Advances in Fresh-Cut Fruits and Vegetables Processing

FOOD PRESERVATION TECHNOLOGY SERIES

Series Editor
Gustavo V. Barbosa-Cánovas

Advances in Fresh-Cut Fruits and Vegetables Processing
Editors: Olga Martín-Belloso and Robert Soliva-Fortuny

Cereal Grains: Properties, Processing, and Nutritional Attributes
Sergio O. Serna-Saldivar

Water Properties of Food, Pharmaceutical, and Biological Materials
Maria del Pilar Buera, Jorge Welti-Chanes, Peter J. Lillford, and Horacio R. Corti

Food Science and Food Biotechnology
Editors: Gustavo F. Gutiérrez-López and Gustavo V. Barbosa-Cánovas

Transport Phenomena in Food Processing
Editors: Jorge Welti-Chanes, Jorge F. Vélez-Ruiz, and Gustavo V. Barbosa-Cánovas

Unit Operations in Food Engineering
Albert Ibarz and Gustavo V. Barbosa-Cánovas

Engineering and Food for the 21st Century
Editors: Jorge Welti-Chanes, Gustavo V. Barbosa-Cánovas, and José Miguel Aguilera

Osmotic Dehydration and Vacuum Impregnation: Applications in Food Industries
Editors: Pedro Fito, Amparo Chiralt, Jose M. Barat, Walter E. L. Spiess, and Diana Behsnilian

Pulsed Electric Fields in Food Processing: Fundamental Aspects and Applications
Editors: Gustavo V. Barbosa-Cánovas and Q. Howard Zhang

Trends in Food Engineering
Editors: Jorge E. Lozano, Cristina Añón, Efrén Parada-Arias, and Gustavo V. Barbosa-Cánovas

Innovations in Food Processing
Editors: Gustavo V. Barbosa-Cánovas and Grahame W. Gould

Advances in Fresh-Cut Fruits and Vegetables Processing

EDITED BY

Olga Martín-Belloso
University of Lleida
Lleida, Spain

Robert Soliva-Fortuny
University of Lleida
Lleida, Spain

CRC Press
Taylor & Francis Group
Boca Raton London New York

CRC Press is an imprint of the
Taylor & Francis Group, an **informa** business

CRC Press
Taylor & Francis Group
6000 Broken Sound Parkway NW, Suite 300
Boca Raton, FL 33487-2742

First issued in paperback 2019

© 2011 by Taylor and Francis Group, LLC
CRC Press is an imprint of Taylor & Francis Group, an Informa business

No claim to original U.S. Government works

ISBN-13: 978-1-4200-7121-4 (hbk)
ISBN-13: 978-0-367-38350-3 (pbk)

Library of Congress Cataloging-in-Publication Data

Advances in fresh-cut fruits and vegetables processing / edited by Olga Martín-Belloso, Robert Soliva-Fortuny.
    p. cm. -- (Food preservation technology series)
    Summary: "Taking a multidisciplinary approach, this work explores the basics and the more recent innovations in fresh-cut fruit and vegetable processing. It addresses scientific progress in the fresh-cut area and discusses the industry and the market for these commodities. In addition, the book covers the regulations that affect the quality of the final products and their processing as well as consumers' attitude and sensory perceptions. The design of plants and equipment is also presented, taking into account engineering aspects, safety, and HACCP guidelines. Finally, innovations with regard to healthy and attractive products are examined"-- Provided by publisher.
    Includes bibliographical references and index.
    ISBN 978-1-4200-7121-4 (hardback)
    1. Fruit--Processing. 2. Vegetables--Processing. 3. Fruit--Preservation. 4. Vegetables--Preservation. 5. Food contamination--Prevention. I. Martín-Belloso, Olga. II. Soliva-Fortuny, Robert. III. Title. IV. Series.

TP440.A28 2011
664'.8--dc22
                                                                2010034918

Visit the Taylor & Francis Web site at
http://www.taylorandfrancis.com

and the CRC Press Web site at
http://www.crcpress.com

Contents

Preface

The fresh-cut fruit and vegetable market is clearly expanding worldwide. In developed countries, those commodities are provided by the food industry, while in the rest of the countries, these products are prepared under uncontrolled conditions that may pose a risk for consumers. Conscientious of the growing interest in these kinds of products, researchers are increasing efforts to offer adequate technologies and practices to processors in order to assure safety while keeping the highest nutritional properties and best sensory properties of the fresh fruits or vegetables. This has led to a significant increase in the amount of new scientific data available. However, this information needed to be presented in a critical and feasible way.

This book is the result of the valuable contribution of experts from industry, research centers, and academia working on different topics regarding fresh-cut produce. We are sincerely thankful to all of them.

The Editors

Olga Martín-Belloso holds a PhD in chemical sciences. She belonged to the National Technical Center of Canned Vegetables from 1984 to 1992 when she joined University of Lleida, Spain. She is presently a professor of Food Science and Technology and head of the research unit on New Technologies for Food Processing.

Her research interests are focused on the development of ready-to-eat, safe, and healthy products by combining already existing processing technologies with novel techniques, as well as the valorization of wastes generated by the fruits and vegetables processing industries.

Pulsed electric fields and intense pulsed light treatments, edible coatings, modified atmosphere packaging, as well as the use of natural antimicrobial and antioxidant substances are among the key technologies developed by her research group.

She has authored more than 200 research papers, several books, book chapters, and patents. She also belongs to the editorial board of recognized journals and is a member of several executive committees of international scientific organizations, such as the Nonthermal Processing Division of the Institute of Food Technologists (NPD-IFT) and the European Federation of Food Science and Technology (EFFoST). In addition, she has been invited as a speaker in numerous international meetings and courses.

Robert Soliva-Fortuny holds a PhD in food technology. He worked on research and development projects for a fruit processing company from 2002 to 2005. In 2005 he was awarded by the Spanish government with a research fellowship. He is currently associate professor at the Department of Food Technology at University of Lleida, Spain, and member of the research unit on New Technologies for Food Processing.

His research activities are focused on food processing and product development. He has authored more than 70 peer-reviewed research papers and several book chapters.

The development of high-quality, safe, and healthy ready-to-eat products by combining the already existing processing technologies with novel techniques is one of his main research activities. He is actively participating in several research projects dealing with the application of nonthermal processing technologies such as high-intensity pulsed electric fields or intense pulsed light treatments.

Contributors

Ana Allende
Research Group on Quality, Safety, and
    Bioactivity of Plant Foods
Department of Food Science and
    Technology
Centro de Edafologia y Biología
    Aplicada del Segura
Consejo Superior de Investigaciones
    Científicas
Murcia, Spain

Begoña de Ancos
Department of Plant Foods Science and
    Technology
Instituto del Frío
Consejo Superior de Investigaciones
    Científicas
Ciudad Universitaria
Madrid, Spain

J. Fernando Ayala-Zavala
Coordinación de Tecnología de
    Alimentos de Origen Vegetal
Centro de Investigación en
    Alimentación y Desarrollo, AC
Hermosillo, Sonora, Mexico

Leen Baert
Department of Food Safety and Food
    Quality
Ghent University
Ghent, Belgium

Jinhe Bai
USDA-ARS
Citrus and Subtropical Products
    Laboratory
Winter Haven, Florida

Elizabeth A. Baldwin
USDA-ARS
Citrus and Subtropical Products
    Laboratory
Winter Haven, Florida

John C. Beaulieu
United States Department of
    Agriculture
Agricultural Research Service
Southern Regional Research Center
New Orleans, Louisiana

M. Pilar Cano
Department of Plant Foods Science and
    Technology
Instituto del Frío
Consejo Superior de Investigaciones
    Científicas
Ciudad Universitaria
Madrid, Spain

María del Milagro Cerdas-Araya
Postharvest Technology Laboratory
Center for Agronomic Research
University of Costa Rica
San José, Costa Rica

Frank Devlieghere
Department of Food Safety and Food
    Quality
Ghent University
Ghent, Belgium

Edward Garner
Kantar Worldpanel
London, United Kingdom

José M. Garrido
Vega Mayor S.L.
Milagro
Navarra, Spain

Maria Isabel Gil
Research Group on Quality, Safety, and
    Bioactivity of Plant Foods
Department of Food Science and
    Technology
Centro de Edafologia y Biología
    Aplicada del Segura
Consejo Superior de Investigaciones
    Científicas
Murcia, Spain

Nathalie Gontard
UMR 1208 Agropolymers Engineering
    and Emerging Technologies
Montpellier SupAgro
France

Gustavo A. González-Aguilar
Coordinación de Tecnología de
    Alimentos de Origen Vegetal
Centro de Investigación en
    Alimentación y Desarrollo
AC, Hermosillo
Sonora, Mexico

Valérie Guillard
UMR 1208 Agropolymers Engineering
    and Emerging Technologies
Montpellier SupAgro
France

Carole Guillaume
UMR 1208 Agropolymers Engineering
    and Emerging Technologies
Montpellier SupAgro
France

Liesbeth Jacxsens
Department of Food Safety and Food
    Quality
Ghent University
Ghent, Belgium

Olga Martín-Belloso
Department of Food Technology
University of Lleida
Lleida, Spain

Peter McClure
Unilever Safety and Environmental
    Assurance Centre
Bedford, United Kingdom

Marta Montero-Calderón
Postharvest Technology Laboratory
Center for Agronomic Research
University of Costa Rica
San José, Costa Rica

Gemma Oms-Oliu
Department of Food Technology
University of Lleida
Lleida, Spain

Lucía Plaza
Department of Plant Foods Science and
    Technology
Instituto del Frío
Consejo Superior de Investigaciones
    Científicas
Ciudad Universitaria
Madrid, Spain

Peter Ragaert
Department of Food Safety and Food
    Quality
Ghent University
Ghent, Belgium

M. Alejandra Rojas-Graü
Department of Food Technology
University of Lleida
Lleida, Spain

Concepción Sánchez-Moreno
Department of Plant Foods Science and
    Technology
Instituto del Frío
Consejo Superior de Investigaciones
    Científicas
Ciudad Universitaria
Madrid, Spain

Maria Victoria Selma
Research Group on Quality, Safety, and
    Bioactivity of Plant Foods
Department of Food Science and
    Technology
Centro de Edafologia y Biología
    Aplicada del Segura
Consejo Superior de Investigaciones
    Científicas
Murcia, Spain

Robert Soliva-Fortuny
Department of Food Technology
University of Lleida
Lleida, Spain

Alessandro Turatti
Turatti SrL
Cavarzere
Venezia, Italy

Isabelle Vandekinderen
Department of Food Safety and Food
    Quality
Ghent University
Ghent, Belgium

Menno van der Velde
Law and Governance Group
Wageningen University
The Netherlands

1 The Fresh-Cut Fruit and Vegetables Industry
Current Situation and Market Trends

M. Alejandra Rojas-Graü, Edward Garner, and Olga Martín-Belloso

CONTENTS

1.1 INTRODUCTION

Fresh-cut fruit and vegetables, initially called minimally processed or lightly processed products, can be defined as any fresh fruit or vegetable that has been physically modified from its original form (by peeling, trimming, washing, and cutting) to obtain 100% edible product that is subsequently bagged or prepackaged and kept in refrigerated storage (IFPA, 2005). Fresh-cut produce includes any kind of fresh commodities and their mixtures in different cuts and packaging. Items such as bagged salads, baby carrots, stir-fry vegetable mixes, and fresh-cut apples, pineapple, or melon are only some examples of this type of product.

The production and consumption of fresh-cut commodities is not new. According to the International Fresh-Cut Produce Association (IFPA), fresh-cut products have been available to consumers since the 1930s in retail supermarkets. However, the fresh-cut industry was first developed to supply hotels, restaurants, catering services, and other institutions. For the food service industry and restaurants, fresh-cut produce presents a series of advantages, including a reduction in the need of manpower for food preparation, reduced need of special systems to handle waste, and the possibility to deliver in a short time, specific forms of fresh-cut products (Watada et al., 1996). Yet it has not been until the past two decades that fresh-cut fruit and vegetable products have gained popularity and penetration in the produce business as a result of a general trend to increase fresh fruit and vegetable consumption (Mayen and Marshall, 2005). The fresh-cut fruit and vegetable industry is constantly growing mainly due to the consumers' tendency to consume healthy and convenient foods and their interest in the role of food in improving human well-being (Gilbert, 2000; Ragaert et al., 2004). In fact, organizations such as the World Health Organization (WHO), Food and Agriculture Organization (FAO), United States Department of Agriculture (USDA), and European Food Safety Authority (EFSA) recommended an increase of fruit and vegetable consumption to decrease the risk of cardiovascular diseases and cancer (Allende et al., 2006). In countries such as the United States, the consumption of fresh whole fruit increased from 282.1 to 284.6 lb/year per capita during the last decade of the 20th century (USDA, 2003), probably as a consequence of an increased public awareness regarding the importance of healthy eating habits.

On the other hand, fresh-cut products are a very convenient way to supply consumers with ready-to-eat foods. Washed, bite-size, and packaged fresh fruit and vegetables allow consumers to eat healthy on the run and to save time on food preparation. For instance, the availability of fresh-cut fruits in vending machines in schools and at workplaces would constitute an excellent strategy to improve the nutritional quality of snacks and convenience foods in a time when obesity and nutrition-related illnesses affect large percentages of the population (Olivas and Barbosa-Cánovas, 2005). In addition to the convenience, there are other reasons for the success of fresh-cut produce, such as the absence of waste material. Waste is generated in peeling and coring fruit. However, when utilizing fresh-cut produce, 100% is consumable, and there is a substantial decrease in labor required for home produce preparation and waste disposal (Garcia and Barrett, 2005).

A study conducted by the IFPA revealed that 76% of surveyed households buy fresh-cut produce at least once a month, and 70% buy fresh-cut fruit every few months (IFPA, 2003). About 30% of consumers prefer fresh-cut fruits and vegetables to their unprocessed equivalents. In addition, Sonti et al. (2003) indicated that women are more likely to buy fresh-cut fruit than men, and as the income level increases, the probability of consuming fresh-cut fruits also increases.

Fresh-cut fruits and vegetables, prepackaged salads, locally grown items, and exotic produce as well as hundreds of new varieties and processed products have been introduced or expanded since the early 1980s. Supermarket produce departments carry over 400 produce items today, up from 150 in the mid-1970s and from 250 in the late 1980s. Also, the number of ethnic, gourmet, and natural food stores that highlight fresh-cut produce continues to rise. Some fresh-cut produce currently available in supermarkets is included in Table 1.1. Because of their convenience and consistent quality, packaged salads continue to be the most popular fresh-cut product. Today, packaged salads account for about 7% of all produce department sales. In fact, in countries such as the United States, consumers have made packaged salads the second-fastest-selling item in grocery stores, trailing only bottled water (Bhagwat, 2006). According to Garrett (2002), organically grown fruits and vegetables are another segment of the fresh produce industry that experienced strong growth in the 1990s, including both whole commodities and fresh-cut products.

TABLE 1.1

Prepared Fruit, Leafy Salad, and Mixed-Tray Salads*

Prepared Fruits	Leafy Salads	Mixed-Tray Salads
Classic salad	Sweet + crunchy salad	Potato + egg salad
Pineapple chunks	Watercress	Sweet + crunchy salad
Luxury fruit salad	Crispy salad	Lettuce + tomato + cucumber + celery
Melon medley	Iceberg lettuce	Mediterranean salad
Melon + grape	Italian salad	Fresh + crispy
Pineapple pieces	Rocket salad	Ribbon salad lettuce + cucumber
Sliced melon selection	Rocket	Mixed pepper salad
Tropical fruit salad	Baby leaf salad	Prawn + pasta salad
Fruit salad	Spinach + watercress + rocket salad	Tuna niçoise
Pineapple slices	Mixed salad	Pasta + cheese salad
Fresh fruit salad	Alfresco salad	Mixed salad white + red cabbage
Mango chunks	Bistro salad	Sweet pepper salad
Grape + kiwi + pineapple	Caesar salad	Oriental edamame soya bean
Apple + grape	Italian leaf salad	Greek salad
Pomegranates	Herb salad	Crunchy lettuce salad + cucumber
Fruit fingers	Ruby salad	Crisp mixed salad
Fruit selection	French style salad	King prawn + pasta salad
Rainbow fruit salad	Crisp mixed salad	Tuna + pasta
Mango + lime wedge	Fine cut salad	Egg salad
Mixed fruit salad	Crispy leaf salad	Poto + peas + bean salad
Fruit cocktail	Leaf salad	Pasta + pepper salad
Seasonal melon medley	Watercress salad	Salmon + potato
Mango pieces	Four leaf salad	Chicken + bacon Caesar salad
Grape + melon	Seasonal baby leaf salad	Avocado spinach + tomato
Apple segments	Mixed leaf salad	King prawn noodle salad
Melon selection	Crunchy mixed salad	Tomato + cheese pasta salad
Summer berry medley	Santa plum tomato salad	
Apple slices + grapes	Tender leaf salad	
Fruit medley	Watercress + spinach + rocket	
Red grape		

Source: Garner, E. 2008. European trends in fresh-cut convenience. Conference presented at the Freshconex, Berlin, Germany

* 52-week ranking ending December 2007.

1.2 GLOBAL MARKET TENDENCIES

Today, there are more fresh-cut fruit and vegetables being consumed as people seek to replace unhealthy snack foods with healthier fruit and vegetable products. This trend has led the fresh-cut industry to increase investment in research and development to address issues regarding raw product supply, packaging technology, processing equipment, and refrigeration. After its popularity in the fast food sector, fresh-cut produce became available at a retail level. This led the way for expansion in the industry, which continues, including more recent additions of fresh-cut fruits at quick-service restaurants and in retail stores.

The production and commercialization of fresh-cut fruits has grown rapidly in recent years, but fresh-cut vegetables, salad in particular, dominate the production of minimally processed foods. According to Mayen and Marshall (2005), the emerging fresh-cut fruit sector will probably overshadow salad sales in the future, because fresh-cut fruits are more attractive to young consumers and aging baby boomers and in general are more likely to be consumed as snack products. In addition, fresh-cut fruits on average have higher margins than bagged salads from retail, which will result in ample space for display in the stores.

1.2.1 AMERICAN TRENDS

Recently, there has been a boom of fresh-cut produce all over the world, especially in many American countries; however, the main production and consumption are concentrated in North America, with the United States as the leader.

In the United States, fresh-cut produce first appeared in retail markets in the 1940s, but second-quality, misshapen produce was used, quality was unpredictable, and shelf life was limited. In the mid-1970s, fast food chains were using shredded fresh-cut lettuce and chopped onions. In the mid-1980s, salad bars opened, and fresh-cut produce start replacing canned products (Garrett, 2002). In fact, the main expansion of fresh-cut fruits and vegetables in the United Stated occurred in the food service sector. In the 1980s, fast food restaurants like McDonald's and Burger King were booming in the United States, so fresh-cut products used in their salad bars and ready-to-eat salads, especially fresh-cut lettuce, were the more required items. In effect, in 2006, in the United States alone, McDonald's used 80 million pounds of salad greens (including spring mix), 100 million pounds of leaf lettuce and iceberg lettuce on sandwiches, 30 million pounds of tomatoes, 54 million pounds of apples for apple dippers and fruit and walnut salad, and 6.5 million pounds of grapes (McDonald's, 2006). Nowadays, fresh-cut produce is one of the fastest growing food categories in U.S. supermarkets, with packaged salads the most important item sold (Figure 1.1). Additionally, the most popular Stock Keeping Unit (SKU)—washed, peeled, and packaged—in the United States is mini carrots, available in a number of sizes (USDA, 2003).

Fresh-cut fruit and vegetable sales have grown to approximately $15 billion per year in the North American food service and retail market and account for nearly 15% of all produce sales. According to the United Fresh Produce Association (2007), the largest portion of U.S. fresh-cut produce sales at retail is fresh-cut salads, with sales of $2.7 billion per year. However, the fast food sector is increasing the demand for packaged fresh-cut fruits by offering healthier choices on their menus. Scott (2008) reported that the U.S. sales of fresh-cut fruit items increased for every product, ranging from 7% to 54% growth. Melons were the segment with a faster growth. This trend is expected to continue at least during the next few years. A number of consumer market research reports have predicted that the demand for fresh-cut fruit products will continually increase, with food service establishments and school

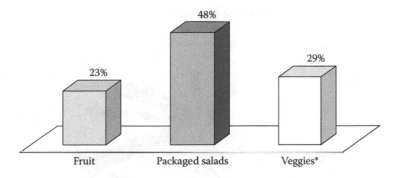

FIGURE 1.1 Fresh-cut produce sales via supermarket channels, 52-week sales ending June 30, 2007, $6 billion total. *Carrots = 45% of vegetables. (Adapted from Cook, R. 2008. The dynamic U.S. fresh produce industry: an industry in transition. *Fresh Fruit and Vegetable Marketing and Trade Information*, University of California, Davis.)

lunch programs being major customers (Anonymous, 2000; Gorny, 2003). According to a study reported by the Perishables Group in 2008, mixed fruits and vegetables, watermelon, pineapple, carrots, and mushrooms have the most important sales and volume in U.S. fresh-cut items (Figure 1.2).

Currently in the United States, the most important company in fresh-cut fruit sales is Ready Pac, with a market share of 23%, followed by private-label store brands (31%), Del Monte (13%), Country Fresh (7%), Club Fresh (3%), and Fresh Express (1%), according to Information Resources, Inc. (2003). Many of these companies' fresh-cut fruit includes products such as pineapple, melons, grapes, citrus, apples, and kiwi. In the case of fresh-cut vegetables, Fresh Express and Dole reached shares of 42% and 48%, respectively, of retail packaged salad sales in 2005, followed by Ready Pac (8%) and other companies (4%) (PMA, 2006). Many kinds of lettuces, such as radicchio, arugula, and red oak, have gained in popularity in past years because of their inclusion in fresh-cut salad mixes and on upscale restaurant menus. In fact, salad blends reached a share of 37% of the salads market in 2005, followed by iceberg lettuce (13%), romaine lettuce (11%), garden salads (8%), salad kits (6%), organic salad blends (4%), premium garden blends (83%), shredded lettuce (3%), spinach (3%), and other (12%) (PMA, 2006).

1.2.2 EUROPEAN TRENDS

In Europe, fresh-cut products were introduced in France in the early 1980s by Florette Group. It was the first production unit of fresh-cut vegetables in Europe which subsequently started various activities to export to other countries such as the United Kingdom, Italy, and Switzerland. Fresh-cut products have been adapted to each country according to consumer preferences, production, distribution, and legislation. In Spain, for instance, fresh-cut products were introduced by Vega Mayor, which were present on the Spanish and Portuguese markets since 1989. Twenty years later, Vega Mayor was acquired by the Florette Group and at the moment is the Spanish leader in the fresh-cut vegetables market. Currently, the main manufacturers and

Fresh-cut Fruit Share ($)

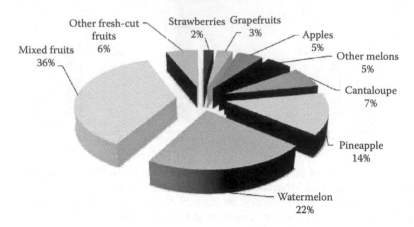

Fresh-cut Vegetables Share ($)

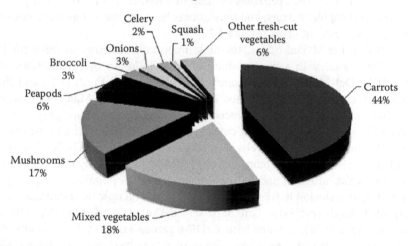

FIGURE 1.2   Fresh-cut vegetables and fruits share in the United States (in dollars). (Adapted from Perishables Group. 2008. U.S. fresh cut produce trends. Conference presented at the Freshconex, Berlin, Germany.)

traders operating in the Spanish market are Vega Mayor, Verdifresh, Kernel, Tallo Verde, Sosegol, and Primaflor (Figure 1.3). However, other producers in the food industry have started to commercialize their new fresh-cut products in the last years. For instance, Vitacress, which is the second firm leading in sales in the English market, after their introduction and success in Portugal, is now beginning their entry into the Spanish market through Vitacress Iberia. Another example is the company Cofrusa, which offers fresher products, such as salads, ready-to-eat vegetables, and, more recently, individual portions of fresh-cut fruits.

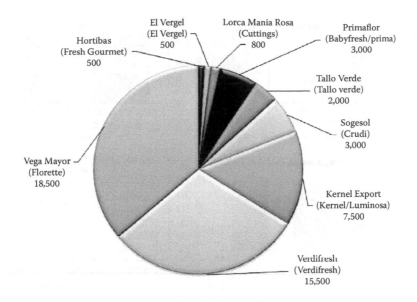

FIGURE 1.3 Ranking of fresh-cut fruits and vegetables manufacturers in Spain. Data correspond to 2004 or estimates for 2005 in tons (t). (Adapted from Alimarket, www.alimarket. es.)

The European markets for fresh-cut fruits and vegetables vary between countries. In some, there is a wide range available in supermarkets, and in others, ready-to-eat food is practically a novelty.

The fresh-cut industry is rising in many European countries with the United Kingdom, France, and Italy as share leaders. Over the last decade, ready-to-eat mixed salad packs have been one of the greatest successes of the UK food industry. The United Kingdom is the leader of the sector, supplying 120,000 tons of fresh-cut salads in 2004, equal to €700 M ($840 M U.S.); France followed with 77,000 tons considering fresh-cut and grilled/steamed vegetables. In Italy, the sales exceeded 42,000 tons of production, corresponding to €375 M ($450 M U.S.) in 2004 (Nicola et al., 2006). Currently, the countries with higher growth in the fresh-cut fruits and vegetables market are Germany, The Netherlands, Spain, and United Kingdom (Figure 1.4) (Garner, 2008). According to the latest data released by Afhorla (the Spanish Association of Washed and Ready-to-Use Fruit and Vegetables) between January and April 2006, Spanish sales had reached 14,675 tonnes, 18.5% more than the same months in 2005. Fresh-cut products represent 5% of all fruit and vegetables consumed in Spain. Some studies indicate that this segment could grow by more than 25% annually.

The average European consumes up to 3 kilos of fresh-cut products a year, but the differences are quite substantial within Europe. For instance, in the United Kingdom the rate is 12 kg per capita per year, France comes second consuming half that of its neighbor with 6 kg per capita, and Italians consume around 4 kg. Other countries where fresh-cut foods are well established, although far less than those already mentioned, are Belgium, The Netherlands, and Germany. In the countries

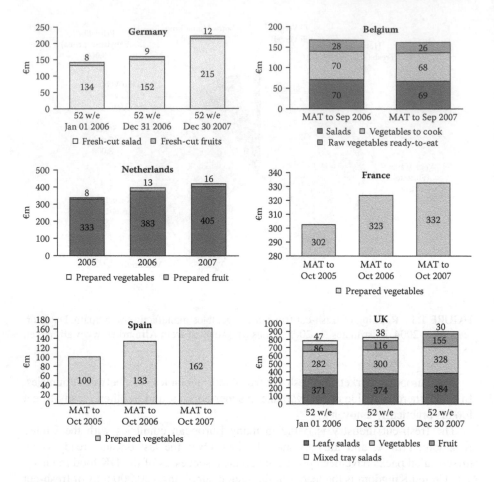

FIGURE 1.4 Fresh-cut fruit and vegetable European market trends. (Garner, 2008.)

of Eastern Europe, with increasingly healthier economies, they are beginning to see great growth in this sector, a development that has not been ignored by the large international holding companies.

1.2.3 ASIATIC TRENDS

Fresh-cut products were introduced in Korea and Japan in the 1990s and 1980s, respectively. Initially, in both countries the main user of fresh-cut products was the food service industry for school meals and restaurants, but in recent years, the consumption has expanded to retail markets (Kim and Jung, 2006). In China, the market for fresh-cut products has been growing since the late 1990s, with more Western fast food industries entering and developing in the Chinese market.

Korea is one of the fastest-growing markets in Asia, with a wide variety of products in the retail segment. In 2006, there were 102 companies producing fresh-cut produce. No fresh-cut fruits were found on supermarkets until the late 1990s

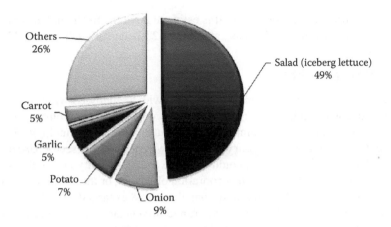

FIGURE 1.5   Proportion of fresh-cut vegetable items based on processing amount in Korea in 2005. (Adapted from Kim, J.G. 2007. Fresh-cut market potential and challenges in Far-East Asia. *Acta Horticulturae* 746: 33–38.)

in Korea. However, the fresh-cut fruits industry has recently enjoyed double-digit growth rates, reaching an estimated $50 million in 2006 (Kim, 2007). Despite this growh, fresh-cut vegetables continue to dominate the production of fresh-cut items, with salad made from iceberg lettuce the most popular fresh-cut produce, comprising 48.7% of total fresh-cut vegetables (Figure 1.5).

According to the Korean Fresh-cut Produce Association (KFPA), in Korea, the fresh-cut produce market reached approximately $1.1 billion in 2006, up from $530 million in 2003. In Japan, the sales of fresh-cut produce have grown from approximately $1 billion in 1999 to $2.6 billion in 2005, which is about 10% of total fresh produce sales (Izumi, 2007).

In Japan, the food service sector, which supplies produce to restaurants, fast-food outlets, and school meals, makes up about 66% of the total fresh-cut market. Sales of fresh-cut produce in the retail sector, including supermarkets and convenience stores, are $0.9 billion, 34% of the total market (Kim, 2007). According to a study published by the Association of Minimally Processed Fruits and Vegetables (AMPFV) in 1999, there are 161 enterprises producing fresh-cut products in Japan (AMPFV, 2000).

Iceberg lettuce, onion, cabbage, Japanese radish, edible burdock, potato, Chinese cabbage, pumpkin, sweet pepper, cucumber, carrot, watermelon, pineapple, and melon are the most popular fresh-cut vegetables and fruits in Japan. In this country, manufacturers such as Dole, for example, offer cut lettuce, cabbage mixes, coleslaw, bean salad, tomato salad, onion salad, corn salad, Caesar salad, and a wide variety of specialty mixes, which make up the bagged category. According to the AMPFV (2004), in Japan, the total input of vegetables to fresh-cut production was reported to be 92,672 tonnes in 2002.

Although no exact data show the scale of the fresh-cut fruits and vegetables market in China, China will become the largest consumer of these products in the future. For example, in Beijing, a new fresh-cut factory was established 3 years ago, and sales reached 3900 tons by 2006. It is estimated that the fresh-cut fruits and vegetables market in China will increase at a rate of 20% annually (Zhang, 2007).

1.3 FINAL REMARKS

Fresh-cut fruits and vegetables are commodities with a rapidly growing sector in the food industry, with both retail and food service outlets. At the moment, the main factor that has promoted and maintained fresh-cut sales is the technology. However, permanent innovations are necessary to drive new growth in this sector. Use of innovative packaging technology that could improve product quality and shelf life, new fruit mixtures with more variety, incorporation of flavors, or the use of steamer bags for vegetables, are just a few considerations that could expand the markets of fresh-cut products. Worldwide, there is a wide range of vegetables that could be used to broaden and increase the product offerings in the market. However, in many countries, it is necessary to improve preparation and preservation techniques with the purpose of keeping the product safe and of high quality long enough to make the distribution of fresh-cut commodities feasible and achievable.

REFERENCES

Allende, A., F.A. Tomás-Barberán, and M.I. Gil. 2006. Minimal processing for healthy traditional foods. *Trends in Food Science and Technology* 17:513–519.

AMPFV, Association of Minimally Processed Fruits and Vegetables Industries, Japan. 2000. A survey report on the fresh-cut vegetables in 1999.

AMPFV, Association of Minimally Processed Fruits and Vegetables Industries, Japan. 2004. A survey report on the fresh-cut vegetables in 2002.

Anonymous. 2000. Fresh sliced apples: waiting to boom? *Fresh Cut* 8: 18–22.

Bhagwat, A.A. 2006. Microbiological safety of fresh-cut produce: where are we now? In: *Microbiology of fresh produce.* Edited by: K.R. Matthews. ASM Press, Washington, pp. 121–166.

Cook, R. 2008. The dynamic U.S. fresh produce industry: an industry in transition. Fresh Fruit and Vegetable Marketing and Trade Information, University of California, Davis.

Garcia, E., and D.M. Barrett. 2005. Fresh-cut fruits. In: *Processing fruits science and technology.* Edited by: D.M. Barrett, L. Somogyi, and H. Ramaswamy. CRC Press, Boca Raton, FL, pp. 53–72.

Garner, E. 2008. European trends in fresh-cut convenience. Conference presented at the Freshconex, Berlin, Germany.

Garrett, E.H. 2002. Fresh-cut produce: tracks and trends. In: *Fresh-cut fruits and vegetables: science, technology and market.* Edited by: O. Lamikanra. CRC Press, Boca Raton, FL, pp. 1–10.

Gilbert, L.C. 2000. The functional food trend: what's next and what Americans think about eggs. *Journal of the American College of Nutrition* 19: 507–512.

Gorny, J.R. 2003. New opportunities for fresh-cut apples. *Fresh Cut* 11: 14–15.

IFPA, International Fresh-cut Produce Association. 2003. http://www.fresh-cuts.org/fcf-html. (accessed July 17, 2008).

IFPA, International Fresh-cut Produce Association. 2005. The convenience, nutritional value and safety of fresh-cut produce. http://www.gov.on.ca/GOPSP/en/graphics/053125.pdf (accessed August 5, 2008).

Information Resources, Inc. 2003. Fresh-cut fruit market shares.

Izumi, H. 2007. Current status of the fresh-cut produce industry and sanitizing technologies in Japan. *Acta Horticulturae* 746: 45–52.

Kim, J.G. 2007. Fresh-cut market potential and challenges in Far-East Asia. *Acta Horticulturae* 746: 33–38.

Kim, J.G., and J.W. Jung. 2006. Status of fresh-cut industry in foreign countries. *Postharvest Horticulture* 14: 4–19.

Mayen, C., and M.I. Marshall. 2005. Opportunities in the fresh-cut fruit sector for Indiana melon growers. Purdue New Ventures. http://www.agecon.purdue.edu/newventures (accessed August 1, 2008).

McDonald's. 2006. Food quality at McDonald's. Fact sheet. http://www.mcdonalds.com/corp/about/factsheets.pdf (accessed September 12, 2008).

Nicola, S., E. Fontana, C. Torassa, and J. Hoeberechts. 2006. Fresh-cut produce: postharvest critical issues. *Acta Horticulturae* 712: 223–230.

Olivas, G.I., and G.V. Barbosa-Cánovas. 2005. Edible coatings for fresh-cut fruits. *Critical Review in Food Science and Nutrition* 45: 657–670.

Perishables Group. 2008. U.S. fresh cut produce trends. Conference presented at the Freshconex, Berlin, Germany.

PMA, Produce Marketing Association. 2006. Fresh-cut produce industry. http://www.pma.com (accessed August 1, 2008).

Ragaert, P., W. Verbeke, F. Devlieghere, and J. Debevere. 2004. Consumer perception and choice of minimally processed vegetables and packaged fruits. *Food Quality and Preference* 15: 259–270.

Scott, C. 2008. Fresh-cut growth trend continues. *Fresh Cut*. www.freshcut.com/pages/arts.php?ns=794 (accessed July 2, 2010).

Sonti, S., W. Prinyawiwatkul, J.M. Gillespie, K.H. McWatters, and S.D. Bhale. 2003. Analysis of consumer perception of fresh-cut fruits and vegetables and edible coating. Paper presented at the Institute of Food Technologist Annual Meeting, Chicago.

United Fresh Produce Association. 2007. Available at http://www.unitedfresh.org. Accessed March 31, 2007.

USDA. Economic Research Service United States Department of Agriculture. 2003. http://www.ers.usda.gov/publications/Agoutlook/AOTables/.Statisticalindicators.

Watada, A.E., N.P. Ko, and D.A. Minott. 1996. Factors affecting quality of fresh-cut horticultural products. *Postharvest Biology and Technology* 9:115–125.

Zhang, X. 2007. New approaches on improving the quality and safety of fresh cut fruits and vegetables. *Acta Horticulturae* 746: 97–102.

2 Regulatory Issues Concerning the Production of Fresh-Cut Fruits and Vegetables

Menno van der Velde

CONTENTS

2.1 INTRODUCTION

2.1.1 THE TYPES OF LAW

The introduction of fresh-cut fruits and vegetables (FcFV) on the market as a new type of product extended the production chain of human food. It added some new challenges for producers, but it did not introduce a new type of law.

FcFV are subjected to the food law of the country where they are grown, harvested, processed, transported, and sold to consumers by caterers and retailers. The applicable food law multiplies when they are traded internationally. The food laws of all participating countries are then applicable, as is international food law, if only in an effort to harmonize the laws of the different legal systems. The Codex Alimentarius Commission (CAC) is the international intergovernmental organization for food standards, guidelines, and recommended practices. An intergovernmental organization can only make recommendations to the governments of the member states, but the Codex Alimentarius, the collection of CAC food law, nevertheless, has much authority.

National and international food laws makes use of many instruments: binding legislation and regulations, along with several voluntary instruments, such as good practices, especially Good Agricultural Practices (GAPs) and Good Manufacturing Practices (GMPs), for the successive stages in the food production chain with other good practices for the remaining stages.

GHP follow the produce from the beginning to the end of the production chain. Systems based on the Hazard Analysis and Critical Control Point (HACCP) principles are equally applicable throughout the production chain. Governments publish guidance documents to explain their legislation; sometimes they prefer the assistance of voluntary codes made by the food businesses. Some of these voluntary codes are approved by public authorities. Letters from public law authorities remind the food business operators of the prescribed levels of hygiene, and the regulatory powers that could be used if the industry continues to fail to live up to the expectations that are also legal requirements.

In short, there is a varied array of binding and nonbinding law. Although the term *binding law* may seem to be a tautology, and the term *nonbinding law* an internal contradiction, both types of law exist. The nonbinding legal instruments are very important in food law.

The simple structure of binding rules, usually made in legislation for food safety, more specific legislation for food hygiene, still more specific rules for fresh produce, and finally detailed legislation for fresh-cut fruits and vegetables does not exist. It is complicated by the practice of mixing legislation with nonbinding law. Mixing takes place at two levels: legally binding instruments are combined with voluntary instruments that can be used as an alternative, under certain conditions set by the law. On a deeper level, legally binding instruments take the essential elements of a voluntary instrument such as a GHP and prescribe it as the law. Enforcement of these binding rules is then an alternative to voluntary good practices. The result is a complicated collection of law, soft and hard, voluntarily accepted and binding, national and international, intergovernmental and supranational, governmental and private. A selection of all of these different kinds of relevant law is presented in this chapter. It is representative but by no means complete.

2.1.2 THE PRODUCE

FcFV are the results of an extended food production chain. Final preparation of this food has been transferred from consumers and caterers* to a new set of producers who depend on new production methods for ready-to-eat food and the logistics of transports, distribution, and wholesale to get the FcFV in time to the retailers and caterers where the consumers will buy them.

The production of this type of ready-to-eat food requires not only additional steps in the production process but also additional food hygiene measures. Preparation of FcFV removes their natural protection against desiccation and contamination. It increases the contact surfaces between the produce and the oxygen in the atmosphere, frees the moisture, and presents abundant nutrition and ideal living conditions to a host of unwanted creatures. It also excludes the use of production processes that could eliminate this contamination, such as freezing or heating to lethal levels, and sweetening, acidifying, and other countermeasures to deal with microbial and other contamination.

The production of FcFV requires increased control over fruits and vegetables, especially backwards in the primary production stage where many potential causes for fresh-cut contamination attach themselves easily to the raw material and are difficult to remove.

The second major concern is the part of the production chain after the FcFV have been made: ways have to be found to preserve their freshness during transport,

---

* The word *caterer* is used here to indicate restaurants, canteens, schools, hospitals, and similar institutions where food is offered for immediate consumption. Compare the phrase "foods for catering purposes" in Codex Alimentarius Commission, CODEX STAN 1-1985 Labelling of Prepackaged Foods, http://www.codexalimentarius.net/web/more_info.jsp?id_sta=322.

distribution, catering, or retail, and even beyond the production chain, because food safety has to be guaranteed by the producers for the entire shelf life—that is, either the period preceding the "use by" date or the "minimum durability date," as defined in Articles 9 and 10 of Directive 2000/13/EC.*

There are several items for which the law for fresh-cut produce has to be different from the law for fresh produce.

2.1.3 EXAMPLES OF BINDING AND NONBINDING LAW
ON FRESH-CUT FRUITS AND VEGETABLES

FcFV have entered food hygiene law in various ways and to various degrees. EC legislation on microbiological contamination is an example of binding law. The EC has set food safety criteria for *Salmonella* in precut fruits and vegetables[†] and for *Listeria monocytogenes* in relation to three ready-to-eat foods.[‡] A food that does not satisfy a food safety criterion is banned from the market and banned from export and import. It also means that food business operators have to install a testing program and conduct studies to investigate compliance with the criteria throughout the shelf life of their products.

Process hygiene criteria have been set for *Escherichia coli* in relation to precut ready-to-eat fruits and vegetables.[§] Food business operators have to ensure that the process hygiene criteria are met in the supply, handling, and processing of raw materials and foodstuffs under their control. Failure to meet the food safety criteria means the duty to destroy the produce when no alternative use is possible. Failure to meet the process hygiene criteria creates the obligation to improve the processing or the raw material.

FcFV have also attracted nonbinding law.

The CAC made voluntary rules in the "Annex for Ready-to-Eat Fresh Pre-cut Fruits and Vegetables" to the "Code of Hygienic Practice for Fresh Fruits and Vegetables."[¶]

---

* Article 2(f) European Community, Commission Regulation (EC) No. 2073/2005 of 15 November 2005 on microbiological criteria for foodstuffs, OJ L 338, 22.12.2005, p. 1–26) http://eur-lex.europa.eu/ LexUriServ/LexUriServ.do?uri=CONSLEG:2005R2073:20071227:EN:PDF and Directive 2000/13/ EC of the European Parliament and of the Council of 20 March 2000 on the approximation of the laws of the member states relating to the labeling, presentation, and advertising of foodstuffs, OJ L 109, 6.5.2000, p. 29, http://eur-lex.europa.eu/LexUriServ/LexUriServ.do?uri=CONSLEG:2000L0013:200 71129:EN:PDF.

† Food category 1.19, Chapter 1. Food safety criteria. Annex I Microbiological criteria for foodstuffs, Commission Regulation (EC) No. 2073/2005.

‡ Food categories 1.1, 1.2, and 1.3, Chapter 1. Food safety criteria. Annex I Microbiological criteria for foodstuffs, Commission Regulation (EC) No. 2073/2005.

§ Food category 2.5.1, 2.5 Vegetables, fruits, and products thereof, Chapter 2. Process hygiene criteria. Annex I Microbiological criteria for foodstuffs, Commission Regulation (EC) No. 2073/2005.

¶ Codex Alimentarius Commission, CAC/RCP 53-2003 Code of Hygienic Practice for Fresh Fruits and Vegetables, Annex I, Annex for Ready-to-eat Fresh Pre-cut Fruits and Vegetables, http://www.codex-alimentarius.net/web/more_info.jsp?id_sta=10200.

The United States Food and Drug Administration (FDA) published the "Guide to Minimize Microbial Food Safety Hazards of Fresh-cut Fruits and Vegetables" as voluntary guidance against the background of FDA's binding regulatory powers.[*]

2.2 NONBINDING LAW

2.2.1 THE CODEX ALIMENTARIUS

The CAC decision-making process is structured to involve as many governments as possible and to prepare decisions in as many steps as are necessary to achieve consensus. Of course, other international intergovernmental organizations can use the Codex in their practice, and a significant development is the use of the Codex as the source of law for decisions in the dispute settlement procedures of the World Trade Organization.[†] The Codex is also used by nongovernmental organizations, food business operators, consumers, and standardization and certification organizations. Three Codex documents are basic to FcFV.

The "Recommended International Code of Practice—General Principles of Food Hygiene" is the basic document on food hygiene law for the whole world. It was published in 1969 and is known as the General Principles of Food Hygiene.[‡] The CAC presents and recommends in the same document a HACCP-based approach for the entire food production chain as a means to enhance food safety.[§]

The more specific CAC "Code of Hygienic Practice for Fresh Fruits and Vegetables" was published in 2003. It has an "Annex for Ready-to-eat Fresh Pre-cut Fruits and Vegetables."

The three Codex documents are cumulative.[¶] The "Code of Hygienic Practice for Fresh Fruits and Vegetables" accepts the "General Principles of Food Hygiene" and adds the specific rules for the fresh fruits and vegetables. The "Annex for Ready-to-eat Fresh Pre-cut Fruits and Vegetables" in its turn accepts the "General Principles" and the "Code of Hygienic Practice" and adds the specifics for fresh-cut produce. Table 2.1 presents an overview of the three basic Codex Alimentarius law documents for FcFV.

There are many other Codex documents on specific aspects of fruits and vegetables. The "Recommended International Code of Practice for Packaging and Transport of Fresh Fruit and Vegetables" was made in 1995 for tropical fresh fruit

---

[*] U.S. Department of Health and Human Services, Food and Drug Administration and Center for Food Safety and Applied Nutrition, Guidance for Industry: Guide to Minimize Microbial Food Safety Hazards of Fresh-cut Fruits and Vegetables, February 2008. http://www.cfsan.fda.gov/~dms/prodgui4. html.

[†] Bernd van der Meulen and Menno van der Velde, European Food Law Handbook, Wageningen 2008, pp. 468–472.

[‡] Codex Alimentarius Commission CAC/RCP 1-1969, Rev. 4-2003 Recommended International Code of Practice—General Principles of Food Hygiene including Annex on Hazard Analysis and Critical Control Point (HACCP) system and Guidelines for its Application. The Code was adopted by the Codex Alimentarius Commission in 1969, revision 4 made in 2003 is the latest revision to date, http://www.codexalimentarius.net/web/more_info.jsp?id_sta=23.

[§] Codex Alimentarius Commission CAC/RCP 1-1969, Rev. 4-2003, p. 3.

[¶] CAC/RCP 53-2003, 2.2 Use, see note 6 supra.

TABLE 2.1

The Codex Alimentarius Law for Fresh-Cut Fruits and Vegetables

CAC/RCP 1-1969, Rev. 4-2003 General Principles of Food Hygiene	CAC/RCP 53-2003 Code of Hygienic Practice for Fresh Fruits and Vegetables	CAC/RCP 53-2003 Annex for Ready-to-Eat Fresh Precut Fruits and Vegetables
Section I Objectives	**Introduction**	**Introduction**
1.1 The Codex General Principles of Food Hygiene	1. Objectives of the Code	1. Objective
Section II Scope, Use, and Definition	**2. Scope, Use, and Definitions**	**2. Scope, Use, and Definitions**
2.1 Scope	2.1 Scope	2.1 Scope
2.1.1 The food chain		
2.1.2 Roles of governments, industry, and consumers		
2.2 Use	2.2 Use	2.2 Use
2.3 Definitions	2.3 Definitions	2.3 Definitions
Section III Primary Production	**3. Primary Production**	**3. Primary Production**
3.1 Environmental Hygiene	3.1 Environmental Hygiene	
3.2 Hygienic Production of Food Sources	3.2 Hygienic Primary Production of Fresh Fruits and Vegetables	
	3.2.1 Agricultural input requirements	
	3.2.1.1 Water for primary production	
	3.2.1.1.1 Water for irrigation and harvesting	
	3.2.1.1.2 Water for fertilizers, pest control, and other agricultural chemicals	
	3.2.1.1.3 Hydroponic water	
	3.2.1.2 Manure, biosolids, and other natural fertilizers	
	3.2.1.3 Soil	
	3.2.1.4 Agricultural chemicals	
	3.2.1.5 Biological control	
	3.2.2 Indoor facilities associated with growing and harvesting	
	3.2.2.1 Location, design, and layout	
	3.2.2.2 Water supply	
	3.2.2.3 Drainage and waste disposal	

TABLE 2.1 (Continued)
The Codex Alimentarius Law for Fresh-Cut Fruits and Vegetables

CAC/RCP 1-1969, Rev. 4-2003 General Principles of Food Hygiene	CAC/RCP 53-2003 Code of Hygienic Practice for Fresh Fruits and Vegetables	CAC/RCP 53-2003 Annex for Ready-to-Eat Fresh Precut Fruits and Vegetables
	3.2.3 Personnel health, hygiene, and sanitary facilities	
	3.2.3.1 Personnel hygiene and sanitary facilities	
	3.2.3.2 Health status	
	3.2.3.3 Personal cleanliness	
	3.2.3.4 Personal behavior	
	3.2.4 Equipment associated with growing and harvesting	
3.3 Handling, Storage, and Transport	3.3 Handling, Storage, and Transport	
	3.3.1 Prevention of cross-contamination	
	3.3.2 Storage and transport from the field to the packing facility	
3.4 Cleaning, Maintenance, and Personnel Hygiene at Primary Production	3.4 Cleaning, Maintenance, and Sanitation	
	3.4.1 Cleaning programs	
	3.4.2 Cleaning procedures and methods	
	3.4.3 Pest control systems	
	3.4.4 Waste management	
Section IV Establishment: Design and Facilities	**4. Packing Establishment: Design and Facilities**	**4. Establishment: Design and Facilities**
4.1 Location		
4.1.1 Establishments		
4.1.2 Equipment		
4.2 Premises and Rooms		
4.2.1 Design and layout permit Good Hygiene Practices (GHPs), protection against cross-contamination		
4.2.2 Internal structures and fittings		
4.2.3 Temporary /mobile premises and vending machines		
4.3 Equipment		
4.3.1 General		

TABLE 2.1 (Continued)
The Codex Alimentarius Law for Fresh-Cut Fruits and Vegetables

CAC/RCP 1-1969, Rev. 4-2003 General Principles of Food Hygiene	CAC/RCP 53-2003 Code of Hygienic Practice for Fresh Fruits and Vegetables	CAC/RCP 53-2003 Annex for Ready-to-Eat Fresh Precut Fruits and Vegetables
4.3.2 Food control and monitoring equipment		
4.3.3 Containers for waste and inedible substances		4.4 Facilities
4.4 Facilities		
4.4.1 Water supply		
4.4.2 Drainage and waste disposal	3.2.2.3 Drainage and waste disposal	4.4.2 Drainage and waste disposal
4.4.3 Cleaning		
4.4.4 Personnel hygiene facilities and toilets	3.2.3.1 Personnel hygiene and sanitary facilities	
4.4.5 Temperature control		
4.4.6 Air quality and ventilation		
4.4.7 Lighting		
4.4.8 Storage		
Section V Control of Operation	**5. Control of Operation**	**5. Control of Operation**
5.1 Control of Food Hazards	5.1 Control of Food Hazards	5.1 Control of Food Hazards
Food business operators should control food hazards using systems such as Hazard Analysis and Critical Control Point (HACCP);		Suppliers (growers, harvesters, packers and distributors) have to minimize contamination of the raw material and adopt CAC/RCP 53-2003.
Identify operation steps critical to food safety;		Certain pathogens, *Listeria monocytogenes* and *Clostridium botulinum* present specific food safety problems
Implement effective control procedures at those steps;		
Monitor control procedures to ensure effectiveness;		
Review control procedures periodically.		
5.2 Key Aspects of Hygiene Control Systems	5.2 Key Aspects of Hygiene Control Systems	5.2 Key Aspects of Control Systems
5.2.1 Time and temperature control	5.2.1 Time and temperature control	
5.2.2 Specific process steps	5.2.2 Specific process steps	5.2.2 Specific process steps
Chilling	5.2.2.1 Postharvest water use	5.2.2.1 Receipt and inspection of raw materials
Thermal processing	5.2.2.2 Chemical treatments	5.2.2.2 Preparation of raw material before processing
Irradiation	5.2.2.3 Cooling of fresh fruits and vegetables	
Drying		
Chemical preservation		

TABLE 2.1 (Continued)
The Codex Alimentarius Law for Fresh-Cut Fruits and Vegetables

CAC/RCP 1-1969, Rev. 4-2003 General Principles of Food Hygiene	CAC/RCP 53-2003 Code of Hygienic Practice for Fresh Fruits and Vegetables	CAC/RCP 53-2003 Annex for Ready-to-Eat Fresh Precut Fruits and Vegetables
5.2.2 Specific process steps (Continued) Vacuum or modified atmospheric packaging	5.2.2.4 Cold storage	5.2.2.3 Washing and microbiological decontamination 5.2.2.4 Precooling fresh fruits and vegetables 5.2.2.5 Cutting, slicing, shredding, and similar precut processes 5.2.2.6 Washing after cutting, slicing, shredding, and similar precut processes 5.2.2.7 Cold storage
5.2.3 Microbiological and other specifications	5.2.3 Microbiological and other specifications	
5.2.4 Microbial cross-contamination	5.2.4 Microbial cross-contamination	
5.2.5 Physical and chemical contamination	5.2.5 Physical and chemical contamination	
5.3 Incoming Material Requirements	5.3 Incoming Material Requirements	
5.4 Packaging	5.4 Packaging	
5.5 Water	5.5 Water Used in the Packing Establishment	
5.5.1 In contact with food		
5.5.2 As an ingredient		
5.5.3 Ice and steam		
5.6 Management and Supervision	5.6 Management and Supervision	
5.7 Documentation and Records	5.7 Documentation and Records In addition, keep current all information, keep records of processing much longer than shelf life Growers: production site, suppliers' information, use and lot numbers of agricultural inputs, irrigation practices, water quality data, pest control, cleaning schedules	5.7 Documentation and Records

TABLE 2.1 (Continued)
The Codex Alimentarius Law for Fresh-Cut Fruits and Vegetables

CAC/RCP 1-1969, Rev. 4-2003 General Principles of Food Hygiene	CAC/RCP 53-2003 Code of Hygienic Practice for Fresh Fruits and Vegetables	CAC/RCP 53-2003 Annex for Ready-to-Eat Fresh Precut Fruits and Vegetables
5.7 Documentation and Records (continued)	Packers: each lot, incoming materials, information from growers, lot numbers, data on quality processing water, pest control programs, cooling and storage temperatures, chemicals used in postharvest treatments, cleaning schedules	5.7 Documentation and Records
5.8 Recall Procedures	5.8 Recall Procedures In addition: keep information to trace products from the distributor to the field.	5.8 Recall Procedures

Section VI Establishment: Maintenance and Sanitation
6.1 Maintenance and Cleaning
6.1.1 General
6.1.2 Cleaning procedures and methods
6.2 Cleaning Programs
6.3 Pest Control Systems
6.3.1 General
6.3.2 Preventing access
6.3.3 Harborage
6.3.4 Monitoring and detection
6.3.5 Eradication
6.4 Waste Management
6.5 Monitoring Effectiveness

Section VII Establishment: Personal Hygiene
7.1 Health Status
7.2 Illness and Injuries
7.3 Personal Cleanliness
7.4 Personal Behavior
7.5 Visitors

Section VIII Transportation
8.1 General
8.2 Requirements
8.3 Use and Maintenance

TABLE 2.1 (Continued)
The Codex Alimentarius Law for Fresh-Cut Fruits and Vegetables

CAC/RCP 1-1969, Rev. 4-2003 General Principles of Food Hygiene	CAC/RCP 53-2003 Code of Hygienic Practice for Fresh Fruits and Vegetables	CAC/RCP 53-2003 Annex for Ready-to-Eat Fresh Precut Fruits and Vegetables
Section IX Product Information and Consumer Awareness 9.1 Lot Identification 9.2 Product Information 9.3 Labeling 9.4 Consumer Education		
Section X Training 10.1 Awareness and Responsibilities	**10 Training**	**10 Training**
	Personnel for growing and harvesting must know the Good Agricultural Practices (GAPs), GHPs, their role and responsibility in hygiene. Knowledge and skills for agricultural activities and handling fresh fruits and vegetables (FFV) Personnel for packing know Good Manufacturing Practices (GMPs), GHPs. Knowledge and skills to handle FFV and minimize microbial, chemical, and physical contamination Personnel handling cleaning chemicals or other potentially hazardous chemicals: instructed in safe handling Aware of their role and responsibility	Refer to CAC/RCP 1-1969 Rev 4 2003 and to CAC/RCP 53-2003 Required level of training Packaging systems used for FFV with risks of contamination or microbiological growth Importance of temperature control and GMPs
10.2 Training Programs Nature of the food, ability to sustain growth of pathogenic or spoilage microorganisms Manner of handling and packaging, with contamination probability	10.2 Training Programs Nature of the food, ability to sustain growth of pathogenic microorganisms. Agricultural techniques and inputs and the associated probability of microbial, chemical, and physical contamination.	10.2 Training Programs Additional: The packaging systems used for fresh precut fruits and vegetables, including the risks of contamination or microbiological growth involved with this method;

TABLE 2.1 (Continued)
The Codex Alimentarius Law for Fresh-Cut Fruits and Vegetables

CAC/RCP 1-1969, Rev. 4-2003 General Principles of Food Hygiene	CAC/RCP 53-2003 Code of Hygienic Practice for Fresh Fruits and Vegetables	CAC/RCP 53-2003 Annex for Ready-to-Eat Fresh Precut Fruits and Vegetables
Extent and nature of processing or further preparation before final consumption	Task with hazards and controls.	The importance of temperature control and GMPs
Storage conditions	Processing and packaging manners for FFV.	
Expected length of time before consumption	Extent and nature of processing or further preparation by consumer.	
	Importance of good health and hygiene for personal health and food safety.	
	Importance of hand washing and hand-washing techniques.	
	Importance of using sanitary facilities to reduce contaminating fields, produce, other workers, and water supplies.	
	Techniques for hygienic handling and storage of FFV by transporters, distributors, storage handlers, and consumer.	
10.3 Instruction and Supervision		
10.4 Refresher Training		

and vegetables. The scope of this Code of Practice was enlarged to packaging and transport of all fresh fruits and vegetables by an amendment in 2004 that eliminated the word "tropical" from the title and where appropriate from the text.*

There are hundreds of national and international agrarian or marketing quality standards for individual fresh fruits and vegetables sold as whole fresh produce.

2.2.2 Good Practices

Good practices are developed by active and knowledgeable persons and organizations with no government power or authority to introduce these practices. They have

---

* CAC/RCP 44-1995, AMD. 1-2004, Recommended International Code of Practice for Packaging and Transport of Fresh Fruit and Vegetables, http://www.codexalimentarius.net/web/more_info.jsp?id_sta=322.

to rely on the force of reason and proven success to convince others that a particular good practice is a valuable instrument for specific objectives.

Several organizations promote these good practices. When an international organization like the CAC goes beyond mere promotion and invests in lengthy negotiations to produce recommended practices, something is added to the good practice as it stood before the CAC published its nonbinding instrument. For one, dozens of government representatives have contributed to the discussions about the right formulation of the text. This process involves many consultations in the member states when stakeholders and other interested persons and organizations are asked to contribute ideas to determine the positions that the national representative will take.*

Reworking a nonbinding good practice into a nonbinding Codex document seemingly has little legal relevance. But governments change the legal character of the Codex Alimentarius recommendations and the good practices when they participate in CAC decision making. This is reinforced by explicit statements in their own legislation that they will contribute to the construction of international food law and will take this law into consideration when they make their own legislation. The EC is an example: it became a member of the CAC in 2003 and refers to the Codex at several places in its extensive collection of food regulations.†

2.2.3 DIFFERENT TYPES OF GOOD PRACTICES

Good practices are made by all kinds of organizations for activities in the food production chain. Some of these practices are made for particular stages of the food production chain in the way that a GAP will be followed by a GMP, and other practices for packaging and transport may be made corresponding to the CAC recommended international Codes of Practice.

2.2.4 THE CODEX GENERAL PRINCIPLES OF FOOD HYGIENE

The General Principles of Food Hygiene deal with all aspects of food hygiene in ten sections. These are presented with their subject matter in Table 2.1. The additional rules for fresh fruits and vegetables and for fresh-cut fruits and vegetables are also indicated.

Sections III to IX follow the food chain from primary production to the consumer. Section III begins with the selection of a suitable site for food production. Environmental hygiene and public health require that harmful substances from the environment cannot become part of the food in unacceptable levels. Production of

---

* See the U.S. Federal Register with prescribed information on Codex activities at http://www.fsis.usd. gov/Codex_Alimentarius/Related_Federal_Register_Notices/index.asp. See also the Food Safety Inspection Service that organizes public meetings with U.S. Delegates to Codex committees before their committee meetings to inform the public about the meeting agenda and proposed U.S. positions on the issues, http://www.fsis.usda.gov/codex_alimentarius/public_meetings/index.asp.

† Article 13 on international standards in Regulation (EC) No. 178/2002; Recital (18) in Regulation (EC) No. 852/2004: "This Regulation takes account of international obligations laid down in the WTO Sanitary and Phytosanitary Agreement and the international food safety standards contained in the Codex Alimentarius." See Bernd van der Meulen and Menno van der Velde, European Food Law Handbook, Wageningen 2008, pp. 467–482.

food can take place only where there are no such substances or when changes can be made that will prevent this contamination.

The hygienic production of food requires that the practices of primary production do not contaminate the food. Producers have to test the way they operate to identify the most likely activities or circumstances where contamination can occur and take measures to prevent contamination or reduce it to acceptable levels. The HACCP system is recommended as a method for this work. Producers are to take measures to

> Control contaminations from air, soil, water, feedstuffs, fertilizers (including natural fertilizers), pesticides, veterinary drugs or any other agent used in primary production; control plant and animal health so that it does not pose a threat to human health through food consumption, (…) and protect food sources from faecal and other contamination.*

The rules on handling, storage, and transport as part of primary production call for sorting the harvested material to reject the parts that are unfit for food. The rejected material must be disposed of hygienically, and the food has to be protected from the negative influences of pests, and chemical, physical, and microbiological contaminants.†

Section IV Establishment: Design and Facilities deals with the infrastructure, location, and buildings where food is processed after the primary stage. These establishments must be suited to good food hygiene practices.

Section V Control of Operation deals with food hazards. It recommends to use a HACCP system, and deals with time and temperature and specific process steps under the heading "Key Aspects of Hygiene Control Systems."

Section VI addresses the maintenance and sanitation of the establishment.

Personnel and personal hygiene are important items of several sections. Facilities have to be present for any necessary cleaning and maintenance and for personal hygiene in the primary production phase.‡ In any building or area where food is handled, facilities like an adequate supply of potable water, drainage, and waste disposal systems have to be present.§

Personnel hygiene requires that the management of the establishment provides the means needed for hygienically washing and drying hands, for lavatories, and for changing rooms.¶ Personal hygiene is one of the subjects of Section VII as part of a set of rules on the consequences of illness and injuries, personal cleanliness, and behavior. Persons who suffer from, or are carriers of, an illness that can be transmitted through food have to be prevented from contaminating it. The symptoms have to be recognized as early as possible. It is the responsibility of the inflicted person to

---

* CAC/RCP 1-1969, Rev. 4-2003, Section III Primary production, 3.2 Hygienic production of food sources.
† CAC/RCP 1-1969, Rev. 4-2003, Section III Primary production, 3.3 Handling, storage and transport.
‡ CAC/RCP 1-1969, Rev. 4-2003, Section III Primary production, 3.4 Cleaning, maintenance and personnel hygiene at primary production.
§ CAC/RCP 1-1969, Rev. 4-2003, Section IV Establishment: Design and Facilities, 4.4 Facilities, 4.4.1 Water supply; 4.4.2 Drainage and waste disposal; 4.4.3 Cleaning.
¶ CAC/RCP 1-1969, Rev. 4-2003, Section IV Establishment: Design and Facilities, 4.4.4 Personnel hygiene facilities and toilets.

report this condition to the management, and it is the management's responsibility to recognize the symptoms as early as possible and act.

The Recommended International Code of Practice deals with the whole food production chain and presents, without using the term, a GHP.

2.2.5   THE CODEX ALIMENTARIUS CODE OF HYGIENIC PRACTICE FOR FRESH FRUITS AND VEGETABLES

The "Code of Hygienic Practice for Fresh Fruits and Vegetables" gives a more detailed application of the General Principles of Food Hygiene to fresh fruits and vegetables. It recommends explicitly GAPs and GMPs. That combines well with the GHP of the General Principles. The main additions are in Section 3 dealing with the conditions for hygienic primary production of fresh fruits and vegetables,* and Section 5 dealing with control of food hazards and key aspects of hygiene control systems. Microbial and chemical contaminants in agricultural inputs must remain below the levels that affect the safety of fresh fruits and vegetables.[†] Personal health and hygiene of the people who work on the farms are of the greatest importance.[‡] The equipment and facilities must be easy to clean and kept clean systematically by cleaning procedures and methods.[§] Pests have to be kept at bay and rigorously suppressed when they nevertheless show up.[¶] Wastes have to be managed to avoid direct contamination.[**]

2.2.6   COMMODITY-SPECIFIC ANNEXES TO THE CODE OF HYGIENIC PRACTICE FOR FRESH FRUITS AND VEGETABLES

The Codex Committee on Food Hygiene (CCFH) formulated an extensive request for scientific advice on the microbiological hazards associated with fresh produce.[††]

A Food and Agriculture Organization (FAO)/World Health Organization (WHO) Expert Meeting established the following criteria to rank the commodities of concern with regard to microbiological hazards associated with fresh produce:

---

[*] CAC/RCP 53-2003, 3.1 Environmental Hygiene, 3.2 Hygienic Primary Production of Fresh Fruits and Vegetables, 3.3 Handling, Storage and Transport, and 3.4 Cleaning, Maintenance and Sanitation.

[†] CAC/RCP 53-2003, 3.2 Hygienic Primary Production of Fresh Fruits and Vegetables, 3.2.1 Agricultural input requirements.

[‡] CAC/RCP 53-2003, 3.2 Hygienic Primary Production of Fresh Fruits and Vegetables, 3.2.1 Personnel health, hygiene and sanitary facilities.

[§] CAC/RCP 53-2003, 3.2 Hygienic Primary Production of Fresh Fruits and Vegetables, 3.2.4 Equipment associated with growing and harvesting, and 3.4 Cleaning, Maintenance and Sanitation.

[¶] CAC/RCP 1-1969, Rev. 4-2003, Section VI—Establishment: Maintenance and Sanitation. 6.3 Pest Control Systems, 6.3.1 General, 6.3.5 Eradication: Pest infestations should be dealt with immediately. Also: CAC/RCP 53-2003, 3. Primary production, 3.4 Cleaning, Maintenance and Sanitation, 3.4.3 Pest control systems: General Principles of Food Hygiene, section 6.3 should be followed.

[**] CAC/RCP 53-2003, 3. Primary production, 3.4 Cleaning, Maintenance and Sanitation, 3.4.4 Waste management.

[††] Codex Committee on Food Hygiene, Terms of Reference for an FAO/WHO Expert Consultation to Support the Development of Commodity-Specific Annexes for the Codex Alimentarius "Code of Hygienic Practice for Fresh Fruits and Vegetables" (ALINORM 07/30/13, Appendix VI).

- Frequency and severity of disease
- Size and scope of production
- Diversity and complexity of the production chain and industry
- Potential for amplification of foodborne pathogens through the food chain
- Potential for control
- Extent of international trade and economic impact*

The commodities of concern were ranked into three priority groupings with leafy greens as the highest priority, level 1.[†] Leafy greens are grown and exported in large volume, have been associated with multiple outbreaks with high numbers of illnesses in at least three regions of the world, and are grown and processed in diverse and complex ways, ranging from in-field packing to precut and bagged product. The ranking is reflected in the work priorities of the Codex Committee on Food Hygiene. It has prepared a draft Annex on fresh leafy vegetables, including leafy herbs, to the Code of Hygienic Practice for Fresh Fruits and Vegetables.[‡]

2.2.7 THE CODEX ANNEX FOR READY-TO-EAT FRESH PRECUT FRUITS AND VEGETABLES

The law on FcFV has to do without a generally accepted formal definition of the concept of *fresh-cut produce* and the related concept of *fresh produce*. Each jurisdiction can have its own definition or rules for fresh-cut produce without a definition or no rules for FcFV.

The CAC, the custodian of the international common consensual fund of basic food law concepts, uses the terms *fresh-cut fruits and vegetables* and *fresh fruits and vegetables*. It makes separate rules for each group in addition to the prescriptions for all foodstuffs. But it does this without a definition of the characteristics that make these foodstuffs special categories with their own special law. It is not uncommon for lawmakers to define several auxiliary concepts and leave the central concept undefined. This is a legal technique based on the implicit assumption that the set of rules made for an undefined subject is in a larger sense its legal definition.

The Annex deals with the additional precautions and production steps for FcFV. The raw material has to be inspected as it arrives at the production facility.[§] Animal and plant debris, metal, and other foreign material have to be removed.[¶] Water used for final rinses must have potable quality, particularly when the products will not be

---

* Food and Agriculture Organization of the United Nations and World Health Organization, Microbiological hazards in fresh fruits and vegetables, Expert Meeting Report, 2008.
† A definite ranking of the Level 2 Priorities (berries, green onions, melons, sprouted seeds, and tomatoes) and the Level 3 Priorities (carrots, cucumbers, almonds, baby corn, sesame seeds, onions, garlic, mango, paw paw, celery, and maimai) requires more information.
‡ Codex Alimentarius Commission, Joint FAO/WHO Food Standards Programme, Codex Committee on Food Hygiene, Fortieth Session, Guatemala, October 2008. Proposed Draft Annex on Fresh Leafy Vegetables Including Leafy Herbs to the Code of Hygienic Practice for Fresh Fruits and Vegetables at Step 3, CX/FH 08/40/7, October 2008.
§ 5.2.2.1 Annex I CAC/RCP 53-2003 p. 18.
¶ 5.2.2.2 Annex I CAC/RCP 53-2003 p. 18.

washed again before consumption.* Contamination during cutting, slicing, shredding, and similar processes must be minimized,† and cut produce must be washed to reduce some of the cellular fluids that were released during the cutting process. This lowers the level of nutrients for microbiological growth. Wash water must be replaced frequently to prevent the buildup of organic material and prevent cross-contamination. Antimicrobial agents may be used where their use is in line with GHPs. The levels of these agents must be monitored and controlled. Application of antimicrobial agents must ensure that chemical residues do not exceed the recommended Codex levels.‡ Precut fresh fruits and vegetables must be kept at low temperatures at all stages, from cutting through distribution, to minimize microbiological growth.§

Records must be made to provide information about the product and the processing operations. These records must be kept much longer than the shelf life of the product to facilitate recalls and foodborne illness investigations.¶

The training programs must deal with packaging systems for fresh precut fruits and vegetables, including the risks of contamination or microbiological growth. The importance of temperature control and GMPs must be stressed.**

The Codex texts have become part of several guides, practices, schemes, standards, and legislation. A continuous set of practices from primary production to manufacturing and from there to packaging and transport is especially important for fresh and fresh-cut fruits and vegetables, because the conditions for microbial contamination are determined mainly by primary production and packaging.

2.2.8 CODEX GUIDELINES ON GOOD HYGIENE PRACTICES TO CONTROL *LISTERIA MONOCYTOGENES* IN READY-TO-EAT FOODS

The CAC has formulated guidelines on how to apply the General Principles of Food Hygiene to control of *Listeria monocytogenes* in Ready-to-Eat Foods.††

These guidelines are in addition to the CAC RCP 1969 General Principles of Food Hygiene. They are necessary, because *L. monocytogenes* has found a permanent foothold in production and consumer premises. This presence of *L. monocytogenes* in establishments that process ready-to-eat food is caused by the following factors:

1. Inadequate separation of raw and finished product areas
2. Poor control of employees or equipment traffic
3. Inability to properly clean and disinfect equipment and premises due to poor layout or design and areas inaccessible to cleaning
4. Use of spray-cleaning procedures that make the bacteria airborne

---

* 5.2.2.3 Annex I CAC/RCP 53-2003 p. 18.
† 5.2.2.5 Annex I CAC/RCP 53-2003 p. 18.
‡ 5.2.2.6 Annex I CAC/RCP 53-2003 p. 18.
§ 5.2.2.7 Annex I CAC/RCP 53-2003 p. 18.
¶ 5.7 Annex I CAC/RCP 53-2003 p. 18.
** 10.2 Annex I CAC/RCP 53-2003 p. 18.
†† Guidelines on the Application of General Principles of Food Hygiene to the Control of Listeria monocytogenes in Ready-to-Eat Foods (CAC/GL 61-2007), ftp://ftp.fao.org/codex/ccfh40/fh40_05e.pdf.

5. Inability to properly control ventilation to minimize condensate formation on surfaces in food processing plants

The countermeasures follow from these causes. CAC/GL 61-2007 presents the following examples of guidelines on equipment and on temperature control:

4.3   Equipment
4.3.1 General
Due to the ability of *L. monocytogenes* to exist in biofilms and persist in harbourage sites for extended periods, processing equipment should be designed, constructed and maintained to avoid, for example, cracks, crevices, rough welds, hollow tubes and supports, close fitting metal-to-metal or metal-to-plastic surfaces, worn seals and gaskets or other areas that cannot be reached during normal cleaning and disinfection of food contact surfaces and adjacent areas.

5.2.1 Time and temperature control
The risk assessments done by the U.S. FDA/FSIS and FAO/WHO on *L. monocytogenes* in ready-to-eat foods demonstrated the tremendous influence of storage temperature on the risk of listeriosis associated with ready-to-eat foods that support *L. monocytogenes* growth. It is therefore necessary to control the time/temperature combination used for storage. Product temperature should not exceed 6°C (preferably 2°C–4°C).

The Codex Committee on Food Hygiene is developing microbiological criteria for *L. monocytogenes* in Ready-to-Eat Foods.*

2.2.9 FOODBORNE VIRUSES

The FAO and the WHO urge governments and food business operators to pay more attention to the risks posed by foodborne viruses.[†]

Fresh produce belong to the main routes for the introduction of viruses in high-risk commodities. The sources for virus contamination in fresh produce are contaminated water used for irrigation, agrochemical application, or wash water; the use of human sewage as fertilizer; and manual handling during harvest and postharvest. The relative contribution of each is not known.

GAPs, GHPs, GMPs, and guidelines for the quality of irrigation water have to address this.

---

* Codex Alimentarius Commission, Joint FAO/WHO Food Standards Programme, Codex Committee on Food Hygiene, Fortieth Session, Guatemala, October 2008. Proposed Draft Microbiological Criteria for Listeria monocytogenes in Ready-to-Eat Foods at Step 5/8 (ALINORM 09/32/13 para. 69 and Appendix II), http://www.codexalimentarius.net/web/archives.jsp?year=09.
† Food and Agriculture Organization of the United Nations (FAO) and the World Health Organization (WHO), Viruses in Food: Scientific Advice to Support Risk Management Activities Meeting Report, 2008, Microbiological Risk Assessment Series 13, http://www.who.int/foodsafety/publications/micro/Viruses_in_food_MRA.pdf.

2.3 BINDING LAW

Law can be binding as public law in human rights, treaties, customary law, legislation, regulation, taxation, and administrative and judicial decisions. It can be binding as private law in contracts, organizations, and properties.

It is not necessary to create a new branch of law for every extension of the food production chain. Existing law can be adequate, ranging from the very basic such as contract and industrial property law, to general food law and more specialized food law for fresh fruits and vegetables (FFV).

2.3.1 FOOD LEGISLATION

Horizontal food legislation contains rules on issues that apply to all food. Essential rules on food safety and food hygiene are examples of this type of legislation. Vertical food legislation is made to provide rules for specific issues that are not covered by the horizontal legislation with the necessary detail. The most general of all general food law rules determines the responsibilities of food business operators with the prescript that only safe food is brought to the market. The CAC has made this statement in the "Recommended International Code of Practice—General Principles of Food Hygiene" in paragraph 2.1.2 on the roles of governments, industry, and consumers.*

The European Community has made that statement in Regulation (EC) No. 178/2002, the foundation of its food legislation better known as the General Food Law (GFL).† Its central rule on food safety is Article 14 that requires that food will not be placed on the market if it is unsafe. It indicates two categories of unsafe food: produce that is injurious to health and produce that is unfit for human consumption.‡ The broad reach of these rules is assured by Article 17 that establishes the responsibilities of food business operators to make sure that their products meet the requirements of food law. The category of persons that carries the primary responsibility is defined by the concept "food business operator":

> The natural or legal persons responsible for ensuring that the requirements of food law are met within the food business under their control.

---

* CAC/RCP 1-1969, Rev. 4-2003, p. 4.
† Regulation (EC) No. 178/2002 of the European Parliament and of the Council of 28 January 2002 laying down the general principles and requirements of food law, establishing the European Food Safety Authority and laying down procedures in matters of food safety, OJ 1.2.2002 L 31/1, http://eur-lex.europa.eu/Result.do?T1=V2&T2=2002&T3=178&RechType=RECH_consolidated&Submit=Search.
‡ Article 14(1) and (2) Regulation (EC) No. 178/2002 (GFL). The remainder of this article lists the factors that have to be taken into consideration to establish whether food is unsafe, injurious to health, or unfit for human consumption.

"Food business" is defined as

> Any undertaking, whether for profit or not and whether public or private, carrying out any of the activities related to any stage of production, processing and distribution of food.*

The statement that food is not placed on the market if it is unsafe functions only when it has the backup of other parts of the law.

The EC does not have the power to use criminal law as a backup. Instead, it has to rely on the obligation of the member states to lay down the rules on measures and penalties applicable to infringements of food law. After the first paragraph about the responsibilities of the food business operators, the second paragraph of Article 17 GFL continues with the enforcement of food law by the member states. They have to maintain a system of official controls and other appropriate activities, and, finally, to lay down the rules on measures and penalties applicable to infringements of food law. "The measures and penalties have to be effective, proportionate and dissuasive."[†]

The United States also has legislation that prescribes that food must be fit for consumption by humans:

> A food shall be deemed to be adulterated (a) Poisonous, insanitary, or deleterious ingredients, (…) (4) if it has been prepared, packed, or held under insanitary conditions whereby it may have become contaminated with filth, or whereby it may have been rendered injurious to health; (…)."[‡]

The federal government can enforce this requirement with the assistance of criminal law. That takes three steps:

1. Food produced under unsanitary conditions that make it injurious to health is adulterated food.[§]
2. It is prohibited to introduce adulterated food into interstate commerce.[¶]
3. The punishment for this violation of the Federal Food, Drug, and Cosmetic Act is imprisonment for no more than 1 year, a fine of not more than $1,000, or both. The punishment for a violation after a previous final conviction, or committed with the intent to defraud or mislead is imprisonment for not more than 3 years, a fine not more than $10,000, or both.[**]

---

* Articles 3(3) and 3(2) Regulation (EC) No. 178/2002 (GFL).
† Article 17 Regulation (EC) No. 178/2002 (GFL).
‡ 21 U.S.C. 342(a)(4).
§ §402(a)(4) of the Federal Food, Drug, and Cosmetic Act.
¶ U.S. Federal Food, Drug and Cosmetic Act Chapter III: Prohibited Acts and Penalties, Section 301 (21 USC §331) 1a.
** U.S. Federal Food, Drug and Cosmetic Act Section 303, at http://www.fda.gov/opacom/laws/fdcact/fdcact3.htm.

2.3.2 SPECIALIZED FOOD LEGISLATION: EC REGULATION 852/2004 ON THE HYGIENE OF FOODSTUFFS

Regulation 852/2004 is the bridge between the GFL and more specific EC hygiene legislation.* It has four functions: It provides the basic food hygiene prescripts for all stages in the production chain.† It creates the obligation for food business operators to put in place, implement, and maintain a permanent procedure based on the HACCP principles.‡ It also promotes the development, assessment, dissemination, and periodical review of national and Community guides to good practice for hygiene or for the application of HACCP principles.§ The fourth function is to prescribe for which subjects additional legislation has to be made.¶ These more specific hygiene regulations deal with food of animal origin,** official controls,†† microbiological criteria for foodstuffs,‡‡ and several other subjects.

2.3.3 PRIMARY PRODUCERS AND FOOD HYGIENE

Regulation (EC) No. 852/2004 separates primary producers from food business operators in the other stages of the food chain. Primary producers do not have to use procedures based on the HACCP principles.§§ They are subjected to a rudimentary set of hygiene rules for primary production and associated operations specified in Annex I.¶¶

The general hygiene duty is to protect primary products, as much as possible, from contamination. This duty is qualified by any processing in later stages.

Food business operators have to apply Community and national legislative provisions relating to the control of hazards in primary production and associated operations, including measures to control contamination arising from the air, soil, water, feed, fertilizers, veterinary medicinal products, plant protection products, biocides, and the storage, handling, and disposal of waste.***

Food business operators producing or harvesting plant products have to keep clean and, where necessary, disinfect facilities, equipment, containers, crates, vehicles, and vessels; use potable water, or clean water; ensure that staff handling foodstuffs are in good health and undergo training on health risks; prevent animals and pests

---

* Regulation (EC) No. 852/2004 of the European Parliament and of the Council of 29 April 2004 on the hygiene of foodstuffs. OJ L 139, 30 April 2004. Corrected version in OJ L 226, 25 June 6, 2004, http://eur-lex.europa.eu/LexUriServ/LexUriServ.do?uri=OJ:L:2007:204:0026:0026:EN:PDF.
† Articles 3 and 4(1) with Annex I for primary production and 4(2) with Annex II for all subsequent production Regulation (EC) No. 852/2004.
‡ Article 5 Regulation (EC) No. 852/2004.
§ Articles 7 to 9 Regulation (EC) No. 852/2004.
¶ Article 4(3) Regulation (EC) No. 852/2004.
** Article 4(1) and (2) Regulation (EC) No. 852/2004.
†† Article 6 Regulation (EC) No. 852/2004.
‡‡ Article 4(3)(a) Regulation (EC) No. 852/2004. Commission Regulation (EC) No. 2073/2005 of 15 November 2005 on microbiological criteria for foodstuffs (text with EEA relevance), (OJ L 338, 22.12.2005, p. 1).
§§ Article 5(3) Regulation (EC) No. 852/2004.
¶¶ Article 4(1) and Annex I Regulation (EC) No. 852/2004.
*** §2 and 3(a) II. Hygiene provisions Part A Annex I Regulation (EC) No. 852/2004.

from causing contamination; and use plant protection products and biocides correctly. They have to take appropriate remedial action when informed of problems identified during official controls.*

Food business operators have to keep records of measures to control hazards and give this information to the competent authority and receiving food business operators on request.[†]

Especially important is information on any use of plant protection products and biocides, any occurrence of pests or diseases that may affect the safety of products of plant origin, and the results of analyses of samples which are relevant for human health.[‡]

2.3.4 ALL PRODUCERS AFTER PRIMARY PRODUCTION AND FOOD HYGIENE

The food business operators active in the stages after primary production have to apply the food hygiene prescriptions of Annex II.[§] This is an elaborate set of rules organized in twelve chapters which deal with all aspects of food hygiene. Table 2.2 presents the titles of these chapters.

TABLE 2.2

Food Hygiene Prescriptions of Annex II to EC Regulation 852/2004 on the Hygiene of Foodstuffs for Food Business Operators Active in the Stages after Primary Production

	Annex II General Hygiene Requirements for All Food Business Operators (Except when Annex I Applies) EC Regulation 852/2004 on the Hygiene of Foodstuffs
Chapter	**Title**
	Introduction
I	General requirements for food premises (other than those specified in Chapter III)
II	Specific requirements in rooms where foodstuffs are prepared, treated, or processed
III	Requirements for movable and/or temporary premises and premises used primarily as a private dwelling-house
IV	Transport
V	Equipment requirements
VI	Food waste
VII	Water supply
VIII	Personal hygiene
IX	Provisions applicable to foodstuffs
X	Provisions applicable to the wrapping and packaging of foodstuffs
XI	Heat treatment
XII	Training

---

* §5 II. Hygiene provisions Part A. Annex I Regulation (EC) No. 852/2004.
[†] §7 III. Record-keeping Part A. Annex I Regulation (EC) No. 852/2004.
[‡] §9 III. Record-keeping Part A. Annex I Regulation (EC) No. 852/2004.
[§] Article 4(2) and Annex II Regulation (EC) No. 852/2004.

The following is an example of the food hygiene legislation for food business operators from Chapter I about the general requirements for food premises:

1. Food premises are to be kept clean and maintained in good repair and condition.
2. The layout, design, construction, siting and size of food premises are to:
   (a) permit adequate maintenance, cleaning and/or disinfection, avoid or minimise airborne contamination, and provide adequate working space to allow for the hygienic performance of all operations;
   (b) be such as to protect against the accumulation of dirt, contact with toxic materials, the shedding of particles into food and the formation of condensation or undesirable mould on surfaces;
   (c) permit good food hygiene practices, including protection against contamination and, in particular, pest control; and
   (d) where necessary, provide suitable temperature-controlled handling and storage conditions of sufficient capacity for maintaining foodstuffs at appropriate temperatures and designed to allow those temperatures to be monitored and, where necessary, recorded.*

A second example of the rules from Annex II is the complete Chapter V Equipment requirements:

1. All articles, fittings and equipment with which food comes into contact are to:
   (a) be effectively cleaned and, where necessary, disinfected. Cleaning and disinfection are to take place at a frequency sufficient to avoid any risk of contamination;
   (b) be so constructed, be of such materials and be kept in such good order, repair and condition as to minimise any risk of contamination;
   (c) with the exception of non-returnable containers and packaging, be so constructed, be of such materials and be kept in such good order, repair and condition as to enable them to be kept clean and, where necessary, to be disinfected; and
   (d) be installed in such a manner as to allow adequate cleaning of the equipment and the surrounding area.
2. Where necessary, equipment is to be fitted with any appropriate control device to guarantee fulfilment of this Regulation's objectives.
3. Where chemical additives have to be used to prevent corrosion of equipment and containers, they are to be used in accordance with good practice.

The general food hygiene requirements function as the necessary legislation on the responsibilities of the food business operator. The operator has the obligation to apply the HACCP system and is encouraged to use GHPs to prevent problems with food hygiene. These instruments have to make recourse to the legislation about specific acceptable food hygiene conditions unnecessary.

---

* http://eur-lex.europa.eu/LexUriServ/LexUriServ.do?uri=CONSLEG:2004R0852:20081028:EN:PDF

2.3.5 GUIDES BY GOVERNMENTS

The EC food hygiene legislation is accompanied by a set of explanations made by the executive power. Regulation (EC) No. 178/2002 is explained by a guidance document made by the Standing Committee of Representatives of the Commission and the Member States on the civil servant level.* The following three examples explain elements of Regulation (EC) No. 852 on the hygiene of food:

> European Community Commission, Health and Consumer Protection Directorate-General, Guidance document on the implementation of certain provisions of Regulation (EC) No 852/2004 On the hygiene of foodstuffs, Brussels, 21 December 2005, pp. 1–17.
>
> European Community Commission, Health and Consumer Protection Directorate-General, Guidance document on the implementation of procedures based on the HACCP principles, and on the facilitation of the implementation of the HACCP principles in certain food businesses, Brussels, 16 November 2005, pp. 1–26.
>
> European Community Commission, Health and Consumer Protection Directorate-General, Guidance document on certain key questions related to import requirements and the new rules on food hygiene and on official food controls, Brussels, 5.1.2006, pp. 1–28.

2.3.6 NATIONAL AND COMMUNITY GUIDES MADE BY THE ORGANIZATIONS OF FOOD BUSINESS OPERATORS

Regulation 852/2004 on the hygiene of foodstuffs offers organizations of food business operators the opportunity to take the initiative to make national and Community guides to good food hygiene practice or to apply HACCP principles. These guides are a major instrument to achieve the food hygiene objectives of Regulation 852/2004. The guides are voluntary: food business operators do not have to participate in the drafting of the guides. When guides are made by other food business organizations, they remain voluntary. The guides do not replace the legal obligations of the food business operators but help them to fulfill these obligations. The member states have to encourage organizations of food business operators to draft national guides. The Commission has to do the same on the EC level, but only after a consultation of the member states has established the usefulness of Community guides and their scope and subject matter.

The relevant Codes of Practice of the Codex Alimentarius have to be used for the preparation of the national and Community guides. The EC wants to contribute to the development of international food law.†

The guides are made according to similar procedures. National guides are made by organizations of food business operators. Their drafts are assessed by the member states to verify that the guides are made in consultation with the representatives of

---

* European Community, Standing Committee on the Food Chain and Animal Health, Guidance on the Implementation of Articles 11, 12, 16, 17, 18, 19, and 20 of Regulation (EC) No. 178/2002 on General Food Law, Conclusions of the Standing Committee on the Food Chain and Animal Health, 20 December 2004.
† Article 13 Regulation (EC) No. 178/2002 GFL.

organizations whose interests are substantially affected, such as other food business operators, consumer organizations, and public authorities. The member states send approved national guides to the European Commission to place them in a registration system available to the member states. The national guides are disseminated by the organizations of food business operators that made them.

The Community guides are prepared with the assistance of the Commission in a comparable procedure. The Community guides must be assessed in consultation with the member states and will be recognized by the Commission after this approval.*

The Commission has published guidelines on how to make Community guides.[†] The title and references of community guides are published in the C series of the *Official Journal of the European Union*. The guides will be publicly available on a dedicated page of the Commission's Health and Consumers Directorate-General (DG SANCO) Web site.

The references of national guides are published in the Register of National Guides to Good Hygiene Practice.[‡] It presents a table of thirty-eight pages with an overview of guides in the member states with the following information: title (original), title (English), country, language, publisher, edition, ISBN, ISSN, Internet or other contact, and key word.

An example is the following:

> Manuale di corretta prassi igienica per i centri di lavorazione e confezionamento dei prodotti ortofrutticoli freschi, surgelati, di IV gamma, degli agrumi della frutta a guscio ed essiccata. Title (English): Guide to Good Hygiene Practice for processing and packaging centres for fresh, frozen, washed-cut-and-packed fruits and vegetables, citrus fruits, nuts and dried fruit. Published by the Associazione nazionale esportatori i, portatori ortofrutticoli e agrumari (ANEIOA) in June 1999 with www.ministerosalute.it as internet contact and Fruit and Vegetables as key word.[§]

The European Commission published guidelines on the registration of national guides.[¶]

2.3.7 SPECIFIC LEGISLATION ON MICROBES

Although most microbiological hazards are related to food of animal origin, some microbes cause hazards in the production of FcFV. They are subjected to stringent

---

* See Articles 7 to 9 Regulation 852/2004 and Bernd van der Meulen and Menno van der Velde, European Food Law Handbook, Wageningen 2008, pp. 353–358.
† European Community, Commission of the European Communities, Health and Consumers Directorate-General, Guidelines for the Development of Community Guides to Good Practice for Hygiene or for the Application of the HACCP Principles, in Accordance with Article 9 of Regulation (EC) No. 852/2004 on the Hygiene of Foodstuffs and Article 22 of Regulation (EC) No. 183/2005 Laying Down Requirements for Feed Hygiene. http://ec.europa.eu/food/food/biosafety/hygienelegislation/guidelines_good_practice_en.pdf.
‡ http://ec.europa.eu/food/food/biosafety/hygienelegislation/register_national_guides_en.pdf.
§ Page 28 of the Register.
¶ European Community, Commission of the European Communities, Health and Consumers Directorate-General, Guidelines on the registration of national guides to good practice in accordance with Article 8 of Regulation (EC) No. 852/2004, Brussels, 15 July 2008, http://ec.europa.eu/food/food/biosafety/hygienelegislation/national_guides_register_15-07-2008.pdf.

measures because they can be present in ready-to-eat food. One of these organisms is the bacterium *Listeria monocytogenes*—it survives low temperatures and kills one of every five persons it infects.

The reduction of microbiological hazards is an objective of different parts of food law such as hygiene codes, good practices, general hygiene legislation, and specialized legislation. Commission Regulation (EC) No. 2073/2005 on microbiological criteria for foodstuffs* is an example of legislation that adds the determining details to the general requirements of hygiene Regulation EC No. 852/2004.[†] It does this in the context of the implementation of good hygiene practice and application of procedures based on hazard analysis and critical control point (HACCP) principles. "The safety of foodstuffs is mainly ensured by a preventive approach."[‡] This approach and the prescription of microbiological criteria can reinforce each other by using the criteria to verify the good functioning of HACCP procedures and good hygiene practice. However, the microbiological criteria are also used to ban failing food from the market and for hygiene control.

The preventive approach calls for effective good hygiene practices.

2.3.8 SPECIFIC LEGISLATION ON MICROBIAL CONTAMINATION

The CAC has made guidelines on microbiological criteria for foods and the conduct of microbiological risk assessment.[§]

The EC legislation follows the Codex guideline for microbiological criteria.[¶] It is also based on reports and opinions of EC committees and panels about the hazards of several types of microbiological contamination of fruits and vegetables eaten raw.[**] One of these types is *Listeria monocytogenes*. Opinions of the risks related to this microbe were first given for food of animal origin[††] and later extended to all

---

[*] European Commission Regulation (EC) No. 2073/2005 of 15 November 2005 on microbiological criteria for foodstuffs (text with EEA relevance). OJ L 338, 22.12.2005, pp. 1–26. Consolidated version: http://eur-lex.europa.eu/LexUriServ/LexUriServ.do?uri=CONSLEG:2005R2073:20071227:EN:PDF.

[†] Article 4(3)(a) Regulation (EC) No. 852/2004.

[‡] Recital 5 Commission Regulation (EC) No. 2073/2005.

[§] Codex Alimentarius Commission. CAC/GL 21-1997 Principles for the Establishment and Application of Microbiological Criteria for Foods, http://www.codexalimentarius.net/web/more_info.jsp?id_sta=394. Codex Alimentarius Commission CAC/GL-30 (1999) Principles and Guidelines for the Conduct of Microbiological Risk Assessment, http://www.codexalimentarius.net/web/more_info.jsp?id_sta=357.

[¶] Recital (18) Commission Regulation (EC) No. 2073/2005.

[**] European Community, Commission of the European Communities, Health and Consumer Protection Directorate-General, Directorate C—Scientific Opinions, C2—Management of Scientific Committees II; Scientific Cooperation and Networks. Scientific Committee on Food, Risk Profile on the Microbiological Contamination of Fruits and Vegetables Eaten Raw. Report of the Scientific Committee on Food (adopted on the 24th of April 2002), SCF/CS/FMH/SURF/Final, 29 April 2002. At http://ec.europa.eu/food/fs/sc/scf/out125_en.pdf.

[††] European Community, Commission of the European Communities, Health and Consumer Protection Directorate-General, Directorate B—Scientific Health Opinions, Unit B3—Management of Scientific Committees II. Opinion of the Scientific Committee on Veterinary Measures relating to Public Health on *Listeria monocytogenes*, 23 September 1999. http://ec.europa.eu/food/fs/sc/scv/out25_en.pdf.

food.* These opinions have been updated by an opinion of the competent panel of the European Food Safety Authority EFSA.[†]

Commission Regulation (EC) No. 2073/2005 gives food safety criteria that impose limits to the presence of specified microorganisms and their toxins and metabolites in twenty-seven different food categories.[‡] The regulation introduces two types of microbiological criteria, the process hygiene criteria that indicate the acceptable functioning of the production process, and the food safety criteria that define the acceptability of products on the market.

Process hygiene criteria prescribe food business operators to take measures based on their HACCP method and good hygiene practice to carry out the supply, handling, and processing of raw materials and foodstuffs.[§]

Food business operators have to make sure that their products meet the food safety criteria throughout the shelf life.[¶] This responsibility covers all stages following manufacturing: distribution, storage, and retail. It covers use of the produce and reaches into the household of the consumer whose refrigerator temperature is often too high and a major cause for microbial contamination. The food safety criteria take this into account. The responsibility is limited to the reasonably foreseeable conditions.

Food business operators in the manufacturing stage of the production chain have to conduct studies to investigate whether their products meet the criteria throughout the shelf life. This obligation applies in particular to ready-to-eat foods that are able to support the growth of *Listeria monocytogenes* and may pose a risk for public health.** The European Commission developed a guidance document,[††] and the Community Reference Laboratory for *Listeria monocytogenes* prepared a technical guidance document for laboratories conducting shelf-life studies.[‡‡]

Food business operators have to perform tests against the microbiological criteria of Annex I when they verify the correct functioning of the HACCP-based procedure

---

* European Community, Commission of the European Communities, Health and Consumer Protection Directorate-General, Directorate C Scientific Opinions, Directorate C—Scientific Opinions, C3—Management of Scientific Committees II; Scientific Cooperation and Networks, Opinion of the Scientific Committee on Food in respect of Listeria monocytogenes expressed on 22 June 2000. http://ec.europa.eu/food/fs/sc/scf/out63_en.pdf.

† Scientific Opinion of the Panel on Biological Hazards on a request from the European Commission on Request for updating the former SCVPH opinion on *Listeria monocytogenes* risk related to ready-to-eat foods and scientific advice on different levels of *Listeria monocytogenes* in ready-to-eat foods and the related risk for human illness. The EFSA Journal (2007) 599, 1–42. http://www.efsa.europa.eu/cs/BlobServer/Scientific_Opinion/biohaz_op_ej599_listeria_en.pdf?ssbinary=true.

‡ Annex I Commission Regulation (EC) No. 2073/2005.

§ Article 3(1)(a) and Chapter 2 Annex I Commission Regulation (EC) No. 2073/2005.

¶ Article 3(1)(b) and Chapter 1 Annex I Commission Regulation (EC) No. 2073/2005.

** Article 3(2) and Annex II Commission Regulation (EC) No. 2073/2005.

†† European Community, Commission of the European Communities, Guidance Document on *Listeria monocytogenes* shelf-life studies for ready-to-eat foods, under Regulation (EC) No. 2073/2005 of 15 November 2005 on microbiological criteria for foodstuffs, Commission Staff Working Document, SANCO/1628/2008 ver. 9.3 (26112008) Brussels 2008, http://ec.europa.eu/food/food/biosafety/salmonella/docs/guidoc_listeria_monocytogenes_en.pdf.

‡‡ Agence Française de sécurité sanitaire des aliments, AFSSA, Laboratoire d'Etudes et de Recherches sur la Qualité des Aliments et sur les Procédés Agro-alimentaires, Technical Guidance Document on shelf-life studies for *Listeria monocytogenes* in ready-to-eat foods, Version 2 – November 2008, http://ec.europa.eu/food/food/biosafety/salmonella/docs/shelflife_listeria_monocytogenes_en.pdf.

and the good hygiene practice they use. They have to decide the appropriate sampling frequencies in the context of this procedure and practice. They also have to take into account what the instructions for use of the foodstuff will be.*

Commission Regulation (EC) No. 2073/2005 specifies the obligations for food business operators to secure that their products meet the microbiological criteria and prescribes the detection methods they have to use. The detection methods refer to specific International Organization for Standardization (ISO) standards for food samples. Samples have to be taken from processing areas and equipment when necessary, and then ISO Standard 18593 is the reference method.[†] The number of sample units is specified in Annex I, but the food business operator can reduce the number if the producer can demonstrate by historical documentation that effective HACCP-based procedures are used.[‡] Food business operators may use other sampling and testing procedures when they can convince the competent authority that these procedures provide at least equivalent guarantees.[§] Alternative analytical methods are accepted when they are validated against the reference method specified in Annex I and when the operator uses a proprietary method certified by a third party in accordance with the protocol of ISO standard 16140 or other internationally accepted similar protocols.[¶] Other analytical methods are accepted when they are validated against internationally accepted protocols and authorized by the competent authority.[**]

FcFV are on the Annex I food safety criteria list in relation to two microorganisms. A food safety criterion is set for precut fruits and vegetables (ready-to-eat) in relation to *Salmonella*.[††] Food safety measures are taken against *Listeria monocytogenes* in relation to three food categories.[‡‡] The first category is ready-to-eat foods intended for infants and special medical purposes. In this food category, regular testing is not required in normal circumstances for fresh, uncut, and unprocessed vegetables and fruits. By implication, regular testing is required for fresh-cut produce. The two remaining ready-to-eat food categories are for food not intended for infants or special medical purposes. Food category 1.2, the ready-to-eat foods able to support the growth of *Listeria monocytogenes,* is to be tested regularly. The test result of this category is satisfactory when the food business operator is able to demonstrate that the product will not exceed the limit of 100 cfu/g throughout the shelf life.[§§] When the operator cannot demonstrate this condition, the following test results apply before the food has left the immediate control of the producer—a satisfactory result if all observed values indicate the absence of the bacterium, and an unsatisfactory result if its presence is detected in any of the sample units. The third category

---

* Article 4(2) Commission Regulation (EC) No. 2073/2005.
† Article 5(2) Commission Regulation (EC) No. 2073/2005.
‡ Article 5(3) Commission Regulation (EC) No. 2073/2005.
§ Article 5(5) Commission Regulation (EC) No. 2073/2005.
¶ Article 5(5)(§3) Commission Regulation (EC) No. 2073/2005.
** Article 5(5)(§4) Commission Regulation (EC) No. 2073/2005.
†† Food category 1.1, 1.2, and 1.3 Chapter 1 Food safety criteria. Annex I Commission Regulation (EC) No. 2073/2005.
‡‡ Food category 1.19 Chapter 1 Food safety criteria. Annex I Commission Regulation (EC) No. 2073/2005.
§§ Cfu/g stands for colony forming units per gram.

is the ready-to-eat foods unable to support the growth of this bacterium. Products with a specified acidity and water activity or a shelf life of less than 5 days belong automatically to this category.* Other categories of products can be added when scientifically justified. Regular testing is required for fresh-cut produce, but the test results are less stringent. They are satisfactory if all the values observed are lower than or equal to the limit of 100 cfu/g.

2.3.9   THE CONSEQUENCES OF FAILURES TO MEET THE FOOD SAFETY CRITERIA AND PROCESS HYGIENE CRITERIA

The food business operator has to withdraw or recall the product or batch when the tests against the food safety criteria specified in Chapter 1 of Annex I are unsatisfactory.[†]

The food safety criteria define the requirements that apply to the product in a definitive manner: Unsatisfactory test results mean that the product is not allowed—it is unsafe according to Article 14 GFL. Food placed on the market or imported into the EC that does not meet the food safety criteria has to be removed.

Process hygiene criteria are set for ready-to-eat precut fruits and vegetables in relation to *E. coli* at the manufacturing stage. Unsatisfactory test results have to lead the food business operator to improve production hygiene and selection of raw materials.[‡]

2.3.10   THE REGULATORY STYLE OF THE U.S. FOOD AND DRUG ADMINISTRATION

The FDA has its own toolkit to promote food safety. It can use its regulatory powers to legislate detailed rules, publish guides, give advice, and investigate. Usually, several of these instruments are combined.

2.3.11   CURRENT GOOD MANUFACTURING PRACTICE IN MANUFACTURING, PACKING, OR HOLDING HUMAN FOOD (CGMP)

The "Current Good Manufacturing Practice in Manufacturing, Packing, or Holding Human Food" is an example of very detailed legislation. It is made by the U.S. Food and Drug Administration (FDA) based on the regulatory powers given by the Federal Food, Drug, and Cosmetic Act and the Federal Public Health Act.[§]

The name is confusing, because it is not a GMP from the collection of voluntary good practices. It is binding regulatory law. This is one example of how legislation changes the significance of good practices by referring to them. It can be said to be

---

* Products with pH ≤ 4.4 or a_w ≤ 0.92, products with pH ≤ 5 and a_w ≤ 0.94, Food category 1.3 Chapter 1 Food safety criteria. Annex I Commission Regulation (EC) No. 2073/2005.

† Article 7 Commission Regulation (EC) No. 2073/2005 and Article 19 Regulation (EC) No. 178/2002, GFL.

‡ Food category 2.5.1, 2.5 Vegetables, fruits and products thereof, Chapter 2 Process hygiene criteria. Annex I Commission Regulation (EC) No. 2073/2005.

§ United States of America, Code of Federal Regulations (CFR) Title 21: Food and Drugs, Part 110. Current Good Manufacturing Practice in Manufacturing, Packing, or Holding Human Food. (Abbreviated to 21 CFR Part 110), http://ecfr.gpoaccess.gov/cgi/t/text/text-idx?c=ecfr;sid=03c0f0ea8b 688b533744ae63a89f6aa4;rgn=div5;view=text;node=21%3A2.0.1.1.10;idno=21;cc=ecfr.

an extreme example, because the legislation in this case does not merely refer to a settled practice or incorporate a part of it, but it determines it word for word, sentence by sentence as binding legislation.

These food safety practices have to be applied by processors who manufacture, process, pack, or hold human food.* This means that this regulation does not cover the production processes on the farm. It does cover the production of FcFV.

The CGMP prescribes the duties of the persons who manage food production to secure the highest possible levels of food hygiene. It provides rules for all aspects of food processing after harvesting for all persons and all materials that can influence food hygiene. The CGMP is applied in the buildings or other facilities where human food is manufactured, packed, labeled, or held. All reasonable measures and precautions have to be taken to ensure that the personnel working in those facilities report their health condition, maintain cleanliness, and are educated and trained in proper food handling and food protection principles.

Disease control can serve as an example of the nature of the requirements and the degree of detail. It is formulated as follows:

> Any person who, by medical examination or supervisory observation, is shown to have, or appears to have, an illness, open lesion, including boils, sores, or infected wounds, or any other abnormal source of microbial contamination by which there is a reasonable possibility of food, food-contact surfaces, or food-packaging materials becoming contaminated, shall be excluded from any operations which may be expected to result in such contamination until the condition is corrected. Personnel shall be instructed to report such health conditions to their supervisors.[†]

Cleanliness is obligatory for all persons working in direct contact with food, food-contact surfaces, and food-packaging materials. They have to conform to hygienic practices at work to the extent necessary to protect against contamination of food.[‡] The responsibility for assuring compliance by all personnel with all these requirements has to be clearly assigned to competent supervisory personnel.[§]

The duties of the responsible operator with regard to buildings, grounds, other facilities, sanitary operations, sanitary facilities, equipment and utensils, process controls, storage, and distribution are regulated in the same manner.

The paragraph on sanitary operations opens with the following statement:

> (a) General maintenance. Buildings, fixtures, and other physical facilities of the plant shall be maintained in a sanitary condition and shall be kept in repair sufficient to prevent food from becoming adulterated within the meaning of the act. Cleaning and sanitizing of utensils and equipment shall be conducted in a manner that protects against contamination of food, food-contact surfaces, or food-packaging materials.[¶]

---

* Code of Federal Regulations (CFR) Title 21: Food and Drugs, Part 110. Current Good Manufacturing Practice in Manufacturing, Packing, or Holding Human Food. (Abbreviated to 21 CFR Part 110).
† 21 CFR Part 110 Subpart A. General Provisions § 110.10 Personnel (a) Disease control
‡ 21 CFR Part 110 Subpart A. General Provisions § 110.10 Personnel (b) Cleanliness.
§ 21 CFR Part 110 Subpart A. General Provisions § 110.10 Personnel (d) Supervision.
¶ 21 CFR Part 110. §110.35(a) Sanitary operations.

The production and process controls are regulated in seventeen items with Point 13 on protection against contamination formulated as follows:

Filling, assembling, packaging, and other operations shall be performed in such a way that the food is protected against contamination. Compliance with this requirement may be accomplished by any effective means, including:

   (i) Use of a quality control operation in which the critical control points are identified and controlled during manufacturing.
  (ii) Adequate cleaning and sanitizing of all food-contact surfaces and food containers.
 (iii) Using materials for food containers and food-packaging materials that are safe and suitable, as defined in §130.3(d) of this chapter.
  (iv) Providing physical protection from contamination, particularly airborne contamination.
   (v) Using sanitary handling procedures.*

The complete section on production and process controls does not mention the HACCP system as such, but Point 13(i) refers to the use of the critical control points.

2.3.12   U.S. Guides

The 2008 U.S. "Guide to Minimize Microbial Food Safety Hazards of Fresh-Cut Fruits and Vegetables" is made by the FDA as a set of nonbinding recommendations for the industry. It is one of the instruments used by the FDA to deal with the food safety hazards of fresh produce. Other instruments are the "Guide to Minimize Microbial Food Safety Hazards for Fresh Fruits and Vegetables" made in 1998 (USA FFV Guide 1998)† and the "Letter to California Firms That Grow, Pack, Process, or Ship Fresh and Fresh-cut Lettuce" sent and published in 2005.‡

Both guides are not binding and do not create rights or obligations, not for the producers and not for the FDA or the federal government. They reflect only the latest considered knowledge of the FDA. Producers with alternatives are invited to contact the FDA to discuss their ideas.§ The guides are suggestions of food safety practices that processors of fresh-cut produce can follow to meet the mandatory CGMP requirements. The guides can deal in greater detail with the issues that are of particular significance for the producers of FcFV. The regulatory CGMP is made for all kinds of products. One can imagine the CGMP, a text of some six thousand words, as the regulatory core founded on federal legislation and surrounded by several guides explaining the regulation for different groups of produce. The 2008 guide

---

* 21 CFR Part 110. Subpart E. Production and Process Controls. § 110.80. Processes and controls.
† U.S. Department of Agriculture, Centers for Disease Control and Prevention, and U.S. Department of Health and Human Services, Food and Drug Administration, Center for Food Safety and Applied Nutrition (CFSAN), Guidance for Industry. Guide to Minimize Microbial Food Safety Hazards for Fresh Fruits and Vegetables, October 26, 1998. http://www.foodsafety.gov/~dms/prodguid.html.
‡ United States of America, Food and Drug Administration, CFSAN/Office of Plant and Dairy Foods, Letter to California Firms That Grow, Pack, Process, or Ship Fresh and Fresh-cut Lettuce, November 4, 2005, http:www.cfsan.fda.gov/~dms/prodltr2.html.
§ USA FcFV Guide 2008, text box preceding 1. Introduction.

on fresh-cut produce uses fifteen thousand words to explain the CGMP, and the 1993 guide on fresh produce uses some thirty-two thousand words.

In addition to providing tailor-made advice to producers, the two guides also play a role in the regulatory style developed by the FDA to achieve the required level of food safety by a combination of instruments.

The FFV Guide USA of 1998 was followed by letters of the FDA to growers of fresh lettuce that was processed into fresh-cut lettuce to inform them that significant food safety incidents were caused by food hygiene failures in their area. The products could not be traced back to individual companies. The letters urged the producers to apply the mandatory hygiene prescriptions, and conveyed the agency's "serious concern with the continuing outbreaks of foodborne illness associated with the consumption of fresh and fresh-cut lettuce and other leafy greens." The agency used the letters to outline its plans and to specify the actions it expects from the industry to enhance the safety of these products. It kindly reminds the producers that it is investigating its regulatory options and will consider enforcement actions against firms and farms that grow, pack, or process fresh lettuce and leafy greens under unsanitary conditions.

The use of the HACCP system is a second example of this "hands-off" approach. The FDA notes that U.S. law does not require FcFV producers to apply the HACCP system. At the same time, it refers approvingly to the United Fresh Produce Association's recommendation to use HACCP, and the information that this is done by most producers.*

2.3.13 STANDARDS

Legislation can promote the use of standards directly by referring to them as technical norms that have to be applied to carry out the law. Legislation can promote the use of standards indirectly by introducing concepts like *due diligence* into food law. The government of the United Kingdom did this after a series of major food safety problems. The due diligence concept changes the burden of proof. It means that in any proceeding for an offense under the Food Safety Act 1990, a food business operator can defend himself by proving "that he took all reasonable precautions and exercised all due diligence to avoid the commission of the offence by himself or by a person under his control."†

This worked as a stimulus to pay more systematic attention to food safety issues, geared to the prevention and collection of verifiable records of good practices. It led the British Retail Consortium (BRC) to introduce the Food Technical Standard for retailers to evaluate the work done by the manufacturers of their own brand of food products. That proved to be only the first of a series of BRC food-related standards. The most recent representatives are the Global Standard for Food Safety and the Global Standard for Packaging and Packaging Materials, both published in 2008.

---

* USA FcFV Guide 2008, Section II Fresh-cut Produce and HACCP Systems.
† United Kingdom Food Safety Act 1990, 1990 Chapter 16, Part II Main Provisions 21, http://www.opsi.gov.uk/acts/acts1990/ukpga_19900016_en_3#pt2-pb4-l1g21.

The International Organization for Standardization published the ISO 22000 standard for "Food safety management systems—Requirements for organisations throughout the food chain." It applies the HACCP system in every stage of the production chain.

2.3.14 THE GLOBAL G.A.P. STANDARD

A sophisticated combination of instruments for food safety and quality was developed by the GlobalG.A.P. organization, the globalized successor of EurepGAP, an initiative of a group of leading European retailers to transmit the preferences of their customers, and some of their own, to primary producers.

GlobalG.A.P. is many things at the same time—a GAPractice, an application of the HACCP system to primary production, a standard for certification, and, finally, a farm insurance contract.*

The farmers and growers who participate in the scheme have to check the way they work at least twice a year against the GlobalG.A.P. standard. First as self-assessment, and then in the certification procedure with an independent certification body, selected by the GlobalG.A.P. organization.

The control points and compliance criteria are organized in three frameworks arranged according to the specialization in the production processes. The first framework is Section AF, the "All Farm Base," covering the general aspects of primary production.

The second framework contains several sections of larger product groups such as CB Crops Base," LB "Livestock Base," and AB "Aquaculture Base." Each of these groups is subdivided into a number of specific products that form the third framework.

Section FV "Fruit and Vegetables" is the first specific product group in the "Crops Base." The 236 potential control points for fruits and vegetables are found by collecting them from the "All Farm Base," "Crops Base," and FV "Fruit and Vegetable Section."

They begin with record keeping and a reference system for each field, orchard, and greenhouse. That provides the necessary information for traceability, one of the major concerns of the retailers. A few examples of the control points and their compliance criterion are as follows:

Section AF, the "All Farm Base"

AF.1 Record Keeping and Internal Self-Assessment/Internal Inspection
AF.1.1 Control point: Are all records requested during the external inspection accessible and kept for a minimum period of time of two years, unless a longer requirement is stated in specific control points?

Compliance criterion: Producers keep up to date records for a minimum of two years from the date of first inspection, unless legally required to do so for a longer period.

AF.2. Site History and Site Management

---

* For the Internet addresses of the freely available GlobalG.A.P. documents, see the GlobalG.A.P. entries in the References.

AF.2.1.2. Control point: Is a reference system for each field, orchard, greenhouse, yard, plot, livestock building or other area/location used in production established and referenced on a farm plan or map?

Compliance criterion: Compliance must include visual identification in the form of a physical sign at each field/greenhouse/plot/livestock building/pen or other farm, or a farm plan or map that could be cross referenced to the identification system.

Section CB Crops Base

CB.1 Traceability. Traceability facilitates the withdrawal of foods and enables customers to be provided with targeted and accurate information concerning implicated products.
CB.1.1 Control point: Is GLOBALGAP (EUREPGAP) registered product traceable back to and trackable from the registered farm (and other relevant registered areas) where it has been grown?

Compliance criterion: There is a documented identification and traceability system that allows GLOBALGAP (EUREPGAP) registered product to be traced back to the registered farm or, in a Farmer Group, to the registered farms of the group, and tracked forward to the immediate customer. Harvest information must link a batch to the production records or the farms of specific producers. (Refer to General Regulations Part III for information on segregation in Option 2). Produce handling must also be covered if applicable.

CB.6 Quality of Irrigation Water
CB.6.3 Control point: Has the use of untreated sewage water for irrigation/fertigation been banned?

Compliance criterion: Untreated sewage water is not used for irrigation/fertigation. Where treated sewage water is used, water quality complies with the WHO published Guidelines for the Safe Use of Wastewater and Excreta in Agriculture and Aquaculture 1989. Also, when there is doubt if water is coming from a possibly polluted source (because of a village upstream, etc.) the grower has to demonstrate through analysis that the water complies with the WHO guideline requirements or the local legislation for irrigation water. See Table 3 in Annex AF.1 for Risk Assessments.

Section FV Fruit and Vegetables

FV.4 Harvesting
FV.4.1.1 Control point: Has a hygiene risk analysis been performed for the harvest and pre-farm gate transport process?

Compliance criterion: There is a documented and up to date (reviewed annually) risk analysis covering physical, chemical and microbiological contaminants and human transmissible diseases, customised to the products. It must also include FV.4.1.2 to FV.4.1.9. The risk analysis shall be tailored to the scale of the farm, the crop, and the technical level of the business.

FV.5.2 Personal Hygiene
FV.5.2.1 Control point: Have workers received basic instructions in hygiene before handling

Compliance criterion: There must be evidence that the workers received training regarding transmission of communicable diseases, personal cleanliness and clothing, i.e. hand washing, wearing of jewellery and fingernail length and cleaning, etc.; personal behaviour, i.e. no smoking, spitting, eating, chewing, perfumes, etc.

FV.5.6 Rodent and Bird Control
Control point: FV.5.6.4 Are detailed records of pest control inspections and necessary actions taken, kept?

Compliance criterion: Records of pest control inspections and follow up action plan(s). The producer can have his own records. Inspections must take place whenever there is evidence of presence of pests. In case of vermin, the producer must have a contact number of the pest controller or evidence

FV.5.8 Post-Harvest Treatments
FV.5.8.3 Control point: Are only any biocides, waxes and plant protection products used on harvested crop destined for sale in the European Union that are not banned in the European Union?

Compliance criterion: The documented post harvest biocide, wax and crop protection product application records confirm that no biocides, waxes and crop protection products that have been used within the last 12 months on the harvested crop grown under GLOBALGAP (EUREPGAP) destined for sale within the E.U., have been prohibited by the E.U. (under EC Prohibition Directive List - 79/117/EC.)

GlobalG.A.P. is a business-to-business standard. Retailers developed it to pass consumers' preferences on to the producers who can produce the preferred primary produce.

Retailers and caterers can use the same instrument with its existing and still to be formulated control points and compliance criteria to meet their requirements with regard to FcFV right from the start. Prevention of microbiological contamination is one of those issues that depend on the development of the best available production methods. GlobalG.A.P can be one instrument that primary producers share with retailers and caterers to stimulate this development.

2.4 CONCLUSION

The law that applies to FcFV is mainly general food law and food hygiene law. The Codex Alimentarius Commission will increasingly be the forum where different national laws will be harmonized to the level that will be required for trade and food safety. Concerns about foodborne diseases and the potential role of fresh-cut produce as a medium that transmits the microbes involved, have already produced some specific fresh-cut legislation and guidance documents on these issues. It is probably just the beginning of far more specific food safety law making.

REFERENCES

Associazione nazionale esportatori i, portatori ortofrutticoli e agrumari (ANEIOA), Manuale di corretta prassi igienica per i centri di lavorazione e confezionamento dei prodotti ortofrutticoli freschi, surgelati, di IV gamma, degli agrumi della frutta a guscio ed essiccata, Title (original) and Title (English), Guide to Good Hygiene Practice for processing and packaging centres for fresh, frozen, washed-cut-and-packed fruits and vegetables, citrus fruits, nuts and dried fruit.

Codex Alimentarius Commission, Joint FAO/WHO Food Standards Program, Codex Committee on Food Hygiene, Fortieth Session, Guatemala, October 2008. Proposed Draft Annex on Fresh Leafy Vegetables Including Leafy Herbs to the Code of Hygienic Practice for Fresh Fruits and Vegetables at Step 3, CX/FH 08/40/7, October 2008, http://www.codexalimentarius.net/web/archives.jsp?year=09.

Codex Alimentarius Commission, CAC/RCP 53-2003 Code of Hygienic Practice for Fresh Fruits and Vegetables, Annex I, Annex for Ready-to-Eat Fresh Pre-cut Fruits and Vegetables, http://www.codexalimentarius.net/web/more_info.jsp?id_sta=10200.

Codex Alimentarius Commission, CAC/GL-30 (1999) Principles and Guidelines for the Conduct of Microbiological Risk Assessment, http://www.codexalimentarius.net/web/more_info.jsp?id_sta=357.

Codex Alimentarius Commission, CAC/GL 21-1997 Principles for the Establishment and Application of Microbiological Criteria for Foods, http://www.codexalimentarius.net/web/more_info.jsp?id_sta=394.

Codex Alimentarius Commission, CAC/RCP 44-1995, Amended 1-2004 Recommended International Code of Practice for Packaging and Transport of Fresh Fruit and Vegetables, http://www.codexalimentarius.net/web/more_info.jsp?id_sta=322.

Codex Alimentarius Commission, CODEX STAN 1-1985 Labelling of Prepackaged Foods, http://www.codexalimentarius.net/web/more_info.jsp?id_sta=322.

Codex Alimentarius Commission, CAC/RCP 1-1969, Rev. 4-2003 Recommended International Code of Practice—General Principles of Food Hygiene Including Annex on Hazard Analysis and Critical Control Point (HACCP) System and Guidelines for Its Application, http://www.codexalimentarius.net/web/more_info.jsp?id_sta=23.

Codex Alimentarius Committee on Food Hygiene, Terms of Reference for an FAO/WHO Expert Consultation to Support the Development of Commodity-Specific Annexes for the Codex Alimentarius "Code of Hygienic Practice for Fresh Fruits and Vegetables" (ALINORM 07/30/13, Appendix VI.), http://www.codexalimentarius.net/web/archives.jsp?year=07.

European Community, Commission of the European Communities, Overview: Register of National Guides to Good Hygiene Practice, 2009, http://ec.europa.eu/food/food/biosafety/hygienelegislation/register_national_guides_en.pdf.

European Community, Commission of the European Communities, Health and Consumers Directorate-General, Guidelines on the registration of national guides to good practice In accordance with Article 8 of Regulation (EC) No. 852/2004, Brussels, 15 July 2008, http://ec.europa.eu/food/food/biosafety/hygienelegislation/national_guides_register_15-07-2008.pdf.

European Community, Commission of the European Communities, Health and Consumer Protection Directorate-General, Guidance document on certain key questions related to import requirements and the new rules on food hygiene and on official food controls, Brussels, 1 May 2006, pp. 1–28, http://ec.europa.eu/food/international/trade/interpretation_imports.pdf.

European Community, Commission of the European Communities, Health and Consumer Protection Directorate-General, Guidance document on the implementation of certain provisions of Regulation (EC) No. 852/2004 On the hygiene of foodstuffs, Brussels, 21 December 2005, pp. 1–17, http://ec.europa.eu/food/food/biosafety/hygienelegislation/guidance_doc_852-2004_en.pdf.

European Community, Commission of the European Communities, Health and Consumer Protection Directorate-General, Guidance document on the implementation of procedures based on the HACCP principles, and on the facilitation of the implementation of the HACCP principles in certain food businesses, Brussels, 16 November 2005, pp. 1–26, http://ec.europa.eu/food/food/biosafety/hygienelegislation/guidance_doc_haccp_en.pdf.

European Community, Commission of the European Communities, Commission Regulation (EC) No. 2073/2005 of 15 November 2005 on microbiological criteria for foodstuffs OJ L 338, 22.12.2005, pp. 1–26), http://eur-lex.europa.eu/LexUriServ/LexUriServ.do?uri=CONSLEG:2005R2073:20071227:EN:PDF.

European Community, Commission of the European Communities, Guidelines for the Development of Community Guides to Good Practice for Hygiene or for the Application of the HACCP Principles, in Accordance with Article 9 of Regulation (EC) No. 852/2004 on the Hygiene of Foodstuffs and Article 22 of Regulation (EC) No. 183/2005 Laying Down Requirements for Feed Hygiene, http://ec.europa.eu/food/food/biosafety/hygienelegislation/guidelines_good_practice_en.pdf.

European Community, Commission of the European Communities, Health and Consumer Protection Directorate-General, Directorate C—Scientific Opinions, C2—Management of Scientific Committees II; Scientific Cooperation and Networks. Scientific Committee on Food, Risk Profile on the Microbiological Contamination of Fruits and Vegetables Eaten Raw. Report of the Scientific Committee on Food (adopted on the 24 April 2002), SCF/CS/FMH/SURF/Final, 29 April 2002, http://ec.europa.eu/food/fs/sc/scf/out125_en.pdf.

European Community, Commission of the European Communities, Health and Consumer Protection Directorate-General, Directorate C—Scientific Opinions, Directorate C—Scientific Opinions, C3—Management of Scientific Committees II; Scientific Cooperation and Networks, Opinion of the Scientific Committee on Food in respect of *Listeria monocytogenes* expressed on 22 June 2000, http://ec.europa.eu/food/fs/sc/scf/out63_en.pdf.

European Community, Commission of the European Communities, Health and Consumer Protection Directorate-General, Directorate B—Scientific Health Opinions, Unit B3—Management of Scientific Committees II. Opinion of the Scientific Committee on Veterinary Measures Relating to Public Health on *Listeria monocytogenes,* 23 September 1999, http://ec.europa.eu/food/fs/sc/scv/out25_en.pdf.

European Community, Directive 2000/13/EC of the European Parliament and of the Council of 20 March 2000 on the approximation of the laws of the Member States relating to the labelling, presentation and advertising of foodstuffs, OJ L 109, 5 June 2000, p. 29, 02000L0013-20071129, http://eur-lex.europa.eu/LexUriServ/LexUriServ.do?uri=CONSLEG:2000L0013:20071129:EN:PDF.

European Community, Regulation (EC) No. 852/2004 of the European Parliament and of the Council of 29 April 2004 on the hygiene of foodstuffs, OJ L 139, 30, 4 April 2004. Corrected version in Official Journal L 226, 25, 6 June 2004, http://eur-lex.europa.eu/LexUriServ/LexUriServ.do?uri=OJ:L:2007:204:0026:0026:EN:PDF.

European Community, Regulation (EC) No. 178/2002 of the European Parliament and of the Council of 28 January 2002 laying down the general principles and requirements of food law, establishing the European Food Safety Authority and laying down procedures in matters of food safety, OJ L 31, 2 January 2002, pp. 1–24, http://eur-lex.europa.eu/pri/en/oj/dat/2002/l_031/l_03120020201en00010024.pdf.

European Community, Standing Committee on the Food Chain and Animal Health, Guidance on the Implementation of Articles 11, 12, 16, 17, 18, 19, and 20 of Regulation (EC) No. 178/2002 on General Food Law, Conclusions of the Standing Committee on the Food Chain and Animal Health, 20 December 2004, http://ec.europa.eu/food/food/foodlaw/guidance/guidance_rev_7_en.pdf.

Food and Agriculture Organization of the United Nations (FAO) and the World Health Organization (WHO), Microbiological hazards in fresh fruits and vegetables, Meeting Report, 2008, http://www.who.int/foodsafety/publications/micro/MRA_FruitVeges.pdf.

Food and Agriculture Organization of the United Nations (FAO) and the World Health Organization (WHO), Viruses in food: scientific advice to support risk management activities, Meeting Report, 2008, Microbiological Risk Assessment Series 13, http://www.who.int/foodsafety/publications/micro/Viruses_in_food_MRA.pdf.

Food and Agriculture Organization of the United Nations (FAO) and the World Health Organization (WHO), Codex Alimentarius, Fresh Fruits and Vegetables, First edition, Rome, 2007.

GLOBALG.A.P. (EUREPGAP), Control Points and Compliance Criteria Integrated Farm Assurance: Introduction, English Versions V.3.0-3_Apr09 Valid from 29 April 2009; All Farm Base, English Version V3.0-2_Sep07 Valid from 30 September 2007; http://www.globalgap.org/cms/upload/The_Standard/IFA/English/CPCC/GG_EG_IFA_CPCC_Intro-AF_ENG_V3_0_3_Apr09.pdf; Crops Base, English Version V.3.0-3_Feb09 Valid from 16 February 2009; http://www.globalgap.org/cms/upload/The_Standard/IFA/English/CPCC/GG_EG_IFA_CPCC_CB_ENG_V3_0_3_Feb09.pdf; Fruits and Vegetables, English Version V3.0-2 Sep07 Valid from 30 September 2007; http://www.globalgap.org/cms/upload/The_Standard/IFA/English/CPCC/GG_EG_IFA_CPCC_FV_ENG_V3_0_2_Sep07.pdf.

Meulen, Bernd van der, and Velde, Menno van der. *European Food Law Handbook*, Wageningen Academic Publishers, Wageningen, 2008.

United Kingdom, Food Safety Act 1990, http://www.opsi.gov.uk/acts/acts1990/ukpga_19900016_en_3#pt2-pb4-l1g21.

United States of America, Department of Health and Human Services, Food and Drug Administration and Center for Food Safety and Applied Nutrition, Guidance for Industry: Guide to Minimize Microbial Food Safety Hazards of Fresh-cut Fruits and Vegetables, February 2008, http://www.cfsan.fda.gov/~dms/prodgui4.html.

United States of America, Food and Drug Administration, CFSAN/Office of Plant and Dairy Foods, Letter to California Firms That Grow, Pack, Process, or Ship Fresh and Fresh-cut Lettuce, 4 November 2005, http://www.cfsan.fda.gov/~dms/prodltr2.html.

United States of America, Food and Drug Administration, CFSAN/Office of Plant and Dairy Foods, Produce Safety. From Production to Consumption: 2004 Action Plan to Minimize Foodborne Illness Associated with Fresh Produce Consumption, October 2004, http://www.cfsan.fda.gov/~dms/prodpla2.html.

United States of America, Food and Drug Administration, and U. S. Department of Agriculture, Centers for Disease Control and Prevention, and U.S. Department of Health and Human Services, Food and Drug Administration, Center for Food Safety and Applied Nutrition (CFSAN), Guidance for Industry. Guide to Minimize Microbial Food Safety Hazards for Fresh Fruits and Vegetables, 26 October 1998, http://www.foodsafety.gov/~dms/prodguid.html.

United States of America, Code of Federal Regulations (CFR) Title 21: Food and Drugs, Part 110. Current Good Manufacturing Practice in Manufacturing, Packing, or Holding Human Food. (Abbreviated to 21 CFR Part 110), http://ecfr.gpoaccess.gov/cgi/t/text/text-idx?c=ecfr;sid=03c0f0ea8b688b533744ae63a89f6aa4;rgn=div5;view=text;node=21%3A2.0.1.1.10;idno=21;cc=ecfr.
United States of America, Federal Food, Drug and Cosmetic Act Section 303, http://www.fda.gov/opacom/laws/fdcact/fdcact3.htm.

3 Microbiological and Safety Aspects of Fresh-Cut Fruits and Vegetables

Peter Ragaert, Liesbeth Jacxsens,
Isabelle Vandekinderen, Leen Baert,
and Frank Devlieghere

CONTENTS

3.1   SPOILAGE PATTERN AND SPOILAGE MICROORGANISMS RELATED TO FRESH-CUT FRUITS AND VEGETABLES

3.1.1   INTRODUCTION

In spoilage of fresh-cut fruits and vegetables, two major patterns can be defined, which are different but influence each other. A distinction needs to be made between physiological spoilage (due to enzymatic and metabolic activity of the living plant tissue) and microbiological spoilage (due to proliferation of microorganisms). Processing fresh fruits and vegetables removes the natural protection of the epidermis and destroys the internal compartmentalization that separates enzymes from substrates. Consequently, plant tissues suffer physical damages that make them much more perishable than when the original product is intact (Artés et al. 2007). Moreover, processing results in a stress response by the produce characterized by an increased respiration rate (wound respiration) and ethylene production, leading to faster metabolic rates (Rosen and Kader 1989; Howard et al. 1994; Watada et al. 1996; Saltveit 1999; Kang and Saltveit 2002; Knee and Miller 2002; Laurila and Ahvenainen 2002; Surjadinata and Cisneros-Zevallos 2003; Artés et al. 2007). In addition to these changes in metabolic rates, damage of the plant tissue leads to exposure to air, desiccation, and the bringing together of enzymes with substrates, all leading to quality degradation (Klein 1987; King and Bolin 1989; Roura et al. 2000; Knee and Miller 2002). Next to physiological processes, the release of nutrients on the cut surfaces allows the growth of microorganisms. Under natural conditions, the outer layer of the plant tissue consists of a hydrophobic surface providing a natural barrier for microorganisms (Lund 1992). Due to damage of the surface, nutrients are released from the plant tissue, which can be used by microorganisms as shown by Mercier and Lindow (2000) after inoculating *Pseudomonas fluorescens* on different leaves of vegetable plants. Densities of microorganisms were directly correlated with the amount of sugars present on the surface of leaves, and these sugars were the limiting factor with regard to colonization. Related to the presence of damaged areas, Babic et al. (1996) found that microbiological populations were situated on cut surfaces of spinach after 12 days of storage at 10°C. Brocklehurst and Lund (1981) found after inoculation of celery, that soft rot could not be caused on unwounded tissue, possibly due to limited proliferation. Damaged spots on plant tissue thus provide a better substrate for microbiological growth by providing nutrients (King et al. 1991; Zagory 1999). Some of the microorganisms produce pectinolytic enzymes degrading texture and as such provide more nutrients for microbiological activity (e.g., during soft rotting, spoilage of leafy vegetables such as fresh-cut lettuce). Commodities that are susceptible to a high degree of nutrient release will result in intense microbiological proliferation (e.g., during spoilage of intensively cut fruits and vegetables such as cucumber cubes, zucchini slices, and melon parts). Moreover, high microbial loads can induce increasing respiration rates of the produce as shown by Saftner et al. (2006) in the case of fresh-cut melon. Yeasts and molds, which are less sensitive to a low pH, will develop on fruits with lower pH (e.g., strawberries, raspberries, and nectarines), while on the more pH-neutral vegetables, bacterial growth can be detected.

The above-mentioned intrinsic properties (e.g., pH of the tissue and nutrient availability) determine the growth rate and the type of microorganisms developing on the produce and, consequently, the type of spoilage pattern. Extrinsic properties (e.g., storage temperature and gas atmosphere) also influence the spoilage behavior of fresh-cut fruits and vegetables. With regard to microbiological processes, Zagory (1999) reviewed some experiments with fresh-cut vegetables, illustrating that the effect of modified atmosphere on microbiological growth is not consistent and that storage temperature rather than gas composition controls microbiological growth. Moreover, Nguyen-the and Carlin (1994) stated that the benefits of modified atmosphere for quality are not consistently related to a reduction in the growth of mesophilic flora. Bennik et al. (1998) reported that the physiological state of the product, rather than the inhibition of soft-rot bacteria, plays an important role in the beneficial effects of controlled atmosphere storage of vegetables. However, too low O_2 concentrations or too high CO_2 concentrations could induce physiological disorders independent of microbiological processes (Giménez et al. 2003). Rocha et al. (1995) stated that in the case of fresh-cut oranges, the microbial population does not cause spoilage, at least when appropriate temperatures are employed. Rather, some factors in the fruit together with factors acquired during processing (e.g., interference with enzymes) may mediate spoilage. Similar observations were found by Pretel et al. (1998). The typical gas conditions applied for packaging fresh-cut fruits and vegetables, 3% to 5% O_2, combined with 5% to 10% CO_2 (and balanced by N_2) are not directly affecting the microbiological growth on produce. The positive effect of modified atmosphere on the quality and spoilage retardation of fresh-cut produce is due to the inhibition of metabolic activity of the living plant tissue. In this way, plant tissue is keeping for a longer period its natural strength and protection (Jacxsens et al. 2002a, 2002b, 2003). Spoilage of fresh-cut fruits and vegetables will thus be determined by the rate of physiological and microbiological processes, as there is a possible interaction between these two processes (as reviewed by Ragaert et al. 2007) in the case of fresh-cut vegetables.

3.1.2 OVERVIEW OF THE MOST IMPORTANT MICROORGANISMS RELATED TO SPOILAGE

Both fresh-cut fruits and vegetables are susceptible to many different contamination sources, such as seed, soil, irrigation water, animals, manure/sewage sludge use, harvesting, processing, and packaging. Total counts of microbiological populations on fresh-cut vegetables after processing range from 3 to 6 log CFU/g. The bacterial population during storage mainly consists of species belonging to *Pseudomonadaceae* (especially *P. fluorescens*) and *Enterobacteriaceae* (especially *Erwinia herbicola* and *Rahnella aquatilis*) in addition to some species belonging to lactic acid bacteria (especially *Leuconostoc mesenteroides*) (Lund 1992; Nguyen-the and Carlin 1994; Vankerschaver et al. 1996; Bennik et al. 1998). Unlike bacteria, many different yeast species of comparable numerical importance have been identified in fresh-cut vegetables: *Candida* sp., *Cryptococcus* sp., *Rhodotorula* sp., *Trichosporon* sp., *Pichia* sp., and *Torulaspora* sp. (Nguyen-the and Carlin 1994).

Molds are not so important in the case of fresh-cut vegetables due to the intrinsic properties favoring the growth of bacteria and yeasts (Magnuson et al. 1990; King et al. 1991; Lund 1992; Moss 1999; Giménez et al. 2003). In the case of fresh-cut fruits, intrinsic properties favor growth of yeasts, molds, and in some cases, lactic acid bacteria mainly due to the lower pH value compared to vegetables (Beaulieu and Gorny, 2002). With regard to processed fruits, Tournas et al. (2006) found on a majority of 38 fruit salad samples (cantaloupe, citrus fruits, honeydew, pine-apple, cut strawberries and mixed fruit salads) yeast levels ranging from less than 2 to 9.72 log CFU/g. The most common yeasts were *Pichia* sp., *Rhodotorula* sp., *Candida pulcherrima, C. lambica, C. sake,* and *Debaryomyces polymorphus.* A wide range of mold species can be present on fruits. Molds reported on berries are *Botrytis cinerea, Rhizopus stolonifer, Mucor piriformis, Rhizoctonia solani,* and *Phytophtora cactorum.* Overripe or damaged berries can be invaded by *Penicillium* and *Cladosporium* species (Pitt and Hocking 1997). However, on intact strawberry and raspberry fruit, bacterial survival is possible and even growth on the calyx of strawberries (Sziro et al. 2006).

3.1.3 PROCESS OF DETERIORATION

Microbiological proliferation is characterized by different processes: Production of enzymes and production of metabolites results in visual and textural defects and off-odors. In this, the interaction with physiological processes is of great impor-tance. For instance, on fresh-cut vegetables sensitive to enzymatic browning, con-ditions decreasing the rate of this physiological process determine if browning influences the sensorial shelf life of the commodities. Lower O_2 concentrations slow enzymatic browning; therefore, sensorial quality factors other than browning become important with regard to sensorial shelf life (e.g., microbiological produc-tion of off-odors) (Jacxsens et al. 2002b). Grated celeriac, for example, became unacceptable due to high sensitivity to browning, before off-odors could have been detected (Jacxsens et al. 2003). Similarly, the visual appearance of whitening and loss of fruity aroma on fresh-cut carrots occurred prior to the presence of off-odors (Lavelli et al. 2006).

Composition of fruits and vegetables will also determine the type of spoilage. In the case of sugar-rich products such as carrots, bell peppers, and most of the fresh-cut fruits, growth of yeasts and lactic acid bacteria is favored, resulting in off-odors caused by microbial proliferation and the production of acids such as lactic acid, ace-tic acid, malic acid, succinic acid, and pyruvic acid (Carlin et al. 1989; Kakiomenou et al. 1996). It can be seen that the detection of off-odors is often accompanied with a bacterial count exceeding 8 log CFU/g or a yeast count exceeding 5 log CFU/g. This was stated by Carlin et al. (1989), Barry-Ryan and O'Beirne (1998), and Hao et al. (1999) in the case of fresh-cut carrots. Production of organic acids on shred-ded mixed bell peppers and grated celeriac was detected when the psychrotrophic count exceeded 8 log CFU/g, dominated by lactic acid bacteria (7 to 8 log CFU/g). In the case of bell peppers, yeasts exceeded 5 log CFU/g at the moment of spoilage (Jacxsens et al. 2003). Sourness is often accompanied by water loss, which is often associated with the proliferation of *L. mesenteroides.*

Besides organic acids, bacteria and yeasts are capable of producing ethanol as demonstrated by Jacxsens et al. (2003) on a simulation medium of mixed-lettuce agar. Production of ethanol is often associated with the production of other volatile organic compounds such as 2-methyl-1-propanol, 2-methyl-1-butanol, and 3-methyl-1-butanol, which could be detected from 8 log CFU/ cm² (bacteria) and 5 log CFU/ cm² (yeasts) (Ragaert et al. 2006a). Smyth et al. (1998) reported inedible lettuce at 20°C, because of severe fermentation and bacterial proliferation (no counts given). Among the volatile organic compounds, 2-methyl-1-butanol, 3-methyl-1-butanol, and 3-methyl-1-pentanol were detected. It was not clear if the production of these compounds was from physiological or microbiological origin. In the case of fresh-cut fruits, which contain high concentrations of fermentable sugars, spoilage is often characterized by off-odors due to yeast growth resulting in the production of volatile organic compounds such as ethanol. Interestingly, yeasts inoculated on a simulation medium of strawberries produced the same compounds as reported on mixed-lettuce agar (ethanol, 2-methyl-1-propanol, 2-methyl-1-butanol, and 3-methyl-1-butanol), also from 5 log CFU/cm² (Ragaert et al. 2006b). Increasing amounts of ethanol were observed on fresh-cut apples with yeast growing to 5.5 log CFU/g (Rojas-Graü et al. 2007). In some cases (e.g., temperature abuse, high initial contamination of molds), molds determine spoilage, resulting in visual spoilage before yeasts or bacteria can produce off-odors (Nielsen and Leufvén 2008). Metabolite production is often accompanied by sugar consumption by microorganisms, as observed by Tassou and Boziaris (2002), Jacxsens et al. (2003), Piga et al. (2003), Soliva-Fortuny et al. (2004), and Shah and Nath (2008) for litchis, different sugar-rich vegetables, carrots, apples, and litchis, respectively.

Production of fermentative metabolites such as ethanol is also possible due to fermentation reactions of the produce when stored at too low O_2 concentrations or too high CO_2 concentrations independent of the microbiological counts. This was shown by López-Gálvez et al. (1997) in the case of packaged salad products (ethanol and acetaldehyde) and by Smyth et al. (1998) in the case of cut iceberg lettuce (ethanol, acetaldehyde, ethyl acetate). Smyth et al. (1999) suggested the presence of ethanol on fresh-cut carrots as to be a wounding response. Mechanical wounding of fruits and vegetables (e.g., cutting) enhances a diverse array of enzymatic pathways, associated in many cases with the generation of volatiles (Toivonen 1997). Purvis (1997) stated that fermentative metabolism can be enhanced in fruits by several factors, including environmental (chilling injury temperature, hypoxic conditions), biotic (microbial infections), and internal (ripening, senescence) factors.

Metabolite production could also affect physiological processes, as demonstrated during storage experiments with strawberries where it was found that ethanol was converted by the physiological processes of the strawberries to ethyl acetate, which limited the sensorial shelf life of strawberries. Increases in ethanol were detected when yeast counts were above 5 log CFU/g (Ragaert et al. 2006c). The conversion of added compounds such as ketones, aldehydes, and alcohols to acetate and butyrate esters by strawberries was also demonstrated by Pesis and Avissar (1990), Hamilton-Kemp et al. (1996), and Yu et al. (2000), indicating that microbiological metabolites can be metabolized by strawberries.

Products with lower sugar concentrations such as lettuce will mainly be spoiled by soft-rot bacteria such as *Pseudomonas* sp. resulting in both textural and visual defects, although these bacteria could also produce metabolites, resulting in off-odors as described above. Besides physiological degradation of texture (Artés et al. 2007), different microorganisms produce pectinolytic enzymes that can influence textural changes in fresh-cut produce by degrading the pectin molecule in the middle lamella and the primary cell wall. This production has been reported for different species of bacteria (Juven et al. 1985; Membré and Burlot 1994; Membré et al. 1995; Fraaije et al. 1997; Liao et al. 1997). The most frequently isolated pectinolytic bacteria regarding fresh-cut vegetables are species of *Erwinia* and *Pseudomonas*. Moreover, a whole range of yeasts have been reported as producers of pectinolytic enzymes, mainly endopolygalacturonases (Blanco et al. 1999), which could be important to consider during storage of fresh-cut fruits (Nakagawa et al. 2004; Restuccia et al. 2006). The softening of the tissue due to microbiological activity is located where nutrients are available (e.g., on the cut surfaces). There seems to be inconsistency in the counts necessary for causing textural decay. Bacterial populations of 7 to 7.7 log CFU/g were found on celery segments with pectinolytic *Pseudomonas* sp. predominating (Robbs et al. 1996a). This was accompanied by soft or macerated tissue. Babic et al. (1996) detected a loss of texture in fresh-cut spinach, when the microbiological count exceeded 8 log CFU/g, dominated by *Pseudomonas fluorescens*. However, Robbs et al. (1996b) found on fresh-cut celery bacterial populations of 7.1 log CFU/g but no spoilage with regard to texture and color, although predominant flora were also pectinolytic *Pseudomonas* sp. Pectinolytic bacteria isolated from fresh-cut (grated) carrots did not influence spoilage on these carrots, possibly due to the dominating lactic acid bacteria and yeasts (Carlin et al. 1989). Different factors play a role in microbiological breakdown of texture; Robbs et al. (1996a) mentioned in the case of cut celery that a complex mixture of bacteria rather than a single pathogenic species will initiate decay, while Robbs et al. (1996b) stated that a possible loss in plant resistance to microbiological attack could lead to development of decay. It should also be mentioned that physiological processes or the activity of pectinolytic microorganisms resulting in "soft-rot" spots or other visual symptoms possibly leads to unacceptable visual defects before textural deviations can be observed. Moreover, reports combining different sensorial evaluations, including texture, found during storage experiments with soft red fruits and different fresh-cut vegetables that visual defects or off-odor and off-flavors determined sensorial shelf life (López-Gálvez et al. 1997; Van der Steen et al. 2002; Giménez et al. 2003; Jacxsens et al. 2003; Sziro et al. 2006). Gas composition can also alter the texture of fruits and vegetables as demonstrated by different publications (Larsen and Watkins 1995; Harker et al. 2000; Giménez et al. 2003; Pelayo et al. 2003; Allende et al. 2004). However, it is not possible from these publications to draw conclusions on which gas composition is optimal, possibly due to the dependence on different factors such as type of product, cultivar, and tolerance to high CO_2 and low O_2 concentrations.

With regard to visual defects, different publications mention that these defects become visible from a microbiological count of 8 log CFU/g (Nguyen-the and Prunier 1989; King et al. 1991; Li et al. 2001b; Giménez et al. 2003). This is,

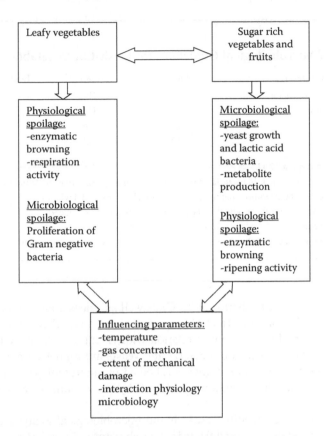

FIGURE 3.1 Overview of dominating mechanisms of spoilage and influences on spoilage of leafy vegetables versus sugar-rich fruits and vegetables.

however, not a sufficient condition as there were leaves with large population densities that remained healthy, showing that interaction with the plant material is also important (Jacques and Morris 1995). The effect of processing conditions on microbiological growth was demonstrated by Barry-Ryan and O'Beirne (1998), reporting that fresh-cut carrot slices having undergone less stressful conditions during processing (sharper blades) remained visually acceptable, due to maintaining the integrity of the tissue. Related to this, total aerobic count never exceeded 8 log CFU/g. Also, visual defects can occur before reaching 8 log CFU/g as demonstrated in the case of fresh-cut bunched onions (10 cm stalks), celery sticks, and kohlrabi slices, which can be attributed to faster physiological processes (Hong and Kim 2004; Gòmez and Artés 2005; Escalona et al. 2006). An overview of different, dominating mechanisms of spoilage typically associated with leafy vegetables, sugar-rich vegetables and sugar-rich fruits, respectively, is given in Figure 3.1.

In shelf-life studies, the evolution of yeast count of fresh-cut produce is often ignored (Canganella et al. 1998; Viljoen et al. 2003; Restuccia et al. 2006). Total

TABLE 3.1
Overview of Microbiological Guidelines for Fresh-Cut Vegetables (CFU/g)

Parameter	Target[c]	Tolerance[d]	Best before Date[e]
Total aerobic psychrotrophic count[a]	10^5	10^6	10^8
Lactic acid bacteria[b]	10^3	10^4	10^7
Yeasts	10^3	10^4	10^5
Molds	10^3	10^4	10^4

[a] Incubated for 5 days at 22°C.
[b] When the number of lactic acid bacteria on the best before dates greater than 10^7/g the food product can only be rejected on the condition that there are unacceptable sensorial deviations.
[c] Target is the guideline for the production day, in the best conditions produced.
[d] Tolerance is the maximum guideline for the production day.
[e] Best before date is the end of the shelf life, and above these guidelines, notable spoilage will occur.

psychrotrophic count (incubation at 22°C) as well as yeasts count should be evaluated when investigating the effect of microorganisms on quality. Moreover, when evaluating microbiological count in relation with visual defects, it should be considered that surface densities of bacteria and yeasts are higher on cut surfaces or damaged surfaces, which is not always reflected by microbiological analysis due to the dilution of such high-contaminated spots by less-contaminated nondamaged surfaces.

Based on intensive scientific research and microbiological analysis of different types of fresh-cut vegetables and fruits in the Laboratory of Food Microbiology and Food Preservation (LFMFP), Ghent University, Belgium, specific microbiological guidelines are proposed for spoilage causing microorganisms (Tables 3.1 and 3.2).

TABLE 3.2
Overview of Microbiological Guidelines for Fresh-Cut Fruits (CFU/g)

Parameter	Target[c]	Tolerance[d]	Best before Date[e]
Total aerobic psychrotrophic count[a]	10^5	10^6	10^7
Lactic acid bacteria[b]	10^3	10^4	10^7
Yeasts	10^3	10^4	10^5
Molds	10^2	10^3	10^3

[a] Incubated for 5 days at 22°C.
[b] When the number of lactic acid bacteria on the best before date is greater than 10^7/g the food product can only be rejected on condition that there are unacceptable sensorial deviations.
[c] Target is the guideline for the production day, in best conditions produced.
[d] Tolerance is the maximum guideline for the production day.
[e] Best before date is the end of the shelf life, above these guidelines notable spoilage will occur.

3.2 PATHOGENIC MICROORGANISMS RELATED TO FRESH-CUT FRUITS AND VEGETABLES

3.2.1 ASSOCIATION OF PATHOGENS WITH CONSUMPTION OF (FRESH-CUT) FRUITS AND VEGETABLES

Fresh-cut fruits and vegetables are often considered as relatively safe food products from a microbiological point of view compared to food from animal origin or other ready-to-eat food products. However, some recent foodborne outbreaks have resulted in the fact that fresh-cut fruits and vegetables are considered as possible vehicles of foodborne pathogens by scientific and commercial stakeholders (e.g., norovirus outbreak due to frozen raspberry fruit in Scandinavian countries in 2006, outbreak of *Escherichia coli* O157:H7 associated with fresh spinach in the United States in 2007, and *Salmonella* outbreak in the United States with tomatoes in 2008). The raw consumption of the products, the application of mild processing techniques, and a subsequent storage period have presented indigenous and pathogenic microorganisms with new ecosystems and potential infection vehicles. Because of the excellent marketing properties of the products, fresh-cut fruits and vegetables have quickly gained a large market share, often without extensive evaluation of the safety aspects of this type of product. Pathogens may be present on the raw vegetables or due to cross-contamination during processing (Nguyen-the and Carlin 1994; Beuchat 1996; Seymour 2001). Mechanisms by which fresh produce can become contaminated with pathogenic microorganisms and serve as vehicles of human disease are demonstrated in Figure 3.2. Typical transmission routes in primary production are water, soil, or sewage that are contaminated with fecal (enteric) pathogens such as *Salmonella* spp., *E. coli*, and viruses. Also during harvesting and further processing, (fecal)

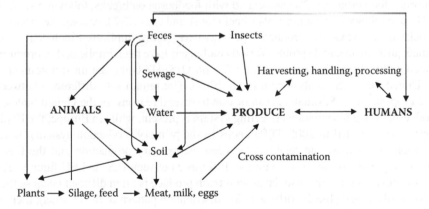

FIGURE 3.2 Mechanisms by which fresh produce can become contaminated with pathogenic microorganisms and serve as vehicles of human disease. (After Beuchat, L. 1996. Pathogenic microorganisms associated with fresh produce. *Journal of Food Protection* 59:204–216; Tauxe, R., Kruse, H., Hedberg, C., Potter, M., Madden, J., and Wachsmugh, K. 1997. Microbial hazards and emerging issues associated with produce. A preliminary report to the national advisory committee on microbiologic criteria for foods. *Journal of Food Protection* 60:1400–1408. With permission.)

contamination can occur through humans and materials. It is generally thought that these enteric pathogens do not fit in the plant habitat as native plant–associated bacteria, but recent research has demonstrated that they have the ability to grow and persist on crop plants (Brandt 2006).

The association of fresh-cut fruits and vegetables with food poisoning is emerging due to their increased consumption, increased consumers' awareness of the importance of fruits and vegetables in their diet, and convenience. Moreover, a wider variety of fresh produce is available due to the globalization of the fresh produce supply chain, which is resulting in their availability throughout the year. This globalization also implicates larger periods between harvest, processing, and consumption and demands the implementation of well-controlled storage and long-distance transport conditions. By applying processing techniques such as cutting, slicing, and shredding, natural plant protection barriers are being removed, providing a suitable medium for microbial growth. These processing steps are also introducing more possibilities for cross-contamination.

Generally, comparing different sources of foodborne outbreaks, fresh-cut fruits and vegetables are only rarely involved in Europe (European Food Safety Authority [EFSA] 2007; Rapid Alert System for Food and Feed [RASFF] 2007). From the annual Rapid Alert reports in the EU, it can be deduced that the relative portion of alert notifications for fruits and vegetables remained stable during the period of 2000 to 2005 at approximately 9%. In 2006, 72 alert notifications were reported with fruits and vegetables. But the majority was related to the presence of mycotoxins or pesticide residues. Only at third place were pathogenic microorganisms, especially *Salmonella* (RASFF 2007), found. The number of reported produce-related outbreaks in the United States increased from 2 per year in the 1970s to 7 per year in the 1980s and to 16 per year in the 1990s. In the United States, fresh produce is considered as the second most important food product (22%) associated with foodborne outbreaks, following seafood (33%), and followed by poultry (18%), beef (16%), and eggs (13%). Among the produce-associated outbreaks, the produce types most frequently implicated included salad, lettuce, juice, melon, and sprouts. Various pathogens have been implicated in produce-related outbreaks, with *Salmonella* and *E. coli* O157:H7 being the most dominant in the United States (Sivapalasingam et al. 2004). The remarkable difference between the European and U.S. situations can be due to several reasons. In the United States, a good, established information collecting system is present, while in Europe, the Rapid Alert System started in 2005. Differences in the primary production systems, water management systems, and relations between the primary production and the fresh-cut industry may also partly explain the difference between U.S. and EU figures. The investigation of outbreaks associated with fresh produce is often difficult because they are typically geographically diffuse (wide distribution pattern of the product); there is a low attack rate (low level of contamination); and due to its short shelf life and rapid turnover, the implied product or even product originating from the same batch is rarely available by the time an outbreak has been identified (Tauxe et al. 1997).

In Europe, as reported by EFSA, foodborne viruses (namely, adenovirus, norovirus [NoV], enterovirus, hepatitis A virus [HAV], rotavirus) are responsible for 593 outbreaks of a total of 5,710 reported foodborne outbreaks, which means 10.2% of the reported outbreaks in 2006 over all food products. In total, 13.345 persons were involved, which made foodborne viruses the second most important agent after *Salmonella* (EFSA 2007). In the United States, NoV, HAV, rotavirus, and astrovirus are included in the thirteen major foodborne pathogens identified by the Centers for Disease Control and Prevention (CDC) (Mead et al. 1999). These viruses make up approximately 67% of all foodborne illnesses in the United States, with NoV by far the greatest contributor. Outbreaks associated with coleslaw, tossed salad (White et al. 1986), green salads (Griffin et al. 1982), green onions (Wheeler et al. 2005), fresh-cut fruit (Herwaldt et al. 1994), small soft fruits such as raspberries (Gaulin et al. 1999; Korsager et al. 2005; Cotterelle et al. 2005; Calder et al. 2003), and potato salad (Patterson 1997) are well known.

The largest foodborne outbreaks due to protozoan parasites were caused by *Cyclospora cayetanensis* during the late 1990s in North America. Fresh raspberries imported from Guatemala were indicated as the source of the outbreak (Herwaldt 2000). Although *Cryptosporidium parvum* and *Giardia lamblia* are more likely to cause waterborne outbreaks, outbreaks associated with green onions (Quinn et al. 1998) and basil (Lopez 2001) are documented.

3.2.2 OVERVIEW OF THE MOST IMPORTANT PATHOGENS RELATED TO FRESH-CUT FRUITS AND VEGETABLES

Numerous pathogens have been isolated from a wide variety of fresh-cut fruits and vegetables, although not all of them could be associated with foodborne outbreaks. Practically any bacterial pathogens are potentially present, but only a few are associated with fresh (cut) produce (Table 3.3). Enteric pathogens are mainly associated with outbreaks in fresh-cut vegetables (e.g., *E. coli* O157:H7 and *Salmonella*

TABLE 3.3

Pathogenic Organisms of Concern or Potential Concern in Fresh-Cut Produce

Pathogens of Concern	Pathogens of Possible Concern
Listeria monocytogenes	Nonproteolytic *Clostridium botulinum* types B, E, F
Escherichia coli (O157:H7)	*Aeromonas hydrophila/caviae*
Shigella spp.	*Bacillus cereus*
Salmonella spp.	*Yersinia enterocolitica*
Parasites	*Campylobacter* spp.
Viruses	

Sources: After Brackett, R. 1992. Shelf stability and safety of fresh produce as influenced by sanitation and disinfection. *Journal of Food Protection* 55:808–814; Nguyen-the, C., and Carlin, F. 1994. The microbiology of minimally processed fresh fruits and vegetables. *Critical Reviews in Food Science and Nutrition* 34:371–401; and Francis, G. A., Thomas, C., and O'Beirne, D. 1999. The microbiological safety on minimally processed vegetables. *International Journal of Food Science and Technology* 34:1–22. With permission.

---

or *Shigella* are associated with imported products from Mediterranean or African countries). Recently, *Campylobacter* was detected on fresh produce, due to cross-contamination in farms with a poultry staple. Numerous types of viruses and parasites can be present on fresh produce, using the produce as a transfer medium to humans. *Listeria monocytogenes,* an emerging and widely spread pathogen in nature and the food processing environment, should also be considered. Plant material can be contaminated by nature with spore-forming pathogens such as *Bacillus cereus* or *Clostridium* spp.

3.2.2.1 *Salmonella* and *Shigella*

Salmonella and *Shigella* spp. are mesophilic fecal-associated pathogens. Their growth on fresh-cut vegetables is mostly provoked by temperature abuse (T > 10 to 12°C). The minimal growth temperatures of *Shigella sonnei* and *S. flexneri* were reported to be 6°C and 7°C, respectively (ICMSF 1996). The low pH of fruits is limiting their growth (except melons that have a high pH and are associated with outbreaks of *Salmonella*). Although relatively fragile, at least some strains of *Shigella* are able to tolerate acidic conditions with pH 4.5 (Bagamboula et al. 2001) and pH 5 (ICMSF 1996) as the lowest reported pH for growth. The pathogen has a very low infectious dose of less than 100 cells. Shigellosis is usually transmitted from person to person but may also occur through consumption of contaminated water and foods, particularly foods such as salad vegetables. Imported food products from endemic regions where hygienic standards are insufficient have become a potential source of *Shigella* contaminated foods (Smith 1987; ICMSF 1996). Laboratory studies revealed that *S. sonnei* can survive on shredded cabbage at 0 to 6°C for 3 days without decrease in number. At 24°C, a fast reduction in number was observed because of the pH drop of cabbage due to the fast outgrowth of spoilage microorganisms (Satchell et al. 1990). However, *Shigella* spp. survived for several days at both ambient (22°C) and refrigerator temperatures (5 and 10°C) when inoculated onto various commercially prepared salads and vegetables (carrots, coleslaw, radishes, broccoli, cauliflower, lettuce, and celery) (Rafii et al. 1995). *S. flexneri* was able to survive at 4°C for at least 11 days on coleslaw, carrot salad, and potato salad (Rafii and Lunsford 1997). Experiments conducted at 12°C demonstrated that *Shigella flexneri* and *S. sonnei* were able to proliferate on fresh-cut mixed lettuce and shredded carrots at 12°C, while no growth occurred on shredded bell peppers due to their lower pH. At 7°C only a survival or a rare decrease in numbers was detected (Bagamboula et al. 2002). No difference in growth, survival, or die-off was noticed between air conditions and low O_2 and low CO_2 atmospheres.

The genus *Salmonella* is frequently present on raw vegetables and fruits, as reviewed by Doyle (1990), Beuchat (1996), Tauxe et al. (1997), and Francis et al. (1999). Vegetables sampled in the field or retail outlets were contaminated with *Salmonella* spp. at frequencies of 7.5% in Spain and 8% to 63% in the Netherlands (Nguyen-the and Carlin 1994; Pirovani et al. 2000). The growth rate is substantially reduced at less than 15°C, and growth is prevented at less than 7°C (ICMSF 1996). Piagentini et al. (1997) observed a survival of inoculated *S. hadar* on shredded cabbage after 10 days of storage at 4°C, and an increase was detected at 12 and 20°C. The cabbage was packaged in permeable packaging material resulting in a steady-state atmosphere around 1.5% O_2 and 8 to 10% CO_2. On the surfaces of tomatoes, no significant change in the population of *S. montevideo* was found at 10°C (18 days storage), but at 20°C, growth occurred (7 days storage) (Zhuang et al. 1995). Finn and Upton (1997) found an absence (after preenrichment step) of inoculated *S. typhimurium* after 2 days of storage at 7°C on shredded carrots and cabbage (<1% O_2 and >25% CO_2). It was demonstrated by Lin and Wei (1997) that Salmonellae, present at the surface of tomatoes and melons, were spread in the tissue by cutting, provoking a possible hazard in the precut products. *S. enteriditis, S. infantis,* and *S. typhimurium* were reported to be capable of growth in chopped cherry tomatoes. These microorganisms, together with many other *Salmonella* spp., were able to grow at low pH (3.99 to 4.37) under certain conditions (Asplund and Nurmi 1991; Wei et al. 1995; ICMSF 1996). Refrigeration is the best preservation method to prevent an outgrowth of these mesophilic pathogens.

3.2.2.2 *Escherichia coli* O157:H7

Although raw and undercooked foods of bovine origin are the main food sources for *Escherichia coli* O157:H7, *E. coli* O157:H7 infections associated with contaminated fruits, vegetables, and water have increased (Park et al. 1999). *E. coli* O157:H7 was isolated in sprouts (alfalfa, mung bean, radish) and in vegetable salads (Lin et al. 1996). Survival and growth patterns of *E. coli* O157:H7 were dependent on vegetable type, package atmosphere, storage temperature, and bacterial strain (Francis and O'Beirne 2001). *E. coli* O157:H7 was able to grow on sprouting alfalfa sprouts stored 2 days at room temperature (Castro-Rosas and Escartín 2000; Charkowski et al. 2002). *E. coli* O157:H7 exponentially grew in surface wounds on different apple cultivars stored at 24°C for 6 days (Janisiewicz et al. 1999; Dingman 2000). At 4°C, the *E. coli* O157:H7 population in shredded lettuce declined approximately 1 log throughout a 14-day storage (Chang and Fang 2007). At a higher temperature (22°C), populations of the same *E. coli* O157:H7 strain increased with about 3 log within 3 days. Similarly, populations of *E. coli* O157:H7 in lettuce stored at 5°C decreased with about 1 log in 18 days but increased with about 3 log when lettuce was stored at 15°C (Li et al. 2001a). In another report, it was observed that *E. coli* O157:H7 could survive for 14 days on freshly peeled Hamlin orange of which the pH at the surface was 6 to 6.5 (Pao et al. 1998). *E. coli* O157:H7 was able to multiply on shredded lettuce, sliced cucumber, and shredded carrots at 12 to 21°C, and on melon cubes at 25°C. At 5°C, *E. coli* O157:H7 could survive on melon cubes for 34 hours and up to 14 days on shredded lettuce, sliced cucumber, and shredded carrots (Abdul-Raouf et al. 1993; Del Rosario and Beuchat 1995). The combination of the capacity of *E. coli* O157:H7 to grow on produce at higher temperatures and to survive at refrigerated temperatures and its low infectious dose (10 to 100 CFU/g) makes the presence of this pathogen in produce an important risk for public health (Chang and Fang 2007). Packaging fresh produce under modified atmospheric conditions does not have an univocal effect on the survival and growth of *E. coli* O157:H7. Although modified atmosphere packaging had beneficial effects on the shelf life of shredded lettuce, the extended shelf life allowed *E. coli* O157:H7 to grow to higher numbers within the shelf-life period compared to air-held shredded lettuce (Diaz and Hotchkiss 1996).

Gunes and Hotchkiss (2002) observed that *E. coli* O157:H7 survived in fresh-cut apples but was inhibited in modified atmospheres with high carbon dioxide concentrations at abusive temperatures. Next to the traditional intrinsic and extrinsic parameters, the growth of *E. coli* O157:H7 on fresh-cut produce is also influenced by the degree of processing such as the degree and the way of cutting (Gleeson and O'Beirne 2005).

3.2.2.3 Campylobacter spp.

Campylobacter is the most common cause of bacterial gastroenteritis worldwide, yet the etiology of this infection remains only partly explained (Evans et al. 1997). Next to traditional risk factors like eating poultry meat and contact with animals, less evident risk factors such as the consumption of raw vegetables (tomato, cucumber) and drinking bottled water were recognized as risk factors for *Campylobacter* infections (Evans et al. 2003). Many studies deal with the prevalence of *Campylobacter* in fresh produce. *Campylobacter* spp. was not detected during an evaluation of five types of fresh produce obtained at the retail level (Thunberg et al. 2002) and was also not isolated from imported lettuce samples (Little et al. 1999), prepared salads and vegetables (Whyte et al. 2004), and Vietnamese vegetables (Dao and Yen 2006). In another report, the prevalence of *Campylobacter* spp. in a popular Malaysian salad dish was analyzed, originating from both a traditional wet market and two modern supermarkets (Chai et al. 2007). In this case, the average prevalence of *Campylobacter* spp. in raw vegetables from these locations ranged between 29.4% and 67.7%. Although, to a lesser degree, *Campylobacter jejuni* was also isolated from 1.5% of the investigated fresh mushrooms (Doyle and Schoeni 1986) and from 3.57% of Indian vegetable samples (Kumar et al. 2001). Thermotolerant *Campylobacters* were detected in spinach (3.3%), lettuce (3.1%), radish (2.7%), green onions (2.5%), parsley (2.4%), and potatoes (1.6%) sold at farmers' outdoor markets, whereas the samples sold in supermarkets were all negative in Canada (Park and Sanders 2002). The presence of *Campylobacter* in produce originates from cross-contamination at the holding and packaging stage in supermarkets or in the kitchen, poor hygiene of the food handler or contamination by contact with natural fertilizers or contaminated water (Kumar et al. 2001; De Cesare et al. 2003; Kärenlampi and Hänninen 2004; Chai et al. 2007).

Despite fastidious growth requirements, *Campylobacter* spp. survive at refrigeration temperatures for extended periods within nutrient-limited environments. This property, combined with the low infective dose and their microaerophilic nature indicate the potential significance of *Campylobacter* with respect to refrigerated produce (Francis et al. 1999). Furthermore, ready-to-eat vegetables in modified atmosphere packaging are kept at high relative humidity, refrigerated (<10°C) conditions with a relatively low level of oxygen (<5%). As survival of *Campylobacter* is longest at low oxygen concentrations, modified atmosphere packaging may prolong the shelf life of this pathogen (Federighi et al. 1999). Although the survival of *Campylobacter* in foods of animal origin (milk, eggs, meat) was extensively studied, the survival of *Campylobacter* on fresh produce is rather poorly characterized. Kärenlampi and Hänninen (2004) were the sole researchers examining the survival of *Campylobacter jejuni* on fresh produce, like fresh-cut iceberg lettuce, cantaloupe pieces, cucumber slices, grated carrot, and strawberries. At 7°C, the mean death rates (day^{-1}) were varying between 0.41 and 1.02 log depending on the type of produce (Kärenlampi and Hänninen 2004). The corresponding death rates at 21°C ranged between 1.52 and 8.74 day^{-1}. The death rate in strawberries was significantly higher than in other produce, probably due to the low pH (3.4) of strawberries in comparison with the other tested produce (5.8 to 6.8) (Kärenlampi and Hänninen 2004). Based on these results, Kärenlampi and Hänninen (2004) stated that after a contamination of fresh produce, including strawberries, *Campylobacter jejuni* may survive long enough to pose a risk for the consumer.

3.2.2.4 *Listeria monocytogenes*

Listeria monocytogenes has increasingly been considered as a cause of foodborne disease, sometimes responsible for large outbreaks of infection. Lettuce and other raw vegetables have been identified as potential vectors of listeriosis (Ho et al. 1986; Sizmur and Walker 1988). A screening of imported vegetables in the United Kingdom revealed that on the 151 samples no *L. monocytogenes* was isolated (Little et al. 1999). The same results were obtained in Italy for the period 1989 to 1999 (Messi et al. 2000) and in Canada on 100 samples of lettuce, celery, radishes, and tomatoes (Farber et al. 1989). The organism was rarely detected on vegetables prior to processing (Fenlon et al. 1996). But, in Spain, 30% of 70 samples of mixed lettuce were found to be positive on the presence of *L. monocytogenes* (Garcia-Gimeno et al. 1996). Fresh-cut vegetables showed an incidence of *L. monocytogenes* varying from 0% to 19% in Europe (Carlin and Nguyen-the 1994). A contamination level of fresh vegetables for *L. monocytogenes* was reviewed by Francis et al. (1999) as being 0% to 44%. Farber et al. (1998) assessed the potential of *L. monocytogenes* to survive and grow at refrigeration temperatures (4 and 10°C) on various retail and wholesale packaged fresh-cut produce. Carlin and Nguyen-the (1994) studied the fate of *L. monocytogenes* on green leafy vegetables. Several investigations demonstrated possible growth of *L. monocytogenes* on modified atmosphere packaged fresh-cut vegetables, although the results depended on the type of vegetables and the storage temperature (Berrang et al. 1989b; Beuchat and Brackett 1990b; Farber, 1991; Omary et al. 1993; Carlin et al. 1995; Carlin et al. 1996a, 1996b; Zhang and Farber 1996; Juneja et al. 1998; Bennik et al. 1999; Jacxsens et al. 1999; Liao and Sapers 1999; Thomas et al. 1999; Castillejo-Rodriguez et al. 2000).

L. monocytogenes is an important human pathogen associated with fresh-cut produce because the pathogen is widespread in the natural environment of fruits and vegetables, the pathogen is psychrotrophic from nature (minimal temperature for growth is between 0 and 4°C), the minimal pH is 4.5 to 5 (ICMSF 1996), and it is not influenced by the modified atmospheres applied for fresh-cut vegetables and fruits (Berrang et al. 1989a; Beuchat and Brackett 1990b; Kallander et al. 1991; Bennik et al. 1995; Carlin et al. 1996a; Francis and O'Beirne 1997; Jacxsens et al. 1999). The composition of the produce could have an effect as demonstrated for carrots, which have an antilisterial effect (Beuchat and Brackett 1990a; Nguyen-the and Lund 1991; Farber et al. 1998; Jacxsens et al. 1999).

3.2.2.5   Spore-Forming Pathogens: *Clostridium botulinum* and *Bacillus cereus*

In the early days of modified atmosphere (MA) packaging of vegetables, attention was focused primarily on anaerobic pathogens, especially proteolytic *Clostridium botulinum*, which produces a lethal heat-unstable toxin but does not grow below 10°C, although nonproteolytic types B, E, and F have been recorded as growing and producing toxins at temperatures as low as 3.3°C (Francis et al. 1999). The growth of aerobic spoilage microflora rapidly decreases the redox potential of the food, improving conditions for the growth of *C. botulinum* (Francis et al. 1999). These findings stimulated the FDA to recommend the puncturing of packaging films applied to high respiring mushrooms (Doyle 1990). Before a real hazard to food safety, growth to higher counts is necessary (10^5 to 10^6 CFU/g), and toxin production must occur. This growth is mostly inhibited by the natural microflora present in fresh-cut fruits and vegetables. Moreover, toxin production is not favored in these conditions. However, they are often isolated from plant materials as these spores are present in soil.

3.2.2.6   Viruses

Enteric viruses are another group of microorganisms potentially posing a health problem for fresh-cut vegetables and fruits. Foodborne viruses can be classified according to the disease that they cause: viruses causing gastroenteritis such as noroviruses and viruses causing hepatitis such as hepatitis A virus (Koopmans and Duizer 2004). They are characterized by a low infectious dose, and high numbers of viruses are usually present in feces (Carter 2005). Enteric viruses are obligate intracellular parasites; as a consequence, foodborne viruses cannot grow on foods, unlike bacterial pathogens. However, survival on fruits, vegetables, and soil is reported. Badawy et al. (1985) studied the survival of rotavirus on lettuce, radishes, and carrots stored at 4°C and room temperature. The virus survived for 25 to 30 days at 4°C and 5 to 25 days at room temperature. The greatest survival was observed on lettuce. No decline of poliovirus after 2 weeks storage of strawberries was noticed by Kurdziel et al. (2001). Less than 1 log reduction of MS2 coliphage on several produce types such as lettuce, cabbage, and carrots occurred after 7 days at 4°C (Dawson et al. 2005). Similarly, no reduction of HAV was observed on lettuce stored 9 days (Croci et al. 2002). Even at low temperatures (3 to 10°C), a long survival of minimally 90 days was found of poliovirus and coxsackie virus in soils (Brackett 1992). This might be long enough to contaminate short-season vegetables.

Viral contamination is possible at the preharvest, harvest, and postharvest stages. Preharvest applications such as the use of fecal contaminated sewage or irrigation water on the field can introduce enteric viruses to fresh produce (Stine et al. 2005). Mechanical harvesting is not feasible for certain kinds of produce, such as raspberries or strawberries. The perishable berries need to be picked manually, and this imposes additional risk if fruit pickers do not follow good hygienic practices. Outbreaks of viral gastroenteritis have also been documented in which food handlers became ill after preparing or serving implicated foods (Beuchat 1996; Bidawid et al. 2000).

3.2.2.7 Parasites

The main parasitic protozoa of concern to the food production industry are *Giardia duodenalis* (sometimes referred to as *G. intestinalis* or *G. lamblia*), *Cryptosporidium parvum,* and *Cyclospora cayetanensis* that cause intestinal infections. Tissue protozoa *Toxoplasma gondii* causing fetus malformations also have the potential to cause food- and waterborne toxoplasmosis (Dawson 2005). This group of foodborne pathogens has received little attention in developed countries, although the problem is of increasing concern (Rose and Slifco 1999). A survey undertaken in Norway between August 1999 and January 2001 showed the presence of *Cryptosporidium* or *Giardia* in several vegetables. Mung bean sprouts was the most important contaminated product, and lettuce, dill, radish sprouts, and strawberries were also found to be contaminated (Robertson and Gjerde 2001). The increasing number of susceptible persons, more extensive trade in produce across the borders, advancement in food-processing technology, and changes in national and international policies concerning food safety are the main issues in the observed increase. Detection of parasites from fruit and vegetables is generally inadequate, with low and variable recovery efficiencies. However, Robertson and Gjerde (2000) proposed new methods for the isolation and enumeration of several parasites from fruits and vegetables. Survival studies are scarce and mainly focused on watery environments (Erickson and Ortega 2006). It is reported that *C. parvum* oocysts inoculated on iceberg lettuce showed 90% inactivation after 3 days storage at 4°C, whereas 100% inactivation was observed after 3 days at 22°C. Survival was greater on textured leaves (50% inactivation/4 days on Rav Baby Leaf Head lettuce) than on smoother leaves (e.g., iceberg lettuce) (Warnes and Keevil 2003).

Vehicles of contamination include sewage effluent, surface water, or contaminated irrigation water. *G. lamblia* and *C. parvum* have both animal and human reservoirs, and *C. cayentanensis* is thought to be transmitted by only human sewage (Dawson 2005). In addition to water, the food handler may be an important source of contamination of fresh produce with protozoan parasites during picking or handling fresh produce.

3.2.3 Setting Microbiological Criteria for Pathogens Associated with Fresh-Cut Fruits and Vegetables

The safety of food must be assured by a preventive approach based on the application of hygiene measures such as Good Agricultural Practices (GAP), the Prerequisite Programs (PRPs), and Hazard Analysis Critical Control Point (HACCP) at different stages of the supply chain. The agricultural sector, the food industry, and the distribution network must work together in order to improve the microbiological safety of fresh-cut fruits and vegetables. In order to verify the food chain, microbial sampling can be performed; therefore, criteria must be available. It is recognized that microbiological criteria are widely used in the food industry, but they have been published rarely (Stannard 1997) because they are mainly internal criteria applied by food business operators. However, with the publication of the EU Regulation 2073/2005 on microbiological criteria for foodstuffs, a big step forward is made for the safety of fresh-cut vegetables and fruits because criteria are now included both for *L. monocytogenes* as a food safety criteria as well as for *Salmonella* spp. and *E. coli* as process hygiene criteria in Europe. Fresh, uncut, and unprocessed vegetables and fruits, excluding sprouted seeds, are considered as ready-to-eat products, but regular testing for *L. monocytogenes* is not useful in normal circumstances as stated in this regulation. However, due to the fact that *L. monocytogenes* can be present on the raw materials entering the fresh-cut food industries, it is advisable to check specific raw materials for *L. monocytogenes*.

Fresh-cut fruits and vegetables, ready-to-eat, must be considered as "ready-to-eat foods able to support growth of *L. monocytogenes*," where the food safety criteria arc set as maximum 100 CFU/g at the end of the shelf life, an indication that during the entire shelf life, a maximum of 100 CFU/g may not be exceeded. However, depending on the shelf life of the fresh-cut vegetables (less than 5 days of shelf life) and the pH of the produce (pH below 4.4, as will be the case for the majority of fruits), they can also be considered as "ready-to-eat foods unable to support growth of *L. monocytogenes*." When growth on the food product is possible, the objective or target value at the day of production should be absent in 25 g and a tolerance value can be defined as long as the maximum of 100 CFU/g at the end of the shelf life is not exceeded. It is up to the food business operator to investigate how *L. monocytogenes* will behave during shelf life. In relation to the potential of *L. monocytogenes* to grow in the concerned food product in the prescribed storage conditions within the prescribed shelf-life period, the tolerance value on the day of production can be adjusted. This growth can be quantified (e.g., by means of challenge tests, in which the pathogen is artificially inoculated on the product), through scientific literature or through predictive modeling. If this growth cannot be substantiated by the food business operator, the tolerance value of absent in 25 g should be used.

The performance of a challenge test is encouraged where the growth or survival of the pathogen in the specific product is quantified in order to be able to set specifications of the product on the day of production (Norrung et al. 1999; Norrung 2000). Distinction can be made between (Table 3.4) group 1—fresh-cut fruits and vegetables, where growth during shelf life can be expected which can provide risk for public health by exceeding the criteria of 100 CFU/g (e.g., 2 log growth during 7 days of shelf life for MA packaged fresh-cut produce at 7°C); group 2—fresh-cut fruits and vegetables where a limited growth can be expected during shelf life (e.g., 1 log growth during 7 days of shelf life for MA packaged fresh-cut produce at 7°C); and group 3—fresh-cut fruits and vegetables where no growth is possible during shelf life (of greater than 5 days) or product with a shelf life less than 5 days (according to EU Regulation 2073/2005).

In non-European-Community countries, there is often another point of view regarding *L. monocytogenes*: For example, the United States and Canada introduced a zero tolerance for some foods (absence of *L. monocytogenes* in 25 g), especially foods that are supportive of growth and have extended shelf lives. In these countries, decontamination techniques are often allowed in the production chain in order to reduce the bacterial load and avoid the presence of pathogens. In Table 3.5, an

TABLE 3.4

Classification of Fresh-Cut Fruits and Vegetables in Possible Risk of Outgrowth of *Listeria monocytogenes* during Shelf Life

Group 1	Group 2	Group 3
• Fruit: melon parts • Sugar-rich vegetables with cut surfaces (e.g., cubes of zucchini, pumpkin, cucumber slices, eggplant cubes)	• Leafy fresh-cut vegetables (e.g., lettuce, celery, leek, chicory endive, spinach) • Red fruits with green crown part (e.g., strawberry, tomato)	• Fresh-cut produce with pH <4.4 (e.g., most fresh-cut fruit products) • Fresh-cut produce with shelf life <5 days • Mixed bell peppers due to low pH • Grated carrots due to antilisterial compound in carrots

overview is given of the microbiological criteria for *L. monocytogenes* on fresh-cut vegetables and fruits, according to European legislation 2073/2005.

With regard to other pathogens, the absence of *Salmonella* in 25 g is generally recommended for fresh-cut produce in France (Nguyen-the and Carlin, 1994), in the United Kingdom (Stannard 1997; Francis et al. 1999), in the United States (Francis et al. 1999), and in Germany (Francis et al. 1999) and is followed in the European regulation regarding criteria for foodstuffs (EU Regulation 2073/2005) but is defined as a process hygiene indicator (Table 3.5).

A number of bacteria can be monitored as hygiene indicators for the production of food products. The presence of a hygiene indicator beyond a certain limit indicates an insufficiently hygienic production process in general and a possible fecal contamination in particular. The presence of *E. coli* in addition indicates the possible presence of ecologically similar pathogens (e.g., *Shigella*, *Salmonella*). *Escherichia coli* is a true indicator of fecal origin whose presence is linked to the possible presence of other fecal pathogens (taxonomically, ecologically, and physiologically). *E. coli* can be used as a hygiene indicator both in the production chain and during storage. In EU Regulation 2073/2005, *E. coli* is used as a hygiene indicator for fresh-cut vegetables and fruits (Table 3.5).

Another hygiene indicator often applied is *Staphylococcus aureus* as an indication of personal hygiene. It is a typical bacterium that is present on the hands and in the mouth and nose. *S. aureus* is also a food intoxicant but only poses a risk when it grows to high numbers (10^5 to 10^6 CFU/g) (ICMSF 1996). Low numbers can be tolerated (Table 3.5). Because this bacterium cannot grow at a temperature less than 10°C and its development on raw products is inhibited because of competition by a large accompanying flora, high numbers can never be reached when the cold chain is kept intact. Excessive values indicate temperature abuse or a local severe postcontamination due to personnel handling.

TABLE 3.5

Overview of Legal Microbiological Criteria (Based on EU Regulation 2073/2005) and Microbiological Guide Values (Expressed as CFU/g) for Fresh-Cut Ready-to-Eat Fruits and Vegetables

Parameter	Target[h]	Tolerance[i]	Best Before Date[j]
Escherichia coli[a]	10^2	10^3	10^3
Staphylococcus aureus	10^2	10^3	10^3
Salmonella spp.[b,c]	Absent/25 g	Absent/25 g	Absent/25 g
Listeria monocytogenes[d]	Absent/25 g[e]	Absent/x g[e,f] $100/g^g$	10^2

[a] Legal base for *E. coli:* EU Regulation 2073/2005 process hygiene criteria category 2.5.1 "pre-cut vegetables ready-to-eat."

[b] Legal base for *Salmonella*: EU Regulation 2073/2005: food safety criteria category 1.19 "pre-cut vegetables ready-to-eat."

[c] If it concerns a vegetable that is specifically destined to be heated before consumption (e.g., mixture of vegetables for cooking and soup vegetables), this criteria needs to be considered as a process parameter.

[d] Legal base for *Listeria monocytogenes*: EU Regulation 2073/2005: food safety criteria categories 1.2 and 1.3 (ready-to-eat food products).

[e] Depending on the growth potential—determined by the intrinsic and extrinsic factors of the food product and the shelf life—the tolerance value needs to be adjusted in such a way that the guide value of 100 CFU/g can still be guaranteed at the end of the shelf life. See three groups of fresh-cut vegetables and fruits that are defined in Table 3.4. In group 1 where approximately 2 log growth is possible, this tolerance must be set at "absence in 1 g" in order to reach a maximum of 100 CFU/g at the end of the shelf life. In group 2 where approximately 1 log growth is possible, this tolerance must be set as "absence in 10 g."

[f] If the food business operator is not able to demonstrate to what extent growth of *Listeria monocytogenes* can occur during the shelf life of the product concerned and in the predetermined storage conditions, then the tolerance value, absence in 25 g, must be adhered to and no tolerance is possible during the day of production.

[g] This tolerance value is valid only if *Listeria monocytogenes* cannot grow because of intrinsic and extrinsic factors. Also, according to EU Regulation 2073/2005, this applies to products with a pH $\leq$ 4.4 or a_w $\leq$ 0.92; products with a pH $\leq$ 5 and a_w $\leq$ 0.94; and products with a shelf life that is shorter than 5 days. Other categories of products can also be assigned to this category if there are scientific reasons for this (see Table 3.4).

[h] Target is the guideline for the production day, in the best conditions produced.

[i] Tolerance is the maximum guideline for the production day.

[j] Best before date is the end of the shelf life. Above these guidelines, notable spoilage will occur.

3.3 DETERMINATION OF THE SHELF LIFE OF FRESH-CUT FRUITS AND VEGETABLES: COMBINING SPOILAGE AND SAFETY ASPECTS

Each food processor is responsible for setting the shelf-life date on products put on the market. Regarding fresh-cut fruits and vegetables, this shelf-life date will be relatively short, so a "best before" date should be mentioned on each package with an indication of the day and month and, preferably, the year. In order to set a correct shelf-life date, different aspects of fresh-cut fruits and vegetables have to be taken into account. As previously discussed, fresh-cut vegetables and fruits are subjected to both physiological (metabolic) and microbiological spoilage. When physiological activity can proceed, spoilage will occur altering its sensorial properties (e.g., discoloration, enzymatic browning). Also, in the case of extensive growth of spoilage microorganisms, deterioration can occur resulting in a decline of the sensorial quality (e.g., off odor and off flavor, loss of texture). Food safety must also be considered: fresh-cut fruits and vegetables placed on the market may not be harmful for consumers. Consequently, when defining the shelf-life date of fresh-cut fruits and vegetables, food safety (and possible growth of pathogens) and food quality (microbiological and physiological deterioration) should be considered and evaluated. Interactions between these should be investigated during the shelf-life study, because fresh-cut vegetables and fruits, different from other food products, are still living tissues with a specific metabolism. Attention has to be given to the selection of the correct parameters that will be followed and evaluated during the shelf-life studies.

There are several reasons why fresh-cut produce is relatively safe when compared to other foods. Conditions used with fresh produce are usually unfavorable for the growth of most pathogens (refrigeration temperatures, relatively low nutrients available in some types of vegetables, e.g., leafy vegetables, low pH of fruits, short shelf life). The spoilage microorganisms in refrigerated produce are usually psychrotrophic and therefore have a competitive advantage over most pathogens. Sometimes this competition prevents the growth of pathogens (Hotchkiss and Banco 1992; Brackett 1994; Carlin et al. 1996b; Francis et al. 1999). In other cases, the food simply spoils before it is eaten. Nevertheless, foodborne disease can and does occur with consumption of fruits and vegetables, especially when fresh-cut produce is modified atmosphere packaged, as it increases the shelf life of the products, and pathogens have more time to develop to infectious numbers before the product is notably spoiled.

When setting up shelf-life studies, temperature abuse should be considered as well. In most European countries, these products are stored between 4 and 8°C; higher temperatures will induce physiological spoilage and microorganisms will be able to develop more quickly.

Below is a case study presenting the results of the evaluation of shelf life for different types of fresh-cut vegetables (after Jacxsens et al. 1999) (Table 3.6).

TABLE 3.6
Overall Shelf Life (Days) of Fresh-Cut Produce Stored under Modified Atmosphere (MA) or in Air at 7°C and Outgrowth of Pathogens during the Shelf Life

Vegetable Type	Atmosphere	Sensorial Shelf Life (Days)	Microbial Shelf Life (Days)	Growth of Listeria monocytogenes
Brussels sprouts (trimmed)	MA	7 (discoloration of cut surfaces, freshness)	—[a]	1 log CFU/g reduction/7 days
	Air	<1 (discoloration of cut surfaces)	—[a]	—[b]/<1 day
Carrots (grated)	MA	7 (acetic taste and odor)	6 (total count, lactic acid bacteria)	2 log CFU/g reduction/6 days
	Air	4 (dry taste and appearance)	4 (total count)	—[c]
Chicory endive (0.5 cm)	MA	4 (discoloration of cut surfaces)	—[a]	<1 log CFU/g growth/4 days
	Air	<1 (discoloration of cut surfaces)	—[a]	—[b]/<1 day
Iceberg lettuce (1 cm)	MA	5 (discoloration of cut surfaces)	—[a]	<1 log CFU/g growth/5 days
	Air	3 (discoloration of cut surfaces)	—[a]	—[b]/3 days

[a] No microbial limit exceeded in 6 (chicory endives) or 7 days (Brussels sprouts and iceberg lettuce) storage at 7°C.
[b] Survival (no growth).
[c] Contamination under detection level (<2 log CFU/g).

Storage experiments were conducted to follow the behavior of pathogens on fresh-cut vegetables (trimmed Brussels sprouts, grated carrots, shredded iceberg lettuce, and shredded chicory endives) packaged under equilibrium-modified atmosphere (2% to 3% O_2, 2% to 3% CO_2, and 94% to 96% N_2) MA (modified atmosphere) and stored at 7°C. As a comparison, fresh-cut vegetables were also packaged in a macro-perforated high-barrier film (air conditions) and stored at 7°C.

In a first step, the shelf life of the vegetables in the two kinds of packages was determined by evaluating the microbiological quality as well as the sensorial quality (appearance, taste, and odor). The end of the microbiological shelf life is reached when the limit (CFU/g) of a particular group of spoilage microorganisms is exceeded, and thus, the fresh-cut produce is no longer consumable based on the microbiological quality (Table 3.1 and Table 3.2). After each microbiological sampling, the fresh-cut vegetables were scored for visual and organoleptical properties by four to six members of a trained sensorial panel. In general, sensorial properties were faster limiting the shelf life than microbiological criteria if visual properties limited the shelf life. The shelf life of the vegetables stored under MA was extended by 40% or more, compared to the air-stored vegetables (see Table 3.6).

In a second storage experiment, the four fresh-cut vegetables were inoculated with a cocktail of psychrotrophic strains of *Listeria monocytogenes* before packaging under MA and air at 7°C. Growth of the inoculated pathogens was more influenced by the type of vegetable than by the type of atmosphere. No growth was detected on the Brussels sprouts, and the antilisterial component of carrots prevented *Listeria* from growing on this commodity (see Table 3.6). Packaging fresh-cut vegetables under an EMA (2% to 3% O_2, 2% to 3% CO_2, 94% to 96% N_2) had a positive effect on the sensorial properties of the packaged vegetables. Suppression of enzymatic discoloration and a longer retention of the rigid structure by retardation of the respiration rate/transpiration losses are caused by storing the vegetables in a lower oxygen atmosphere (<5% O_2). The influence of the EMA on microbiological quality was not always obvious. The beneficial effect of EMA storage of vegetables is, for the most part, the result of the physiological state of the produce rather than the inhibition of soft-rot bacteria (Bennik et al. 1995; Carlin et al. 1996a; Bennik et al. 1998).

However, an extension of the lag phase could be postulated for spoilage microorganisms on shredded chicory endives and iceberg lettuce. When the sensorial quality of the packaged vegetables decreased, the outgrowth of the spoilage microorganisms reached the same level as the microbiological counts for the air-stored vegetables at 7°C. The sensorial properties limited the shelf life of the EMA and the air-stored produce faster than the microbiological criteria. Grated carrots were an exception: EMA conditions favored growth of lactic acid bacteria.

Psychrotrophic pathogens, such as *L. monocytogenes*, were able to grow under an EMA at 7°C and were more influenced by the type of vegetable than by the type of atmosphere. Based on the determination of the shelf life, safety criteria could be defined on the day of production for *Listeria monocytogenes*: absence in 0.1 g for shredded iceberg lettuce and shredded chicory endives. For grated carrots and trimmed Brussels sprouts, absence in 0.01 g on the day of production could be postulated, as *L. monocytogenes* did not grow on these fresh-cut vegetables.

REFERENCES

Abdul-Raouf, U. M., Beuchat, L. R., and Ammar, M. S. 1993. Survival and growth of *Escherichia coli* O157:H7 on salad vegetables. *Applied and Environmental Microbiology* 59:1999–2006.

Allende, A., Luo, Y. G., Mcevoy, J. L., Artes, F., and Wang, C. Y. 2004. Microbial and quality changes in minimally processed baby spinach leaves stored under super atmospheric oxygen and modified atmosphere conditions. *Postharvest Biology and Technology* 33:51–59.

Artés, F., Gómez, P. A., and Artés-Hernandéz, F. 2007. Physical, physiological and microbial deterioration of minimally fresh processed fruits and vegetables. *International Food Science and Technology* 13:177–188.

Asplund, K., and Nurmi, E. 1991. The growth of salmonellae in tomatoes. *International Journal of Food Microbiology* 13:177–182.

Babic, I., Roy, S., Watada, A. E., and Wergin, W. P. 1996. Changes in microbial populations on fresh cut spinach. *International Journal of Food Microbiology* 31:107–119.

Badawy, A., Gerba, C., and Kelly, L. 1985. Survival of rotavirus SA-11 on vegetables. *Food Microbiology* 2:199–205.

Bagamboula, C., Uyttendaele, M., and Debevere, J. 2001. Acid tolerance of *Shigella sonnei* and *flexneri*. Proceedings of Nizo Dairy Conference on Food Microbes, 13–15 June 2001, Ede, the Netherlands, p. 127.

Bagamboula, C., Uyttendaele, M., and Debevere, J. 2002. Growth and survival of *Shigella sonnei* and *S. flexneri* in minimal processed vegetables packed under equilibrium modified atmosphere and stored at 7°C and 12°C. *Food Microbiology* 19:529–536.

Barry-Ryan, C., and O'Beirne, D. 1998. Quality and shelf-life of fresh cut carrot slices as affected by slicing method. *Journal of Food Science* 63:851–856.

Beaulieu, J. C., and Gorny, J. R. 2002. Fresh-cut fruits. In *The commercial storage of fruits, vegetables, and florist and nursery stocks,* USDA Handbook No. 66, 3rd Edition, ed. K. C. Gross, C. Y. Wang, and M. E. Saltveit, 1–17. Washington, DC: Agricultural Research Service.

Bennik, M., Smid, E., Rombouts, F., and Gorris, L. 1995. Growth of psychrotrophic foodborne pathogens in a solid surface model system under the influence of carbon dioxide and oxygen. *Food Microbiology* 12:509–519.

Bennik, M. H. J., Vorstman, W., Smid, E. J., and Gorris, L. G. M. 1998. The influence of oxygen and carbon dioxide on the growth of prevalent *Enterobacteriaceae* and *Pseudomonas* species isolated from fresh and controlled-atmosphere-stored vegetables. *Food Microbiology* 15:459–469.

Bennik, M., Van Overbeek, W., Smid, E., and Gorris, L. 1999. Biopreservation in modified atmosphere stored mungbean sprouts: the use of vegetable-associated bacteriogenic lactic acid bacteria to control the growth of *Listeria monocytogenes*. *Letters in Applied Microbiology* 28:226–232.

Berrang, M., Brackett, R., and Beuchat, L. 1989a. Growth of *Aeromonas hydrophila* on fresh vegetables stored under a controlled atmosphere. *Applied and Environmental Microbiology* 55:2167–2171.

Berrang, M., Brackett, R., and Beuchat, L. 1989b. Growth of *Listeria monocytogenes* on fresh vegetables stored under a controlled atmosphere. *Journal of Food Protection* 52:702–705.

Beuchat, L. 1996. Pathogenic microorganisms associated with fresh produce. *Journal of Food Protection* 59:204–216.

Beuchat, L., and Brackett, R. 1990a. Inhibitory effect of raw carrots on *L. monocytogenes*. *Applied and Environmental Microbiology* 56:1734–1742.

Beuchat, L., and Brackett, R. 1990b. Survival and growth of *L. monocytogenes* on lettuce as influenced by shredding, chlorine treatment, modified atmosphere packaging and temperature. *Journal of Food Science* 55:755–758, 870.

Bidawid, S., Farber, J., and Sattar, S. 2000. Contamination of foods by food handlers; experiments on Hepatitis A virus transfer to food and its interruption. *Applied and Environmental Microbiology* 66:2759–2763.

Blanco, P., Sieiro, C., and Villa, T. G. 1999. Production of pectic enzymes in yeasts. *FEMS Microbiology Letters* 175:1–9.

Brackett, R. 1992. Shelf stability and safety of fresh produce as influenced by sanitation and disinfection. *Journal of Food Protection* 55:808–814.

Brackett, R. 1994. Microbiological spoilage and pathogens in minimally processed refrigerated fruits and vegetables. In *Minimally processed refrigerated fruits and vegetables*, ed. R. Wiley, 269–312. New York: Chapman and Hall.

Brandt, M. 2006. Fitness of human enteric pathogens on plants and implications for food safety, *Annual Review of Phytopathology* 44:367–392.

Brocklehurst, T. F., and Lund, B. M. 1981. Properties of Pseudomonads causing spoilage of vegetables stored at low-temperature. *Journal of Applied Bacteriology* 50:259–266.

Calder, L., Simmons, G., Thornley, C., Taylor, P., Pritchard, K., Greening, G., and Bishop, J. 2003. An outbreak of hepatitis A associated with consumption of raw blueberries. *Epidemiology and Infection* 131:745–751.

Canganella, F., Ovidi, M., Paganini, S., Vettraino, A. M., Bevilacqua, L., and Trovatelli, L. D. 1998. Survival of undesirable micro-organisms in fruit yoghurts during storage at different temperatures. *Food Microbiology* 15:71–77.

Carlin, F., Nguyen-the, C., Cudennec, P., and Reich, M. 1989. Microbiological spoilage of fresh, ready-to-use grated carrots. *Sciences des Aliments* 9:371–386.

Carlin, F., and Nguyen-the, C. 1994. Fate of *Listeria monocytogenes* on four types of minimally processed green salads. *Letters in Applied Microbiology* 18:222–226.

Carlin, F., Nguyen-the, C., and Abreu da Silva, A. 1995. Factors affecting the growth of *L. monocytogenes* on minimally processed fresh endive. *Journal of Applied Bacteriology* 78:636–646.

Carlin, F., Nguyen-the, C., Abreu da Silva, A., and Cochet, C. 1996a. Effects of carbon dioxide on the fate of *L. monocytogenes*, of aerobic bacteria and on the development of spoilage in minimally processed fresh endive. *International Journal of Food Microbiology* 32:159–172.

Carlin, F., Nguyen-the, C., and Morris, C. 1996b. The influence of the background microflora on the fate of *Listeria monocytogenes* on minimally processed fresh broad leaved endive. *Journal of Food Protection* 59:698–703.

Carter, M. J. 2005. Enterically infecting viruses: pathogenicity, transmission and significance for food and waterborne infection. *Journal of Applied Microbiology* 98:1354–1380.

Castillejo-Rodriquez, A., Barco-Alcala, E., Garcia-Gimeno, R., and Zurera-Cosano, G. 2000. Growth modelling of *Listeria monocytogenes* in packaged fresh green asparagus. *Food Microbiology* 17:421–427.

Castro-Rosas, J., and Escartín, E. F. 2000. Survival and growth of *Vibrio cholerae* O1, *Salmonella typhi*, and *Escherichia coli* O157:H7 in alfalfa sprouts. *Journal of Food Science* 65:162–165.

Chai, L. C., Robin, T., Ragavan, U. M., Gunsalam, J. W., Bakar, F. A., Ghazali, F. M., Radu, S., and Kumar, M. P. 2007. Thermophilic *Campylobacter* spp. in salad vegetables in Malaysia. *International Journal of Food Microbiology* 117:106–111.

Chang, J.-M., and Fang, T. J. 2007. Survival of *Escherichia coli* O157:H7 and *Salmonella enterica* serovars *Typhimurium* in iceberg lettuce and the antimicrobial effect of rice vinegar against *E. coli* O157:H7. *Food Microbiology* 24:745–751.

Charkowski, A. O., Barak, J. D., Sarreal, C. Z., and Mandrell, R. E. 2002. Differences in growth of *Salmonella enterica* and *Escherichia coli* O157:H7 on alfalfa sprouts. *Applied and Environmental Microbiology* 68:3114–3120.

Cotterelle, B., Drougard, C., Rolland, J., Becamel, M., Boudon, M., Pinede, S., Traoré, O., Balay, K., Pothier, P., and Espié, E. 2005. Outbreak of Norovirus infection associated with the consumption of frozen raspberries, France, March 2005, *Euro Surveillance*, no. 10 (April): pii 2690. http://www.eurosurveillance.org/ew/2005/050428.asp#1.

Croci, L., De Medici, D., Scalfaro, C., Fiore, A., and Toti, L. 2002. The survival of hepatitis A virus in fresh produce. *International Journal of Food Microbiology* 73:29–34.

Dao, H. T. A., and Yen, P. T. 2006. Study of *Salmonella*, *Campylobacter*, and *Escherichia coli* contamination in raw food available in factories, schools, and hospital canteens in Hanoi, Vietnam. *Impact of Emerging Zoonotic Diseases on Animal Health* 1081:262–265.

Dawson, D. 2005. Foodborne protozoan parasites. *International Journal of Food Microbiology* 103:207–227.

Dawson, D. J., Paish, A., Staffell, L. M., Seymour, I. J., and Appleton, H. 2005. Survival of viruses on fresh produce, using MS2 as a surrogate for norovirus. *Journal of Applied Microbiology*, 98:203–209.

De Cesare, A., Sheldon, B. W., Smith, K. S., and Jaykus, L. A. 2003. Survival and persistence of *Campylobacter* and *Salmonella* species under various organic loads on food contact surfaces. *Journal of Food Protection* 66:1587–1594.

Del Rosario, B. A., and Beuchat, L. R. 1995. Survival and growth of enterohemorrhagic *Escherichia coli* O157:H7 in cantaloupe and watermelon. *Journal of Food Protection* 58:105–107.

Diaz, C., and Hotchkiss, J. H. 1996. Comparative growth of *Escherichia coli* O157:H7, spoilage organisms and shelf life of shredded iceberg lettuce stored under modified atmospheres. *Journal of the Science of Food and Agriculture* 70:433–438.

Dingman, D.W. 2000. Growth of *Escherichia coli* O157: H7 in bruised apple (*Malus domestica*) tissue as influenced by cultivar, date of harvest, and source. *Applied and Environmental Microbiology*, 66:1077–1083.

Doyle, M. 1990. Fruit and vegetable safety—Microbiological considerations. *HortScience* 25:1478–1482.

Doyle, M. P., and Schoeni, J. L. 1986. Isolation of *Campylobacter jejuni* from retail mushrooms. *Applied and Environmental Microbiology* 51:449–450.

Dziezak, J. 1986. Preservatives: antioxidants, the ultimate answer to oxidation. *Food Technology* 9:94–102.

European Commision. 2005. European Commission Regulation 2073/2005 on microbiological criteria in foodstuffs.

European Food Safety Authority (EFSA). 2007. The community summary report on trends and sources of zoonoses, zoonotic agents, antimicrobial resistance and foodborne outbreaks in the European Union in 2006. *The EFSA Journal* 130. http://www.efsa.europa.eu/EFSA/DocumentSet/Zoon_report_2006_en,0.pdf.

European Food Safety Authority (EFSA). 2007. Annual report on Zoonoses, 2006. www.efsa.europe.eu/EFSA.

Erickson, M. C., and Ortega, Y. R. 2006. Inactivation of protozoan parasites in food, water, and environmental systems. *Journal of Food Protection* 69:2786–2808.

Escalona, V. H., Aguayo, E., and Artés, F. 2006. Metabolic activity and quality changes of whole and fresh-cut kohlrabi (*Brassica oleracea* L. gongylodes group) stored under controlled atmospheres. *Postharvest Biology and Technology* 41:181–190.

Evans, M. R., Ribeiro, C. D., and Salmon, R. L. 2003. Hazards of healthy living: bottled water and salad vegetables as risk factors for Campylobacter infection. *Emerging Infectious Diseases* 9:1219–1225.

Evans, R., Russell, N., Gould, G., and McClure, P. 1997. The germinability of spores of a psychrotolerant, non-proteolytic strain of *C. botulinum* is influenced by their formation and storage temperature. *Journal of Applied Microbiology* 83:273–280.

Farber, J. 1991. Microbiological aspects of modified atmosphere packaging technology—a review. *Journal of Food Protection* 54:58–70.

Farber, J., Sanders, G., and Johnston, M. 1989. A survey of various foods for the presence of *Listeria* species. *Journal of Food Protection* 52:456–458.

Farber, J., Wang, S., Cai, Y., and Zhang, S. 1998. Changes in populations of *Listeria monocytogenes* inoculated on packaged fresh cut vegetables. *Journal of Food Protection* 61:192–195.

Federighi, M., Magras, C., Pilet, M. F., Woodward, D., Johnson, W., Jugiau, F., and Jouve, J. L. 1999. Incidence of thermotolerant *Campylobacter* in foods assessed by NF ISO 10272 standard: results of a two-year study. *Food Microbiology* 16:195–204.

Fenlon, D., Wilson, J., and Donachie, W. 1996. The incidence and level of *L. monocytogenes* contamination of food sources at primary production and initial processing. *Journal of Applied Bacteriology* 81:641–650.

Finn, M.J., and Upton, M.E. 1997. Survival of pathogens on modified-atmosphere-packaged shredded carrot and cabbage. *Journal of Food Protection* 60:1347–1350.

Fraaije, B. A., Bosveld, M., Vandenbulk, R. W., and Rombouts, F. M. 1997. Analysis of conductance responses during depolymerization of pectate by soft rot *Erwinia* spp. and other pectolytic bacteria isolated from potato tubers. *Journal of Applied Microbiology* 83:17–24.

Francis, G., and O'Beirne, D. 1997. Effect on gas atmosphere, antimicrobial dip and temperature on the fate of *L. innocua* and *L. monocytogenes* on minimally processed lettuce. *International Journal of Food Science and Technology* 32:141–151.

Francis, G. A., Thomas, C., and O'Beirne, D. 1999. The microbiological safety on minimally processed vegetables. *International Journal of Food Science and Technology* 34:1–22.

Francis, G.A., and O'Beirne, D. 2001. Effects of vegetable type, package atmosphere and storage temperature on growth and survival of *Escherichia coli* O157: H7 and *Listeria monocytogenes*. *Journal of Industrial Microbiology & Biotechnology* 27:111–116.

Garcia-Gimeno, R., Zurera-Cosano, G., and Amaro-Lopez, M. 1996. Incidence, survival and growth of *Listeria monocytogenes* in ready-to-use mixed vegetable salads in Spain. *Journal of Food Safety* 16:75–86.

Gaulin, C. D., Ramsay, D., Cardinal, P., and D'Halevyn, M. A. 1999. Viral gastroenteritis epidemic associated with the ingestion of imported raspberries. *Canadian Journal of Public Health* 90:37–40.

Giménez, M., Olarte, C., Sanz, S., Lomas, C., Echávarri, J. F., and Ayala, F. 2003. Relation between spoilage and microbiological quality in minimally processed artichoke packaged with different films. *Food Microbiology* 20:231–242.

Gleeson, E., and O'Beirne, D. 2005. Effects of process severity on survival and growth of *Escherichia coli* and *Listeria innocua* on minimally processed vegetables. *Food Control* 16:677–685.

Gómez, P. A., and Artés, F. 2005. Improved keeping quality of minimally fresh processed celery sticks by modified atmosphere packaging. *Lebensmittel-Wissenschaft und Technologie* 38:323–329.

Griffin, M. R., Surowiec, J. J., Mccloskey, D. I., Capuano, B., Pierzynski, B., Quinn, M., Wojnarski, R., Parkin, W. E., Greenberg, H., and Gary, G. W. 1982. Foodborne Norwalk virus. *American Journal of Epidemiology* 115:178–184.

Gunes, G. G., and Hotchkiss, J. H. 2002. Growth and survival of *Escherichia coli* O157:H7 on fresh-cut apples in modified atmospheres at abusive temperatures. *Journal of Food Protection* 65:1641–1645.

Hamilton-Kemp, T. R., Archbold, D. D., Loughrin, J. H., Collins, R. W., and Byers, M. E. 1996. Metabolism of natural volatile compounds by strawberry fruit. *Journal of Agricultural and Food Chemistry* 44:2802–2805.

Hao, Y. Y., Brackett, R. E., Beuchat, L. R., and Doyle, M. P. 1999. Microbiological quality and production of botulinal toxin in film-packaged broccoli, carrots, and green beans. *Journal of Food Protection* 62:499–508.

Harker, F. R., Elgar, H. J., Watkins, C. B., Jackson, P. J., and Hallett, I. C. 2000. Physical and mechanical changes in strawberry fruit after high carbon dioxide treatments. *Postharvest Biology and Technology* 19:139–146.

Herwaldt, B. L. 2000. Cyclospora cayetanensis: A review, focusing on the outbreaks of cyclosporiasis in the 1990s. *Clinical Infectious Diseases* 31:1040–1057.

Herwaldt, B. L., Lew, J. F., Moe, C. L., Lewis, D. C., Humphrey, C. D., Monroe, S. S., Pon, E. W., and Glass, R. I. 1994. Characterization of a variant strain of Norwalk virus from a food-borne outbreak of gastroenteritis on a cruise ship in Hawaii. *Journal of Clinical Microbiology* 32:861–866.

Ho, J., Shands, K., Freidland, G., Eckind, P., and Fraser, D. 1986. An outbreak of type 4b *Listeria monocytogenes* infection involving patients from eight Boston hospitals. *Archives Internal Medicine* 146:520–524.

Hong, S. I., and Kim, D. 2004. The effect of packaging treatment on the storage quality of minimally processed bunched onions. *International Journal of Food Science and Technology* 39:1033–1041.

Hotchkiss, J., and Banco, M. 1992. Influence of new packaging technologies on the growth of microorganisms in produce. *Journal of Food Protection* 55:815–820.

Howard, L. R., Yoo, K. S., Pike, L. M., and Miller, G. H. 1994. Quality changes in diced onions stored in film packages. *Journal of Food Science* 59:110–112, 117.

ICMSF (International Commission on Microbiological Specifications for Foods). 1996. *Microorganisms in foods 5*, ed. T. Roberts, A. Baird-Parker, and R. Tompkin. London: Blackie Academic and Professional.

Jacques, M.A., and Morris, C.E. 1995. Bacterial population-dynamics and decay on leaves of different ages of ready-to-use broad-leaved endive. *International Journal of Food Science and Technology* 30:221–236.

Jacxsens, L., Devlieghere, F., Falcato, P., and Debevere, J. 1999. Behaviour of *Listeria monocytogenes* and *Aeromonas* spp. on fresh-cut produce packaged under equilibrium modified atmosphere. *Journal of Food Protection* 62:1128–1135.

Jacxsens, L., Devlieghere, F., and Debevere, J. 2002a. Predictive modelling for packaging design: equilibrium modified atmosphere packages of fresh-cut vegetables subjected to a simulated distribution chain. *International Journal of Food Microbiology* 73:331–341.

Jacxsens, L., Devlieghere, F., and Debevere, J. 2002b. Temperature dependence of shelf-life as affected by microbial proliferation and sensorial quality of equilibrium modified atmosphere packaged fresh produce. *Postharvest Biology and Technology* 26:59–73.

Jacxsens, L., Devlieghere, F., Ragaert, P., Vanneste, E., and Debevere, J. 2003. Relation between microbiological quality, metabolite production and sensory quality of equilibrium modified atmosphere packaged fresh-cut produce. *International Journal of Food Microbiology* 83:263–280.

Janisiewicz, W.J., Conway, W.S., Brown, M.W., Sapers, G.M., Fratamico, P., and Buchanan, R.L. 1999. Fate of *Escherichia coli* 0157: H7 on fresh-cut apple tissue and its potential for transmission by fruit flies. *Applied and Environmental Microbiology* 65:1–5.

Juneja, V., Martin, S., and Sapers, G. 1998. Control of *L. monocytogenes* in vacuum-packaged pre-peeled potatoes. *Journal of Food Science* 63:911–914.

Juven, B. J., Lindner, P., and Weisslowicz, H. 1985. Pectin degradation in plant-material by *Leuconostoc mesenteroides*. *Journal of Applied Bacteriology* 58:533–538.

Kakiomenou, K., Tassou, C., and Nychas, G. J. 1996. Microbiological, physicochemical and organoleptic changes of shredded carrots stored under modified storage. *International Journal of Food Science and Technology* 31:359–366.

Kallander, K., Hitchins, A., Lancette, G., Schmieg, J., Garcia, G., Solomon, H., and Sofos, J. 1991. Fate of *L. monocytogenes* in shredded cabbage stored at 5 and 25°C under a modified atmosphere. *Journal of Food Protection* 54:302–304.

Kang, H. M., and Saltveit, M. E. 2002. Antioxidant capacity of lettuce leaf tissue increases after wounding. *Journal of Agricultural and Food Chemistry* 50:7536–7541.

Kärenlampi, R., and Hänninen, M.-L. 2004. Survival of *Campylobacter jejuni* on various fresh produce. *International Journal of Food Microbiology* 97:187–195.

King, A. D., and Bolin, H. R. 1989. Physiological and microbiological storage stability of minimally processed fruits and vegetables. *Food Technology* 43:132–135, 139.

King, A. D., Magnuson, J. A., Torok, T., and Goodman, N. 1991. Microbial flora and storage quality of partially processed lettuce. *Journal of Food Science* 56:459–461.

Klein, B. P. 1987. Nutritional consequences of minimal processing of fruits and vegetables. *Journal of Food Quality* 10:179–193.

Knee, M., and Miller, A. R. 2002. Mechanical injury. In *Fruit quality and its biological basis*, ed. M. Knee, 157–179. Sheffield: Sheffield Academic Press.

Koopmans, M., and Duizer, E. 2004. Foodborne viruses: an emerging problem. *International Journal of Food Microbiology* 90:23–41.

Korsager, B., Hede, S., Bøggild, H., Böttiger, B., and Mølbak, K. 2005. Two outbreaks of norovirus infections associated with the consumption of frozen raspberries Denmark, May–June 2005. *Euro Surveillance*, 10: pii 2729. http://www.eurosurveillance.org/ew/2005/050623.asp#1.

Kumar, A., Agarwal, R. K., Bhilegaonkar, K. N., Shome, B. R., and Bachhil, V. N. 2001. Occurrence of *Campylobacter jejuni* in vegetables. *International Journal in Food Microbiology* 67:153–155.

Kurdziel, A. S., Wilkinson, N., Langton, S., and Cook, N. 2001. Survival of poliovirus on soft fruit and salad vegetables. *Journal of Food Protection* 64:706–709.

Larsen, M., and Watkins, C. B. 1995. Firmness and concentrations of acetaldehyde, ethyl acetate and ethanol in strawberries stored in controlled and modified atmospheres. *Postharvest Biology and Technology* 5:39–50.

Laurila, E., and Ahvenainen, R. 2002. Minimal processing in practice: fresh fruits and vegetables. In *Minimal processing technologies in the food industry*, ed. T. Ohlsson and N. Bengtsson, 219–244. Cambridge: Woodhead.

Lavelli, V., Pagliarini, E., Ambrosoli, R., Minati, J. L., and Zanoni, B. 2006. Physicochemical, microbial, and sensory parameters as indices to evaluate the quality of minimally-processed carrots. *Postharvest Biology and Technology* 40:34–40.

Li, Y., Brackett, R. E., Chen, J., and Beuchat, L. R. 2001a. Survival and growth of *Escherichia coli* O157:H7 inoculated onto cut lettuce before or after heating in chlorinated water, followed by storage at 5 or 15°C. *Journal of Food Protection* 64:305–309.

Li, Y., Brackett, R. E., Shewfelt, R. L., and Beuchat, L. R. 2001b. Changes in appearance and natural microflora on iceberg lettuce treated in warm, chlorinated water and then stored at refrigeration temperature. *Food Microbiology* 18:299–308.

Liao, C. H., Sullivan, J., Grady, J., and Wong, L. J. C. 1997. Biochemical characterization of pectate lyases produced by fluorescent Pseudomonads associated with spoilage of fresh fruits and vegetables. *Journal of Applied Microbiology* 83:10–16.

Liao, C., and Sapers, G. 1999. Influence of soft rot bacteria on growth *of Listeria monocytogenes* on potato slices. *Journal of Food Protection* 62:343–348.

Lin, C., and Wei, C. 1997. Transfer of *S. Montevideo* onto the interior surfaces of tomatoes by cutting. *Journal of Food Protection* 60:858–863.

Lin, C.-M., Fernando, S. Y., and Wei, C. 1996. Occurence of *Listeria monocytogenes*, *Salmonella* spp., *Escherichia coli* and *E. coli* O157:H7 in vegetable salads. *Food Control* 3:135–140.

Little, C., Roberts, D., Youngs, E., and De Louvois, J. 1999. Microbiological quality of retail imported unprepared whole lettuces: a PHLS food working group study. *Journal of Food Protection* 62:325–328.

Lopez, A. S., Dodson, D. R., Arrowood, M. J., Orlandi, P. A., Da Silva, A. J., Bier, J. W., Hanauer, S. D., Kuster, R. L., Oltman, S., Baldwin, M. S., Won, K. Y., Nace, E. M., Eberhard, M. L., and Herwaldt, B. L. 2001. Outbreak of cyclosporiasis associated with basil in Missouri in 1999. *Clinical Infectious Diseases* 32:1010–1017.

López-Gálvez, G., Peiser, G., Nie, X. L., and Cantwell, M. 1997. Quality changes in packaged salad products during storage. *Zeitschrift für Lebensmittel-Untersuchung und-Forschung* 205:64–72.

Lund, B. M. 1992. Ecosystems in vegetable foods. *Journal of Applied Bacteriology* 73:S115–S126.

Magnuson, J. A., King, A. D., and Torok, T. 1990. Microflora of partially processed lettuce. *Applied and Environmental Microbiology* 56:3851–3854.

Mead, P. S., Slutsker, L., Dietz, V., Mccaig, L. F., Bresee, J. S., Shapiro, C., Griffin, P. M., and Tauxe, R. V. 1999. Food-related illness and death in the United States. *Emerging Infectious Diseases* 5:607–625.

Membré, J. M., and Burlot, P. M. 1994. Effects of temperature, pH, and NaCl on growth and pectinolytic activity of *Pseudomonas marginalis*. *Applied and Environmental Microbiology* 60:2017–2022.

Membré, J. M., Goubet, D., and Kubaczka, M. 1995. Influence of salad constituents on growth of *Pseudomonas marginalis*—a predictive microbiology approach. *Journal of Applied Bacteriology* 79:603–608.

Mercier, J., and Lindow, S. E. 2000. Role of leaf surface sugars in colonization of plants by bacterial epiphytes. *Applied and Environmental Microbiology* 66:369–374.

Messi, P., Casolari, C., Fabio, A., Fabio, G., Gibertoni, C., Menziani, G., and Quaglio, P. 2000. Occurrence of *Listeria* in food matrices. *Industrie Alimentari* 39:151–157.

Moss, M. O. 1999. Spoilage problems/problems caused by fungi. In *Encyclopedia of food microbiology*, ed. R. K. Robinson, C. A. Batt, and P. D. Patel, 2056–2062. London: Academic Press.

Nakagawa, T., Nagaoka, T., Taniguchi, S., Miyaji, T., and Tomizuka, N. 2004. Isolation and characterization of psychrophilic yeasts producing cold-adapted pectinolytic enzymes. *Letters in Applied Microbiology* 38:383–387.

Nguyen-the, C., and Carlin, F. 1994. The microbiology of minimally processed fresh fruits and vegetables. *Critical Reviews in Food Science and Nutrition* 34:371–401.

Nguyen-the, C., and Lund, B. 1991. The lethal effect of carrot on *Listeria* species. *Journal of Applied Bacteriology* 70:479–488.

Nguyen-the, C., and Prunier, J. P. 1989. Involvement of Pseudomonads in deterioration of ready-to-use salads. *International Journal of Food Science and Technology* 24:47–58.

Nielsen, T., and Leufvén, A. 2008. The effect of modified atmosphere packaging on the quality of Honeoye and Korona strawberries. *Food Chemistry* 107:1053–1063.

Norrung, B. 2000. Microbiological criteria for *L. monocytogenes* in foods under special consideration of risk assessment approaches. *International Journal of Food Microbiology* 62:217–221.

Norrung, B., Andersen, J., and Schlundt, J. 1999. Incidence and control of *L. monocytogenes* in foods in Denmark. *International Journal of Food Microbiology* 53:195–203.

Omary, M., Testin, R., Barefoot, S., and Rushing, J. 1993. Packaging effects on growth of *Listeria innocua* in shredded cabbage. *Journal of Food Science* 58:623–626.

Pao, S., Brown, G. E., and Schneider, K. R. 1998. Challenge studies with selected pathogenic bacteria on freshly peeled Hamlin orange. *Journal of Food Science* 63:359–362.

Park, C. E., and Sanders, G. W. 1992. Occurrence of thermotolerant Campylobacters in fresh vegetables sold at farmers outdoor markets and supermarkets. *Canadian Journal of Microbiology* 38:313–316.

Park, S., Worobo, R. W., and Durst, R. A. 1999. *Escherichia coli* O157:H7 as an emerging foodborne pathogen: a literature review. *Critical Reviews in Food Science and Nutrition* 39:481–502.

Patterson, W., Haswell, P. T., and Green, J. 1997. Outbreak of small round structured virus gastroenteritis arose after kitchen assistant vomited. *Canada Communicable Disease Report* 27:R101–R103.

Pelayo, C., Ebeler, S.E., and Kader, A.A. 2003. Postharvest life and flavor quality of three strawberry cultivars kept at 5 degrees C in air or air + 20 kPa CO_2. *Postharvest Biology and Technology* 27:171–183.

Pesis, E., and Avissar, I. 1990. Effect of postharvest application of acetaldehyde vapor on strawberry decay, taste and certain volatiles. *Journal of the Science of Food and Agriculture* 52:377–385.

Piagentini, A., Pirovani, M., Guëmes, D., Di Pentima, J., and Tessi, M. 1997. Survival and growth of *Salmonella hadar* on minimally processed cabbage as influenced by storage abuse conditions. *Journal of Food Science* 62:616–618, 631.

Piga, A., Del Caro, A., Pinna, I., and Agabbio, M. 2003. Changes in ascorbic acid, polyphenol content and antioxidant activity in minimally processed cactus pear fruits. *Lebensmittel-Wissenschaft und-Technologie* 36:257–262.

Pirovani, M., Güemes, D., Di Pentima, J., and Tessi, M. 2000. Survival of *Salmonella hadar* after washing disinfection of minimally processed spinach. *Letters in Applied Microbiology* 31:143–148.

Pitt, J. I., and Hocking, A. D. 1997. Spoilage of fresh and perishable foods. In *Fungi and food spoilage*, ed. J. I. Pitt, and A. D. Hocking. London: Blackie Academic and Professional.

Pretel, M. T., Fernández, P. S., Romojaro, F., and Martínez, A. 1998. The effect of modified atmosphere packaging on "ready-to-eat" oranges. *Lebensmittel Wissenschaft und Technologie* 31:322–328.

Purvis, A. C. 1997. The role of adaptative enzymes in carbohydrate oxidation by stressed and senescing plant tissues. *Hortscience* 32:1165–1168.

Quinn, K., Baldwin, G., Stepak, P., Thorburn, K., Bartleson, C., Goldoft, M., Kobayashi, J., and Stehr-Green, P. 1997. Civilian outbreak of adenovirus acute respiratory disease–South Dakota (Reprinted from *MMWR*, vol 47, pg 565–567, 1998). *Journal of the American Medical Association* 280:595–596.

Rafii, F., Holland, M., Hill, W., and Cerniglia, C. 1995. Survival of *Shigella flexneri* on vegetables and detection by polymerase chain reaction. *Journal of Food Protection* 58:727–732.

Rafii, F., and Lunsford, P. 1997. Survival and detection of *Shigella flexneri* on vegetables and commercial prepared salads. *Journal of AOAC International* 80:1191–1197.

Ragaert, P., Devlieghere, F., and Debevere, J. 2007. Role of microbiological and physiological spoilage mechanisms during storage of minimally processed vegetables: a review. *Postharvest Biology and Technology* 44:185–194.

Ragaert, P., Devlieghere, F., Devuyst, E., Dewulf, J., Van Langenhove, H., and Debevere, J. 2006a. Volatile metabolite production of spoilage micro-organisms on a mixed-lettuce agar during storage at 7°C in air and low oxygen atmosphere. *International Journal of Food Microbiology* 112:162–170.

Ragaert, P., Devlieghere, F., Loos, S., Dewulf, J., Van Langenhove, H., and Debevere, J. 2006b. Metabolite production of yeasts on a strawberry-agar during storage at 7°C in air and low oxygen atmosphere. *Food Microbiology* 23:154–161.

Ragaert, P., Devlieghere, F., Loos, S., Dewulf, J., Van Langenhove, H., Foubert, I., Vanrolleghem, P.A., and Debevere, J. 2006c. Role of yeast proliferation in the quality degradation of strawberries during refrigerated storage. *International Journal of Food Microbiology* 108:42–50.

Rapid Alert System for Food and Feed (RASFF). 2007. Annual Report of the Rapid Alert System 2005–2005 by European Commission. http://ec.europa.eu/food/food/rapidalert/index_en.htm.

Restuccia, C., Randazzo, C., and Caggia, C. 2006. Influence of packaging on spoilage yeast population in minimally processed orange slices. *International Journal of Food Microbiology* 109:146–150.

Robbs, P. G., Bartz, J. A., Mcfie, G., and Hodge, N. C. 1996a. Causes of decay of fresh-cut celery. *Journal of Food Science* 61:444–448.

Robbs, P. G., Bartz, J. A., Sargent, S. A., Mcfie, G., and Hodge, N. C. 1996b. Potential inoculum sources for decay of fresh-cut celery. *Journal of Food Science* 61:449–452, 455.

Robertson, L. J., and Gjerde, B. 2001. Occurrence of parasites on fruits and vegetables in Norway. *Journal of Food Protection* 64:1793–1798.

Robertson, L. J., and Gjerde, B. 2000. Isolation and enumeration of *Giardia* cysts, *Cryptosporidium* oocysts, and *Ascaris* eggs from fruits and vegetables. *Journal of Food Protection* 63:775–778.

Rocha, A. M. C. N., Brochado, C. M., Kirby, R., and Morais, A. M. M. B. 1995. Shelf-life of chilled cut orange determined by sensory quality. *Food Control* 6:317–322.

Rojas-Graü, M. A., Raybaudi-Massilia, R. M., Soliva-Fortuny, R. C., Avena-Bustillos, R. J., McHugh, T. H., and Martín-Belloso, O. 2007. Apple puree-alginate edible coating as carrier of antimicrobial agents to prolong shelf-life of fresh-cut apples. *Postharvest Biology and Technology* 45:254–264.

Rose, J. B., and Slifko, T. R. 1999. *Giardia*, *Cryptosporidium*, and *Cyclospora* and their impact on foods: a review. *Journal of Food Protection* 62:1059–1070.

Rosen, J. C., and Kader, A. A. 1989. Postharvest physiology and quality maintenance of sliced pear and strawberry fruits. *Journal of Food Science* 54:656–659.

Roura, S. I., Davidovich, L. A., and Del Valle, C. E. 2000. Quality loss in minimally processed Swiss chard related to amount of damaged area. *Lebensmittel-Wissenschaft und-Technologie* 33:53–59.

Saftner, R., Abbott, J. A., Lester, G., and Vinyard, B. 2006. Sensory and analytical comparison of orange-fleshed honeydew to cantaloupe and green-fleshed honeydew for fresh-cut chunks. *Postharvest Biology and Technology* 42:150–160.

Saltveit, M. E. 1999. Effect of ethylene on quality of fresh fruits and vegetables. *Postharvest Biology and Technology* 15:279–292.

Satchell, F., Stephenson, P., Andrews, W., Estala, L., and Allen, G. 1990. The survival of *Shigella sonnei* in shredded cabbage. *Journal of Food Protection* 53:558–562.

Seymour, I. J., and Appleton, H. 2001. Foodborne viruses and fresh produce. *Journal of Applied Microbiology* 91:759–773.

Shah, N. S., and Nath, N. 2008. Changes in qualities of minimally processed litchis: effect of antibrowning agents, osmo-vacuuming drying and moderate vacuum packaging. *Lebensmittel Wissenschaft und Technologie* 41:660–668.

Sivapalasingam, S., Friedman, C., Cohen, L., and Tauxe, R. 2004. Fresh produce: a growing cause of outbreaks of foodborne illness in the United States, 1973 through 1997. *Journal of Food Protection* 67:2342–2353.

Sizmur, K., and Walker, C. 1988. *Listeria* in prepacked salads. *The Lancet* 5:1167.

Smith, J. 1987. *Shigella* as a foodborne pathogen. *Journal of Food Protection* 50:788–801.

Smyth, A. B., Song, J., and Cameron, A. C. 1998. Modified atmosphere packaged cut iceberg lettuce: effect of temperature and O_2 partial pressure on respiration and quality. *Journal of Agricultural and Food Chemistry* 46:4556–4562.

Smyth, A. B., Talasila, P. C., and Cameron, A. C. 1999. An ethanol biosensor can detect low-oxygen injury in modified atmosphere packages of fresh-cut produce. *Postharvest Biology and Technology* 15:127–134.

Soliva-Fortuny, R. C., Elez-Martinez, P., and Martin-Belleso, O. 2004. Microbiological and biochemical stability of fresh cut apples preserved by modified atmosphere packaging. *Innovative Food Science and Emerging Technologies* 5:215–224.

Stannard, C. 1997. Development and use of microbiological criteria for foods. *Food Science and Technology Today* 11:137–177.

Stine, S. W., Song, I. H., Choi, C. Y., and Gerba, C. P. 2005. Application of microbial risk assessment to the development of standards for enteric pathogens in water used to irrigate fresh produce. *Journal of Food Protection* 68:913–918.

Surjadinata, B. B., and Cisneros-Zevallos, L. 2003. Modeling wound-induced respiration of fresh-cut carrots (*Daucus carota* L.). *Journal of Food Science* 68:2735–2740.

Sziro, I., Devlieghere, F., Jacxsens, L., Uyttendaele, M., and Debevere, J. 2006. The microbial safety of strawberry and raspberry fruit packaged in high oxygen and equilibrium modified atmospheres compared to air storage. *International Journal of Food Science and Technology* 41:93–103.

Tassou, C. C., and Boziaris, J. S. 2002. Survival of *Salmonella enteritidis* and changes in pH and organic acids in grated carrots inoculated or not inoculated with *Lactobacillus* sp. and stored under different atmospheres at 4°C. *Journal of the Science of Food and Agriculture* 82:1122–1127.

Tauxe, R., Kruse, H., Hedberg, C., Potter, M., Madden, J., and Wachsmugh, K. 1997. Microbial hazards and emerging issues associated with produce. A preliminary report to the national advisory committee on microbiologic criteria for foods. *Journal of Food Protection* 60:1400–1408.

Thomas, C., Prior, O., and O'Beirne, D. 1999. Survival and growth of *Listeria* species in a model ready-to-use vegetable product containing raw and cooked ingredients as affected by storage temperature and acidification. *International Journal of Food Science and Technology* 34:317–324.

Thunberg, R. L., Tran, T. T., Bennett, R. W., Matthews, R. N., and Belay, N. 2002. Microbial evaluation of selected fresh produce obtained at retail markets. *Journal of Food Protection* 64:677–682.

Toivonen, P. M. A. 1997. Non-ethylene, non-respiratory volatiles in harvested fruits and vegetables: their occurrence, biological activity and control. *Postharvest Biology and Technology* 12:109–125.

Tournas, V. H., Heeres, J., and Burgess, L. 2006. Moulds and yeasts in fruit salads and fruit juices. *Food Microbiology* 23:684–688.

Van der Steen, C., Jacxsens, L., Devlieghere, F., and Debevere, J. 2002. Combining high oxygen atmospheres with low oxygen modified atmosphere packaging to improve the keeping quality of strawberries and raspberries. *Postharvest Biology and Technology* 26:49–58.

Vankerschaver, K., Willocx, F., Smout, C., Hendrickx, M., and Tobback, P. 1996. The influence of temperature and gas mixtures on the growth of the intrinsic micro-organisms on cut endive: predictive versus actual growth. *Food Microbiology* 13:427–440.

Viljoen, B. C., Khouri, A. R., and Hattingh, A. 2003. Seasonal diversity of yeasts associated with white-surface mould-ripened cheeses. *Food Research International* 36:275–283.

Warnes, S., and Keevil, C. W. 2003. Survival of *Cryptosporidium parvum* in faecal wastes and salad crops. Available at http://www.teagasc.ie/publications/2003/conferences/cryptosporidiumparvum/paper02.htm. Accessed on 9 June 2008.

Watada, A. E., Ko, N. P., and Minott, D. A. 1996. Factors affecting quality of fresh-cut horticultural products. *Postharvest Biology and Technology* 9:115–125.

Wei, C., Huang, T., Kim, J., Lin, W., Tamplin, M., and Bartz, J. 1995. Growth and survival of *S. montevideo* on tomatoes and disinfection with chlorinated water. *Journal of Food Protection* 58:829–836.

Wheeler, C., Vogt, T. M., Armstrong, G. L., Vaughan, G., Weltman, A., Nainan, O. V., Dato, V., Xia, G. L., Waller, K., Amon, J., Lee, T. M., Highbaugh-Battle, A., Hembree, C., Evenson, S., Ruta, M. A., Williams, I. T., Fiore, A. E., and Bell, B. P. 2005. An outbreak of hepatitis A associated with green onions. *New England Journal of Medicine* 353:890–897.

White, K. E., Osterholm, M. T., Mariotti, J. A., Korlath, J. A., Lawrence, D. H., Ristinen, T. L., and Greenberg, H. B. 1986. A foodborne outbreak of Norwalk virus gastroenteritis—evidence for post-recovery transmission. *American Journal of Epidemiology* 124:120–126.

Whyte, P., McGill, K., Madden, R. H., Moran, L., Scates, P., Carroll, C., O'Leary, A., Fanning, S., Collins, J. D., McNamara, E., Moore, J. E., and Cormican, M. 2004. Occurrence of *Campylobacter* in retail foods in Ireland. *International Journal of Food Microbiology* 95:111–118.

Yu, K. S., Hamilton-Kemp, T. R., Archbold, D. D., Collins, R. W., and Newman, M. C. 2000. Volatile compounds from *Escherichia coli* O157:H7 and their absorption by strawberry fruit. *Journal of Agricultural and Food Chemistry* 48:413–417.

Zagory, D. 1999. Effects of post-processing handling and packaging on microbial populations. *Postharvest Biology and Technology* 15:313–321.

Zhang, S., and Farber, J. 1996. The effects of various disinfectants against *L. monocytogenes* on fresh-cut vegetables. *Food Microbiology* 13:311–321.

Zhuang, R., Beuchat, L., and Angulo, F. 1995. Fate of *S. montevido* on and in raw tomatoes as affected by temperature and treatment with chlorine. *Applied and Environmental Microbiology* 61:2127–2131.

4 Physiology of Fresh-Cut Fruits and Vegetables

Elizabeth A. Baldwin and Jinhe Bai

CONTENTS

4.1 INTRODUCTION

The idea to preprocess fruits and vegetables in the fresh state started with fresh-cut salads and now has expanded to fresh-cut fruits and other vegetables. The fresh-cut portion of the fresh produce industry includes fruits, vegetables, sprouts, mushrooms, and even herbs that are cut, cored, sliced, peeled, diced, or shredded, but not heated or altered from their fresh state in any way (Shewfelt, 1987). Physiologically this is, in fact, wounding living tissue which starts a cascade of metabolic reactions that can result in texture changes, accelerated ripening and senescence, off flavors, discoloration, and other undesirable events that can render the product unmarketable. Microbiologically, removing the protective peel of fresh produce leaves a cut surface that is awash with cell contents, which makes the surface attractive to plant pathogens (King and Bolin, 1989). The extra handling and processing of the produce result in increased ethylene, a gaseous ripening plant hormone, which results in genetic signals that promote ripening and senescence (Karakurt and Huber, 2007). The handling and processing also increase the respiration rate, using up sugar and acid substrates and shortening shelf life. Finally, the extra handling, removal of the peel, and altering of the normal microbial ecology of fresh-cut products allow for possible contamination by and growth of human pathogens (Breidt and Fleming, 1997).

Many processing techniques (Artés and Allende, 2005; Palumbo et al., 2007; Reyes et al., 1995) have been developed to counteract the reactions of the cut produce to ethylene production and other wound responses, including use of precooling (Vigneault et al., 2008) to reduce respiration; prequality assessment for microbial stability (Fan and Song, 2008), modified atmosphere packaging (MAP) and other packaging to reduce respiration and ethylene production (Barmore, 1987; Day, 1994; Jacxsens et al., 2000; Myers, 1989; Reyes et al., 1995), 1-methylcyclopropene (1-MCP, marketed commercially as Smartfresh) to minimize ethylene action (Toivonen, 2008a), various sanitation or decontamination procedures (Artés et al., 2007; Gómez-López et al., 2008; Gómez-López et al., 2009), use of antimicrobial agents to reduce spoilage (Ayala-Zavala et al., 2008; Brecht et al., 1993; Breidt and Fleming, 1997; King and Bolin, 1989), use of surface treatments and edible coatings to reduce water loss, dehydration, and browning (Baldwin et al., 1995; King and Bolin, 1989; Olivas et al., 2008; Vargas et al., 2008), use of ethylene absorbents to retard ripening and senescence (Abe and Watada, 1991), and calcium for firmness and preservation (Martín-Diana et al., 2007). All the above have been the subject of numerous reviews, as cited. Other general reviews for extending the shelf life and quality of fresh-cut produce include those by Ahvenainen (1996), Shewfelt (1987), Soliva-Fortuny and Martín-Belloso (2003), and Watada and Qi (1999). Guerzoni et al. (1996) reviewed the topic of modeling for shelf-life prediction. Reviews for Asian fresh-cut fruits have also been published (Rattanapanone et al., 2000).

The consequences of processing fresh produce into fresh-cut products have also been reviewed, including the effect on general quality (Hu and Jiang, 2007; Rico et al., 2007; Shewfelt, 1987), warnings (Brody, 1998), physiology (Parkin, 1987; Rolle and Chism, 1987; Varoquaux et al., 1995), effect on flavor (Forney, 2008), browning (He and Luo, 2007; Salcini and Massantini, 2005), physiological deterioration (Artés et al., 2007; King and Bolin, 1989), microbial deterioration (Artés et al., 2007; Brackett, 1987; King and Bolin, 1989), respiration (Watada et al., 1996), as well as markets and distribution (Huxsoll and Bolin, 1989). Generally, fresh-cut fruit deteriorate much faster than their intact fruit counterpart. In a study where fresh-cut fruit was compared to the intact fruit counterpart, the fruit were stored up to 9 days at 5°C. Of the fruit tested (pineapples, mangoes, cantaloupes, watermelons, strawberries, and kiwifruits), only fresh-cut watermelon and mango pieces were still marketable at the end of the storage period (Gil et al., 2006). Losses of vitamin C and carotenoids was 5 to 25% and 10 to 25%, respectively, in some of the cut fruit products, but negligible in the intact fruit. However, for intact versus fresh-cut spinach during 1 week at 4°C, there were no changes in vitamin C, and flavonoids and color were not altered (Bottino et al., 2009), so the effect of minimal processing depends on the commodity.

In light of the many reviews published to date on fresh-cut products, this chapter will summarize the literature for the past 5 years or so that concerns fresh-cut physiology. Much research continues to be published in this area which relates to quality and microbial stability, although food safety will not be covered here except in cases

TABLE 4.1
Ripening Responses Initiated by Ethylene

Unripe Fruit			Ripe Fruit	
Appearance	Chemical Cause	Involved Enzymes	Chemical Cause	Appearance
Green	Chlorophyll	Hydrolase	Anthocynin, carotenoids	Yellow, blue, red
Firm	Insoluble pectin	Pectinases	Soluble pectin	Soft
Sour	Acid	Kinase	Neutral	Less sour
Dry	Starch	Amylase	Sugar	Sweet + juicy
Odorless	Large organic cpds	Hydrolases and ester synthase	Small organic cpds	Aromatic

Source: Modified from Koning (1994).

where there are reports of impacts on fresh-cut produce physiology. Physiological changes due to wounding, such as occurs with fresh-cut commodities, impact quality and spoilage; thus, this chapter may overlap some other chapters but will focus on the physiological reasons and metabolic mechanisms that cause quality and microbial problems. Plant wound responses fall into several categories: increased ethylene production, increased respiration, increased production of secondary metabolites, accelerated ripening and senescence, cut surface discoloration, texture changes, and off-flavors. To understand the physiology of fresh-cut produce, some basic plant physiological information is first presented, particularly as pertains to ripening (Table 4.1), wounding (Table 4.2), ethylene, and respiration.

4.2   PLANT STRESS AND WOUNDING

Injured or stressed plants produce signals that induce a wide range of genes whose products are meant to help repair the plant, defend against pathogens (Hillwig et al., 2008), or signal cell death to wall off infection (Dangl and Jones, 2001; Kessler and Baldwin, 2002). Oligogalacturonides (OGAs) (Viña et al., 2007), which come from the cell wall; jasmonic acid (JA) and methyl jasmonate (MeJA) (Howe, 2004), which comes from membranes (plastids, linolenic acid) via lipoxygenase (13-LOX); salicylic acid and methyl salicylate (wintergreen) from benzoic acid (Mur et al., 1997) and abscisic acid (ABA) (Hillwig et al., 2008; Siqueira-Júnior et al., 2008; Wasternack et al., 2006) regulate signaling pathways that induce wound-responsive genes. The wound-induced JA pathway is in turn regulated by other signals including the protein systemin, OGAs, ethylene, and other elicitors (Howe, 2004; Ryan and Moura, 2002). These signals are thought to be transduced to a lipase, which causes the release of linolenic and linoleic acids from membrane lipids. A full-length cDNA clone encoding *Capsicum annum* GDSL-lipase 1 (CaGL1) expression was triggered by MeJA and wound stress, for example (Kim et al., 2008b). These signals also give rise to pathogenesis-related (PR) proteins and proteinase inhibitors (Fallico et al., 1996). Ethylene has been shown to be a signal that mediates wound

TABLE 4.2

Wound Responses

- Increased ethylene production
  Cell wall degrading enzymes (PG)—softening
  Chlorophyll degradation (chlorophyllase)—degreening
  Membrane leakage (lipoxygenase)—water soaking
  Carotenoid synthesis, anthocyanin synthesis—color development
  Phenolic metabolism
  Volatile synthesis (?)
  Susceptibility to pathogens
  Reduced shelf life
- Phenolic metabolism (polyphenyloxidase [PPO], peroxidase [POD], and phenylalanine ammonia-lyase [PAL])
  Browning
  Secondary metabolites like lignin (toughening of asparagus) or coumarin (bitterness in carrots)
- Increased respiration
  Use up sugars, acids—poor flavor
  Use up ascorbic acid—reduced nutrition
  Generated heat
  Reduced shelf life

responses (O'Donnell et al., 1996). JA induced by wounding, systemin, and OGAs acts in concert with ethylene (O'Donnell et al., 1996) and hydrogen peroxide (Katsir et al., 2008; Orozco-Cardenas et al., 2001) to positively regulate the expression of downstream target genes (Howe, 2004).

Fresh-cut fruits and vegetables are wounded tissue, so all of the above could be going on in these products. Karakurt and Huber (2003) established that cell walls and membranes played a role in the rapid deterioration of fresh-cut papaya, resulting in softening, breakdown of pectin, increased polygalacturonase (PG, releases OGAs), α- and β-galactosidase, lipoxygenase, and phospholipase D activities (release linolenic and linoleic acids). These likely produce wound signals. They then studied signaling and expression of wound-regulated cDNAs in fresh-cut papaya compared to intact fruit (Karakurt and Huber, 2007). The cDNAs showed homology to signaling pathway genes, membrane proteins, cell-wall enzymes, proteases, ethylene biosynthetic enzymes, and enzymes involved in plant defense responses.

Abiotic stresses and wounding can also result in an increase in antioxidant phytochemicals. Antioxidants scavenge reactive oxygen species (ROS) that can damage plant cells, are generated as part of normal metabolism as by-products, and are involved in the signaling and function of antioxidant systems to detoxify ROS. ROS may be involved as signal messengers after wounding (Orozco-Cardenas et al., 2001) and are involved with synthesis of lignin and suberin that is synthesized during wound healing (Reyes et al., 2007). Phenolic compounds in fruits and vegetables result in elevated antioxidant activity (Mahattanatawee et al., 2006) and their induced accumulation could relate to health benefits. In cut apple, however, the cut surface browning, due to synthesis of phenolic compounds induced by cutting, is considered undesirable and can be inhibited by surface treatments described later in this chapter. Nevertheless, the stress of cutting may result in increased antioxidants in some fresh-cut products which could translate into health benefits.

The way in which fruit is cut can result in more or less wound responses, which in turn affects shelf life. Generally, respiration rates increase with increased cutting (wounding). In a study on the processing conditions of carrots (Iqbal et al., 2008), the respiration rate was highest for shredded (75 mL O_2/kg h) carrots and lowest for whole carrots (26 mL O_2/kg h) of the various cuts attempted. For fresh-cut papaya fruit, different cut types made a difference in the physiochemical and microbiological properties. Papaya spheres resulted in less color change, firmness loss and higher titratable acidity, maintenance of soluble solids and less weight loss, higher vitamin C content and lower microbial counts compared to papaya cubes (Argañosa et al., 2008). For fresh-cut lemons, wedges, slices, and half and quarter slices were studied while stored at different temperatures. The quarter slice did not maintain its sensory attributes as well as the larger cuts (Artés-Hernández et al., 2007).

4.3 ETHYLENE PRODUCTION

Ethylene is a gaseous plant hormone and biologically active at low concentrations (parts per billion to parts per million) (Abeles et al., 1992; Saltveit, 1999). Ethylene production occurs in all plant tissues at some minimal level. It is necessary for growth (usually thickening or lateral growth such as with etiolated pea seedlings), inhibits longitudinal growth, and promotes seed germination, degreening, adventitious root formation, abscission, ripening, and senescence (Baldwin, 2004; Reid, 1985). Ethylene is considered autocatalytic when ethylene stimulates its own synthesis and autoinhibitory when it turns off continued synthesis (Mattoo and White, 1991). System-1 ethylene, engaged during vegetative growth, is thought to be autoinhibitory, and system-II ethylene, engaged during ripening of climacteric fruits and senescence of some flowers, is generally autocatalytic. Each seems to be regulated in a unique way (Barry et al., 2000). System-II ethylene coincides with increased respiration and induces activity of PG (DellaPenna et al., 1986), cellulase, chlorophyllase (Baremore, 1975), polyphenyloxidase (PPO), peroxidase (POD), and phenylalanine ammonia-lyase (PAL) (Kader, 1985; Watada, 1986) which promote softening, color changes, and browning, respectively. Ethylene initiates ripening in fruits (Koning, 1994). It is used commercially to ripen fruits like bananas, mangoes, melons, and tomatoes (100 to 1,000 µL/L) and to degreen citrus (Watada, 1986). Generally, rates of ethylene production are highest when associated with meristematic, stressed, or ripening tissues. For example, ethylene production is high in young fruit or tissues during the period of rapid cell division, declines during cell expansion, and increases again during ripening (fruits) or senescence (flowers) (Abeles et al., 1992; McGlasson, 1985). Climacteric fruit can ripen off the mother plant in response to ethylene (Baldwin, 2004). They can respond either to their own ethylene synthesis, which becomes autocatalytic, or to exogenously applied ethylene, or to both. Ethylene is associated with ripening-related color and flavor development as well as softening in many fruits and vegetables (Kader, 1985; Watada, 1986). Ethylene is also produced at elevated levels in response to plant stress or wounding (Baldwin, 2004; Hyodo, 1991; Morgan and Drew, 1997) as occurs with fresh-cut products. This elevated ethylene, either due to ripening for climacteric fruit or to wounding for both climacteric and nonclimacteric fruit representing system-I and -II ethylene, can be active in fresh-cut products. Ripening or wound response ethylene triggers via signaling and subsequent gene expression many ripening and stress and wound responses (Bailey et al., 2005). These, as mentioned above, include production of browning enzymes, secondary metabolites, chlorophyllase, cell wall degrading enzymes, secondary metabolites, lignin and more ethylene (if autocatalytic), or a reduction of ethylene synthesis (if autoinhibitory). This can result in yellowing of broccoli florets, softening of fruit tissue, abscission of cabbage leaves, pinking and browning of lettuce, browning of cut apple, sprouting of potato, toughening of asparagus stems, and bitterness (isocoumarin) in carrots. For intact fruit, ethylene promotes color changes, softening, and general ripening and senescence. For fresh-cut fruit, ethylene promotes discoloration, off-flavor, softening, ripening, and senescence. Ethylene generally shortens the shelf life of whole or fresh-cut products. Some techniques to prolong the shelf life of intact or fresh-cut produce are targeted at inhibiting or retarding ripening and wound response ethylene production or action.

Ethylene is thought to be synthesized from the amino acid L-methionine, which is converted to S-adenosylmethionine (SAM or adomet) by the enzyme methionine S-adenosyltransferase, which is converted to 1-aminocyclopropane-1-carboxylic acid (ACC) by ACC synthase (ACS), which, in turn, is converted to ethylene by ethylene-forming enzyme (Podoski et al., 1997) now called ACC oxidase (ACO) (Adams and Yang, 1979; Imaseki, 1991; Yang and Hoffman, 1984) (Figure 4.1). The key regulatory enzymes in the pathway are ACS and ACO. Ripening positively modulates ACS and ACO; however, wounding and water stress only positively modulate ACS. 1-MCP negatively modulates ACS and ACO, but aminoethoxyvinylglycine (AVG) and aminooxyacetic acid (AOA) only negatively modulate ACS, and high temperature ($>35°C$) and anaerobiosis only negatively modulate ACO. The enzyme malonyltransferase can catalyze the conjugation of ACC into malonyl ACC (Fallico et al., 1996). Ethylene can turn on or off genes that encode some portion of its biosynthetic pathway (autocatalytic or autoinhibitory, respectively). As mentioned above, it is often autoinhibitory in vegetative and immature reproductive tissues and autocatalytic in mature reproductive tissues like flowers and fruit (Baldwin, 2004; Lelièvre et al., 1997; Saltveit, 1999). Therefore, fresh-cut products may behave differently if made from vegetative, reproductive, climacteric, or nonclimacteric tissue.

A diverse multigene family encodes ACS (Zarembinski and Theologis, 1994), which is the rate-limiting step in ethylene biosynthesis. Ethylene can increase levels of malonyltransferase in various tissues which may account for the autoinhibition that is sometimes observed (Vangronsveld et al., 1988). Synthesis of ethylene requires O_2 and is inhibited by high levels of CO_2, especially if the response is autocatalytic as with climacteric fruits (Abeles et al., 1992; Sisler and Wood, 1988). Thus, MAP is used with fresh-cut products to control ethylene and respiration and thereby extend shelf life. Wound-response ethylene is thought to enhance activity of ACS (Apelbaum and Yang, 1981; Boller and Kende, 1980). To have a biological effect, ethylene must bind to a receptor that is thought to reside in a membrane. If the

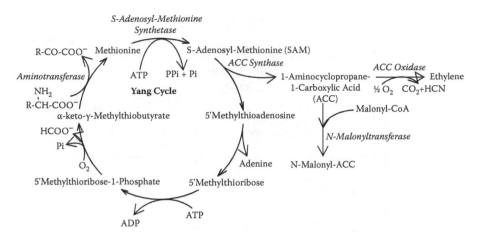

FIGURE 4.1 The ethylene synthesis pathway and Yang cycle. (Adapted from Adams, D.O., and S.F. Yang. 1979. Ethylene biosynthesis: Identification of 1-aminocyclopropane-1-carboxylic acid as an intermediate in the conversion of methionine to ethylene. *Proc. Natl. Acad. Sci. U.S.A.* 76: 170–174; Imaseki, H. 1991. The biochemistry of ethylene biosynthesis, pp. 1–20. In: A.K. Mattoo and J.C. Suttle (eds.). *The Plant Hormone Ethylene*. CRC Press, Boca Raton, FL; and Yang, S.F., and N.E. Hoffman. 1984. Ethylene biosynthesis and its regulation in higher plants. *Ann. Rev. Plant Physiol.* 35: 155–189.)

receptor is blocked, then ethylene will have no effect (inhibition of ethylene action). Silver ions, CO_2, 2,5-norbonradiene and 1-MCP can block the ethylene receptor, and therefore inhibit ethylene-induced effects, including its own synthesis in cases where ethylene is autocatalytic or promote its synthesis where ethylene is autoinhibitory (Abeles et al., 1992; Saltveit, 1999; Sisler et al., 1986). Of these compounds, only 1-MCP is approved for food and thus has been used to control ethylene responses and production in fresh-cut products (Toivonen, 2008a).

Suppression of the ethylene receptor (LeETR4) resulted in early ripening of tomato but did not affect fruit size, yield, or flavor composition, demonstrating that ethylene receptors may act as biological clocks that regulate onset of fruit ripening (Kevany et al., 2008). Ethylene receptors are degraded in the presence of ethylene, likely through a proteasome-dependent pathway; therefore, immature fruits exposed to ethylene undergo a reduction in the amount of receptor protein and earlier ripening (Kevany et al., 2007). In other words, the receptor acts as a negative regulator of ethylene responses, and this suppression is removed when ethylene binds the receptor. Figure 4.2 (Ecker, 2004; Li and Guo, 2007; Schweighofer and Meskiene, 2008) shows that after wounding or other stress perception, plant cells activate protein phosphorylation cascades mediated by kinases. Mitogen-activated protein (MAP) kinases, MKK4, MKK5, and MPK6 phosphorylate target proteins, leading to cell responses. One of the targets is the enzyme responsible for ethylene production, ACC synthase (ACS). Members in the two major subgroups of the ACS family, represented by ACS5 and ACS6, are regulated by different kinase pathways. Phosphorylation stabilizes ACS6 (and possibly ACS5), and interaction of ACS5 with unknown components targets ACS5 for degradation by the proteasome. Fine-tuning

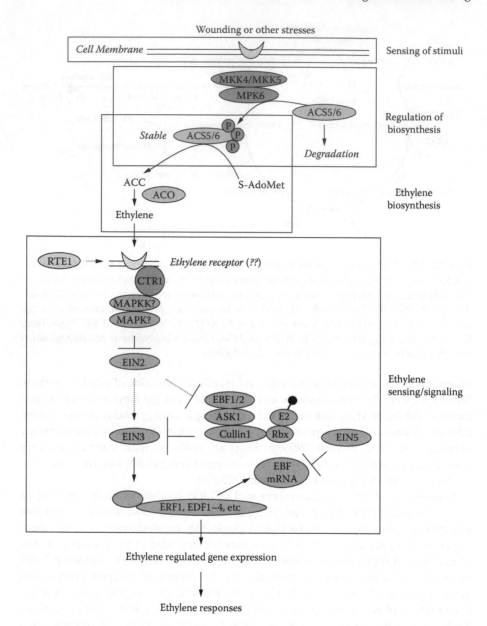

FIGURE 4.2 A simplified model depicting the induction of ethylene biosynthesis and their sensing and signaling pathway in *Arabidopsis*. Arrows and t-bars represent positive and negative regulations, respectively. Solid arrows and t-bars correspond to direct interactions, and dotted lines indicate the likely existence of unidentified elements between upstream and downstream components. (Adapted from Ecker, J.R. 2004. Reentry of the ethylene MPK6 module. *Plant Cell.* 16: 3169–3173; Li, H., and H. Guo. 2007. Molecular basis of the ethylene signaling and response pathway in *Arabidopsis. J. PlantGrowth Regul.* 26: 106–117; and Schweighofer, A., and I. Meskiene. 2008. Regulation of stress hormones jasmonates and ethylene by MAPK pathways in plants. *Molecul. BioSyst.* 4: 799–803.)

of ethylene induction is achieved by tissue-specific expression, gene activation, and the formation of heterodimers between different ACS members. ACC oxidase (ACO) activity can also affect levels of ethylene induction. Ethylene binding leads to the inactivation of ER-localized receptors (ETR1, ETR2, ERS1, ERS2, and EIN4) by an unknown mechanism. A novel membrane protein, RTE1, might specifically enhance the function of the ETR1 receptor. An inactive receptor is incapable of recruiting the negative regulator CTR1 to the ER membrane, which in turn shuts off its activity. EIN2 is then free from inhibition by CTR1 and increases the nuclear accumulation of EIN3 protein by repressing its turnover, which is mediated by SCF complexes containing the F-box proteins EBF1/2. There are two possibilities for how EIN2 stabilizes EIN3: EIN2-derived signal modulates EIN3 directly or inhibits the SCFEBF complex. One of the *EBF* genes, *EBF2*, is induced by ethylene in an EIN3-dependent manner. Thus a negative feedback loop is formed between EIN3 and EBF. EIN5, an exoribonuclease, seems to downregulate the level of *EBF1* and *EBF2* mRNAs without affecting their half-life. The nuclear accumulation of EIN3 induces a large amount of gene expression and ultimately triggers various ethylene responses.

Therefore, inhibition or reduction of ethylene synthesis or action can help to extend the shelf life of fresh-cut products but again may differ depending on the type of product (vegetative, reproductive, harvest maturity, etc.). There are numerous techniques to reduce ethylene production or effects which can be applied to the intact fruit prior to processing or to the processed product directly. Such treatments include genetics (Klee and Clark, 2002), storage temperature, MAP packaging, coatings, ethanol, ethylene absorbents, and 1-MCP. Ethylene synthesis inhibitors such as AVG can be used preharvest to inhibit ethylene production (Schupp and Greene, 2004; Yang, 1985), which can affect postharvest shelf life. Genetic engineering can also be used to reduce fruit ethylene production as has been shown in tomato by downregulating ACS (Oeller et al., 1991), ACO (Hamilton et al., 1990), or enhancing expression of ACC deaminase (Klee et al., 1991). Antisense molecular techniques have also been used to knock out the ethylene receptor (Klee and Clark, 2002), making the tissue nonresponsive to ethylene. For example, tomatoes with downregulated ACS or ACO do not produce much ethylene and do not ripen, and tomatoes with inserted ACC deaminase do not make much ethylene either, because this enzyme degrades ACC to α-ketobutyric acid. If ever approved, these transgenic fruits would be useful as fresh-cut products.

Temperatures above 35°C can cause injury to the ethylene biosynthetic pathway, disrupting ethylene synthesis (Antunes and Sfakiotakis, 2000; Lurie, 1998). Cold temperatures below 2.5°C slow metabolism, including ethylene synthesis (Wang and Adams, 1982) due to reduced ACO activity. The relatively low oxygen (O_2) and high carbon dioxide (CO_2) in MAP or the internal atmosphere of coated fruit (Bai et al., 2001; Bai et al., 2002) inhibit ethylene production because the ethylene biosynthetic pathway requires O_2 and is inhibited by CO_2 (Figure 4.1). Anaerobic conditions inhibit ACO. Ethanol has been shown to inhibit ethylene production in intact tomato fruits (Kelly and Saltveit, 1988; Saltveit and Mencarelli, 1988; Saltveit and Sharaf, 1992) and ethanol vapor was used to inhibit ethylene production in cut apple and mango (Bai et al., 2004; Plotto et al., 2006). Ethanol vapor treatment can be problematic because it sometimes results in altered or off-flavors, but at lower levels this was minimized, and the ethanol treatment of intact mango fruit had the added benefit of reducing the microbial population on the surface of the cut mango (Plotto et al., 2006).

The ethylene action inhibitor, 1-MCP, also reduced ethylene effects by inhibiting ethylene action in cut fruit (Bai et al., 2004; Ergun et al., 2007; Perera et al., 2003; Saftner et al., 2007). This included, in some cases, ethylene synthesis, if it was autocatalytic (Baldwin, 2004). For fresh-cut apple, treatment of intact apple with 1-MCP resulted in reduced ethylene production, respiration for 'Braeburn,' 'Pacific Rose,' and 'Gala' subsequent sliced products and reduced browning for 'Braeburn' cut apple (Bai et al., 2004; Perera et al., 2003). For 'Galia' melon, treatment of intact fruit resulted in reduced softening and watersoaking of the subsequent fresh-cut fruit (Ergun et al., 2007). Treatment of intact pineapple resulted in inhibition of respiration and ethylene production and delayed softening in combination with MAP (Rocculi et al., 2009). Treatment of fresh cut cilantro resulted in reduced respiration (Kim et al., 2007). Low dosage 1-MCP treatments prior to ethylene exposure of intact watermelon prevented ethylene-mediated quality deterioration in subsequent fresh-cut slices under MAP (Saftner et al., 2007). Application of 1-MCP before or after processing of kiwifruit reduced ethylene and softening which was enhanced by a dip of calcium chloride ($CaCl_2$). 1-MCP also reduced softening and browning when applied directly on fresh-cut mango slices. For fresh-cut persimmons, direct application of 1-MCP after processing resulted in higher ethylene production, slower softening, and darkening of color with no effect on respiration (Vilas-Boas and Kader, 2007).

4.4   RESPIRATION

Fruit ripening, senescence, and the wound response are energy-consuming processes. The respiratory quotient of a fruit or vegetable is the ratio of the CO_2 evolved to O_2 consumed. Novel proteins, messenger ribonucleic acids (mRNAs), pigments, and flavor compounds are synthesized in fresh produce, requiring energy and carbon that are supplied by the fruit as with other tissues. Once harvested, however, the fruit has limited resources, which is even more pronounced in fresh-cut products. For example, fresh-cut lemon had two to five times the respiration of whole lemons when stored below 5°C and 12-fold higher respiration if stored at 10°C (Artés-Hernández et al., 2007). Ripening of climacteric fruit and wounding of all fruit produce a characteristic peak of respiratory activity. In general, the higher the respiratory rate, the shorter the shelf life, as respiratory substrates, generally sugars and acids, are consumed (Tucker, 1983). Usually, sugars and acids of fruits are sequestered in the vacuole but are released periodically or maintained in a separate pool for use in respiration (Tucker, 1983; Tucker and Grierson, 1987). Respiratory pathways by which fruit oxidize sugars are glycolysis, oxidative pentose phosphate (OPP) pathway (less important, but may operate in climacteric respiration), and the tricarboxylic acid (TCA) pathway. In glycolysis, glucose-6-phosphate (P) is converted to fructose 6-P which is converted to fructose-1, 6-bisphosphate, and phosphoenol pyruvate to pyruvate. Pyruvate feeds into the the TCA cycle. Malic acid appears to be used as a respiratory substrate via malic enzyme which decarboxylates malate to pyruvate, with the extra carbon being fed into the TCA cycle, while citrate can feed directly into the TCA cycle (Goodenough et al., 1985; Tucker, 1993). The NAD(P)H produced by glycolysis, TCA, or malic enzyme activity is oxidized by plant mitochondria and the energy used for adenosine triphosphate (ATP) synthesis via oxidative phosphorylation, and mediated by membrane-bound dehydrogenases in the mitochondria (Palmer and Moller, 1982; Tucker, 1993).

Through the engulfing of a α-proteobacterium, the mitochondrion, the ancestral eukaryotic cell gained many metabolic capabilities (Sweetlove et al., 2007). The mitochondrion is uniquely sensitive to ATP demands in the cell. The mitochondria also produces ROS, which as mentioned before, are potent intracellular signals for, among other things, abiotic stress and oxidative damage of proteins, so that it acts as an early warning sensor of altered cellular redox balance. ROS are generated in the mitochondrion when overreduction of the respiratory complexes occurs, causing leakage of electrons to molecular oxygen causing superoxide formation (Moller, 2001). This is a result of an imbalance between electrons entering the respiratory chain and the dissipation of the proton gradient by ATP synthesis, or by reactions of electron transport due to altered physiological state (low temperature) (Sweetlove et al., 2007) or wounding, as with fresh-cut produce. What is important is the balance between ROS production and antioxidant capacity. The alternative oxidase (AOX) responds to redox signals, which bypass much of the respiratory chain and significantly reduce proton pumping and ROS production (Maxwell et al., 1999). AOX expression is linked to the TCA cycle, and the TCA cycle intermediates may act as signals for gene expression (Sweetlove et al., 2007).

As for the connection between climacteric ethylene and respiration, a developmental factor seems to be involved in the activation of the respiratory climacteric. For example, honeydew melon needs to achieve a certain maturity for low levels of exogenous ethylene to stimulate climacteric ethylene and respiration. In an experiment using antisense ACO cantaloupe harvested at 20 to 35 days past pollination compared to wild-type fruit, it seemed there was an interaction between developmental factors and ethylene levels in the induction of the respiratory upsurge (Flores et al., 2008).

4.5 STORAGE TEMPERATURE AND MODIFIED ATMOSPHERE

4.5.1 ATMOSPHERE

Storage temperature of fresh-cut commodities affects their respiration rate, which increases with increasing temperature. This effect is exacerbated if the commodity is in a modified atmosphere (MAP package, coating or CA storage) of relatively low O_2 and high CO_2, likely designed for low-temperature (therefore, low respiration) storage (2 to 10°C). If the cold chain is broken, then the respiration rate/O_2 demand increases, and the O_2 level may fall below the extinction point for that particular commodity, and the anaerobic pathway is engaged, producing ethanol and off-flavors. Sometimes the product has acceptable appearance but not acceptable flavor. The product can then quickly deteriorate. The Q_{10} of respiration rates of fresh-cut vegetables ranged from 2 to 7.5 when the temperature was increased from 0 to 10°C

(Watada et al., 1996). High O_2 MAP (70% + O_2) has also been used to inhibit enzymatic discoloration, prevent anaerobic fermentation, and inhibit aerobic and anaerobic microbial growth. Argon and nitrous oxide have also been shown to reduce respiration rates (Day, 1994).

4.5.2 STORAGE TEMPERATURE

The temperature at which products are stored can also affect the nutritional shelf life of a fresh-cut product. In a study of the antioxidant potential of fresh-cut strawberries, temperature affected the anthocyanins and vitamin C content as well as antioxidant capacity of the cut strawberry stored 21 days under 80% O_2, which reduced wounding stress (Odriozola-Serrano et al., 2009). Vitamin C degradation was more susceptible to small temperature increments than the other antioxidant compounds, and a storage temperature of 5°C was most effective in keeping the antioxidant properties of cut strawberry under the high O_2 atmosphere. For fresh-cut tomato, lycopene, vitamin C, and phenolic contents as well as other physiochemical parameters were maintained for 14 days at 5°C in MAP (5% O_2 + 5% CO_2). Raising the storage temperature enhanced lycopene and total phenolics but decreased vitamin C, and shelf life was reduced due to microbial growth at storage temperatures above 10°C (Odriozola-Serrano et al., 2008). One advantage of fresh-cut produce from chilling-sensitive fruits is that the cut product often does not manifest chilling injury, symptoms of which are often exhibited in the intact fruit peel.

Films available for fresh-cut produce often do not have sufficient O_2 and CO_2 transmission rates to reach steady-state gas concentrations before reaching O_2 and CO_2 levels that are too low or too high, respectively (Oms-Oliu et al., 2007). This promotes anaerobic respiration or CO_2 injury that results in fermentive off-flavor and odor, excess CO_2 production, as well as visual defects. In a survey of bagged salads from major supermarket chains, the mean headspace was 1.2% O_2 and 12% CO_2, and the mean ethanol content was 700 parts per million (ppm) (Hagenmaier and Baker, 1998). Fresh-cut sweet potato in medium- or high-permeability film bags stored at 2 or 8°C resulted in anaerobiosis at 2°C and especially 8°C for the low-permeability bags and at 8°C for the medium-permeability bags (Erturk and Picha, 2008). Nevertheless, MA is used to reduce respiration and ethylene production of fresh-cut produce to extend shelf life, generally in the range of 1 to 5% O_2 and 5 to 10% CO_2 (Gorny et al., 2002). This was successfully used for Kohlrabi sticks (2.5% O_2 and 9% CO_2) (Escalona et al., 2007), broccoli and cauliflower florets (1% O_2 and 21% CO_2) (Schreiner et al., 2007), minimally processed bok choy (5% O_2 and 2% CO_2) (Lu, 2007) and fresh-cut pineapple (2% O_2 and 12% CO_2) (Montero-Calderón et al., 2008), while low O_2 reduced browning of cut carambola (Teixeira et al., 2007) in vacuum-sealed bags. MAP and osmotic dehydration worked well for minimally processed guava (Pereira et al., 2004). Studies on the use of superatmospheric O_2 (>70%) were done to prevent anaerobic conditions, although this technique reduced the quality of fresh-cut pears (Oms-Oliu et al., 2007) by promoting oxidative processes (reduced vitamin C, solids, and acids). The high oxygen active packages, however, reduced accumulation of fermentative metabolites compared to typical MAP (relatively low O_2 and high CO_2) and inhibited some spoilage microorganisms

(Oms-Oliu et al., 2007). For cut peppers, 50% to 80% O_2 with less than 20% CO_2 was found to maintain good visual appearance without fermentation or off-odors. Higher levels of CO_2 promoted respiration (Conesa et al., 2007). Coating cut pears with antioxidant compounds (N-acetylcysteine and glutathione at 75%) in high O_2 (70%) atmosphere, low O_2 atmosphere (2.5% O_2 and 7% CO_2 active flush), or passively modified atmosphere showed that the low O_2 with the antioxidant treatments best maintained vitamin C, chlorogenic acid, and antioxidant capacity and reduced browning and ethylene production compared with the high O_2 treatment (Oms-Oliu et al., 2008a). However, Chinese bayberry fruits treated with air or 80% to 100% O_2 exhibited less decay and high levels of total soluble solids, titratable acidity and ascorbic acid contents, and reduced pH compared to air. The high O_2 treatment was less stressful as demonstrated by lower malonaldehyde content and higher catalase, ascorbic acid peroxidase, and peroxidase activities during storage. Elevated O_2 levels may improve the antioxidative defense mechanism and decay resistance (Yang et al., 2009).

4.6 GENETICS AND HARVEST MATURITY

4.6.1 GENETICS

Breeders are beginning to look for ways to breed varieties with characteristics tailored to a fresh-cut product. For example, lettuce breeders are seeking to develop multiuse cultivars. To this end, breeding lines were screened for increased cut lettuce shelf life in modified atmosphere (MA) environments such as encountered with MAP (Hayes and Liu, 2008). Genetic variation was thus determined for cut lettuce shelf life in low O_2 MA environments of 0.2% to 5% O_2 with the balance of N_2 in bags or in CA, to which cultivars could avoid damage from self-generated CO_2 injury or fermentation from low O_2. A novel hybrid muskmelon bred from an ultrafirm parent resulted in a fruit that was larger and had high external and internal firmness, ideal for long-distance shipping (Lester and Saftner, 2008). This hybrid also had high concentrations of sugar, β-carotene, and folic acid.

4.6.2 HARVEST MATURITY

When a product is harvested can affect the shelf life and quality of a fresh-cut product in that the fruit may or may not continue to ripen and soften after cutting. In some studies, relationships between harvest maturity and shelf life have been found (Chiesa et al., 2003; Couture et al., 1993; Watada and Qi, 1999), but this has not been found in others (Hayes and Liu, 2008). For cantaloupe, the length of storage duration decreased as harvest maturity increased for fresh-cut cantaloupe. Vitamin C losses were greater in fresh-cut cubes from more mature harvested fruit (full slip), and firmness losses were faster in more mature fruit (Beaulieu and Lea, 2007). Volatiles were reduced in more immature harvested cantaloupe (1/4 slip) (Beaulieu, 2006). It was therefore recommended to harvest at moderate maturity (1/2 slip) (Beaulieu and Lea, 2007) for a fresh-cut cantaloupe product. On the other hand, pear destined for a fresh-cut product was more acceptable if harvested 1 month later than normal commercial harvest due to less browning potential (lower levels of phenolic compounds in the pulp) and better flavor (sweeter due to lower titratable acidity, higher levels of total aroma volatiles) (Bai et al., 2009). For mango also, later harvest maturity resulted in generally higher levels of desirable aromatic compounds (Beaulieu and Lancaster, 2008). Cut apple prepared from partially ripe intact 'Fuji' apples was most suitable for a fresh-cut product (Rojas-Graü et al., 2007b). However, for 'Granny Smith' apples, fruit harvested 2 weeks or more before induction of climacteric ethylene production (starch index <2.5) had more cut edge browning, despite use of anti-browning dip (Toivonen, 2008b).

4.7 PHYSIOLOGICAL EFFECTS OF SANITIZING TREATMENTS

Some sanitizers have physiological effects on the treated fresh-cut products. Sodium hypochlorite (NaOCl) has been shown to reduce browning of apple and potato, although $CaCl_2$ was found to be more effective (Brecht et al., 1993). Cut tomato treated with hydrogen peroxide (H_2O_2) exhibited reduced microbial populations when stored at 10°C; however, they also had reduced phenolic and antioxidant levels after 7 days of storage and exhibited reduced color and levels of carotenoids. On fresh-cut Chinese water chestnuts, use of H_2O_2 reduced surface discoloration as well as decay. The H_2O_2 also delayed increases in total phenolic contents and reduced PPO, PAL, and POD activities (Peng et al., 2008; Podoski et al., 1997). Heat treatments for sanitation purposes can also cause heat shock and denaturation of metabolic enzymes that result in physiological effects. Hot water treatment of cantalopes to "pasteurize" the surface prior to cutting resulted in reduced microbial populations on the rind and of the subsequent cut melon pieces without affecting the quality of the fresh-cut product (Fan et al., 2008). A mild heat treatment of fresh-cut table grapes (45°C hot water or 55°C hot air for 8 or 5 minutes, respectively) followed by storage at 5°C resulted in reduced O_2 consumption, ethylene production, and decay (Kou et al., 2007).

Irradiation doses of 0.2 to 0.8 kGy can result in a 1 log reduction of bacterial pathogens, while 1 to 3 kGy is necessary to achieve a 1-log reduction of pathogenic viruses and fungi but can impact quality. Doses of 1 kGy or less are tolerated by most fresh-cut fruits and vegetables with little change in appearance, flavor, color, or texture. In some cases, there is loss of firmness and wilting due to an effect on permeability and functionality of cell membranes resulting in electrolyte leakage and loss of tissue integrity, especially at doses above 1 kGy (Fan and Sokorai, 2008). Low-dose irradiation (1 kGy) that can potentially inactivate *Escherichia coli* O157:H7 by 5 logs was tested on 13 common fresh-cut vegetables and was found not to result in significant losses of quality attributes (appearance, texture, flavor, and nutritional composition), with the exception of irradiated green and red leaf lettuce having reduced vitamin C values (Fan and Sokorai, 2008). Low-dose electron beam irradiation on fresh-cut cantalope lowered and stabilized respiration compared to nonirradiated samples while lowering microbial populations (Boynton et al., 2006).

Heat is also used to reduce ripening, senescence, and wound responses of fresh-cut produce (respiration, ethylene, softening, browning). Treatment of fresh-cut peach with hot water at 50°C for 10 minutes followed by MAP storage at 5°C controlled browning and retained firmness during storage, with lower concentrations of CO_2 and ethylene in the package atmosphere (Koukounaras et al., 2008). However, the heat treatment increased total carotenoid loss and reduced chroma color values on the slices compared to nonheated controls, but PPO activity was not affected. Hot dips in $CaCl_2$ or other calcium salts (60°C) helped maintain firmness (increased bound Ca levels in the cell wall), reduced microbial growth, and improved sensory quality compared to unheated controls (Aguayo et al., 2008). Other calcium salts were evaluated with varying amounts of success. Heat-treated cut apple (4 days at 38°C hot air) reduced ethylene synthesis, respiration, and softening, but also reduced aroma volatiles (Bai et al., 2004). Mild heat shock of lettuce leaves had mixed results, reducing browning but causing deleterious surface color and texture effects (Moreira et al., 2006).

UV light can also be used to sanitize fresh-cut produce, but it can also affect quality. Fresh-cut tomatoes, grown hydroponically, when exposed to UV-C light exhibited reduced microbial populations, increased phenolic content, and delayed degradation of vitamin C after 7 days storage at 4 to 6°C. The UV light treatment did not affect appearance, color, or lycopene content when the tomatoes were grown in lower salt concentrations. When the tomatoes were grown at higher salt concentrations and exposed to UV treatment, lycopene and vitamin C contents were increased likely due to stress; however, the UV-C treatment accelerated the decline of phenolic compounds and vitamin C in the subsequent fresh-cut product of these fruit (Kim et al., 2008a). Use of higher UV-C doses (up to 11.35 kJm^{-2}) on fresh-cut spinach resulted in a loss of lightness likely due to superficial tissue damage and an accelerated decrease in total antioxidant activity and polyphenol content during storage, although microbial growth was reduced (Artés-Hernández et al., 2009). UV irradiation of cut cantaloupe enhanced terpenoid production accompanied by some ester losses, which were cultivar and maturity dependent (Beaulieu and Lea, 2007).

Ozonated water is also used as a sanitizer, but it can have physiological effects. Use of ozonated water on the shelf life of fresh-cut lettuce antioxidants, including polyphenols and vitamin C, did not affect these compounds and did not increase respiration. The ozone treatment reduced browning and improved appearance compared to water and chlorine rinses (Beltran et al., 2005). In a comparison of sanitizing methods including sodium hypochlorite, electrolyzed water, and peroxyacetic acid for effect on respiration rates of fresh-cut vegetables stored at 7°C with 3% O_2, sodium hypochlorite increased respiration of iceberg lettuce but not carrot, leek, or cabbage. Peroxyacetic acid did not affect respiration of leek and iceberg lettuce, but it decreased respiration of carrot and cabbage. Electrolyzed water, however, reduced respiration of leek and cabbage (Vandekinderen et al., 2008).

4.8 SURFACE TREATMENTS AND COATINGS

Often, fresh-cut fruit are treated to combat derivative physiological changes such as browning, softening, flavor changes, and losses of nutrients or other healthful compounds and dehydration.

4.8.1 SURFACE TREATMENTS

Browning of cut tissue is generally due to the action of PPO with some involvement in some cases of POD and PAL. POD is a heme-containing enzyme that can perform single-electron oxidation of phenolic compounds in the presence of H_2O_2. Generation of H_2O_2 by PPO when oxidizing some phenolics can induce synergistic action between PPO and POD (He and Luo, 2007). Ascorbic acid or its derivatives can be used as an antioxidant to reduce browning and as an acidulant. Citric, malic, tartaric, oxalic, and succinic acids can also be used to acidify the surface and to act as chelator of browning compounds. Calcium or its derivatives can be used to reduce browning and to increase firmness (He and Luo, 2007). Acidifying the surface helps to inhibit PPO activity due to its pH optimum being quite sensitive (Chisari et al., 2008). PPO isoenzymes have two copper ions at the active site. This enzyme can catalyze the hydroxylation of monophenols and the oxidation of diphenols to quinones, using oxygen as a cosubstrate. The quinones polymerize into brown pigments (Arias et al., 2007; He and Luo, 2007; McEvily et al., 1991; Taylor and Clydesdale, 1987). Other antibrowning agents include thiol-containing compounds cysteine, glutathione, and acetylcysteine, which are thought to form colorless thiol-conjugated o-quinnones (He and Luo, 2007). Calcium ascorbate is also effective (adapted by the fresh-cut industry in the commercial form of Nature Seal), as are sulfates (banned by the U.S. Food and Drug Administration [FDA] due to allergenic reactions). The effect of calcium ascorbate was demonstrated on cut apple (Wang et al., 2007). 4-Hexylresorcinol and ascorbic acid were investigated in pear. It was found that ascorbic acid did not interact directly with PPO but prevented browning by reducing oxidized substrates. 4-Hexylresorcinol, however, interacted with the deoxy form of PPO, thus inactivating the enzyme in the absence of substrates, or competed with substrates for the catalytic site (Arias et al., 2007). POD activity, however, was more associated with browning of cut melon than PPO (Chisari et al., 2008). Oxyresveratrol and an extract from mulberry twigs (mulberroside) in combination with isoascorbic acid, calcium chloride, and acetycysteine delayed browning of cut apple (Li et al., 2007). Ascorbic acid, citric acid, and calcium chloride were used on cut mango to retain color and antioxidant activity (Robles-Sánchez et al., 2009). In cut apple, antibrowning agents ascorbic acid, 4-hexylresorcinol, N-acetylcyteine, and glutathione were tested. N-acetylcysteine and glutathione inhibited PPO activity. Ascorbic acid in combination with 4-hexylresorcinol, N-acetylcysteine, or glutathione inhibited POD activity. Generally, 4-hexylresorcinol and ascorbic acid individually were not very effective at levels tested for browning of cut apple (Rojas-Graü et al., 2008) compared to the sulfur compounds N-acetylcysteine and glutathione. Other inhibitors of PPO and browning from natural sources include kojic acid, arbutin, and glabridin, as well as an extract from the wood of *Artocarpus heterophyllus*, which was effective on cut apple (Zheng et al., 2008). Lychee fruit have a bright red pericarp that browns after harvest due to dehydration, breakdown of anthocyanins, and action of PPO. Therefore, a lychee fresh-cut product was developed where the edible aril was removed from the peel and treated with cysteine, ascorbic acid, and 4-hexylresorcinol along with osmotic vacuum dehydration, which maintained color and sensory characteristics (Shah and Nath, 2008).

4.8.2 COATINGS

One problem of minimal processing of fruits and vegetables is water loss and dehydration. Use of edible coatings can reduce water loss by acting as a barrier to water vapor transfer, or by acting as a sacrificing agent, losing water from the coating before losing water from the product tissue, and thus delaying water loss by the cut product. Coatings can also be used as carriers of the above antioxidants, sanitizers, acidulents, chelators, firming agents, and probiotics. The problem is that fresh-cut products often have high water vapor activity on the cut surface, making use of hydrophobic coatings difficult, although they are best at preventing water loss. Gellan-based films containing antioxidants and viable probiotic bilfidobacteria on cut papaya and apple exhibited good water vapor properties (Tapia et al., 2007). Although hydrophilic protein and polysaccharide films are more compatible with a cut surface, an antioxidant candelilla wax carrying *Aloe vera* reduced weight loss, changes in pH, and firmness when tested on fresh-cut avocados, bananas, and apples (Saucedo-Pompa et al., 2007). Algenate and gellan-based edible coatings on fresh-cut 'Fuji' apples delayed ethylene production, reduced microbial growth, and maintained firmness and color when N-acetylcysteine was added (Rojas-Graü et al., 2007b). Alginate coating with acetylated monoglyceride minimized weight loss of cut 'Gala' apple (Olivas et al., 2007). Apple puree-alginate coatings with antimicrobial essential oils (lemongrass, oregano, and vanillin) and antioxidants reduced respiration, ethylene production, and microbial growth when applied to fresh-cut apples but exhibited some ethanol and acetaldehyde formation in the first week of storage (Rojas-Graü et al., 2007a). Candelilla wax with ellagic acid-reduced color changes. Generally, these films reduced softening of cut avocado, banana, and apple (Saucedo-Pompa et al., 2007). Coatings of sour whey powder, soy protein isolate, and calcium caseinate on cut apple, carrots, and potato resulted in reduced color changes and weight loss (Shon and Haque, 2007). Application of chitosan coatings on fresh-cut mushroom delayed discoloration and reduced enzyme activities of PPO, POD, catalase, PAL, and laccase as well as resulted in lower phenolic content. A chitosan coating also maintained color and reduced microbial growth on cut papaya (González-Aguilar et al., 2009). The coatings also reduced activities of cellulose, total amylase, and α-amylase and inhibited microbial populations (Eissa, 2007). Edible coatings have also been used to reduce dehydration of the surface of peeled carrots, which takes on an undesirable white coloration (white blush) using cellulose (Mei et al., 2002; Sargent et al., 1994), sodium caseinate/stearic acid (Avena-Bustillos et al., 1994), whey protein, or xanthan gum (Mei et al., 2002). Vacuum impregnation of chitosan-based films reduced peeled carrot whitening and resistance to water vapor transmission (Vargas et al., 2008). Application of a chitosan/methyl cellulose film on fresh-cut cantaloupe and pineapple reduced microbial growth when vanillin was added and helped maintain moisture content (Sangsuwan et al., 2008). Alginate, pectin, and gellan-based coatings prevented dehydration of cut melon and inhibited ethylene production (Oms-Oliu et al., 2008b).

4.8.2.1 Firming Agents

Calcium can interlink pectin chains in cell walls and thus improve firmness. Calcium lactate strengthened the structure of fresh-cut 'Fuji' apples stored for 3 weeks at 4°C (Alandes et al., 2006) and counteracted activity of both pectinmethylesterase (PME) and PG. Addition of calcium chloride as a cross-linking agent helped maintain firmness of fresh-cut melon (Oms-Oliu et al., 2008b).

4.9 CONCLUSION

Fresh-cut fruits and vegetables present unique physiological challenges to postharvest physiologists and food scientists. The products are living tissues that have the outer protective peel breached in some way, allowing water loss and providing an entrance for microbes. The processing also induces a range of wound responses, including ethylene production and increased respiration, which in turn induce undesirable color, texture, and flavor changes. Many methods have been developed and continue to be tested to counteract the physiology, including MAP packaging, edible coatings, antioxidants, acidulants, antimicrobial agents, and firming agents. Anti-ethylene production and action agents have also been employed, including 1-MCP, ethanol, and use of temperature. In the end, successful marketing of fresh-cut products is evident judging by the rapidly growing fresh-cut segment of the fresh produce industry.

REFERENCES

Abe, K., and A.E. Watada. 1991. Ethylene absorbent to maintain quality of lightly processed fruits and vegetables. *J. Food Sci.* 56: 1589–1592.

Abeles, F.B., P.W. Morgan, and M.E. Saltveit. 1992. *Ethylene in Plant Physiology,* 2nd ed. Academic Press, San Diego, CA.

Adams, D.O., and S.F. Yang. 1979. Ethylene biosynthesis: Identification of 1-aminocyclopropane-1-carboxylic acid as an intermediate in the conversion of methionine to ethylene. *Proc. Natl. Acad. Sci. U.S.A.* 76: 170–174.

Aguayo, E., V.H. Escalona, and F. Artés. 2008. Effect of hot water treatment and various calcium salts on quality of fresh-cut 'Amarillo' melon. *Postharvest Biol. Technol.* 47: 397–406.

Ahvenainen, R. 1996. New approaches in improving the shelf life of minimally processed fruits and vegetables. *Tren. Food Sci. Technol.* 7: 179–187.

Alandes, L., I. Hernando, A. Quiles, I. Pérez-Munuera, and M.A. Lluch. 2006. Cell wall stability of fresh-cut Fuji apples treated with calcium lactate. *J. Food Sci.* 71: S615–S620.

Antunes, M.D.C., and E.M. Sfakiotakis. 2000. Effect of high temperature stress on ethylene biosynthesis, respiration and ripening of 'Hayward' kiwifruit. *Postharvest Biol. Technol.* 20: 251–259.

Apelbaum, A., and S.F. Yang. 1981. Biosynthesis of stress ethylene induced by water deficit. *Plant Physiol.* 68: 594–596.

Argañosa, A.C.S., M. Filomena, J.R. Paula, C.M.T. Alcina, and M.M.B. Morais. 2008. Effect of cut-type on quality of minimally processed papaya. *J. Sci. Food Agric.* 88: 2050–2060.

Arias, E., J. González, R. Oria, and P. Lopez-Buesa. 2007. Ascorbic acid and 4-hexylresorcinol effects on pear PPO and PPO catalyzed browning reaction. *J. Food Sci.* 72: C422–C429.

Artés, F., and A. Allende. 2005. Processing lines and alternative preservation techniques to prolong the shelf-life of minimally fresh processed leafy vegetables. *Eur. J. Hort. Sci.* 70: 231–245.

Artés, F., P.A. Gomez, and F. Artés-Hernandez. 2007. Physical, physiological, and microbial deterioration of minimally fresh processed fruits and vegetables. *Food Sci. Technol. Intl.* 13: 177–188.

Artés-Hernández, F., V.H. Escalona, P.A. Robles, G.B. Martínez-Hernández, and F. Artés. 2009. Effect of UV-C radiation on quality of minimally processed spinach leaves. *J. Sci. Food Agric.* 89: 414–421.

Artés-Hernández, F., F. Rivera-Cabrera, and A.A. Kader. 2007. Quality retention and potential shelf-life of fresh-cut lemons as affected by cut type and temperature. *Postharvest Biol. Technol.* 43: 245–254.

Avena-Bustillos, R.J., L.A. Cisneros-Zevallos, J.M. Krochta, and M.E. Saltveit Jr. 1994. Application of casein-lipid edible film emulsions to reduce white blush on minimally processed carrots. *Postharvest Biol. Technol.* 4: 319–329.

Ayala-Zavala, F.J., L. del Toro-Sánchez, E. Alvarez-Parrilla, H. Soto-Valdez, O. Martín-Belloso, S. Ruiz-Cruz, and G. González-Aguilar. 2008. Natural antimicrobial agents incorporated in active packaging to preserve the quality of fresh fruits and vegetables. *Stewart Postharvest Rev.* 4: 1–9.

Bai, J., E.A. Baldwin, R.C. Soliva-Fortuny, J.P. Mattheis, R. Stanley, C. Perera, and J.K. Brecht. 2004. Effect of pretreatment of intact 'Gala' apple with ethanol vapor, heat, or 1-methylcyclopropene on quality and shelf life of fresh-cut slices. *J. Amer. Soc. Hort. Sci.* 129: 583–593.

Bai, J., R.D. Hagenmaier, and E.A. Baldwin. 2002. Volatile response of four apple varieties with different coatings during marketing at room temperature. *J. Agric. Food Chem.* 50: 7660–7668.

Bai, J., P. Wu, J. Manthey, K. Goodner, and E. Baldwin. 2009. Effect of harvest maturity on quality of fresh-cut pear salad. *Postharvest Biol. Technol.* 51: 250–256.

Bai, J.H., R.A. Saftner, A.E. Watada, and Y.S. Lee. 2001. Modified atmosphere maintains quality of fresh-cut cantaloupe (*Cucumis melo* L.). *J. Food Sci.* 66: 1207–1211.

Bailey, B.A., M.D. Strem, H. Bae, G.A. de Mayolo, and M.J. Guiltinan. 2005. Gene expression in leaves of 'Theobroma' cacao in response to mechanical wounding, ethylene, and/or methyl jasmonate. *Plant Sci.* 168: 1247–1258.

Baldwin, E., M. Nisperos-Carriedo, and R. Baker. 1995. Use of edible coatings to preserve quality of lightly (and slightly) processed products. *Crit. Rev. Food Sci. Nutr.* 35: 509–524.

Baldwin, E.A. 2004. Ethylene and postharvest commodities. *HortScience.* 39: 1538–1540.

Barmore, C.R. 1975. Effect of ethylene on chlorophyllase activity and chlorophyll content in calmondin rind tissue. *HortScience.* 10: 595.

Barmore, C.R. 1987. Packaging technology for fresh and minimally processed fruits and vegetables. *J. Food Qual.* 10: 207–217.

Barry, C.S., M.I. Llop-Tous, and D. Grierson. 2000. The regulation of 1-aminocyclopropane 1-carboxylic acid synthase gene expression during the transition from system-1 to system-2 ethylene synthesis in tomato. *Plant Physiol.* 123: 979–986.

Beaulieu, J.C. 2006. Volatile changes in cantaloupe during growth, maturation, and in stored fresh-cuts prepared from fruit harvested at various maturities. *J. Amer. Soc. Hort. Sci.* 131: 127–139.

Beaulieu, J.C., and V.A. Lancaster. 2008. Correlating volatile compounds, sensory attributes, and quality parameters in stored fresh-cut cantaloupe. *J. Agric. Food Chem.* 56: 2866–2866.

Beaulieu, J.C., and J.M. Lea. 2007. Quality changes in cantaloupe during growth, maturation, and in stored fresh-cut cubes prepared from fruit harvested at various maturities. *J. Amer. Soc. Hort. Sci.* 132: 720–728.

Beltran, D., M.V. Selma, A. Marin, and M.I. Gil. 2005. Ozonated water extends the shelf life of fresh-cut lettuce. *J. Agric. Food Chem.* 53: 5654–5663.

Boller, T., and H. Kende. 1980. Regulation of wound ethylene synthesis in plants. *Nature.* 286: 259–260.

Bottino, A., E. Degl'Innocenti, L. Guidi, G. Graziani, and V. Fogliano. 2009. Bioactive compounds during storage of fresh-cut spinach: The role of endogenous ascorbic acid in the improvement of product quality. *J. Agric. Food Chem.* 57: 2925–2931.

Boynton, B.B., B.A. Welt, C.A. Sims, M.O. Balaban, J.K. Brecht, and M.R. Marshall. 2006. Effects of low-dose electron beam irradiation on respiration, microbiology, texture, color, and sensory characteristics of fresh-cut cantaloupe stored in modified-atmosphere packages. *J. Food Sci.* 71: S149–S155.

Brackett, R.E. 1987. Microbiological consequences of minimally processed fruits and vegetables. *J. Food Qual.* 10: 195–206.

Brecht, J.K., A.U.O. Sabaa-Srur, S.A. Sargent, and R.J. Bender. 1993. Hypochlorite inhibition of enzymic browning of cut vegetables and fruit. *Acta Hort.* 343: 341–344.

Breidt, F., and H.P. Fleming. 1997. Using lactic acid bacteria to improve the safety of minimally processed fruits and vegetables. *Food Technol.* 51: 44–51.

Brody, A. 1998. Minimally processed foods demand maximum research and education. *Food Technol.* 52: 62, 64, 66, 204, 206.

Chiesa, A., D. Frezza, A. Fraschina, G. Trinchero, S. Moccia, and A. León. 2003. Preharvest factors and fresh-cut vegetables quality. *Acta Hort.* 604: 153–159.

Chisari, M., R.N. Barbagallo, and G. Spagna. 2008. Characterization and role of polyphenol oxidase and peroxidase in browning of fresh-cut melon. *J. Agric. Food Chem.* 56: 132–138.

Conesa, A., B.E. Verlinden, F. Artés-Hernández, B. Nicolaï, and F. Artés. 2007. Respiration rates of fresh-cut bell peppers under superamospheric and low oxygen with or without high carbon dioxide. *Postharvest Biol. Technol.* 45: 81–88.

Couture, R., M.I. Cantwell, D. Ke, and M.E. Saltveit, Jr. 1993. Physiological attributes related to quality attributes and storage life of minimally processed lettuce. *HortScience.* 28: 723–725.

Dangl, J.L., and J.D.G. Jones. 2001. Plant pathogens and integrated defence responses to infection. *Nature.* 411: 826–833.

Day, B.P.F. 1996. High oxygen modified atmosphere packaging for fresh prepared produce. *Postharvest News Information* 7:31N–33N.

DellaPenna, D., D.C. Alexander, and A.B. Bennett. 1986. Molecular cloning of tomato fruit polygalacturonase: Analysis of polygalacturonase mRNA levels during ripening. *Proc. Natl. Acad. Sci. U.S.A.* 83: 6420–6424.

Ecker, J.R. 2004. Reentry of the ethylene MPK6 module. *Plant Cell.* 16: 3169–3173.

Eissa, H.A.A. 2007. Effect of chitosan coating on shelf life and quality of fresh-cut mushroom. *J. Food Qual.* 30: 623–645.

Ergun, M., J. Jeong, D.J. Huber, and D.J. Cantliffe. 2007. Physiology of fresh-cut 'Galia' (*Cucumis melo* var. *reticulatus*) from ripe fruit treated with 1-methylcyclopropene. *Postharvest Biol. Technol.* 44: 286–292.

Erturk, E., and D.H. Picha. 2008. The effects of packaging film and storage temperature on the internal package atmosphere and fermentation enzyme activity of sweet potato slices. *J. Food Process. Preserv.* 32: 817–838.

Escalona, V.H., E. Aguayo, and F. Artés. 2007. Quality changes of fresh-cut kohlrabi sticks under modified atmosphere packaging. *J. Food Sci.* 72: S303–S307.

Fallico, B., M.C. Lanza, E. Maccarone, C.N. Asmundo, and P. Rapisarda. 1996. Role of hydroxycinnamic acids and vinylphenols in the flavor alteration of blood orange juices. *J. Agric. Food Chem.* 44: 2654–2657.

Fan, L., and J. Song. 2008. Microbial quality assessment methods for fresh-cut fruits and vegetables. *Stewart Postharvest Rev.* 4: 1–9.

Fan, X., B.A. Annous, J.C. Beaulieu, and J.E. Sites. 2008. Effect of hot water surface pasteurization of whole fruit on shelf life and quality of fresh-cut cantaloupe. *J. Food Sci.* 73: M91–M98.

Fan, X., and K.J.B. Sokorai. 2008. Retention of quality and nutritional value of 13 fresh-cut vegetables treated with low-dose radiation. *J. Food Sci.* 73: S367–S372.

Flores, F.B., M.C. Martinez-Madrid, and F. Romojaro. 2008. Influence of fruit development stage on the physiological response to ethylene in cantaloupe 'Charentais' melon. *Food Sci. Technol. Intl.* 14: 87–94.

Forney, C.F. 2008. Flavour loss during postharvest handling and marketing of fresh-cut produce. *Stewart Postharvest Rev.* 4: 1–10.

Gil, M.I., E. Aguayo, and A.A. Kader. 2006. Quality changes and nutrient retention in fresh-cut versus whole fruits during storage. *J. Agric. Food Chem.* 54: 4284–4296.

Gómez-López, V., P. Ragaert, J. Debevere, and F. Devlieghere. 2008. Decontamination methods to prolong the shelf-life of minimally processed vegetables, state-of-the-art. *Crit. Rev. Food Sci. Nutr.* 48: 487–495.

Gómez-López, V.M., A. Rajkovic, P. Ragaert, N. Smigic, and F. Devlieghere. 2009. Chlorine dioxide for minimally processed produce preservation: A review. *Tren. Food Sci. Technol.* 20: 17–26.

González-Aguilar, G.A., E. Valenzuela-Soto, J. Lizardi-Mendoza, F. Goycoolea, M.A. Martínez-Téllez, M.A. Villegas-Ochoa, I.N. Monroy-García, and J.F. Ayala-Zavala. 2009. Effect of chitosan coating in preventing deterioration and preserving the quality of fresh-cut papaya lsquoMaradolrsquo. *J. Sci. Food Agric.* 89: 15–23.

Goodenough, P.W., I.M. Prosser, and K. Young. 1985. NADP-linked malic enzyme and malate metabolism in ageing tomato fruit. *Phytochem.* 24: 1157–1162.

Gorny, J.R., B. Hess-Pierce, R.A. Cifuentes, and A.A. Kader. 2002. Quality changes in fresh-cut pear slices as affected by controlled atmospheres and chemical preservatives. *Postharvest Biol. Technol.* 24: 271–278.

Guerzoni, M., A. Gianotti, M. Corbo, and M. Sinigaglia. 1996. Shelf-life modeling for fresh-cut vegetables. *Postharvest Biol. Technol.* 9: 195–207.

Hagenmaier, R.D., and R.A. Baker. 1998. A survey of the microbial population and ethanol content of bagged salad. *J. Food Protect.* 61: 357–359.

Hamilton, A.J., G.W. Lycett, and D. Grierson. 1990. Antisense gene that inhibits synthesis of the hormone ethylene in transgenic plants. *Nature.* 346: 284–287.

Hayes, R.J., and Y.-B. Liu. 2008. Genetic variation for shelf-life of salad-cut lettuce in modified-atmosphere environments. *J. Amer. Soc. Hort. Sci.* 133: 228–233.

He, Q., and Y. Luo. 2007. Enzymatic browning and its control in fresh-cut produce. *Stewart Postharvest Rev.* 3: 1–7.

Hillwig, M., N. LeBrasseur, P. Green, and G. MacIntosh. 2008. Impact of transcriptional, ABA-dependent, and ABA-independent pathways on wounding regulation of RNS1 expression. *Molecul. Genet. Genom.* 280: 249–261.

Howe, G.A. 2004. Jasmonates as signals in the wound response. *J. Plant Growth Regul.* 23: 223–237.

Hu, W., and Y. Jiang. 2007. Quality attributes and control of fresh-cut produce. *Stewart Postharvest Rev.* 3: 1–9.

Huxsoll, C.C., and H. Bolin. 1989. Processing and distribution alternatives for minimally processed fruits and vegetables *Food Technol.* 43: 124–128.

Hyodo, H. 1991. Stress/wound ethylene, pp. 43–46. In: A.K. Mattoo and J.C. Suttle (eds.). *The Plant Hormone Ethylene.* CRC Press, Boca Raton, FL.

Imaseki, H. 1991. The biochemistry of ethylene biosynthesis, pp. 1–20. In: A.K. Mattoo and J.C. Suttle (eds.). *The Plant Hormone Ethylene.* CRC Press, Boca Raton, FL.

Iqbal, T., F.A.S. Rodrigues, P.V. Mahajan, J.P. Kerry, L. Gil, M.C. Manso, and L.M. Cunha. 2008. Effect of minimal processing conditions on respiration rate of carrots. *J. Food Sci.* 73: E396–E402.

Jacxsens, L., F. Devlieghere, T. De Rudder, and J. Debevere. 2000. Designing equilibrium modified atmosphere packages for fresh-cut vegetables subjected to changes in temperature. *Leb. Wiss. Technol.* 33: 178–187.

Kader, A.A. 1985. Ethylene-induced senescence and physiological disorders in harvested horticultural crops. *HortScience.* 20: 52–54.

Karakurt, Y., and D.J. Huber. 2003. Activities of several membrane and cell-wall hydrolases, ethylene biosynthetic enzymes, and cell wall polyuronide degradation during low-temperature storage of intact and fresh-cut papaya (*Carica papaya*) fruit. *Postharvest Biol. Technol.* 28: 219–229.

Karakurt, Y., and D.J. Huber. 2007. Characterization of wound-regulated cDNAs and their expression in fresh-cut and intact papaya fruit during low-temperature storage. *Postharvest Biol. Technol.* 44: 179–183.

Katsir, L., H.S. Chung, A.J.K. Koo, and G.A. Howe. 2008. Jasmonate signaling: A conserved mechanism of hormone sensing. *Curr. Opin. Plant Biol.* 11: 428–435.

Kelly, M.O., and M.E. Saltveit, Jr. 1988. Effect of endogenously synthesized and exogenously applied ethanol on tomato fruit ripening. *Plant Physiol.* 88: 143–147.

Kessler, A., and I.T. Baldwin. 2002. Plant responses to insect herbivory: The emerging molecular analysis. *Ann. Rev. Plant Biol.* 53: 299–328.

Kevany, B.M., M.G. Taylor, and H.J. Klee. 2008. Fruit-specific suppression of the ethylene receptor *LeETR4* results in early-ripening tomato fruit. *Plant Biotechnol. J.* 6: 295–300.

Kevany, B.M., D.M. Tieman, M.G. Taylor, V.D. Cin, and H.J. Klee. 2007. Ethylene receptor degradation controls the timing of ripening in tomato fruit. *Plant J.* 51: 458–467.

Kim, H.-J., J.M. Fonseca, C. Kubota, M. Kroggel, and J.-H. Choi. 2008a. Quality of fresh-cut tomatoes as affected by salt content in irrigation water and post-processing ultraviolet-C treatment. *J. Sci. Food Agric.* 88: 1969–1974.

Kim, J.G., Y. Luo, and Y. Tao. 2007. Effect of the sequential treatment of 1-methylcyclopropene and acidified sodium chlorite on microbial growth and quality of fresh-cut cilantro. *Postharvest Biol. Technol.* 46: 144–149.

Kim, K.-J., J.H. Lim, M.J. Kim, T. Kim, H.M. Chung, and K.-H. Paek. 2008b. GDSL-lipase1 (CaGL1) contributes to wound stress resistance by modulation of CaPR-4 expression in hot pepper. *Biochem. Biophysic. Res. Communic.* 374: 693–698.

King, A.D., and H.R. Bolin. 1989. Physiological and microbiological storage stability of minimally processed fruits and vegetables. *Food Technol.* 43: 132–135, 139.

Klee, H.J., and D.G. Clark. 2002. Manipulation of ethylene synthesis and perception in plants: The ins and the outs. *HortScience.* 37: 450–452.

Klee, H.J., M.B. Hayford, K.A. Kretzmer, G.F. Barry, and G.M. Kishore. 1991. Control of ethylene synthesis by expression of a bacterial enzyme in transgenic tomato plants. *Plant Cell.* 3: 1187–1193.

Koning, R.E. 1994. Ethylene, plant physiology information Web site: http://plantphys.info/ plant_physiology/ethylene.shtml (accessed 25 March 2009).

Kou, L., Y. Luo, D. Wu, and X. Liu. 2007. Effects of mild heat treatment on microbial growth and product quality of packaged fresh-cut table grapes. *J. Food Sci.* 72: S567–S573.

Koukounaras, A., G. Diamantidis, and E. Sfakiotakis. 2008. The effect of heat treatment on quality retention of fresh-cut peach. *Postharvest Biol. Technol.* 48: 30–36.

Lelièvre, J.-M., A. Latchè, B. Jones, M. Bouzayen, and J.-C. Pech. 1997. Ethylene and fruit ripening. *Physiol. Plant.* 101: 727–739.

Lester, G.E., and R.A. Saftner. 2008. Marketable quality and phytonutrient concentrations of a novel hybrid muskmelon intended for the fresh-cut industry and its parental lines: Whole-fruit comparisons at harvest and following long-term storage at 1 or 5°C. *Postharvest Biol. Technol.* 48: 248–253.

Li, H., and H. Guo. 2007. Molecular basis of the ethylene signaling and response pathway in *Arabidopsis. J. Plant Growth Regul.* 26: 106–117.

Li, H., K.-W. Cheng, C.-H. Cho, Z. He, and M. Wang. 2007. Oxyresveratrol as an anti-browning agent for cloudy apple juices and fresh-cut apples. *J. Agric. Food Chem.* 55: 2604–2610.

Lu, S. 2007. Effect of packaging on shelf-life of minimally processed bok choy (*Brassica chinensis* L.). *LWT—Food Sci. Technol.* 40: 460–464.

Lurie, S. 1998. Postharvest heat treatments. *Postharvest Biol. Technol.* 14: 257–269.

Mahattanatawee, K., J.A. Manthey, G. Luzio, S.T. Talcott, K. Goodner, and E.A. Baldwin. 2006. Total antioxidant activity and fiber content of select Florida-grown tropical fruits. *J. Agric. Food Chem.* 54: 7355–7363.

Martín-Diana, A.B., D. Rico, J.M. Frías, J.M. Barat, G.T.M. Henehan, and C. Barry-Ryan. 2007. Calcium for extending the shelf life of fresh whole and minimally processed fruits and vegetables: A review. *Tren. Food Sci. Technol.* 18: 210–218.

Mattoo, A.K., and W.B. White. 1991. Regulation of ethylene biosynthesis, pp. 21–42. In: A.K. Mattoo and J.C. Suttle (eds.). *The Plant Hormone Ethylene*. CRC Press, Boca Raton, FL.

Maxwell, D.P., Y. Wang, and L. McIntosh. 1999. The alternative oxidase lowers mitochondrial reactive oxygen production in plant cells. *Proc. Natl. Acad. Sci. U.S.A.* 96: 8271–8276.

McEvily, A.J., R. Iyengar, and A. Gross. 1991. Compositions and methods for inhibiting browning in foods using resorcinol derivatives. US patent 5059438.

McGlasson, W. 1985. Ethylene and fruit ripening. *HortScience.* 20: 51–53.

Mei, Y., Y. Zhao, J. Yang, and H.C. Furr. 2002. Using edible coating to enhance nutritional and sensory qualities of baby carrots. *J. Food Sci.* 67: 1964–1968.

Moller, I.M. 2001. Plant mitochondria and oxidative stress: Electron transport, NADPH turnover, and metabolism of reactive oxygen species. *Ann. Rev. Plant Physiol. Plant Molecul. Biol.* 52: 561–591.

Montero-Calderón, M., M.A. Rojas-Graü, and O. Martín-Belloso. 2008. Effect of packaging conditions on quality and shelf-life of fresh-cut pineapple (*Ananas comosus*). *Postharvest Biol. Technol.* 50: 182–189.

Moreira, M.d.R., A.G. Ponce, C.E. Del Valle, and S.I. Roura. 2006. Ascorbic acid retention, microbial growth, and sensory acceptability of lettuce leaves subjected to mild heat shocks. *J. Food Sci.* 71: S188–S192.

Morgan, P.W., and M.C. Drew. 1997. Ethylene and plant responses to stress. *Physiol. Plant.* 100: 620–630.

Mur, L.A.J., Y.-M. Bi, R.M. Darby, S. Firek, and J. Draper. 1997. Compromising early salicylic acid accumulation delays the hypersensitive response and increases viral dispersal during lesion establishment in TMV-infected tobacco. *Plant J.* 12: 1113–1126.

Myers, R.A. 1989. Packaging considerations for minimally processed fruits and vegetables *Food Technol.* 43: 129–131.

O'Donnell, P.J., C. Calvert, R. Atzorn, C. Wasternack, H.M.O. Leyser, and D.J. Bowles. 1996. Ethylene as a signal mediating the wound response of tomato plants. *Science.* 274: 1914–1917.

Odriozola-Serrano, I., R. Soliva-Fortuny, and O. Martín-Belloso. 2008. Antioxidant properties and shelf-life extension of fresh-cut tomatoes stored at different temperatures. *J. Sci. Food Agric.* 88: 2606–2614.

Odriozola-Serrano, I., R. Soliva-Fortuny, and O. Martín-Belloso. 2009. Influence of storage temperature on the kinetics of the changes in anthocyanins, vitamin C, and antioxidant capacity in fresh-cut strawberries stored under high-oxygen atmospheres. *J. Food Sci.* 74: C184–C191.

Oeller, P.W., M.W. Lu, L.P. Taylor, D.A. Pike, and A. Theologis. 1991. Reversible inhibition of tomato fruit senescence by antisense RNA. *Science.* 254: 437–439.

Olivas, G.I., D.S. Mattinson, and G.V. Barbosa-Cánovas. 2007. Alginate coatings for preservation of minimally processed 'Gala' apples. *Postharvest Biol. Technol.* 45: 89–96.

Olivas, G.I., J.E. Davila-Avina, N.A. Salas-Salazar, and F.J. Molina. 2008. Use of edible coatings to preserve the quality of fruits and vegetables during storage. *Stewart Postharvest Rev.* 4: 1–10.

Oms-Oliu, G., R. Soliva-Fortuny, and O. Martín-Belloso. 2007. Respiratory rate and quality changes in fresh-cut pears as affected by superatmospheric oxygen. *J. Food Sci.* 72: E456–E463.

Oms-Oliu, G., I. Odriozola-Serrano, R. Soliva-Fortuny, and O. Martín-Belloso. 2008a. Antioxidant content of fresh-cut pears stored in high-O_2 active packages compared with conventional low-O_2 active and passive modified atmosphere packaging. *J. Agric. Food Chem.* 56: 932–940.

Oms-Oliu, G., R. Soliva-Fortuny, and O. Martín-Belloso. 2008b. Edible coatings with anti-browning agents to maintain sensory quality and antioxidant properties of fresh-cut pears. *Postharvest Biol. Technol.* 50: 87–94.

Orozco-Cardenas, M.L., J. Narvaez-Vasquez, and C.A. Ryan. 2001. Hydrogen peroxide acts as a second messenger for the induction of defense genes in tomato plants in response to wounding, systemin, and methyl jasmonate. *Plant Cell.* 13: 179–191.

Palmer, J.M., and I.M. Moller. 1982. Regulation of NAD(P)H dehydrogenases in plant mitochondria. *Tren. Biochem. Sci.* 7: 258–261.

Palumbo, M.S., J.R. Gorny, D.E. Gombas, L.R. Beuchat, C.M. Bruhn, B. Cassens, P. Delaquis, J.M. Farber, L.J. Harris, K. Ito, M.T. Osterholm, M. Smith, and K.M.J. Swanson. 2007. Recommendations for handling fresh-cut leafy green salads by consumers and retail foodservice operators. *Food Protect. Tren.* 27: 892–898.

Parkin, K.L. 1987. A new technique for the long-term study of the physiology of plant fruit tissue slices. *Physiol. Plant.* 69: 472–476.

Peng, L., S. Yang, Q. Li, Y. Jiang, and D.C. Joyce. 2008. Hydrogen peroxide treatments inhibit the browning of fresh-cut Chinese water chestnut. *Postharvest Biol. Technol.* 47: 260–266.

Pereira, L.M., A.C.C. Rodrigues, C.I.G.L. Sarantopoulos, V.C.A. Junqueira, R.L. Cunha, and M.D. Hubinger. 2004. Influence of modified atmosphere packaging and osmotic dehydration on the quality maintenance of minimally processed guavas. *J. Food Sci.* 69: FEP172–FEP177.

Perera, C.O., L. Balchin, E. Baldwin, R. Stanley, and M. Tian. 2003. Effect of 1-methylcyclopropene on the quality of fresh-cut apple slices. *J. Food Sci.* 68: 1910–1914.

Plotto, A., J. Bai, J.A. Narciso, J.K. Brecht, and E.A. Baldwin. 2006. Ethanol vapor prior to processing extends fresh-cut mango storage by decreasing spoilage, but does not always delay ripening. *Postharvest Biol. Technol.* 39: 134–145.

Podoski, B.W., C.A. Sims, S.A. Sargent, J.F. Price, C.K. Chandler, and S.F. Okeefe. 1997. Effects of cultivar, modified atmosphere, and pre-harvest conditions on strawberry quality. *Proc. Fla. State Hort. Soc.* 110: 246–252.

Rattanapanone, N., C. Chongsawat, and S. Chaiteep. 2000. Fresh-cut fruits in Thailand. *HortScience.* 35: 543–546.

Reid, M.S. 1985. Ethylene and abscission. *HortScience.* 20: 45–50.

Reyes, L.F., J.E. Villarreal, and L. Cisneros-Zevallos. 2007. The increase in antioxidant capacity after wounding depends on the type of fruit or vegetable tissue. *Food Chem.* 101: 1254–1262.

Reyes, V.G., L. Simons, and C. Tran. 1995. Using edible coating to enhance nutritional and sensory qualities of baby carrots. Proceedings of the Australasian Postharvest Hort. Conf. *Science and Technology for the Fresh Food Revolution*, Melbourne, Australia 18–22 September, pp. 451–456.

Rico, D., A.B. Martín-Diana, J.M. Barat, and C. Barry-Ryan. 2007. Extending and measuring the quality of fresh-cut fruit and vegetables: A review. *Tren. Food Sci. Technol.* 18: 373–386.

Robles-Sánchez, R.M., M.A. Rojas-Graü, I. Odriozola-Serrano, G.A. González-Aguilar, and O. Martín-Belloso. 2009. Effect of minimal processing on bioactive compounds and antioxidant activity of fresh-cut 'Kent' mango (*Mangifera indica* L.). *Postharvest Biol. Technol.* 51: 384–390.

Rocculi, P., E. Cocci, S. Romani, G. Sacchetti, and M.D. Rosa. 2009. Effect of 1-MCP treatment and N₂O MAP on physiological and quality changes of fresh-cut pineapple. *Postharvest Biol. Technol.* 51: 371–377.

Rojas-Graü, M.A., R.J. Avena-Bustillos, C. Olsen, M. Friedman, P.R. Henika, O. Martín-Belloso, Z. Pan, and T.H. McHugh. 2007a. Effects of plant essential oils and oil compounds on mechanical, barrier and antimicrobial properties of alginate-apple puree edible films. *J. Food Engineer.* 81: 634–641.

Rojas-Graü, M.A., R. Grasa-Guillem, and O. Martín-Belloso. 2007b. Quality changes in fresh-cut Fuji apple as affected by ripeness stage, antibrowning agents, and storage atmosphere. *J. Food Sci.* 72: S036–S043.

Rojas-Graü, M.A., R. Soliva-Fortuny, and O. Martín-Belloso. 2008. Effect of natural antibrowning agents on color and related enzymes in fresh-cut 'Fuji' apples as an alternative to the use of ascorbic acid. *J. Food Sci.* 73: S267–S272.

Rolle, R.S., and G.W.I. Chism. 1987. Physiological consequences of minimally processed fruits and vegetables. *J. Food Qual.* 10: 157–177.

Ryan, C.A., and D.S. Moura. 2002. Systemic wound signaling in plants: A new perception. *Proc. Natl. Acad. Sci. U.S.A.* 99: 6519–6520.

Saftner, R., Y. Luo, J. McEvoy, J.A. Abbott, and B. Vinyard. 2007. Quality characteristics of fresh-cut watermelon slices from non-treated and 1-methylcyclopropene- and/or ethylene-treated whole fruit. *Postharvest Biol. Technol.* 44: 71–79.

Salcini, M.C., and R. Massantini. 2005. Minimally processed fruits: An update on browning control. *Stewart Postharvest Rev.* 1: 1–7.

Saltveit, M. 1999. Effect of ethylene on quality of fresh fruits and vegetables. *Postharvest Biol. Technol.* 15: 279–292.

Saltveit, M., and A. Sharaf. 1992. Ethanol inhibits ripening of tomato fruit harvested at various degrees of ripeness without affecting subsequent quality. *J. Amer. Soc. Hort. Sci.* 117: 793–798.

Saltveit, M.E., and F. Mencarelli. 1988. Inhibition of ethylene synthesis and action in ripening tomato fruit by ethanol vapors. *J. Amer. Soc. Hort. Sci.* 113: 572–576.

Sangsuwan, J., N. Rattanapanone, and P. Rachtanapun. 2008. Effect of chitosan/methyl cellulose films on microbial and quality characteristics of fresh-cut cantaloupe and pineapple. *Postharvest Biol. Technol.* 49: 403–410.

Sargent, S.A., J.K. Brecht, J.J. Zoellner, E.A. Baldwin, and C.A. Campbell. 1994. Edible films reduce surface drying of peeled carrots. *Proc. Fla. State Hort. Soc.* 107: 245–247.

Saucedo-Pompa, S., D. Jasso-Cantu, J. Ventura-Sobrevilla, A. Saenz-Galindo, R. Rodriguez-Herrera, and C.N. Aguilar. 2007. Effect of candelilla wax with natural antioxidants on the shelf life quality of fresh-cut fruits. *J. Food Qual.* 30: 823–836.

Schreiner, M., P. Peters, and A. Krumbein. 2007. Changes of glucosinolates in mixed fresh-cut broccoli and cauliflower florets in modified atmosphere packaging. *J. Food Sci.* 72: S585–S589.

Schupp, J.R., and D.W. Greene. 2004. Effect of aminoethoxyvinylglycine (AVG) on prehar-vest drop, fruit quality, and maturation of 'McIntosh' apples: Concentration and timing of dilute applications of AVG. *HortScience.* 39: 1030–1035.

Schweighofer, A., and I. Meskiene. 2008. Regulation of stress hormones jasmonates and eth-ylene by MAPK pathways in plants. *Molecul. BioSyst.* 4: 799–803.

Shah, N.S., and N. Nath. 2008. Changes in qualities of minimally processed litchis: Effect of antibrowning agents, osmo-vacuum drying and moderate vacuum packaging. *LWT—Food Sci. Technol.* 41: 660–668.

Shewfelt, R.L. 1987. Quality of minimally processed fruits and vegetables. *J. Food Qual.* 10: 143–156.

Shon, J., and Z.U. Haque. 2007. Efficacy of sour whey as a shelf-life enhancer: Use in antioxi-dative edible coating of cut vegetables and fruit. *J. Food Qual.* 30: 581–593.

Siqueira-Júnior, C., B. Jardim, T. Ürményi, A. Vicente, E. Hansen, K. Otsuki, M. da Cunha, H. Madureira, D. de Carvalho, and T. Jacinto. 2008. Wound response in passion fruit (*Passiflora f.* edulis flavicarpa) plants: Gene characterization of a novel chloroplast-targeted allene oxide synthase up-regulated by mechanical injury and methyl jasmonate. *Plant Cell Rept.* 27: 387–397.

Sisler, E.C., and C. Wood. 1988. Interaction of ethylene and CO_2. *Physiol. Plant.* 73: 440–444.

Sisler, E.C., M.S. Reid, and S.F. Yang. 1986. Effect of antagonists of ethylene action on bind-ing of ethylene in cut carnations. *Plant Growth Regul.* 4: 213–218.

Soliva-Fortuny, R.C., and O. Martín-Belloso. 2003. New advances in extending the shelf-life of fresh-cut fruits: A review. *Tren. Food Sci. Technol.* 14: 341–353.

Sweetlove, L.J., A. Fait, A. Nunes-Nesi, T. Williams, and A.R. Fernie. 2007. The mitochon-drion: An integration point of cellular metabolism and signalling. *Crit. Rev. Plant Sci.* 26: 17–43.

Tapia, M.S., M.A. Rojas-Graü, F.J. Rodriguez, A. Carmona, and O. Martin-Belloso. 2007. Alginate and gellan-based edible films for probiotic coatings on fresh-cut fruits. *J. Food Sci.* 72: 190–196.

Taylor, A.J., and F.M. Clydesdale. 1987. Potential of oxidised phenolics as food colourants. *Food Chem.* 24: 301–313.

Teixeira, G.H.A., J.F. Durigan, R.E. Alves, and T.J. O'Hare. 2007. Use of modified atmo-sphere to extend shelf life of fresh-cut carambola (*Averrhoa carambola* L. cv. 'Fwang Tung'). *Postharvest Biol. Technol.* 44: 80–85.

Toivonen, P.M.A. 2008a. Application of 1-methylcyclopropene in fresh-cut/minimal process-ing systems. *HortScience.* 43: 102–105.

Toivonen, P.M.A. 2008b. Influence of harvest maturity on cut-edge browning of 'Granny Smith' fresh apple slices treated with anti-browning solution after cutting. *LWT—Food Sci. Technol.* 41: 1607–1609.

Tucker, G.A. 1983. Introduction: Respiration and energy, pp. 1–51. In: G.B. Seymour, J.E. Taylor, and G.A. Tucker (eds.). *Biochemistry of Fruit Ripening.* Chapman and Hall, London.

Tucker, G.A. 1993. Introduction: Respiration and energy, pp. 1–51. In: G.B. Seymour, J.E. Taylor, and G.A. Tucker (eds.). *Biochemistry of Fruit Ripening.* Chapman and Hall, London.

Tucker, G.A., and D. Grierson. 1987. Fruit ripening, pp. 265–318. In: D.D. Davis (ed.). *The Biochemistry of Plants—A Comprehensive Treatise.* Academic Press, San Diego, CA.

Vandekinderen, I., F. Devlieghere, B. De Meulenaer, K. Veramme, P. Ragaert, and J. Van Camp. 2008. Impact of decontamination agents and a packaging delay on the respira-tion rate of fresh-cut produce. *Postharvest Biol. Technol.* 49: 277–282.

Vangronsveld, J., H. Clijsters, and M. Poucke. 1988. Phytochrome-controlled ethylene biosyn-thesis of intact etiolated bean seedlings. *Planta.* 174: 19–24.

Vargas, M., C. Pastor, A. Chiralt, D.J. McClements, and C. Gonzalez-Martinez. 2008. Recent advances in edible coatings for fresh and minimally processed fruits. *Crit. Rev. Food Sci. Nutr.* 48: 496–511.

Varoquaux, P., G. Albagnac, and B. Gouble. 1995. Physiology of minimally processed fresh vegetables. Australasian Postharvest Hort. Conf., pp. 445–450.

Vigneault, C., T.J. Rennie, and V. Toussaint. 2008. Cooling of freshly cut and freshly harvested fruits and vegetables. *Stewart Postharvest Rev.* 4: 1–10.

Vilas-Boas, E.V.D.B., and A.A. Kader. 2007. Effect of 1-methylcyclopropene (1-MCP) on softening of fresh-cut kiwifruit, mango and persimmon slices. *Postharvest Biol. Technol.* 43: 238–244.

Viña, S.Z., A. Mugridge, M.A. García, R.M. Ferreyra, M.N. Martino, A.R. Chaves, and N.E. Zaritzky. 2007. Effects of polyvinylchloride films and edible starch coatings on quality aspects of refrigerated brussels sprouts. *Food Chem.* 103: 701–709.

Wang, C.Y., and D.O. Adams. 1982. Chilling-induced ethylene production in cucumbers (*Cucumis sativus* L.). *Plant Physiol.* 69: 424–427.

Wang, H., H. Feng, and Y. Luo. 2007. Control of browning and microbial growth on fresh-cut apples by sequential treatment of sanitizers and calcium ascorbate. *J. Food Sci.* 72: M001–M007.

Wasternack, C., I. Stenzel, B. Hause, G. Hause, C. Kutter, H. Maucher, J. Neumerkel, I. Feussner, and O. Miersch. 2006. The wound response in tomato: Role of jasmonic acid. *J. Plant Physiol.* 163: 297–306.

Watada, A., N. Ko, and D. Minott. 1996. Factors affecting quality of fresh-cut horticultural products. *Postharvest Biol. Technol.* 9: 115–125.

Watada, A.E. 1986. Effects of ethylene on the quality of fruits and vegetables. *Food Technol.* 40: 82–85.

Watada, A.E., and L. Qi. 1999. Quality control of minimally-processed vegetables. *Acta Hort.* 483: 209–220.

Yang, S.F. 1985. Biosynthesis and action of ethylene. *HortScience.* 20: 45.

Yang, S.F., and N.E. Hoffman. 1984. Ethylene biosynthesis and its regulation in higher plants. *Ann. Rev. Plant Physiol.* 35: 155–189.

Yang, Z., Y. Zheng, and S. Cao. 2009. Effect of high oxygen atmosphere storage on quality, antioxidant enzymes, and DPPH-radical scavenging activity of Chinese bayberry fruit. *J. Agric. Food Chem.* 57: 176–181.

Zarembinski, T.I., and A. Theologis. 1994. Ethylene biosynthesis and action: A case of conservation. *Plant Molecul. Biol.* 26: 1579–1597.

Zheng, Z.-P., K.-W. Cheng, J.T.-K. To, H. Li, and M. Wang. 2008. Isolation of tyrosinase inhibitors from *Artocarpus heterophyllus* and use of its extract as antibrowning agent. *Molecul. Nutr. Food Res.* 52: 1530–1538.

5 Factors Affecting Sensory Quality of Fresh-Cut Produce

John C. Beaulieu

CONTENTS

5.1   INTRODUCTION

Fresh-cut produce is the fastest growing food category in U.S. supermarkets. Sales trends for fresh-cut salads clearly indicate that consumers will pay for fresh-cut produce, if quality and convenience are perceived to be better than or equal to uncut product. The most important driving force behind fresh-cut product purchases is convenience (Ragaert et al., 2004). Due to commercial difficulties in testing product quality, it is often assumed that "if it looks good, it tastes good." Unfortunately, quality of intact vegetables and fruits is often determined almost exclusively on appearance, sometimes to the exclusion of flavor and texture. Consumers often buy the first time based on appearance, but repeat purchases are driven by expected quality factors such as flavor and texture (Beaulieu, 2006b; Waldron et al., 2003).

It is well recognized that minimal processing and wounding have profound physiological effects on plant tissue, and most consequences of cutting are physiologically deleterious, especially in sensitive fruits (Karakurt and Huber, 2007).

Fresh-cut salads and vegetables are enjoying success in the marketplace because certain important negative effects of processing have been ameliorated or addressed sufficiently. On the other hand, the fresh-cut fruit market has not experienced phenomenal growth similar to its companion fresh-cut salad and vegetable market due to numerous physiological and biochemical phenomena. Unfortunately, and especially in the fresh-cut fruit arena, growth may be limited by low repeat sales and overall consumer dissatisfaction. This is predominately due to the inherent fragile nature of ripe fruit with removed epidermis, and also has been attributed to likely flavor imbalance, or loss (Beaulieu, 2006a).

The goal of this review is to convey recent findings and results related to agents affecting sensorial acceptance by the consumer as well as traditional and recent trends in fresh-cut fruit and vegetable processing that allow maintaining or extending sensory shelf life. Much background material that could be included in this review of sensory quality in fresh-cut produce was previously published in several reviews (Barrett et al., 2010; Beaulieu and Baldwin, 2002; Beaulieu and Gorny, 2004; Brecht et al., 2004; Forney, 2008; Hodges and Toivonen, 2008). Subsequently, only more current literature and significant items related to sensory qualities of fresh-cut produce will be addressed. Focus will be placed on fresh-cut fruits and include only highly relevant vegetable and salad information. Little to no microbiological treatment effects will be covered (please refer to Chapters 3 and 8), unless volatile and sensory appraisal was a main objective of the study.

5.2   FRESH-CUT FRUIT AND VEGETABLE QUALITY COMPONENTS

For fruits and vegetables, the characteristics imparting the most important quality factors may be described by several different attributes, such as color (appearance), aroma and taste (flavor), texture, and nutritional value. Consumers generally evaluate a product based on four attributes in the order specified above. Specifically, visual cues are first, followed by aroma, taste, and texture. Aroma refers to the orthonasal (sniff) smell of a fruit or vegetable product, whereas flavor includes both aroma and retronasal (mouth) taste. Texture is very broad and encompasses everything from squeeze to handling firmness to sensory attributes upon chewing. As chewing proceeds, the perception of textural quality changes as products generally become softer with ripening and storage. Nutritional value is an extremely important quality factor that is impossible to see, taste, or feel. Although nutritional value is a hidden attribute, this quality factor is becoming increasingly valued by consumers, scientists, and the medical profession as phytonutrients, functional foods, and antioxidants become more appreciated. This chapter will touch upon a few salient studies related to nutritional aspects involving sensory analysis. However, the reader is referred to Chapter 6 for a more comprehensive overview.

5.2.1 COLOR AND APPEARANCE

An important consideration during fresh-cut fruit processing is the preservation of normal tissue color and control of discoloration or surface browning. Oxidative browning is usually caused by the enzyme polyphenol oxidase (PPO) that, in the presence of O_2, converts phenolic compounds into dark-colored pigments. Extent of browning varies markedly by commodity, cultivar, and even growing conditions. Several commercial and research strategies used to reduce PPO-mediated discoloration have been outlined elsewhere (Beaulieu and Gorny, 2004; Garcia and Barrett, 2002; Hodges and Toivonen, 2008). Often, a simple visual, and hopefully objective, appraisal of quality and appearance is desired. A comprehensive manner by which trained persons can uniformly perform subjective fresh-cut quality appraisals has been established by well characterized categorized hedonic scales for mango (Beaulieu and Lea, 2003) and cantaloupe (Beaulieu, 2005). A detailed atlas of color quality scales for both whole and fresh-cut produce is also available (Kader and Cantwell, 2007).

5.2.2 SENSORIAL ASPECTS

Fresh-cut vegetables and salads have great consumer appeal due to their convenience, flexibility of use, and possibly because their desirable flavor often comes about via condiments (croutons, spices, or dressing) and cooking, or because numerous products make up a medley mixture. Nonetheless, certain vegetables have specific characteristic aromas (e.g., S-compounds) that must be perceived within their threshold concentration by consumers. On the other hand, consumer acceptance of fresh-cut fruits most often relies upon the inherent flavor and textural quality of the product, seldom with accompaniments. Several examples illustrate clearly that fresh-cut products (e.g., cantaloupe, carrot, honeydew, orange, tomato, watermelon) lose flavor quality prior to visual quality (Abbey et al., 1988; Barry-Ryan and O'Beirne, 1998; Beaulieu et al., 2004; Hakim et al., 2004; Qi et al., 1998; Rocha et al., 1995). Recently, fruit medleys have gained popularity, as well as dips, spreads, and caramel or chocolate dips for certain fresh-cut products. Such commercial tactics have been implemented to offset slow market growth for fresh-cut fruits, likely due to consumers' apprehension of repeatedly purchasing products that lack consistent flavor.

5.2.2.1 Flavor and Aroma

Consumers often buy the first time based on appearance or impulse. However, intrinsic quality factors such as flavor and texture drive repeat sales. Flavor is comprised of taste and aroma relating mainly to sugars (fructose, glucose, and sucrose), salts, acids (citric, malic, and tartaric), bitter compounds (alkaloids and flavanoids), and volatile components (Baldwin et al., 2007; DeRovira, 1996). Aroma compounds are detected by olfactory nerve endings in the nose (e.g., parts per billion). In contrast, taste is the detection of nonvolatile compounds by several types of receptors in the tongue (e.g., parts per hundred). Flavor is a very complex "trait" that is difficult to analyze. Several to hundreds of volatile and semivolatile compounds may be responsible for the characteristic aroma of a fruit or vegetable. Some of these essential compounds will be present in minute concentrations (parts per billion) that require expensive and complex analytical equipment. Then, determining which volatile components actually impact the flavor of a fruit or vegetable requires method-dependent gas chromatography (GC) or gas chromatography/mass spectrometry (GC-MS) separation, gas chromatography/olfactometer (GC-O), or sensory, which again, are complicated, labor intensive, and expensive. Instrumental techniques must determine compound concentrations in a commodity, accompanied by sensory measurement of odor threshold or flavor units to give a measure of the contribution of that specific compound. Subsequently, flavor detection, importance, and considerations are often overlooked within the industry due to practicality.

Flavor volatiles in fruits and vegetables arise from numerous biosynthetic pathways (carbohydrates, amino acids, fatty acids, oxidations, and ß-oxidation) and include a wide range of molecular weight alcohols, aldehydes, esters, furanes, glucosinolates, ketones, lactones, nitrogen- and sulfur-containing compounds, terpenes, and other compounds. Volatile esters often make the major contribution to impact aromas in fruits such as apple, banana, pear, strawberry, and melon. Flavor loss during fresh-cut storage can proceed as a direct consequence of senescence and may be driven by catabolic, metabolic, and diffusional mechanisms (Beaulieu, 2006a, 2007; Forney, 2008). Flavor changes that occur during fresh-cut storage also may affect the consumers' experience and decision as to buy again or not. Although fresh-cut per se was not explored, in-depth reviews of fruit and vegetable flavor and sensory appraisal were recently published (Baldwin et al., 2007; Brückner and Wyllie, 2008). Several concepts and themes presented therein, as well as in Barrett et al. (2010), are relevant and applicable in fresh-cuts.

Several fresh-cut studies have illustrated that a product's sensory attributes may decline prior to the physiological appearance declines. A hypothetical postharvest quality figure was presented by Kader (2003) and was adapted for inclusion herein to generalize how flavor quality compares with visual and textural quality in fresh-cut fruits (Figure 5.1). As examples, an informal taste panel determined that fresh-cut honeydew melon stored in air at 5°C for 6 days lacked acceptable textural characteristics and had flat flavor (Qi et al., 1998). Fresh-cut orange segments having acceptable appearance after 14 days storage were found to have unacceptable flavor quality after 5 days at 4°C (Rocha et al., 1995). Likewise, undesirable flavor was the limiting factor in sliced wrapped watermelon stored 7 days at 5°C, even though aroma was still acceptable and microbial populations were not problematic until after day 8 (Abbey et al., 1988). Subsequently, modified atmosphere packaging (MAP) has also been investigated as a means to prolong the flavor and aroma qualities in fresh-cut produce.

5.2.2.2 Texture

Consumers generally cite flavor as the most important quality attribute for fruits and vegetables, but textural defects and the interaction of flavor and texture are more likely to cause rejection of a fresh product (Harker et al., 2003). Texture is a critical

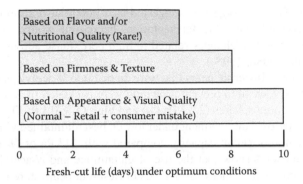

Based on Flavor and/or
Nutritional Quality (Rare!)

Based on Firmness & Texture

Based on Appearance & Visual Quality
(Normal – Retail + consumer mistake)

0   2   4   6   8   10

Fresh-cut life (days) under optimum conditions

FIGURE 5.1 Hypothetical fresh-cut fruit life based on flavor and nutritional quality contrasted against texture and appearance attributes. (Adapted from Kader A.A. 2003. A perspective on postharvest horticulture (1978–2003). *HortSci*. 38(5):1004–1008.)

quality attribute that helps both the industry and consumer determine the acceptability of cut fruits and vegetables. The term *texture* encompasses both structural and mechanical properties of a food, as well as the sensory components in the hand and mouth, all of which can be measured by several destructive and nondestructive instrumental or objective methods (Abbott and Harker, 2004; Barrett et al., 2010; Bourne, 2002). Texture evaluation is most commonly performed by application of a destructive force via puncture or compression. Consistent firmness loss in stored fresh-cut cantaloupe has been documented using a puncture test (Gil et al., 2006; Luna-Guzmán and Barrett, 2000) and compression test (Beaulieu et al., 2004) on the texture analyzer. Due to the empirical nature of these tests, they do not provide an understanding of food microstructure or force-deformation and failure mechanisms at the cellular level (Barrett et al., 2010). The three-dimensional network of plant cell walls is still somewhat unresolved, but it largely dictates the perception of consistency, smoothness, and juiciness, in fruit and vegetable tissues (Waldron et al., 2003). Consumer and panel testing indicates that consumers or panelists are oftentimes more sensitive to small differences in texture than flavor (Beaulieu et al., 2004; Shewfelt, 1999).

5.2.3 NUTRITIONAL CONTENT

After preparing fresh cuts (wounding), the antioxidant capacity of fruit or vegetable tissue may increase (carrot, celery, lettuce, parsnip, purple-flesh potato, sweet potato, and white cabbage) during storage (Kang and Saltveit, 2002; Reyes et al., 2007; Reyes and Cisneros-Zevallos, 2003) or decrease (melon, potato, red cabbage, and zucchini) (Oms-Oliu et al., 2008b; Reyes et al., 2007). During investigations with different initial in-package O_2 and CO_2 concentrations in fresh-cut 'Piel de Sapo' melon, excessively low O_2 (2.5 kPa) and high CO_2 (7 kPa) concentrations stimulated the synthesis of phenolic compounds, reduced vitamin C, and increased peroxidase activity (Oms-Oliu et al., 2008b). This could have been due to the fact that internal package concentrations of O_2 fell below 1% by 9 days storage.

Ultimately, 70 kPa O_2 storage was suggested as a potential optimum atmosphere because it prevented anaerobic fermentation, decreased wound-induced stress, and reduced deteriorative changes related to high peroxidase activity in cut tissue (Oms-Oliu et al., 2008b, 2008c).

Fresh-cut 'Flor de Invierno' pears that were dipped into N-acetylcysteine plus glutathione (0.75% w/v, each) to prevent browning were packaged in active flush MAP delivering low O_2 atmospheres (2.5 kPa O_2) and high O_2 (70 kPa O_2) atmospheres. The synergistic antioxidant treatment under low O_2 best maintained vitamin C, chlorogenic acid, and antioxidant capacity compared with 70 kPa of O_2 (Oms-Oliu et al., 2008a). The results show that the use of glutathione and N-acetylcysteine also enhanced the formation of phenylpropanoids in fresh-cut pears stored under the low O_2 atmosphere (Oms-Oliu et al., 2008a). Similar increases in antioxidant activity have been achieved in fresh-cut apples treated with 1% ascorbic acid and 1% citric acid for 3 min (Cocci et al., 2006).

Vitamin C and carotene degraded very little during short-term (about 1 week) storage in some fresh-cut fruits (Wright and Kader, 1997a, 1997b). However, ascorbic acid and vitamin C content generally decreased after cutting, especially in longer-term fresh-cut storage (Beaulieu and Lea, 2007; Gil et al., 2006; Oms-Oliu et al., 2008b; Reyes et al., 2007). Vitamin C loss after 6 days storage in clamshell containers at 5°C, were 5% in mango, strawberry, and watermelon pieces, 10% in pineapple pieces, 12% in kiwi slices, and 25% in cantaloupe cubes (Gil et al., 2006). Vitamin C content declined significantly after roughly 1 week in stored (4°C) fresh-cut cantaloupe, and the decrease was independent of initial processing maturity (Beaulieu and Lea, 2007). Carotenoids decreased up to 25% in pineapple pieces, and 10% to 15% in cantaloupe, mango, and strawberry pieces after 6 days at 5°C (Gil et al., 2006). Ascorbic acid dips reduced browning and deterioration in fresh-cut 'Ataulfo' mango and maintained vitamin C and ß-carotene contents during 21 days fresh-cut storage at 5°C (González-Aguilar et al., 2008). Total ascorbic acid in processed iceberg lettuce is retained significantly better when there is less cell damage during cutting (Barry-Ryan and O'Beirne, 1999).

5.3 FACTORS AND AGENTS AFFECTING FRESH-CUT SENSORY QUALITY

In reality, research and reviews have illustrated clearly that several important preharvest factors (genetics, cultural practices, environment, harvest maturity) can predict or determine final fruit and vegetable flavor quality (Beaulieu and Baldwin, 2002). However, in practicality, the integration of all aspects from field to fork including selection of variety, horticultural practices, harvest operations, shipping, cooling, washing, processing operations, packaging, cold chain integrity, distribution, marketing, and shelf life of the fresh-cut product all ultimately determine the sensory quality and consumer appraisal. Subsequently, because vertically integrating the whole chain and managing it without flaws is a major challenge, many processors strive to simply ensure the best possible quality produce arriving at their facility.

Then, they process (after passing from insubstantial to stringent in-house quality control checks), package, and distribute what they have. Intrinsic fresh-cut qualities may therefore not always match one's expectation for optimum flavor, aroma, and sensory quality. Use of the "best" cutting and processing treatments, along with packaging and distribution, essentially becomes a game of maintaining the initial quality for the longest rational feasible time period to insure sales, safety, and a satisfied consumer. To this end, there are several factors that come into play regarding the sensorial quality of fresh-cuts. These include variety, maturity, season, precutting treatments (ethanol, 1-MCP, hot water, insect sanitation, heat shock), processing technique and treatments or aids (antibrowning, firmness retention, 1-MCP, edible coatings, microbiological sanitation), combined and synergistic quality retention treatments or hurdles, packaging (containers, form and fill, deli cups, active flush and passive MAP, film-sealed trays, etc.), and temperature management. Within this review, attention will be placed on important color, appearance, texture, flavor and aroma attributes, and nutritional content in recently reported literature for fresh-cuts with the aforementioned technologies.

5.3.1 WHOLE FRUIT OR PRECUTTING TREATMENTS

Treatment efficacy for 1-MCP is rather variable and depends on several factors, such as concentration of 1-MCP used, storage condition and duration, climacteric versus nonclimacteric, and maturity of fruit at time of application. Commercial use for climacteric fruits requires a careful balancing act between cultivar-dependent 1-MCP concentration and exposure periods that will delay but not inhibit ripening (Blankenship and Dole, 2003; Calderon-Lopez et al., 2005; Watkins, 2008). Some additional precutting treatments that have been explored are ethanol vapor and heat or hot water treatments to intact commodities.

Application of 1-MCP to partially ripe fruit before cutting generally suppresses firmness and color loss in fresh-cut kiwi, mango, and persimmon (Mao et al., 2007; Vilas-Boas and Kader, 2007). Fresh-cut apples exposed to 1-MCP have decreased ethylene production, respiration, softening, color change, and synthesis of aroma/volatile compounds (Bai et al., 2004; Calderon-Lopez et al., 2005; Jiang and Joyce, 2002; Perera et al., 2003). Detailed volatile and genetic analysis indicates that marketing of 1-MCP-treated fruit shortly after treatment might result in the delivery of fruit to the consumer with poor likelihood of ester volatile recovery (Ferenczi et al., 2006; Kondo et al., 2005), even when browning is inhibited with NatureSeal™ (Rupasinghe et al., 2005). However, in longer-term storage of apples, volatile recovery has been observed but is variety dependent (Watkins, 2008). Likewise, treatment of whole pears with 1-MCP reduced the production of aromatic volatiles by the fruit but did not dampen the ability to recover volatile synthesis and sensory (texture) acceptability after long-term storage (Moya-León et al., 2006). However, fresh-cuts were not a part of this study. Maturity and variety-dependent effectiveness of 1-MCP is therefore problematic and must be critically evaluated with regard to delivering organoleptically acceptable fresh-cuts.

5.3.2 MATURITY

Both the maturity at harvest and the ripeness stage at cutting will affect the postcutting quality and shelf life of fresh-cut fruit products. Because many fruit are picked before they are fully ripe, the question arises as to what maturity should climacteric fruit be used for fresh-cut processing to optimize product shelf life and eating quality? In order to withstand mechanical damage during postharvest handling and deliver excellent visual quality and acceptance by retailers and consumers, fresh-cuts may be produced with firm fruit that was harvested slightly immature. Consumer preferences within a single fruit type are often cultivar dependent and often defined by the stage of ripeness (Harker et al., 2003). Firmer fruit tends to be less ripe and thus tastes more acidic or "sour" and has a volatile profile rich in aldehydes delivering green grassy aroma and flavor notes. On the other hand, softer fruit is generally much riper, has a lower acidity, and has a volatile profile dominated by esters that deliver typical fruity aroma and flavor notes.

Although bland flavor was reported by the end of storage, sliced, mature-green, tomato fruit ripened normally and attained comparable eating quality to the fruit that was sliced after the whole fruit ripened (Mencarelli and Saltveit, 1988). Yet most fruits do not follow this pattern. Generally, there appears to be a detrimental tradeoff between firmness and acceptable volatiles and flavor/aroma attributes in fresh-cut fruits prepared with less ripe fruit (Beaulieu et al., 2004; Beaulieu and Lea, 2003; Gorny et al., 1998; Gorny et al., 2000; Soliva-Fortuny et al., 2004). For example, mango fruit processed at the firm-ripe stage had lower aroma/smell scores and less ripe odors compared to soft-ripe fruit, yet firm-ripe wedges held under passive MAP for 11 days versus 7 days for soft-ripe wedges (Beaulieu and Lea, 2003).

In sliced pear and peach fruit, some ripening-related phenomena will occur such as softening, but other ripening-related processes such as flavor development and texture seem aberrant if the fruits are processed at an excessively immature stage (Dong et al., 2000; Gorny et al., 1998; Gorny et al., 2000; Mencarelli et al., 1998). Mature green apple slices maintained their initial firmness and color better than partially ripe and ripe slices (Soliva-Fortuny et al., 2002b), and slices of slightly underripe 'Conference' pears exhibited less browning and softening than those from more ripe fruit (Soliva-Fortuny et al., 2004). Also, pear slices processed at partially ripe maturity were the most suitable to obtain fresh-cut products, as opposed to mature green and ripe fruit. Therefore, in climacteric fruits, initial fruit firmness is a good indicator of necessary fruit ripeness required for optimum postcutting quality.

Harvest maturity significantly affects the level of flavor volatiles recovered in fresh-cuts from soft-ripe versus firm-ripe mangos (Beaulieu and Lea, 2003), and 1/4-, 1/2-, 3/4-, and full-slip 'Sol Real' cantaloupe (Beaulieu, 2006b). Harvest maturity significantly affects the level of flavor volatiles extracted from 'Athena' and 'Sol Real' cantaloupe (Beaulieu, 2006b; Beaulieu and Grimm, 2001). Volatile compounds followed highly linear trends by which increasing maturity was associated with increased total compounds, total esters, nonacetate esters, aromatic (benzyl) and sulfur compounds, and decreasing levels of acetates and aldehydes, and maturity-associated volatile trends were conserved through fresh-cut storage (Beaulieu, 2006b).

Cantaloupe fruit harvested at different maturity deliver stored cubes differing significantly in postharvest quality, flavor, sensory, and textural attributes (Beaulieu et al., 2004; Beaulieu, 2006b; Beaulieu and Lancaster, 2007; Beaulieu and Lea, 2007). In general, cubes prepared with less mature fruit, that were excessively firm, lacked flavor volatiles (Beaulieu, 2006b) and had less acceptable sensory scores (Beaulieu et al., 2004). High-quality fresh-cut cantaloupe can be prepared with fruit harvested from at least 1/2-slip, with 3/4-slip optimum, but not from 1/4-slip ripeness (Beaulieu and Lancaster, 2007).

The ratio of the nonacetate esters to acetate esters changed uniformly during fresh-cut cantaloupe storage, which was independent of initial processing maturity (Beaulieu, 2006b). The ester:acetate ratio also changed in a similar manner in stored fresh-cut apple and honeydew (Beaulieu, 2006a). In fresh-cut cantaloupe, further data analysis led to the finding that several correlations with flavor compounds and classes of compounds existed between several quality, textural, and sensory attributes (Beaulieu and Lancaster, 2007). However, descriptive sensory analysis was not correlated with threshold volatile concentrations. Instead, we conducted analyses of variance (year × maturity × day), generated correlation coefficients (e.g., Table 5.1), and visualized the correlation (Figure 5.2 and Figure 5.3) using h-plots (Trosset, 2005). An h-plot displays the Pearson's product-moment correlation coefficient where clusters of positively correlated variables correspond to clusters of radii, and groups of clusters separated by 180° are negatively correlated. Some of the strongest sensory and physiological correlations were attained between sweet taste and a*, b*, a*/b*, and %Brix. Cucurbit, denseness, and water-like were negatively correlated with a*, b*, a*/b*, and %Brix, and positively correlated with L* and desiccation. Young's modulus (firmness) delivered the strongest correlations in the study. It was highly correlated with relative percentage aromatic acetates, total benzyl compounds, and acetates, and negatively correlated with nonacetate esters. In lieu of exhaustively comparing tabulated data for positive and negative correlations (Beaulieu and Lancaster, 2007), they are visualized in Figures 5.2 and 5.3, via tightly or opposing clustered radii. Some of the relationships were even more profound if data were analyzed by pairing daily maturity means, as presented in Beaulieu and Lancaster (2007); however, interaction effects limited the reliability of that interpretation. Evaluating sensory, volatility, and quality data as such would be highly valuable in fresh-cut studies using MAP and synergistic quality retention treatments.

5.3.3 Processing Treatments

Essentially, most fresh-cut processing treatments and dips are dissolved in water or sprayed onto surfaces for antibrowning when phenolic content and PPO are problematic (e.g., apple, lettuce, pear, zucchini), antimicrobial in an attempt to ensure food safety, and as firming agents. A review of possible replacements for chlorine use during processing, as well as alternative washing regimes (mainly antimicrobial) was recently published (Rico et al., 2007), and this topic will not be covered. In almost 20 years in fresh-cut research, several compounds and classes of compounds have been tested alone and in combination plus or minus additional hurdles (e.g., calcium, CA, MAP, thermal, ultraviolet [UV], sanitation). Earlier quality retention treatments

TABLE 5.1

Correlations[a] between Sensory Attributes and Physiological Measurements in Fresh-Cut Cantaloupes Harvested at Various Maturities and Stored up to 14 Days at 4°C (Year by Maturity by Day Means)

Parameters	FTY[b]	CRB	SWA	WTR	SWT	AST	WET	HRD	COH	DEN
L*[c]	-0.55/0.00	**0.73/0.00**	0.19/0.21	0.67/0.00	-0.69/0.00	0.50/0.00	-0.16/0.30	0.54/0.00	0.57/0.00	0.65/0.00
a*	**0.72/0.00**	-0.57/0.00	-0.35/0.02	**-0.76/0.00**	**0.83/0.00**	-0.46/0.00	-0.27/0.07	-0.20/0.18	-0.23/0.12	-0.69/0.00
b*	0.63/0.00	-0.43/0.00	-0.29/0.05	-0.67/0.00	**0.71/0.00**	-0.35/0.02	-0.16/0.29	-0.16/0.28	-0.11/0.45	-0.60/0.00
a*/b*	0.69/0.00	-0.59/0.00	-0.33/0.03	**-0.73/0.00**	**0.81/0.00**	-0.47/0.00	-0.30/0.05	-0.22/0.15	-0.28/0.06	-0.67/0.00
°Brix	0.68/0.00	-0.63/0.00	-0.43/0.00	**-0.84/0.00**	**0.78/0.00**	-0.50/0.00	-0.18/0.21	-0.17/0.23	-0.48/0.00	**-0.73/0.00**
Firmness	-0.01/0.97	0.38/0.01	-0.12/0.42	-0.00/0.99	-0.00/1.00	-0.13/0.38	-0.61/0.00	-0.64/0.00	0.52/0.00	0.12/0.41
Color	-0.21/0.14	0.44/0.00	0.18/0.21	0.34/0.02	-0.27/0.06	0.18/0.20	-0.32/0.03	0.52/0.00	0.49/0.00	0.39/0.01
Edges	0.52/0.00	-0.21/0.14	-0.21/0.14	-0.51/0.00	0.58/0.00	-0.49/0.00	-0.51/0.00	0.27/0.06	0.04/0.77	-0.42/0.00
Desiccation	-0.47/0.00	0.62/0.00	0.23/0.03	0.62/0.00	-0.59/0.00	0.36/0.00	-0.07/1.00	0.54/0.00	0.54/0.00	0.60/0.00
Average	-0.08/0.57	0.36/0.01	0.05/0.73	0.21/0.16	-0.16/0.27	0.09/0.53	-0.38/0.01	0.55/0.00	0.44/0.00	0.25/0.08
Bioforce	-0.27/0.23	0.57/0.01	-0.59/0.01	-0.40/0.07	-0.18/0.43	-0.12/0.60	-0.68/0.00	**0.79/0.00**	0.20/0.39	0.49/0.03
Bioarea	-0.29/0.02	0.58/0.01	-0.58/0.01	-0.42/0.06	-0.22/0.34	-0.13/0.58	-0.66/0.00	**0.79/0.00**	0.18/0.44	0.47/0.03
Slope	-0.23/0.31	0.54/0.01	-0.55/0.01	-0.38/0.09	-0.07/0.77	-0.13/0.56	**-0.70/0.00**	**0.80/0.00**	0.26/0.25	0.49/0.02
TA	-0.24/0.30	0.54/0.01	-0.56/0.01	-0.38/0.09	-0.09/0.70	-0.13/0.56	**-0.71/0.00**	**0.80/0.00**	0.24/0.29	0.48/0.03
YM	0.09/0.71	0.25/0.27	-0.32/0.16	0.02/0.94	0.28/0.22	-0.16/0.49	**-0.71/0.00**	0.58/0.01	0.42/0.06	0.23/0.32

[a] Data reflect the coefficient of correlation, r, followed by the p-value for the test of the null hypothesis, H_o: $r = 0$. Significant correlations are indicated with shaded boxes, and correlations ≥±0.70 are indicated by bold font.

[b] FTY, fruity; CRB, cucurbit; SWA, sweet aromatic; WTR, water-like; SWT, sweet; AST, astringent; WET, wetness; HRD, hardness; COH, cohesiveness; and DEN, denseness.

[c] Instrumental texture measures: force (N) at the bioyield point = bioforce; bioyield area (N*s) = bioarea; slope = force deformation curve (N s^{-1}) until first inflection point (nondestructive elastic deformation); total force area (N*s) = TA; and Young's modulus = YM.

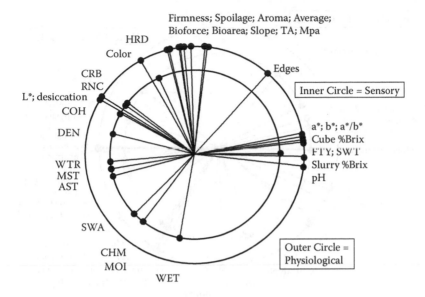

FIGURE 5.2   Sensory versus physiological correlations in fresh-cut cantaloupe ('Sol Real') stored 14 days at 4°C (year by maturity by day means). Sensory attributes, instrumental firmness measures, and subjective quality abbreviations are according to Table 5.1. CHM, MST, RNC, and MOI attributes had no correlation with any physiological measures.

were categorized as physical or chemical and have been summarized for fresh-cut fruits (Beaulieu and Gorny, 2004). The reader should also consult Chapter 9 for more information. Several commercial products are also available and will not be discussed. Recently, treatments have become diversified and synergistic approaches have begun to address quality improvement via combined physical and chemical mechanisms. Herein, focus is placed on recent or novel fresh-cut findings involving flavor and aroma.

5.3.3.1   Antibrowning and Firmness Retention

A vast amount of work has been accomplished in the fresh-cut arena revolving around maintaining or improving storability via technologies and chemical applications (mainly antibrowning, antimicrobial, and firmness agents) (Beaulieu and Gorny, 2004; Brecht et al., 2004). Several investigators are currently focusing on synergistic treatments with MAP, 1-MCP, and coatings and dips. Subsequently, several items will be reviewed in those sections. Several treatments used to reduce browning are complementary in that they contain calcium or salts, which may improve firmness retention. Unfortunately, some treatments used to reduce enzymatic browning or improve texture can impart off-flavors. Calcium chloride concentrations above 0.5% imparted detectable off-flavor in cantaloupe slices, whereas calcium lactate improved firmness without bitter flavor (Luna-Guzmán and Barrett, 2000). Off-odors were detected in potato strips when treated with sodium sulfite and held in passive MAP for 14 days at 4°C (Beltrán et al., 2005), and off-flavor occurred in 'Bartlett' and

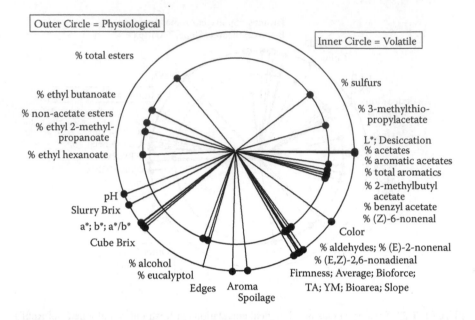

FIGURE 5.3 Physiology versus volatile correlations in fresh-cut cantaloupe ('Sol Real') stored 14 days at 4°C (year by maturity by day means, Y × M × D). Fifty-five compounds were positively identified and quantified, according to Beaulieu and Lancaster (Beaulieu J.C., and V.A. Lancaster. 2007. Correlating volatile compounds, quality parameters, and sensory attributes in stored fresh-cut cantaloupe. *J. Agric. Food Chem.* 55(23):9503–9513). Instrumental firmness measures and subjective quality abbreviations are according to Table 5.1. Color measures a* and b*, and subjective scores for edges, aroma, and spoilage had no correlation with any volatiles.

'Anjou' pear slices only 1 day after treatment with 0.01% 4-hexylresorcinol (Dong et al., 2000). Antibrowning dip solutions (0.1 mol/L isoascorbic acid, and 0.05 mol/L ascorbic acid or acetyl cysteine) applied to pineapple slices resulted in up to 14 days storage at 10°C without off-flavors, even though acetaldehyde and ethanol concentrations increased (González-Aguilar et al., 2004).

5.3.3.2 Ripening Inhibition

A synergistic effect was observed when 1-MCP was combined with $CaCl_2$ (1%) + CA (3 kPa O_2 + 10 kPa CO_2) in strawberry wedges, as it slowed the softening, loss of appearance quality, changes in TA, and microbial growth resulting in a shelf life of 9 days (versus 6 days for the control) at 5°C (Aguayo et al., 2006). Fresh-cut grapes with stems on (1 to 2 mm) that received a hot water dip (55°C for 5 min) had superior quality, sensory scores almost equal to controls, and very little decay evidenced by the lowest population of yeast and mold, lactic acid bacteria, and total mesophilic aerobic bacteria (Kou et al., 2007). Even though these results are very promising, it may not be commercially feasible to manually cut individual berries from grape bunches prior to treating and packaging. Informal sensory attributes (firmness and appearance) were maintained for 6 days in 1-MCP-treated cut papaya versus 2 to 3 days in the controls held at 5°C, and firmness decreased 50% in 1-MCP-treated slices compared with 75% for control (Ergun et al., 2006).

The storage life of fresh-cut 'Galia' melon held at 5°C, based on firmness and sensory evaluations, was extended by 2 to 3 days by 1-MCP treatment (1 μL L^{-1} for 24 h at 20°C). Water soaking and mesocarp softening were significantly reduced or delayed in 1-MCP-treated fresh-cut tissue (Ergun et al., 2007). However, in additional studies, the effect of 1-MCP treatment on the firmness retention and water soaking of other fresh-cut melon cultivars was inconsistent. Subsequently, application of 1-MCP to muskmelons before fresh-cut processing was not recommended (Jeong et al., 2008). According to Huber (2008), because wounding overrides components of ethylene signaling, the benefits conferred by treating whole fruits with 1-MCP may be considerably dampened in the stored fresh-cut tissues (Calderon-Lopez et al., 2005; Ergun et al., 2006; Ergun et al., 2007; Jeong et al., 2008; Mao et al., 2007).

Certain crops, like watermelon, are nonclimacteric yet highly sensitive to ethylene. In an attempt to ameliorate mixed commodity–related and fresh-cut degradation, low dosage 1-MCP (18 h of 0.5 or 1.0 μL L^{-1}) treatments were administered to whole watermelons prior to ethylene exposure (5 days of 10 μL L^{-1}). 1-MCP prevented ethylene-mediated quality deterioration in fresh-cut watermelon slices stored 12 days under passive MAP at 5°C (Saftner et al., 2007).

Exposing mangoes to more than 20 hours of ethanol vapors resulted in stored (7°C) fresh-cuts with higher firmness and acidity as determined by a sensory panel (Plotto et al., 2006). However, off-flavors developed, and similar to previous research, results were affected by harvest maturity (Beaulieu and Saltveit, 1997) and heat treatment. Nevertheless, lower ethanol exposure duration (8 to 10 h) was suggested as a possible safe microbial control in fresh-cut production (Plotto et al., 2006). Use of ethanol vapor treatment on intact apples, to inhibit ethylene production and reduce decay in fresh-cut slices, extended appearance during shelf life, yet also resulted in altered or off-flavor (Bai et al., 2004). Inhibition of senescence in 8-day-old fresh-cut broccoli florets stored at 10°C was achieved by pretreating intact broccoli with ethanol vapor (6 mL kg^{-1} for 5 h). Tissue concentrations of ethanol and acetaldehyde rose sharply, but ethanol declined sufficiently after 1 day so as to not render unpleasant or ethanol flavors (Han et al., 2006).

Fresh-cut cantaloupe cubes prepared from hot water-treated cantaloupes (76°C for 3 min) had no consistent changes in soluble solids, ascorbic acid, fluid loss, and aroma and appearance scores (Fan et al., 2008). Because the hot water treatment significantly reduced total plate count (TPC) and yeast and mold counts on the rinds and resulted in less TPC on cubes as well, it could become one step or hurdle strategy to improve microbiological safety without deleterious effects on cut melon quality. Mild heat pretreatments of intact early maturity (firm ripe) kiwifruit at temperatures below 45°C during less than 25 min improved the firmness quality; however, there was no benefit observed in soft ripe fruit (Beirão-da-Costa et al., 2006) that would likely have superior flavor and consumer attributes.

5.3.3.3   Edible Coatings and Dips

Edible coatings may eventually enhance microbial safety of fresh-cut produce, as carriers of functional ingredients, such as antimicrobial and antioxidant agents. Several challenges exist regarding creating and optimizing various formulations (Lin and Zhao, 2007; Min and Krochta, 2005; Olivas and Barbosa-Cánovas, 2005). These include insufficient moisture barrier, inefficient coating coverage, poor coating adhesion, and fresh-cut surfaces covered with juice can actually dissolve coatings that are absorbed instead of drying to form a uniform boundary layer. Also, there exist several issues relating to potential off-flavors likely imparted by anaerobic respiration that induces synthesis of ethanol and acetaldehyde with various coating materials, or entrapment of volatiles in sealed tissue (Baldwin et al., 1995, 1999). Apple slices coated with soybean oil emulsion or carboxy methylcellulose lost aroma volatiles considered important to the flavor profile (Bai and Baldwin, 2002). On the other hand, application of 1% polysaccharide carboxy methylcellulose ±0.5% maltodextrin coatings on fresh-cut mango cubes resulted in better volatile retention than uncoated fruit, and taste panels did not detect any difference between treatments (Plotto et al., 2004). The lack of ethanol and acetaldehyde accumulation after roughly 6 to 9 days in three cultivars ('Keitt,' 'Kent,' and 'Ataulfo') of stored (5°C) fresh-cut mango was considered a good indicator of the favorable effect of commercial edible coatings (González-Aguilar et al., 2008).

Malic and lactic acids were incorporated into soy protein plus glycerol films to coat fresh-cut cantaloupes that were stored at 5°C without negatively affecting the sensory properties. Sweetness of cubes coated with soy protein plus glycerol and lactic acid was higher than the noncoated and the soy protein plus malic acid coated samples at 7 days at 5°C. However, other sensory attributes of taste and appearance in the coatings on cubes showed no significant difference on days 7 and 14 (Eswaranandam et al., 2006). Subsequently, there appeared to be no substantial beneficial (e.g., flavor preservation) effects. Fresh-cut 'Fuji' apples treated with ascorbic acid plus N-acetylcysteine (10 gL^{-1}, each) for 1 minute displayed a fivefold increase in peroxide levels, indicative of oxidative damage to the tissue. Further analysis needs to be carried out to determine if overall antioxidant potential is adversely affected (Larrigaudière et al., 2008). All edible coating treatments tested (whey protein concentrate and carrageenan) on coated apples slices stored in passive MAP pouches at 3°C for up to 14 days resulted in higher sensory scores (color, firmness, and flavor) for all quality factors appraised (Lee et al., 2003). Composite coatings of whey proteins (hydrophilic phase) and beeswax or carnauba wax (lipid phase) provided antibrowning effects on fresh-cut apples, and incorporation of antioxidants (ascorbic acid, L-cysteine and 4-hexylresorcinol) into the coating reduced browning compared to the use of the antioxidant alone sensory (Perez-Gago et al., 2005, 2006). However, sensory panels were capable of differentiating samples treated with whey protein–based coatings, and when cysteine was included (Perez-Gago et al., 2006).

Edible coatings made from alginate were recently investigated in various fresh-cut fruits. Apple wedges that were coated with three alginate formulations (alginate, alginate-acetylated monoglyceride-linoleic acid, and alginate-butter-linoleic acid) and stored at 5°C had minimized weight loss, maintained firmness, browning inhibition, and no anaerobic respiration (Olivas et al., 2007). Significant changes were reported for certain important flavor-related apple volatiles that the authors believed could be attributed to linoleic and sorbic acid constituents of the coating becoming metabolized; however, no ancillary sensory appraisals were reported (Olivas et al., 2007). Alginate-, pectin-, and gellan-based edible coatings were applied to fresh-cut 'Piel de Sapo' melon stored under passive MAP 15 days at 4°C, with final sensory appraisal on day 7. The pectin coating best maintained quality attributes (7 days), whereas the gellan coating scored the lowest for odor, color, and taste. In general, edible coatings prevented desiccation and maintained firmness, and pectin-based coating best maintained sensory attributes in fresh-cut melon (Oms-Oliu et al., 2008e). The use of alginate- (2% w/v) or gellan-based (0.5% w/v) coating formulations on fresh-cut papaya cylinders stored at 4°C for 8 days in polyvinyl chloride (PVC) cups with airtight PVC lids was also recently reported. Coatings containing 2% glycerol plus 1% ascorbic acid or 1% glycerol plus 1% ascorbic acid slightly improved water barrier properties and firmness, and the addition of ascorbic acid helped maintain nutritional quality throughout storage (Tapia et al., 2008). However, there was no sensory or aroma property reported.

Edible films also offer potential to incorporate synergistic treatments. Recently, essential oils from cinnamon, palmarosa, and lemongrass were incorporated into edible alginate coatings and evaluated on 'Piel de Sapo' fresh-cut melon as natural antimicrobial substances. This study offered good microbial results; however, there were some troubling firmness and sensory issues. Firmness was significantly affected by the incorporation of lemongrass into the edible coating, and cinnamon oil resulted in lower odor and taste characteristics and was not acceptable. Palmarosa oil (0.3% incorporated into the edible coating) was accepted by panelists, maintained quality, inhibited native flora growth, reduced the *Salmonella enteritidis* population, and was suggested as a promising conservation treatment in fresh-cut melon (Raybaudi-Massilia et al., 2008). Plant-derived essential oils (oregano, cinnamon, and lemongrass) incorporated into apple-based antimicrobial edible films have promising antimicrobial effects (Rojas-Graü et al., 2006). However, little work has been accomplished to determine efficacy on cut fruits and to determine if the films will entrap volatiles.

Peeled baby carrots dipped in xanthan gum solution containing 5% calcium lactate plus gluconate and 0.2% α-tocopherol acetate, stored at 2°C, for up to 3 weeks had improved surface color with no deleterious effects (aside from a slightly "slippery" surface) on taste, texture, fresh aroma, and flavor (Mei et al., 2002). Incorporation of vitamin E (0.4% to 0.8% α-tocopherol) into honey-based vacuum impregnation (VI) solutions was recently used to maintain fresh-cut pear quality. Instrumental color analysis and sensory evaluation indicated that VI-fortified pear slices had significantly higher lightness, lower browning index, and higher consumer acceptance rating than unfortified control after 7 days in clamshell containers at 2°C (Lin et al., 2006). Similar results were also attained for VI-treated fresh-cut apples stored at 3°C for up to 14 days (Jeon and Zhao, 2005). Nonetheless, vacuum infiltration technologies may not be readily accepted in the "fresh-cut" arena. Further details concerning coatings and in-depth information can be found in Chapter 11.

5.3.4 VOLATILE AND SENSORY ODDITIES AND DIFFUSIONAL CONSIDERATIONS

Several similarities are reported in the literature where either increasing anaerobic volatile accumulation or marked ester imbalances in fresh-cut MAP apples, durian, and cantaloupe were not associated with development of off-odors or negatively appraised by sensory or human evaluations (Bai et al., 2004; Beaulieu et al., 2004; Beaulieu, 2006b; Bett et al., 2001; Soliva-Fortuny et al., 2005; Voon et al., 2007). In fresh-cut Keitt and Palmer mangoes stored in passive MAP, overall terpene levels decreased roughly 47% and 79% after 7 and 11 days, respectively, and this was associated with loss of desirable aroma. When evaluated with physiological measures, this rendered fresh-cuts unmarketable after 7 days for soft-ripe, and 11 days for firm-ripe cubes (Beaulieu and Lea, 2003). Interestingly, appraisal of aroma/smell was not affected negatively by accumulation of anaerobic volatiles (acetaldehyde, ethanol, ethyl acetate, and ethyl butanoate) in the package headspace atmosphere.

Fresh-cut durian (arils) that were overwrapped with low-density polyethylene (LDPE) film and held at 4°C lost 53% of total volatiles, including 77% of esters after 7 days of storage, but fruity and sweet notes could still be perceived by panelists using quantitative descriptive analysis, even when ester concentrations were not detectable. Only by 14 days, the loss of most esters correlated well with the decrease in intensities of fruity aroma perceived by panelists (Voon et al., 2007).

Fresh-cut apples dipped in 10 g L^{-1} ascorbic acid and 5 g L^{-1} calcium chloride and packaged under different MAP resulted in increased respiration coefficients, which reduced consumer acceptance beyond the second week of storage, even though reduced ethylene and CO_2 production limited accumulation of fermentative metabolites (Soliva-Fortuny et al., 2005). Sensory panels determined that stored 'Gala' apple slices (14 days at 1°C) in barrier bags maintained acceptable flavor and developed no off-flavors or fermented flavors, even though there were significant ester changes (Bett et al., 2001). When 'Gala' apple slices were stored in perforated polyethylene bags at 5.5°C for 7 days, ethyl ester concentration increased 11-fold, while acetate ester concentration declined about 50%, but no significant changes in apple-like flavor were associated with these changes (Bai et al., 2004). Similarly, nonacetate esters increased 87%, and acetate esters decrease 66% in fresh-cut cantaloupe cubes during 14 days of storage at 4°C, but sensory analysis displayed transient increases followed by slight declines for some attributes in flavor and aroma with the exception of mustiness, which increased after 12 days (Beaulieu et al., 2004). As previously mentioned, the ratio change of the nonacetate esters (increased) to acetate esters (decreased) was linear and uniform, independent of maturity, during fresh-cut storage of cantaloupe (Beaulieu, 2006b), and similar trends were observed in fresh-cut apple and honeydew (Beaulieu, 2006a).

In one of the few studies that considered diffusion of volatiles at the cut boundary as a mechanism of flavor loss in fresh-cuts, rapid loss of low molecular weight esters (< C_8) from thin sliced cantaloupe held in open or closed Petri dishes was attributed to off gassing due to the high vapor pressures and low boiling points of the esters (Beaulieu, 2007). There was no evidence for catabolic loss via recovery of esterase-mediated volatiles, and this corresponded with the finding that esterase activity generally declined in thin-sliced cut cantaloupe during storage (Lamikanra and Watson,

2003). Furthermore, there were only secondary volatiles recovered related mainly to lipid oxidation and carotenoid breakdown (Beaulieu, 2007). The boundary layer effect was explored in greater depth by sampling equatorial melon tissue as thin slices (2-mm thick) versus traditional cubes that were washed (1 minute at 5°C) and stored at 5°C. Upon GC-MS volatile determination, cubes were cut so as to mimic the thin-sliced tissue, and also the internal 1 cm³ was sampled after removal of exterior tissue. Table 5.2 indicates that the washed thin-sliced tissue had slightly elevated ester recovery on day 0, followed by significant loss (62%) compared with both tissue types from cubes after 3 days storage. Rinsing the samples provided a reduction in short-term boundary layer ester loss compared to previously reported ester losses in a similar system without washing (Beaulieu, 2007). These data indicate clearly that both internal versus external cube tissue had almost identical ester profiles through storage, whereas esters rapidly diffused away from the thin-sliced tissue during storage, across all molecular weights, with few exceptions. The surface area per unit volume of thin-sliced tissue was roughly four times greater than a typical fresh-cut cube. Subsequently, not only are cutting tool sharpness and relative humidity during storage important for quality retention, but size of tissue cut and surface area to unit volume are important regarding critical volatile diffusion and loss.

5.3.5 PACKAGING AND STORAGE

After lowering product temperature, modified atmosphere packaging (MAP) is considered the second-best method that can serve to further improve keeping quality and shelf life for several fresh-cut commodities. MAP prolongs the shelf life of fresh-cut products by decreasing O_2 and increasing CO_2 concentrations in the package atmosphere, which is accomplished by the interaction between respiratory O_2 uptake and CO_2 evolution of the produce, and gas transfer through the package films (Beaudry, 2007). Theoretically, MAP maintains quality of fresh-cut products by matching the oxygen transmission rate (OTR) of the packaging film to the respiration rate of the product. Often, improper or fluctuating temperatures will hasten product deterioration and upset the normal aroma and flavor profiles in MAP, because films have a fixed OTR, per temperature. This is where one of the most significant problems with MAP occurs: fine-tuning a model to attain equilibrium conditions is almost impossible unless the package is held under closely controlled temperature conditions.

Optimized MAP conditions are common in research studies, yet are often compromised in commercial applications. Unfortunately, there are some possible negative effects of MAP, including water condensation within packages due to temperature fluctuations, abnormal ripening in certain fruits and development of physiological disorders that often lead to anaerobically produced off-flavors and off-odors, and decrease of some bioactive compounds (Artés, 2006). In fact, increased fermentation reactions and undesirable product quality are often observed commercially before completion of the "normal" shelf life because films may not have adequate permeability. The effects of MAP on quality of fresh-cut fruits and vegetables have been reviewed extensively in recent years (Artés, 2006; Brecht, 2006; Erkan and Wang, 2006; Martín-Belloso and Soliva-Fortuny, 2006; Yahia, 2006), and subsequently, this chapter will focus on studies incorporating sensory and flavor aspects. Some

TABLE 5.2
Volatile Changes[a] in Washed Thin-Sliced Cantaloupe Tissue versus External Cube Surfaces and Internal Tissue from the Same Cube

Volatile Compounds Days Stored at 4°C	Washed Slices (2 mm)			Washed Cube Edges (2 mm)			Cube Internal Tissue		
	0	1	3	0	1	3	0	1	3
Ethyl acetate	11.408	13.423	13.276	13.191	16.366	17.220	13.792	16.519	18.067
Ethyl propanoate	2.807	3.213	3.280	3.323	3.591	4.067	3.533	3.534	4.373
Propyl acetate	2.843	2.861	2.220	3.236	3.822	4.187	3.603	4.093	4.584
Ethyl-2-methyl propanoate	0.697	0.006	0.000	0.321	0.480	0.456	0.258	0.375	0.492
2-Methyl propyl acetate	7.336	4.862	3.114	8.940	8.096	8.667	10.181	8.356	9.895
Methyl-2-methyl butanoate	1.109	0.111	0.033	0.429	0.835	1.441	0.413	0.566	1.143
Ethyl butanoate	8.520	0.221	0.027	4.931	6.837	5.960	4.897	5.180	6.824
Butyl acetate	3.463	2.728	2.010	4.170	4.088	4.400	4.454	4.125	4.940
Ethyl-2-methyl butanoate	5.690	0.046	0.003	2.594	3.756	3.517	2.672	3.218	4.410
2- and 3-methyl butyl acetate	10.678	0.133	0.078	5.119	6.636	1.315	3.829	5.065	3.623
Ethyl hexanoate	1.628	0.297	0.035	1.023	1.902	0.855	1.392	2.294	1.911
Hexyl acetate	0.749	0.010	0.002	0.351	0.388	0.052	0.288	0.350	0.061
Benzyl acetate	5.846	0.814	0.000	4.463	4.076	0.154	1.298	2.792	0.098
Ethyl benzoate	0.030	0.020	0.004	0.034	0.054	0.054	0.043	0.069	0.070
Ethyl-2-phenyl acetate	0.030	0.019	0.006	0.029	0.036	0.033	0.042	0.048	0.041
2-Phenylethyl acetate	0.013	0.003	0.000	0.027	0.014	0.019	0.053	0.033	0.056
3-Phenyl-propyl acetate	0.000	0.000	0.000	0.002	0.001	0.003	0.007	0.007	0.008
Total esters	62.85	28.77	24.09	52.18	60.98	52.40	50.76	56.62	60.60
Standard deviation	3.82	3.30	3.19	3.59	4.17	4.37	3.79	4.10	4.63

[a] Counts based on integrated GC-MS target ion abundance (×1,000,000) in compounds authenticated by standards, comparison of mass spectra (NIST Database, v. 1.5, Palisade Corp., Newfield, NY), and an in-house retention index as previously described (Beaulieu, 2006b; Beaulieu J.C. 2007. Effect of UV irradiation on cut cantaloupe: Terpenoids and esters. J. Food Sci. 72(4):S948–S957.), $n = 3$.

proprietary, commercial products exist to compensate for temperature fluctuation (e.g., Breathway™), and "next generation" humidity/temperature devices are under investigation, but these will not be addressed in this review.

5.3.5.1 Fruits

Fresh-cut cantaloupe cubes held best under active MAP (4 kPa O_2 + 10 kPa CO_2) at 5°C for 9 days and maintained mild melon aroma after 12 days compared against passive MAP and perforated films. Meanwhile, the shelf life (5 to 7 days) of cubes held in perforated film packages was limited by tissue translucency and off-odor development (Bai et al., 2001). An intermediate stage of ripeness (slightly underripe) at processing was the most suitable to extend the shelf life of fresh-cut 'Piel de Sapo' melon stored under active MAP (2.5 kPa O_2 + 7 kPa CO_2) at 4°C. In addition, 1% ascorbic acid and 0.5% calcium chloride extended shelf life out to 10 days by preventing alcohol accumulation, color loss, translucency, and flesh softening (Oms-Oliu et al., 2007a). Unfortunately the authors did not report sensory or consumer evaluation. Color and texture, as measured by instruments and trained sensory panels, remained stable for up to 17 (+/−3) days storage at 3°C in fresh-cut cantaloupe receiving electron (E)-beam irradiation (0, 0.5, and 1 kGy) stored under actively flushed MAP (4 kPa O_2 + 10 kPa CO_2) (Boynton et al., 2006). Sensory evaluation rated the 1-kGy sample highest in sweetness and cantaloupe flavor intensity and lowest in off-flavor. Low O_2 and high CO_2 controlled atmospheres (4 kPa O_2 + 15 kPa CO_2, and 21 kPa O_2 + 15 kPa CO_2) reduced the respiration rate and ethylene emission, and resulted in fresh-cut 'Amarillo' melon with better sensory quality than controls held for up to 14 days in 5°C storage. Melon pieces under both CA achieved a shelf life of 14 days, although 4 kPa O_2 + 15 kPa CO_2 was preferable for maintaining firmness (Aguayo et al., 2007). Superatmospheric O_2 (70 kPa O_2) packaging of fresh-cut 'Piel de Sapo' melons predipped in 0.5% calcium chloride maintained instrumental firmness and chewiness and avoided fermentative reactions for 2 weeks at 4°C (Oms-Oliu et al., 2008d). However, superatmospheric O_2 in 'Flor de Invierno' fresh-cut pears dipped in 0.75% N-acetylcysteine and 0.75% glutathione resulted in dramatic modification of some quality attributes, including excessive ethanol accumulation (Oms-Oliu et al., 2007b). Fresh-cut honeydew cubes held 11 days in active MAP (5 kPa O_2 + 5 kPa CO_2) had better visual quality and aroma than passive MAP for both winter and summer season fruit (Bai et al., 2003). Passive MAP (final atmosphere of 4 kPa O_2 + 12 to 13 kPa CO_2) using three commercial films maintained sensorial quality and microbial safety of fresh-cut Amarillo melon for 14 days at 5°C (Aguayo et al., 2003).

Partially ripe fresh-cut 'Conference' pears receiving browning inhibition (10 g L^{-1} ascorbic acid and 5 g L^{-1} calcium chloride for 1 min) and packaged in MAP (2.5 kPa O_2 + 7 kPa CO_2) preserved acceptable sensory quality during 3-week storage at 4°C. However, both low O_2 and high CO_2 were determined to have detrimentally negative effects on quality, and the authors concluded that higher CO_2 and O_2 transmission films are required (Soliva-Fortuny et al., 2007). Antioxidant treatments (0.75% N-acetylcysteine + 0.75% glutathione) in concert with active low O_2 MAP prevented browning and maintained vitamin C, chlorogenic acid, and antioxidant capacity in fresh-cut pears (Oms-Oliu et al., 2008a). Interestingly, the authors were evaluating high O_2 MAP that resulted in inferior quality (color and physiological; unfortunately sensory was not reported) compared with traditional MAP. Low O_2 MAP (2.5 kPa O_2) also best maintained vitamin C and phenolic content during storage of fresh-cut 'Piel de Sapo' melon (Oms-Oliu et al., 2008b). Packaging fresh-cut apple and pear under high N_2 (100 kPa) led to a progressive degradation of product and almost linear texture loss during storage (Soliva-Fortuny et al., 2002a, 2002b).

Fresh-cut pineapple dipped in 0.25% ascorbic acid and 10% sucrose for 2 min and stored under active MAP (4 kPa O_2 + 10 kPa CO_2) maintained sensory quality for 7 days at 4°C, whereas passive MAP developed off-flavor after 3 days storage (Liu et al., 2007). Pineapple chunks that were held in MAP (1.5 kPa O_2 and 11 kPa CO_2) for 14 days at 0°C had no off-flavors and maintained good appearance, yet holding at 5°C resulted in anaerobiosis (<1 kPa O_2 and 15 kPa CO_2), lending to slight off-odor upon opening which quickly dissipated (Marrero and Kader, 2006).

Taste and aroma of fresh-cut tomato slices stored under active (12 to 14 kPa O_2 + 0 kPa CO_2) MAP at 0 or 5°C for up to 10 days, and CA (2.5 kPa O_2 + 5 kPa CO_2) at 1°C were better than air-stored slices (Gil et al., 2002; Hakim et al., 2004). Nonetheless, a slight flavor loss after storage was detected (Gil et al., 2002) which corroborated Mencarelli and Saltveit (1988) who associated the insipid flavor with low soluble solids to titratable acidity ratio and pH. The limit of visual quality acceptance to consumers in pomegranate arils stored in passive MAP was reached after 14 days at 5°C from earlier harvested fruit and 10 days for later harvested fruit (López-Rubira et al., 2005). The use of UV was deemed unjustified, and antioxidant activity of arils did not change significantly between both harvest dates and throughout storage.

5.3.5.2   Vegetables and Salads

Fruity aroma decreased after 5 days storage at 4°C in "julienne" fresh-cut carrot sticks packed in polypropylene trays overwrapped with PVC film (Lavelli et al., 2006). Carrot discs stored in high-permeability microperforated film pouches that developed atmospheres of 5 to 7 kPa O_2 + 12 to 15 kPa CO_2 maintained good aroma and flavor as determined by quantitative descriptive analysis for 6 days at 4°C compared to those stored in conventional polypropylene microperforated polypropylene having low permeability (Cliffe-Byrnes and O'Beirne, 2007). However, extensive surface whitening and moisture loss occurred, which was overcome by Natureseal™ (Cliffe-Byrnes and O'Beirne, 2007).

Increasing delay (4, 8, and 12 hours at 5°C) before packaging fresh-cut Romaine lettuce resulted in progressively decreased fermentative volatile production, off-odor development, and CO_2 injury, yet discoloration increased (Kim et al., 2005). In low OTR polyethylene film bags (8 pmol s^{-1} m^{-2} Pa^{-1}), there was rapid development of off-flavors in fresh-cut salad savoy by day 15 at 5°C, while all other treatments with OTR films ≥16.6 did not develop off-odor until day 20 or 25 (Kim et al., 2004). Snow pea pods held at 5°C in MAP (5 kPa O_2 + 5 kPa CO_2) maintained quality and developed no off-flavors after 28 days, unlike peas held in ambient atmospheres that became unmarketable, and under low O_2 MAP (2.5 kPa +h 5 kPa CO_2) or high CO_2 (10 kPa + 5 kPa O_2) that developed slight off-flavors (Pariasca et al., 2001). Sliced mushrooms and grated celeriac stored in active MAP with high oxygen (95 kPa O_2 + 5 kPa N_2) maintained fresh taste and smell for 7 days at 4°C, whereas storage in passive MAP (3 kPa O_2 + 5 kPa CO_2) was restricted by poor taste and development of off-odors by 3 to 4 days (Jacxsens et al., 2001).

5.3.6 COMBINED OR SYNERGISTIC QUALITY RETENTION TREATMENTS

The above sections already contain several examples where combined treatments have been investigated to improve various aspects of storage quality. Nowadays, food safety has become a major issue, and many analyses (mostly not covered in this chapter) have been addressing efficacy of various agents to reduce loads of microbes and exogenously applied foodborne pathogens in challenge studies. Several of those studies also used synergistic treatments and hurdle technologies. Nonetheless, a few more examples are introduced where traditional agents have been used without MAP, edible coatings and dips, or antimicrobials.

Several studies have illustrated that ascorbic acid, CA, calcium lactate, cysteine, or 4-hexylresorcinol are independently inadequate to effectively control browning in long-term fresh-cut storage of apple and pear (Gorny et al., 2002; Oms-Oliu et al., 2006; Rojas-Graü et al., 2006). However, various combinations of these treatments, with others (reduced glutathione, N-acetyl-L-cysteine), have been highly effective in controlling browning, by direct action on PPO (Rojas-Graü et al., 2008), and in some cases, no negative sensorial consequences occurred as consumer panelists could not distinguish between preservative treated slices and control fruit (Gorny et al., 2002).

The effects of calcium ascorbate, irradiation, and MAP were evaluated in concert in fresh-cut 'Gala' apples. Apple slices were treated with 7% calcium ascorbate followed by irradiation at 0.5 and 1 kGy and storage at 10°C for up to 3 weeks in MAP (film bags). Slices softened during irradiation and storage, yet the firmness decrease was reduced by the calcium treatment (Fan et al., 2005). Overall, the combined technologies enhanced microbial food safety while maintaining fresh-cut quality of the slices. However, sensory evaluations were not performed. Recently, hot water dipping has been used as a hurdle technology to reduce microbial loads and was also synergistically used with firmness retention applications. Calcium propionate (0.9%) and calcium lactate (1.4%) dips applied for 1 min at 60°C maintained fresh-cut melon ('Amarillo') firmness during 8 days storage at 5°C (Aguayo et al., 2008). The calcium dips maintained cube firmness, reduced microbial growth, and improved sensory quality compared to control, and there was not a salty taste in cut melon dipped in 0.5% calcium chloride.

5.4 CONCLUSIONS AND FUTURE DIRECTIONS

In recent years, especially in the United States, there have been several foodborne illnesses involving fresh-cuts. This has created urgency to minimize risk for both the consumer and industry by attempting to rapidly develop food safety treatments that can deliver some additional level of safety while maintaining product quality. Even so, demand for highly convenient fresh-cuts still increases, especially in light of renewed consumer health-conscious attitudes brought on by awareness and appreciation of natural phytochemicals and antioxidants in fruits and vegetables. Flavor and texture are critically important components of fresh-cut fruit and vegetable quality, which are difficult to commercially analyze. Due to commercial difficulties in testing product quality, it is often assumed that "if it looks good, it tastes good." Investigators have determined that some simple instrumental measurements of color, soluble solids, changes in weight or juice leakage, and ascorbate content may be used as indices of quality changes in fresh-cut fruits and vegetables (Barrett et al., 2010).

As reviewed herein, several postharvest chemical and physical treatments used to extend shelf life of fresh-cut produce may negatively affect product flavor. Active and passive MAP, and even tightly sealed packaging, can alter gaseous and flavor compound metabolism leading to beneficial effects, flavor imbalances, inhibition of flavor compounds, or off-flavors. Sensory quality and appreciation were already problematic in fresh-cuts and now will require increased attention as several synergistic treatments and antimicrobial technologies are vigorously researched in concert with packaging. This review highlights much of the recent literature along these lines but, unfortunately, comes up lacking regarding unified solutions to the flavor dilemma, or delivery of the magical "silver bullet" treatment(s) needed to achieve food safety, shelf life, and optimum flavor and sensory quality in stored fresh-cuts. Because inherent variability of produce and all the steps from cultivar selection through the postharvest chain to the consumer can affect flavor and sensory quality, it remains a challenge to design meaningful projects that can answer all the questions. If seed companies and researchers worked together more often, determining critical fresh-cut flavor volatiles, cell wall associations with texture, and elucidating precursor compound pathways, faster genetic or enzymatic regulation could ensue to address the flavor and texture issues. Nonetheless, an exciting future awaits us as more details emerge regarding antimicrobials and packaging (MAP) in concert with temperature and humidity release mechanisms, plant essential oil use, and edible films or coatings.

REFERENCES

Abbey S.D., E.K. Heaton, D.A. Golden, and L.A. Beuchat. 1988. Microbiological and sensory quality changes in unwrapped and wrapped sliced watermelon. *J. Food Protection.* 51:531–533.

Abbott J.A., and F.R. Harker. 2004. Texture. In: Gross, K.C., C.Y. Wang, and M.E. Saltveit (eds.), *USDA Handbook No. 66, 3rd edition. The Commercial Storage of Fruits, Vegetables, and Florist and Nursery Stocks.* Washington, DC: Agricultural Research Service. Accepted (In public review: http://usna.usda.gov/hb66-021texture.pdf).

Aguayo A., A. Allende, and F. Artes. 2003. Keeping quality and safety of minimally fresh processed melon. *Eur. Food Res. Technol.* 216(6):494–499.

Aguayo E., V.H. Escalona, and F. Artés. 2007. Quality of minimally processed *Cucumis melo* var. *saccharinus* as improved by controlled atmosphere. *Eur. J. Hortic. Sci.* 72(1):39–45.

Aguayo E., V.H. Escalona, and F. Artés. 2008. Effect of hot water treatment and various calcium salts on quality of fresh-cut 'Amarillo' melon. *Postharvest Biol. Technol.* 47(3):397–406.

Aguayo E., R. Jansasithorn, and A.A. Kader. 2006. Combined effects of 1-methylcyclopropene, calcium chloride dip, and/or atmospheric modification on quality changes in fresh-cut strawberries. *Postharvest Biol. Technol.* 40:269–278.

Artés F. 2006. Modified atmosphere packaging of fruits and vegetables. *Stewart Postharvest Rev.* 2(5-2):1–13.

Bai J.H., and E.A. Baldwin. 2002. Post-processing dip maintains quality and extends the shelf life of fresh-cut apple. *Proc. Fla. State Hort. Soc.* 115:297–300.

Bai J., E.A. Baldwin, R.C. Soliva-Fortuny, J.P. Mattheis, R. Stanley, C. Perera, and J.K. Brecht. 2004. Effect of pretreatment of intact 'Gala' apple with ethanol vapor, heat, or 1-methylcyclopropene on quality and shelf life of fresh-cut slices. *J. Amer. Soc. Hort. Sci.* 129:583–593.

Bai J.H., R.A. Saftner, and A.E. Watada. 2003. Characteristics of fresh-cut honeydew (*Cucumis xmelo* L.) available to processors in winter and summer and its quality maintenance by modified atmosphere packaging. *Postharvest Biol. Technol.* 28(3):349–359.

Bai J.H., R.A. Saftner, A.E. Watada, and Y.S. Lee. 2001. Modified atmosphere maintains quality of fresh-cut cantaloupe (*Cucumis melo* L.). *J. Food Sci.* 66(8):1207–1211.

Baldwin E.A., T.M.M. Malundo, R.J. Bender, and J.K. Brecht. 1999. Interactive effects of harvest maturity, controlled atmosphere and surface coatings on mango (*Mangifera indica* L.) flavor quality. *HortSci.* 34:514.

Baldwin E.A., M.O. Nisperos-Carriedo, P.E. Shaw, and J.K. Burns. 1995. Effect of coatings and prolonged storage conditions on fresh orange flavor volatiles, degrees Brix, and ascorbic acid levels. *J. Agric. Food Chem.* 43:1321–1331.

Baldwin E.A., A. Plotto, and K. Goodner. 2007. Shelf-life versus flavour-life for fruits and vegetables: how to evaluate this complex trait. *Stewart Postharvest Rev.* 3(1–2):1–10.

Barrett D.M., J.C. Beaulieu, and R.L. Shewfelt. 2010. Color, flavor, texture and nutritional quality of fresh-cut fruits and vegetables: Desirable levels, instrumental and sensory measurement, and effects of processing. *Crit. Rev. Food Sci. Nutr.* 50:369–389.

Barry-Ryan C., and D. O'Beirne. 1998. Quality and shelf-life of fresh-cut carrot slices as affected by slicing method. *J. Food Sci.* 63(5):851–856.

Barry-Ryan C., and D. O'Beirne. 1999. Ascorbic acid retention in shredded iceberg lettuce as affected by minimal processing. *J. Food Sci.* 64(3):498–500.

Beaudry R.M. 2007. MAP as a basis for active packaging. In: Wilson C.L. (ed.), *Intelligent and Active Packaging of Fruits and Vegetables*. Boca Raton, FL: CRC Press, pp. 31–55.

Beaulieu J.C. 2005. Within-season volatile and quality differences in stored fresh-cut cantaloupe cultivars. *J. Agric. Food Chem.* 53(22):8679–8687.

Beaulieu J.C. 2006a. Effect of cutting and storage on acetate and non-acetate esters in convenient, ready to eat fresh-cut melons and apples. *HortSci.* 41(1):65–73.

Beaulieu J.C. 2006b. Volatile changes in cantaloupe during growth, maturation and in stored fresh-cuts prepared from fruit harvested at various maturities. *J. Amer. Soc. Hort. Sci.* 131(1):127–139.

Beaulieu J.C. 2007. Effect of UV irradiation on cut cantaloupe: Terpenoids and esters. *J. Food Sci.* 72(4):S948–S957.

Beaulieu J.C., and E.A. Baldwin. 2002. Flavor and aroma of fresh-cut fruits and vegetables. In: Lamikanra O. (ed.), *Fresh-Cut Fruits and Vegetables. Science, Technology and Market*. Boca Raton, FL: CRC Press, pp. 391–425.

Beaulieu J.C., and J.R. Gorny. 2004. Fresh-cut fruits. In: Gross K.C., C.Y. Wang, and M.E. Saltveit (eds.), *USDA Handbook No. 66, 3rd edition. The Commercial Storage of Fruits, Vegetables, and Florist and Nursery Stocks*. Washington, DC: Agricultural Research Service. Accepted (In public review: http://usna.usda.gov/hb66/146freshcutfruits.pdf).

Beaulieu J.C., and C.C. Grimm. 2001. Identification of volatile compounds in cantaloupe at various developmental stages using solid phase microextraction. *J. Agric. Food Chem.* 49(3):1345–1352.

Beaulieu J.C., D.A. Ingram, J.M. Lea, and K.L. Bett-Garber. 2004. Effect of harvest maturity on the sensory characteristics of fresh-cut cantaloupe. *J. Food Sci.* 69(7):S250–S258.

Beaulieu J.C., and V.A. Lancaster. 2007. Correlating volatile compounds, quality parameters, and sensory attributes in stored fresh-cut cantaloupe. *J. Agric. Food Chem.* 55(23):9503–9513.

Beaulieu J.C., and J.M. Lea. 2003. Volatile and quality changes in fresh-cut mangos prepared from firm-ripe and soft-ripe fruit, stored in clamshell containers and passive MAP. *Postharvest Biol. Technol.* 30(1):15–28.

Beaulieu J.C., and J.M. Lea. 2007. Quality changes in cantaloupe during growth, maturation, and in stored fresh-cuts prepared from fruit harvested at various maturities. *J. Amer. Soc. Hort. Sci.* 132(5):720–728.

Beaulieu J.C., and M.E. Jr. Saltveit. 1997. Inhibition or promotion of tomato fruit ripening by acetaldehyde and ethanol is concentration dependent and varies with initial fruit maturity. *J. Amer. Soc. Hort. Sci.* 122(3):392–398.

Beirão-da-Costa S., A. Steiner, L. Correia, J. Empis, and M. Moldão-Martins. 2006. Effects of maturity stage and mild heat treatments on quality of minimally processed kiwifruit. *J. Food Eng.* 76(4):616–625.

Beltrán D., M.V. Selma, J.A. Tudela, and M.I. Gil. 2005. Effect of different sanitizers on microbial and sensory quality of fresh-cut potato strips stored under modified atmosphere or vacuum packaging. *Postharvest Biol. Technol.* 37(1):37–46.

Bett K.L., D.A. Ingram, C.C. Grimm, S.W. Lloyd, A.M. Spanier, J.M. Miller, K.C. Gross, E.A. Baldwin, and B.T. Vinyard. 2001. Flavor of fresh-cut 'Gala' apples in modified atmosphere packaging as affected by storage time. *J. Food Qual.* 24(2):141–156.

Blankenship S.M., and J.M. Dole. 2003. 1-Methylcyclopropene: A review. *Postharvest Biol. Technol.* 28(1):1–25.

Bourne M. 2002. *Food Texture and Viscosity; Concept and Measurement.* San Diego, CA: Academic Press.

Boynton B.B., B.A. Welt, C.A. Sims, M.O. Balaban, J.K. Brecht, and M.R. Marshall. 2006. Effects of low-dose electron beam irradiation on respiration, microbiology, texture, color, and sensory characteristics of fresh-cut cantaloupe stored in modified-atmosphere packages. *J. Food Sci.* 71(2):S149–S155.

Brecht J.K. 2006. Controlled atmosphere, modified atmosphere and modified atmosphere packaging for vegetables. *Stewart Postharvest Rev.* 2(5):1–6.

Brecht J.K., M.E. Saltveit, S.T. Talcott, K.R. Schneider, K. Felkey, and J.A. Bartz. 2004. Fresh-cut vegetables and fruits. *Hort. Rev.* 30(30):185–251.

Brückner B., and S.G. Wyllie. 2008. *Fruit and Vegetable Flavour: Recent Advances and Future Prospects.* Cambridge: CRC Press, Woodhead.

Calderon-Lopez B., J.A. Bartsch, C.Y. Lee, and C.B. Watkins. 2005. Cultivar effects on quality of fresh cut apple slices from 1-methylcyclopropene (1-MCP)-treated apple fruit. *J. Food Sci.* 70(3):S221–S227.

Cliffe-Byrnes V., and D. O'Beirne. 2007. The effects of modified atmospheres, edible coating and storage temperatures on the sensory quality of carrot discs. *Int. J. Food Sci. Technol.* 42(11):1338–1349.

Cocci E., P. Rocculi, S. Romani, and M. Dalla Rosa. 2006. Changes in nutritional properties of minimally processed apples during storage. *Postharvest Biol. Technol.* 39(3):265–271.

DeRovira D. 1996. The dynamic flavor profile method. *Food Technol.* 50(2):55–60.

Dong X., R.E. Wrolstad, and D. Sugar. 2000. Extending shelf life of fresh-cut pears. *J. Food Sci.* 65(1):181–186.

Ergun M., D.J. Huber, J. Jeong, and J.A. Bartz. 2006. Extended shelf life and quality of fresh-cut papaya derived from ripe fruit treated with the ethylene antagonist 1-methylcyclopropene. *J. Amer. Soc. Hort. Sci.* 131(1):97–103.

Ergun M., J. Jeong, D.J. Huber, and D.J. Cantliffe. 2007. Physiology of fresh-cut Galia (*Cucumis melo* var. *reticulatus*) from ripe fruit treated with 1-methylcyclopropene. *Postharvest Biol. Technol.* 44(3):286–292.

Erkan M., and C.Y. Wang. 2006. Modified and controlled atmosphere storage of subtropical crops. *Stewart Postharvest Rev.* 2(5-4):1–8.

Eswaranandam S., N.S. Hettiarachchy, and J.-F. Meullenet. 2006. Effect of malic and lactic acid incorporated soy protein coatings on the sensory attributes of whole apple and fresh-cut cantaloupe. *J. Food Sci.* 71(3):S307–S313.

Fan X., B.A. Annous, J.C. Beaulieu, and J. Sites. 2008. Effect of hot water surface pasteurization of whole fruit on shelf-life and quality of fresh-cut cantaloupe. *J. Food Sci.* 73(3):M91–M98.

Fan X.T., B.A. Niemira, J.P. Mattheis, H. Zhuang, and D.W. Olson. 2005. Quality of fresh-cut apple slices as affected by low-dose ionizing radiation and calcium ascorbate treatment. *J. Food Sci.* 70(2):S143–S148.

Ferenczi A., J. Song, M. Tian, K. Vlachonasios, D. Dilley, and R.M. Beaudry. 2006. Volatile ester suppression and recovery following 1-methylcyclopropene application to apple fruit. *J. Amer. Soc. Hort. Sci.* 131(5):691–701.

Forney C.F. 2008. Flavour loss during postharvest handling and marketing of fresh-cut produce. *Stewart Postharvest Rev.* 4(3–5):1–10.

Garcia E., and D.M. Barrett. 2002. Preservative treatments for fresh-cut fruits and vegetables. In: Lamikanra O. (ed.), *Fresh-Cut Fruits and Vegetables. Science, Technology, and Market.* Boca Raton, FL: CRC Press, pp. 267–303.

Gil M.I., E. Aguayo, and A.A. Kader. 2006. Quality changes and nutrient retention in fresh-cut versus whole fruits during storage. *J. Agric. Food Chem.* 54(12):4284–4296.

Gil M.I., M.A. Conesa, and F. Artes. 2002. Quality changes in fresh cut tomato as affected by modified atmosphere packaging. *Postharvest Biol. Technol.* 25(2):199–207.

González-Aguilar G.A., J. Celis, R.R. Sotelo-Mundo, L.A. de la Rosa, J. Rodrigo-Garcia, and E. Alvarez-Parrilla. 2008. Physiological and biochemical changes of different fresh-cut mango cultivars stored at 5°C. *Int. J. Food Sci. Technol.* 43(1):1365–2621.

González-Aguilar G.A., S. Ruiz-Cruz, R. Cruz-Valenzuela, A. Rodríguez-Félix, and C.Y. Wang. 2004. Physiological and quality changes of fresh-cut pineapple treated with anti-browning agents. *Lebensm. Wiss. Technol.* 37(3):369–376.

Gorny J.R., R.A. Cifuentes, B. Hess-Pierce, and A.A. Kader. 2000. Quality changes in fresh-cut pear slices as affected by cultivar, ripeness stage, fruit size, and storage regime. *J. Food Sci.* 65(3):541–544.

Gorny J.R., B. Hess-Pierce, R.A. Cifuentes, and A.A. Kader. 2002. Quality changes in fresh-cut pear slices as affected by controlled atmospheres and chemical preservatives. *Postharvest Biol. Technol.* 24(3):271–278.

Gorny J.R., B. Hess-Pierce, and A.A. Kader. 1998. Effects of fruit ripeness and storage temperature on the deterioration rate of fresh-cut peach and nectarine slices. *HortSci.* 33(1):110–113.

Hakim A., M.E. Austin, D. Batal, S. Gullo, and M. Khatoon. 2004. Quality fresh-cut tomatoes. *J. Food Qual.* 27(3):195–206.

Han J., W. Tao, H. Hao, B. Zhang, W. Jiang, T. Niu, Q. Li, and T. Cai. 2006. Physiology and quality responses of fresh-cut broccoli florets pretreated with ethanol vapor. *J. Food Sci.* 71(5):S385–S389.

Harker F.R., F.A. Gunson, and S.R. Jaeger. 2003. The case for fruit quality: An interpretive review of consumer attitudes, and preferences for apples. *Postharvest Biol. Technol.* 28:333–347.

Hodges D.M., and P.M.A. Toivonen. 2008. Quality of fresh-cut fruits and vegetables as affected by exposure to abiotic stress. *Postharvest Biol. Technol.* 48(2):155–162.

Huber D.J. 2008. Suppression of ethylene responses through application of 1-methylcyclopropene: A powerful tool for elucidating ripening and senescence mechanisms in climacteric and nonclimacteric fruits and vegetables. *HortSci.* 43(1):106–111.

Jacxsens L., F. Devlieghere, C. Van der Steen, and J. Debevere. 2001. Effect of high oxygen modified atmosphere packaging on microbial growth and sensorial qualities of fresh-cut produce. *Int. J. Food Microbiol.* 71(2–3):197–210.

Jeon M., and Y. Zhao. 2005. Honey in combination with vacuum impregnation to prevent enzymatic browning of fresh-cut apples. *Int. J. Food Sci. Nutr.* 56(3):165–176.

Jeong J., J.K. Brecht, D.J. Huber, and S.A. Sargent. 2008. Storage life and deterioration of intact cantaloupe (*Cucumis melo* L. var. *reticulatus*) fruit treated with 1-methylcyclopropene and fresh-cut cantaloupe prepared from fruit treated with 1-methylcyclopropene before processing. *HortSci.* 43(2):435–438.

Jiang Y.M., and D.C. Joyce. 2002. 1-Methylcyclopropene treatment effects on intact and fresh-cut apple. *J. Hortic. Sci. Biotechnol.* 77(1):19–21.

Kader A.A. 2003. A perspective on postharvest horticulture (1978–2003). *HortSci.* 38(5):1004–1008.

Kader, A.A., and M. Cantwell. 2007. Produce quality rating scales and color charts. Postharvest Technology Research Information Center. Postharvest Horticulture Series, No. 23B. The University of California, Davis.

Kang H.-M., and M.E. Saltveit. 2002. Antioxidant capacity of lettuce leaf tissue increases after wounding. *J. Agric. Food Chem.* 50(26):7536–7541.

Karakurt Y., and D.J. Huber. 2007. Characterization of wound-regulated cDNAs and their expression in fresh-cut and intact papaya fruit during low-temperature storage. *Postharvest Biol. Technol.* 44(2):179–183.

Kim J.G., Y. Luo, and K.C. Gross. 2004. Effect of package film on the quality of fresh-cut salad savoy. *Postharvest Biol. Technol.* 32(1):99–107.

Kim J.G., Y. Luo, R.A. Saftner, and K.C. Gross. 2005. Delayed modified atmosphere packaging of fresh-cut romaine lettuce: Effects on quality maintenance and shelf-life. *J. Amer. Soc. Hort. Sci.* 130(1):116–123.

Kondo S., S. Setha, D. Rudell, D.A. Buchanan, and J.P. Mattheis. 2005. Aroma volatile biosynthesis in apples affected by 1-MCP and methyl jasmonate. *Postharvest Biol. Technol.* 36(1):61–68.

Kou L., Y. Luo, D. Wu, and X. Liu. 2007. Effects of mild heat treatment on microbial growth and product quality of packaged fresh-cut table grapes. *J. Food Sci.* 72(8):S567–S573.

Lamikanra O., and M.A. Watson. 2003. Temperature and storage duration effects on esterase activity in fresh-cut cantaloupe melon. *J. Food Sci.* 68(3):790–793.

Larrigaudière C., D. Ubach, Y. Soria, M.A. Rojas-Graü, and O. Martín-Belloso. 2008. Oxidative behaviour of fresh-cut 'Fuji' apples treated with stabilising substances. *J. Sci. Food Agric.* 88(10):1770–1776.

Lavelli V., E. Pagliarini, R. Ambrosoli, J.L. Minati, and B. Zanoni. 2006. Physicochemical, microbial, and sensory parameters as indices to evaluate the quality of minimally-processed carrots. *Postharvest Biol. Technol.* 40(1):34–40.

Lee J.Y., H.J. Park, C.Y. Lee, and W.Y. Choi. 2003. Extending shelf-life of minimally processed apples with edible coatings and antibrowning agents. *Lebensmittel-Wissenschaft und-Technologie.* 36(3):323–329.

Lin D., and Y. Zhao. 2007. Innovations in the development and application of edible coatings for fresh and minimally processed fruits and vegetables. *Comp. Rev. Food Sci. Food Safety.* 6(3):60–75.

Lin D.S., S.W. Leonard, C. Lederer, M.G. Traber, and Y. Zhao. 2006. Retention of fortified vitamin E and sensory quality of fresh-cut pears by vacuum impregnation with honey. *J. Food Sci.* 71(7):S553–S559.

Liu C.-L., C.-K. Hsu, and M.-M. Hsu. 2007. Improving the quality of fresh-cut pineapples with ascorbic acid/sucrose pretreatment and modified atmosphere packaging. *Packaging Technol. Sci.* 20(5):337–343.

López-Rubira V., A. Conesa, A. Allende, and F. Artés. 2005. Shelf life and overall quality of minimally processed pomegranate arils modified atmosphere packaged and treated with UV-C. *Postharvest Biol. Technol.* 37(2):174–185.

Luna-Guzmán I., and D.M. Barrett. 2000. Comparison of calcium chloride and calcium lactate effectiveness in maintaining shelf stability and quality of fresh-cut cantaloupes. *Postharvest Biol. Technol.* 19(1):61–72.

Mao L., G. Wang, and F. Que. 2007. Application of 1-methylcyclopropene prior to cutting reduces wound responses and maintains quality in cut kiwifruit. *J. Food Eng.* 78(1):361–365.

Marrero A., and A.A. Kader. 2006. Optimal temperature and modified atmosphere for keeping quality of fresh-cut pineapples. *Postharvest Biol. Technol.* 39(2):163–168.

Martín-Belloso O., and R. Soliva-Fortuny. 2006. Effect of modified atmosphere packaging on the quality of fresh-cut fruits. *Stewart Postharvest Rev.* 2(1–3):1–8.

Mei Y., Y. Zhao, J. Yang, and H.C. Furr. 2002. Using edible coating to enhance nutritional and sensory qualities of baby carrots. *J. Food Sci.* 67(5):1964–1968.

Mencarelli, F., F.G.R. Botondi, and P. Tonutti. 1998. Postharvest physiology of peach and nectarine slices. *Acta Hort.* 465:463–470. Proc. Fourth Intern. Peach Symposium. Bordeaux, France.

Mencarelli F., and M.E. Jr. Saltveit. 1988. Ripening of mature-green tomato fruit slices. *J. Amer. Soc. Hort. Sci.* 113(5):742–745.

Min S., and J.M. Krochta. 2005. Antimicrobial films and coatings for fresh fruit and vegetables. In: Jongen W. (ed.), *Improving the Safety of Fresh Fruit and Vegetables.* Cambridge: Woodhead/Taylor and Francis, pp. 455–492.

Moya-León M.A., M. Vergara, C. Bravo, M.E. Montes, and C. Moggia. 2006. 1-MCP treatment preserves aroma quality of 'Packham's Triumph' pears during long-term storage. *Postharvest Biol. Technol.* 42:185–197.

Olivas G.I., and G.V. Barbosa-Cánovas. 2005. Edible coatings for fresh-cut fruits. *Crit. Rev. Food Sci. Nutr.* 45(7):657–670.

Olivas G.I., D.S. Mattinson, and G.V. Barbosa-Cánovas. 2007. Alginate coatings for preservation of minimally processed 'Gala' apples. *Postharvest Biol. Technol.* 45(1):89–96.

Oms-Oliu G., I. Aguilo-Aguayo, and O. Martin-Belloso. 2006. Inhibition of browning on fresh-cut pear wedges by natural compounds. *J. Food Sci.* 71(3):S216–S224.

Oms-Oliu G., R.C. Soliva-Fortuny, and O. Martín-Belloso. 2007a. Effect of ripeness on the shelf-life of fresh-cut melon preserved by modified atmosphere packaging. *Eur. Food Res. Technol.* 225(3):301–311.

Oms-Oliu G., R.C. Soliva-Fortuny, and O. Martín-Belloso. 2007b. Respiratory rate and quality changes in fresh-cut pears as affected by superatmospheric oxygen. *J. Food Sci.* 72(8):E456–E463.

Oms-Oliu G., I. Odriozola-Serrano, R. Soliva-Fortuny, and O. Martín-Belloso. 2008a. Antioxidant content of fresh-cut pears stored in high-O_2 active packages compared with conventional low-O_2 active and passive modified atmosphere packaging. *J. Agric. Food Chem.* 56(3):932–940.

Oms-Oliu G., I. Odriozola-Serrano, R. Soliva-Fortuny, and O. Martín-Belloso. 2008b. The role of peroxidase on the antioxidant potential of fresh-cut 'Piel de Sapo' melon packaged under different modified atmospheres. *Food Chem.* 106(3):1085–1092.

Oms-Oliu G., R.M. Raybaudi-Massilia Martinez, R. Soliva-Fortuny, and O. Martín-Belloso. 2008c. Effect of superatmospheric and low oxygen modified atmospheres on shelf-life extension of fresh-cut melon. *Food Control.* 19(2):191–199.

Oms-Oliu G., R. Soliva-Fortuny, and O. Martín-Belloso. 2008d. Modeling changes of headspace gas concentrations to describe the respiration of fresh-cut melon under low or superatmospheric oxygen atmospheres. *J. Food Eng.* 85(3):401–409.

Oms-Oliu G., R. Soliva-Fortuny, and O. Martín-Belloso. 2008e. Using polysaccharide-based edible coatings to enhance quality and antioxidant properties of fresh-cut melon. *LWT—Food Sci. Technol.* 41(10):1862–1870.

Pariasca J.A.T., T. Miyazaki, H. Hisaka, H. Nakagawa, and T. Sato. 2001. Effect of modified atmosphere packaging (MAP) and controlled atmosphere (CA) storage on the quality of snow pea pods (*Pisum sativum* L. var. *saccharatum*). *Postharvest Biol. Technol.* 21(2):213–223.

Perera C.O., L. Balchin, E. Baldwin, R. Stanley, and M. Tain. 2003. Effect of 1-methylcyclopropene on the quality of fresh-cut apple slices. *J. Food Sci.* 68(6):1910–1914.

Perez-Gago M.B., M. Serra, M. Alonso, M. Mateos, and M.A. del Río. 2005. Effect of whey protein- and hydroxypropyl methylcellulose-based edible composite coatings on color change of fresh-cut apples. *Postharvest Biol. Technol.* 36(1):77–85.

Perez-Gago M.B., M. Serra, and M.A. del Río. 2006. Color change of fresh-cut apples coated with whey protein concentrate-based edible coatings. *Postharvest Biol. Technol.* 39(1):84–92.

Plotto A., J. Bai, J.A. Narciso, J.K. Brecht, and E.A. Baldwin. 2006. Ethanol vapor prior to processing extends fresh-cut mango storage by decreasing spoilage, but does not always delay ripening. *Postharvest Biol. Technol.* 39(2):134–145.

Plotto A., K.L. Goodner, E.A. Baldwin, and J. Bai. 2004. Effect of polysaccharide coatings on quality of fresh cut mangoes (*Mangifera indica*). *Proc. Fla. State Hort. Soc.* 117:382–388.

Qi, L., A.E. Watada and J.R. Gorny. 1998. In: Gorny J.R. (ed.), *Quality Changes of Fresh-Cut Fruits in CA Storage. 5: Fresh-Cut Fruits and Vegetables and MAP. Postharvest Horticulture Series No. 19. CA '97. Proceedings.* Seventh International Controlled Atmosphere Research Conference, pp 116–121.

Ragaert P., W. Verbeke, F. Devlieghere, and J. Debevere. 2004. Consumer perception and choice of minimally processed vegetables and packaged fruits. *Food Quality and Preference.* 15(3):259–270.

Raybaudi-Massilia R.M., J. Mosqueda-Melgar, and O. Martin-Belloso. 2008. Edible alginate-based coating as carrier of antimicrobials to improve shelf-life and safety of fresh-cut melon. *Int. J. Food Microbiol.* 121(3):313–327.

Reyes L.F., and L. Cisneros-Zevallos. 2003. Wounding stress increases the phenolic content and antioxidant capacity of purple-flesh potatoes (*Solanum tuberosum* L.). *J. Agric. Food Chem.* 51(18):5296–5300.

Reyes L.F., J.E. Villarreal, and L. Cisneros-Zevallos. 2007. The increase in antioxidant capacity after wounding depends on the type of fruit or vegetable tissue. *Food Chem.* 101(3):1254–1262.

Rico D., A.B. Martin-Diana, J.M. Barat, and C. Barry-Ryan. 2007. Extending and measuring the quality of fresh-cut fruit and vegetables: A review. *Trends Food Sci. Technol.* 18(7):373–386.

Rocha A.M.C.N., C.M. Brochado, R. Kirby, and A.M.M.B. Morais. 1995. Shelf-life of chilled cut orange determined by sensory quality. *Food Control.* 6(6):317–322.

Rojas-Graü M.A., R.J. Avena-Bustillos, M. Friedman, P.R. Henika, O. Martín-Belloso, and T.H. McHugh. 2006. Mechanical, barrier, and antimicrobial properties of apple puree edible films containing plant essential oils. *J. Agric. Food Chem.* 54(24):9262–9267.

Rojas-Graü M.A., R. Soliva-Fortuny, and O. Martín-Belloso. 2008. Effect of natural anti-browning agents on color and related enzymes in fresh-cut Fuji apples as an alternative to the use of ascorbic acid. *J. Food Sci.* 73(6):S267–S272.

Rupasinghe H.P.V., D.P. Murr, J.R. DeEll, and J. Odumeru. 2005. Influence of 1-methylcyclopropene and NatureSeal on the quality of fresh-cut 'Empire' and 'Crispin' apples. *J. Food Qual.* 28(3):289–307.

Saftner R., Y. Luo, J. McEvoy, J.A. Abbott, and B.T. Vinyard. 2007. Quality characteristics of fresh-cut watermelon slices from non-treated and 1-methylcyclopropene- and/or ethylene-treated whole fruit. *Postharvest Biol. Technol.* 44(1):71–79.

Shewfelt R.L. 1999. What is quality? *Postharvest Biol. Technol.* 15(3):197–200.

Soliva-Fortuny R.C., N. Alòs-Saiz, A. Espachas-Barroso, and O. Martín-Belloso. 2004. Influence of maturity at processing on quality attributes of fresh-cut conference pears. *J. Food Sci.* 69(7):290–294.

Soliva-Fortuny R.C., N. Grigelmo-Miguel, I. Hernando, M.A. Lluch, and O. Martin-Belloso. 2002a. Effect of minimal processing on the textural and structural properties of fresh-cut pears. *J. Sci. Food Agric.* 82(14):1682–1688.

Soliva-Fortuny R.C., G. Oms-Oliu, and O. Martin-Belloso. 2002b. Effects of ripeness stages on the storage atmosphere, color, and textural properties of minimally processed apple slices. *J. Food Sci.* 67(5):1958–1963.

Soliva-Fortuny R.C., M. Ricart-Coll, P. Elez-Martinez, and O. Martin-Belloso. 2007. Internal atmosphere, quality attributes and sensory evaluation of MAP packaged fresh-cut Conference pears. *Int. J. Food Sci. Technol.* 42(2):208–213.

Soliva-Fortuny R.C., M. Ricart-Coll, and O. Martin-Belloso. 2005. Sensory quality and internal atmosphere of fresh-cut Golden Delicious apples. *Int. J. Food Sci. Technol.* 40(4):369–375.

Tapia M.S., M.A. Rojas-Graü, A. Carmona, F.J. Rodriguez, R. Soliva-Fortuny, and O. Martín-Belloso. 2008. Use of alginate- and gellan-based coatings for improving barrier, texture and nutritional properties of fresh-cut papaya. *Food Hydrocolloids.* 22(8):1493–1503.

Trosset M.W. 2005. Visualizing correlation. *J. Comp. Graph. Stats.* 14(1):1–19.

Vilas-Boas E.V.B., and A.A. Kader. 2007. Effect of 1-methylcyclopropene (1-MCP) on softening of fresh-cut kiwifruit, mango and persimmon slices. *Postharvest Biol. Technol.* 43(2):238–244.

Voon Y.Y., N. Sheikh Abdul Hamid, G. Rusul, A. Osman, and S.Y. Quek. 2007. Volatile flavour compounds and sensory properties of minimally processed durian (*Durio zibethinus* cv. D24) fruit during storage at 4°C. *Postharvest Biol. Technol.* 46(1):76–85.

Waldron K.W., M.L. Parker, and A.C. Smith. 2003. Plant cell walls and food quality. *Comp. Rev. Food Sci. Food Safety.* 2:101–119.

Watkins C.B. 2008. Overview of 1-methylcyclopropene trials and uses for edible horticultural crops. *HortSci.* 43(1):86–94.

Wright K.P., and A.A. Kader. 1997a. Effect of controlled-atmosphere storage on the quality and carotenoid content of sliced persimmons and peaches. *Postharvest Biol. Technol.* 10(1):89–97.

Wright K.P., and A.A. Kader. 1997b. Effect of slicing and controlled-atmosphere storage on the ascorbate content and quality of strawberries and persimmons. *Postharvest Biol. Technol.* 10(1):39–48.

Yahia E.M. 2006. Controlled atmosphere, modified atmosphere and modified atmosphere packaging for vegetables. *Stewart Postharvest Rev.* 2(5–6):1–6.

6 Nutritional and Health Aspects of Fresh-Cut Vegetables

Begoña De Ancos, Concepción Sánchez-Moreno, Lucía Plaza, and M. Pilar Cano

CONTENTS

6.1   INTRODUCTION

The quality of fresh-cut vegetables depends on two types of factors: apparent and nonapparent. Aspects connected with appearance, such as color, texture, aroma, and flavor, are characteristics of fresh-cut vegetables that consumers detect immediately by way of the senses and determine how acceptable or unacceptable they find a pre-cut product. Nonapparent characteristics include microbiological quality, nutritional quality, and health-beneficial properties of fresh-cut vegetables. Microbiological quality and absence of toxic substances are important factors determining the safety of fresh-cut vegetables for public consumption. Safety and sensory quality (color, texture, aroma, and flavor) are the quality characteristics of fresh-cut vegetables which have received the most attention from researchers and processors in the sector (González-Aguilar et al., 2005; Heard, 2005; Hurst, 2005; Jongen, 2005; Balla and Farkas, 2006).

The nutritional quality and health-beneficial properties of plant products have been studied by many research groups because of the growing consumer interest in recent years in including functional foods in their diets. Functional foods, in addition to containing nutrients essential to life, contain other substances that with daily intake as part of a balanced diet help to maintain or improve the general state of health and well-being of consumers. On the basis of this definition, then, plant foods may be considered functional foods as they are, without any need to substitute, add to, or enrich their composition. Moreover, many of the ingredients used in the functional food industry are of vegetable origin (phytosterols, isoflavones, lycopene, etc.) (Hasler et al., 2004).

Numerous epidemiological studies in recent years have shown that there is an inverse relationship between a diet rich in fruit and vegetables and the incidence of chronic degenerative diseases such as certain kinds of cancer and cardiovascular diseases (Steinmetz and Potter, 1996; Ness and Powles, 1997; Tribble, 1998; Lampe, 1999; Slattery et al., 2000; Kris-Etherton et al., 2002; Maynard et al., 2003; Temple and Gladwin, 2003; Trichopoulou et al., 2003). This effect has been attributed to the presence in these foods of compounds with certain biological actions that produce health-beneficial effects and are known as bioactive or phytochemical compounds (Prior and Cao, 2000). The composition of fruits and vegetables therefore includes not only nutrients essential to life (carbohydrates, proteins, fats, vitamins, etc.), but also other substances that can potentially protect against certain degenerative diseases, known as phytochemical or bioactive compounds (carotenoids, phenolics, vitamins A, C, and E, fiber, glucosinolates, organosulfur compounds, sesquiterpenic lactones, etc.), whose biological activity has been studied by means of *in vitro* and *ex vivo* assays and human intervention studies (Eastwood and Morris, 1992; Giovannucci et al., 1995; Ling and Jones, 1995; Knekt et al., 1997; Kohlmeier and Su, 1997; Duthie et al., 2000; Knekt et al., 2000; Le Marchand et al., 2000; Piironen et al., 2000; Skimola and Smith, 2000; Simon et al., 2001; Olmedilla et al., 2001; Lampe and Peterson, 2002; Plaza et al., 2006a, 2006b; Sánchez-Moreno et al., 2006a, 2006b, 2006c).

6.2 MINIMAL PROCESSING OF VEGETABLE PRODUCTS

In the 21st century, the consumption habits of the people and their knowledge about nutrition has changed the lines of research in food technology. We can see the same process in supermarkets, which increasingly offer refrigerated displays with bags or trays of fruits or vegetables that have been selected, washed, and cut ready for consumption (i.e., fresh-cut vegetables). These products offer considerable advantages to present-day consumers in that they are easy and convenient to prepare and retain their original freshness of color, texture, aroma, and flavor without loss of their nutritional and health-beneficial properties. Fresh-cut vegetables are prepared for direct consumption by means of simple processes (selection, washing, peeling, stoning, cutting, hygienizing, etc.); they are packed under plastic film and are stored chilled in modified atmospheres (modified atmosphere packaging [MAP]). Consumers expect foods to be not only microbiologically safe but also nutritious, healthy, and easy to prepare for consumption (González-Aguilar et al., 2005).

Because they have not been subjected to harsh processes such as heat treatments, cut vegetables retain their initial freshness and also their nutritional components and the compounds responsible for their healthy properties. The health-giving potential of fresh cut vegetables should therefore be the same as that of the whole products from which they come; or the concentrations of some bioactive compounds may even be significantly improved depending on the selection of varieties, agricultural practices, physiological state of the plant of origin, and the stress produced by processing.

6.3 PHYTOCHEMICAL COMPOUNDS IN VEGETABLE PRODUCTS

The mechanics of the beneficial effects of consuming fruit and vegetables, either whole or cut, are not fully understood, although they appear to be connected with antioxidant activity, modulation of detoxifying enzymes, stimulation of the immune response, modification of inflammatory processes, reduction of platelet aggregation, disruption of cholesterol metabolism, modulation of the concentration of steroid hormones and hormone metabolism, reduction of blood pressure, and antiviral and antibacterial activity (Lampe, 1999).

Phytochemical compounds may be defined as chemical substances contained in plant foods, which endow the food with physiological properties above the strictly nutritional.

It is now accepted that oxidative stress or oxidative overload is implicit in manifold degenerative diseases (Halliwell, 1987), such as cancer, cardiovascular diseases, atherosclerosis, neurological disorders such as Alzheimer's disease, cataract formation, macular degeneration, inflammatory processes, or the aging process. Most of these phytochemical substances possess antioxidant capacity.

An antioxidant compound may be defined as one that, when present in low concentrations with respect to an oxidizable substrate, delays or inhibits oxidation of that substrate. On a physiological level, the oxidizable substrates are lipids, proteins, and DNA, and the antioxidants may be either endogenous or exogenous (dietary). Among the most important antioxidants present in foods are vitamins C and E, carotenoids, and phenolic compounds. However, some assays involving supplementation with these antioxidants, especially β-carotene, have produced contradictory results. Numerous epidemiological studies in the last 15 years (ATBC, CARET, Women's Health Study) (Heinonen and Albanes, 1994; Hennekens et al., 1996; Mayne, 1996; Lee et al., 1999) have shown that β-carotene supplements, whether alone or in association with vitamin E, do not reduce the incidence of some kinds of cancer or of cardiovascular diseases, and that their use may even be inadvisable in individuals subject to certain physiological conditions, such as smokers. On the other hand, epidemiological studies with fruits and vegetables or derivatives thereof have produced health-beneficial results (Lampe, 1999; Aviram et al., 2000; Kris-Etherton et al., 2002; Temple and Gladwin, 2003).

6.3.1 VITAMIN C

Vitamin C belongs to the group of water-soluble vitamins and is one of the most important micronutrients, 90% of which humans derive from the intake of fruit and vegetables. Structurally, vitamin C is composed of chenodiol conjugated with the carbonyl group in the lactone ring. In the presence of oxygen, ascorbic acid oxidizes to dehydroascorbic acid; this has the same vitamin activity but is more unstable, so that the activity can readily be lost through lactone hydrolysis and formation of 2,3-diketogulonic acid (Figure 6.1).

Numerous epidemiological studies have shown a strong correlation between the healthy effects of consuming fruit and vegetables and the vitamin C content of these (Ness et al., 1996; Block et al., 2002).

Vitamin C (ascorbic acid, ascorbate) is one of the most efficacious and is the least toxic of antioxidants, and it is particularly effective against free radicals. A low intake of vitamin C–rich products results in low levels of vitamin C in the blood (0.3 mg/dL); levels between 0.8 and 1.3 mg/dL, which are considered necessary for good health (Simon et al., 2001), may be attained through a daily intake of 90 mg of vitamin C in adults (Taylor et al., 2000).

In general, the vitamin C (ascorbic acid-AA + dehydroascorbic acid-DAA) content of fruit and vegetables depends on the species, cultivar, climatic conditions, agricultural practices, ripeness, and of course postharvest handling (Lee and Kader, 2000) (Tables 6.1 and 6.2). The vitamin C concentration generally increases in vegetable tissues through the action of light during growth of the plant and tends to decrease in the presence of nitrogen-rich fertilizers. However, the principal cause of vitamin C degradation in vegetables is storage at high or inappropriate postharvest temperatures.

Vitamin C content starts to decline as soon as the product is harvested. Levels depend significantly on the type of vegetable and the processing and storage conditions; the amount of vitamin C that is degraded increases with storage temperature and time (Davey et al., 2000; Lee and Kader, 2000). Fruits with acid pH generally present slower rates of vitamin C degradation than vegetables with high pH. However, among fruits, vitamin C has been found to be more stable in citrus than in berry fruits such as strawberries or raspberries (Davey et al., 2000).

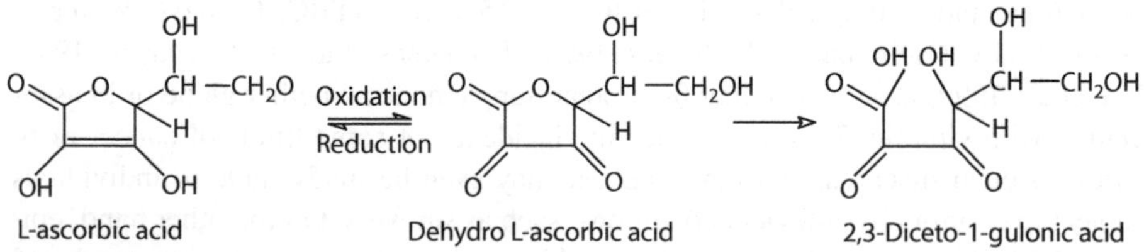

FIGURE 6.1 Oxidation of L-ascorbic to dehydro-L-ascorbic acid followed by evolution into products lacking biological activity.

TABLE 6.1
Vitamin C Content (mg/100 g f.w.) of Some Fruits

Product	Ascorbic Acid	Dehydroascorbic Acid	Total Vitamin C
Banana	15.3	3.3	18.6
Blackberry	18.0	3.0	21.0
Melon	31.3	3.0	34.3
Grape	21.3	2.3	23.6
Kiwi	59.6	5.3	64.9
Lemon	50.4	23.9	74.3
Mandarin	34.0	3.7	37.7
Orange (California)	75.0	8.2	83.2
Orange (Florida)	54.7	8.3	63.0
Persimmon	110.0	100.0	210.0
Raspberry	27.0	2.0	29.0
Strawberry	60.0	5.0	65.0

Source: Lee, S.K.; Kader, A.A. 2000. Preharvest and postharvest factors influencing vitamin C content in horticultural crops. *Postharvest Biol Technol.* 20:207–220. With permission.

TABLE 6.2
Vitamin C Content (mg/100 g f.w.) of Some Vegetables

Products	Ascorbic Acid	Dehydroascorbic Acid	Total Vitamin C
Broccoli	89.0	7.7	96.7
Cabbage	42.3	—	42.3
Cauliflower	54.0	8.7	62.7
Collards	92.7	—	92.7
Peppers (red)	151.0	4.0	155.0
Peppers (green)	129.0	5.0	134.0
Potatoes	8.0	3.0	11.0
Spinach	62.0	13.0	75.0
Swiss chard	—	45.0	45.0
Tomatoes	10.6	1.7	9.7

Source: Lee, S.K.; Kader, A.A. 2000. Preharvest and postharvest factors influencing vitamin C content in horticultural crops. *Postharvest Biol Technol.* 20:207–220. With permission.

6.3.2 CAROTENOID COMPOUNDS

Carotenoids are lipid-soluble plant pigments responsible for the yellow, orange, or red colors of numerous plant products. The carotenoid are isoprenoid compounds (tetraterpenes) generally having 40 carbon atoms. Most carotenoids are structurally arranged as two substituted ionone rings separated by four isoprene units containing nine conjugated double bonds, such as α- and β-carotene, lutein and zeaxanthin, and

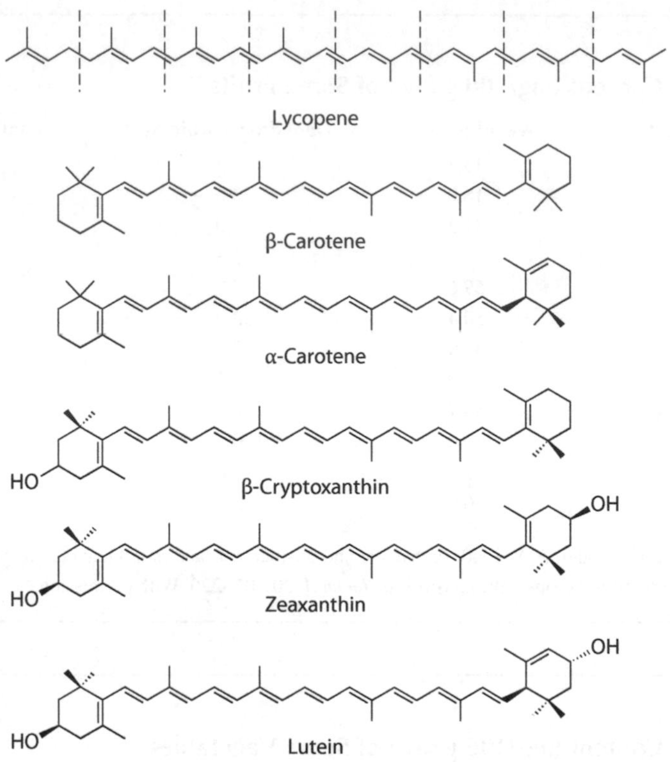

FIGURE 6.2 Majority carotenoids in fruits and vegetables.

α- and β-cryptoxanthin. These carotenoids, along with lycopene, an acyclic biosynthesis precursor of β-carotene, are most commonly consumed and are most prevalent in human plasma (Castenmiller and West, 1998; Van den Berg et al., 2000).

They can be divided into two groups in terms of their chemical composition: *carotenes* that contain only carbon and hydrogen atoms and *xanthophylls* or oxocarotenoids, whose structure contains an oxygen function, such as keto, hydroxyl, or epoxy groups, generally in the terminal rings (Figure 6.2). Because of the presence of double bonds, carotenoids present different geometric shapes (cis/trans), which can be converted from one to another through the action of light or thermal or chemical energy. Therefore, carotenoid compounds are highly sensitive to light, oxygen, and low pH conditions. At present, more than 600 different carotenoids have been identified, although only about two dozen are regularly consumed by humans.

The system of conjugated double bonds influences their physical, biochemical, and chemical properties. Thus, carotenoid compounds like β-carotene, α-carotene, β-cryptoxanthin, lycopene, lutein, and zeaxanthin perform important biological functions such as antioxidant activity, stimulation of intercellular communication, control of cell growth, intercellular differentiation of growth control, cellular differentiation (inhibition of mutagenesis), and modulation of the immune response. But lycopene, the major carotenoid in tomato and its derivatives, is the only one, either ingested or in serum, to have been inversely associated by a wide-ranging epidemiological study with the appearance of prostate cancer (Giovannuci et al., 1995), while others reported better results with derivatives than with unprocessed fresh tomato (Gartner et al., 1997).

Also, high plasma levels of lycopene, lutein, or α-carotene have been inversely associated with risk of coronary disease, myocardial infarction, and atherosclerosis. All these studies have been summarized by Olmedilla et al. (2001). Carotenoid pigments are also of physiological interest in human nutrition, because some are vitamin A precursors, especially β-carotene, α-carotene, and α- and β-cryptoxanthin. Carotenoid intake assessment has been shown to be complicated mainly because of inconsistencies in food composition tables and databases. No formal diet recommendation for carotenoids has yet been established, but some experts suggest intake of 5 to 6 mg/day. In the case of vitamin A, for adult human males, the RDA is 1,000 μg retinyl Eq/day, and for adult females, 800 μg retinyl Eq/day (O'Neill et al., 2001; Trumbo et al., 2003).

6.3.3 PHENOLIC COMPOUNDS

Phenolic compounds are secondary metabolites produced by vegetables, which possess in common a benzene ring in their chemical structure with hydroxyl groups that are responsible for their activity. The polyphenols present a wide variety of structures ranging from simple molecules (monomers and dimers) to polymers (tannins, molecular weights greater than 500 daltons). Among the most abundant are hydrocinnamic acids (C6-C3), benzoic acids (C6-C1), flavonoids (C6-C3-C6), proanthocyanidins (C6-C3-C6)n, stilbenes (C6-C2-C6), lignanes (C6-C3-C3-C6), and lignines (C6-C3)n (Scalbert et al., 2005). Phenolics have been widely studied for their relationship with the quality characteristics of plant foods such as color, because many of them are pigments such as anthocyanins (responsible for the color of grapes, cherries, plums, strawberries, raspberries, etc.). Others, when oxidized enzymatically (polyphenol oxidase [PPO], EC 1.14.18.1), cause enzymatic browning, which is responsible for a large percentage of loss of quality in plant foods during processing and storage. They also contribute to the flavor and aroma of plant foods. For example, condensation of catechols to tannins produces the bitter, astringent taste in unripe apples and in persimmons.

Flavonoids are the most common and widely distributed group of plant phenolics. Over 500 different flavonoids have been described, and they are classified into at least ten chemical groups. Among them, flavones, flavonols, flavanols, flavanones, anthocyanins, and isoflavones are particularly common in fruits and vegetables. Flavonoids have been the most widely studied group of polyphenols in the last few years. Most existing flavonoids in fruits and vegetables have shown to a lower or higher extent some antioxidant and radical scavenger activity in *in vitro* studies (Rice-Evans et al., 1996; Williamson and Manach, 2005).

There are epidemiological studies that associate intake of foods rich in phenolics, and most particularly flavonoids, with a low incidence of cardiovascular diseases and some types of cancer (Derbyshire et al., 1995; Duthie et al., 2000; Skibola and Smith, 2000). The more widely studied foods include tea, spices, and grape derivatives, and

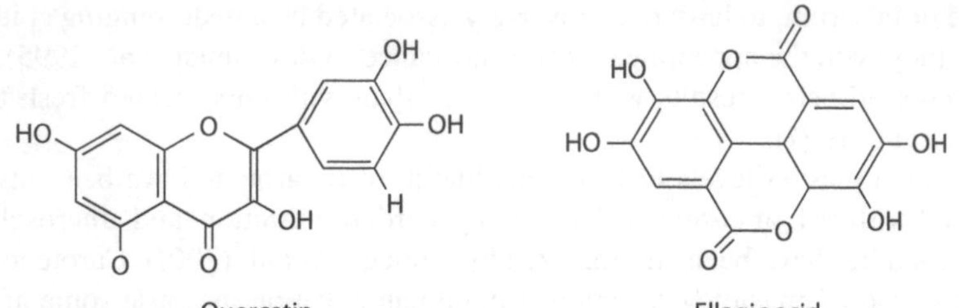

Quercetin   Ellagic acid

FIGURE 6.3 Structures of quercetin and ellagic acid.

also some fruits such as apples and berries (strawberries, raspberries, blackberries, etc.). Flavonoids have been found to possess anti-inflammatory, antiallergic, antiviral, hypocholesterolemic, and anticarcinogenic activities (Middleton, 1996; Knekt et al., 2000; Skibola and Smith, 2000; Wang and Mazza, 2002). It is also important to note another major estrogenic activity found in a group of flavonoids, the isoflavones (genistein and daidzein), large concentrations of which occur in soy and derivatives and which stimulate bone mineralization as well as help prevent atherosclerosis and some kinds of cancer (Ling and Jones, 1995; Liu et al., 2005). Onions are also an important source of the flavonol quercetin (Figure 6.3), which is responsible for the yellow tone of the pulp and the brown of the skin and is believed to be one of the constituents responsible for the protective effects against cancer and cardiovascular diseases which have been associated with the intake of this product (Griffiths et al., 2002). Quercetin is also found in apples and is one of the constituents responsible for the health benefits deriving from their consumption (Boyer and Liu, 2004). We would also note the presence in citrus fruits of glycosidated flavanones (hesperidin and narirutin) that have been associated with health-beneficial properties of oranges and orange juice (De Pascual-Teresa et al., 2007). Red grapes, blackberries, whortleberries, raspberries, and strawberries are all rich in hydroxybenzoic acids, notably ellagic acid that has a protective effect against cancer (Figure 6.3) (Maas and Galleta, 1991). In general, fruits and vegetables are generally good sources of flavonoids, and the more we know about their metabolism, the physiological beneficial effects of their intake are increasingly being attributed to their conjugates and metabolites than to the intact molecule.

6.3.4 PHYTOSTEROLS

Plant sterols have been reported to include over 250 different sterols and related compounds. The most common sterols in fruits and vegetables are β-sitoesterol, and its 22-dehydro analogue stigmasterol, campesterol, and avenasterol. The chemical structure of these sterols is similar to cholesterol differing in the side chain. It has been known for years that sterols, which are found largely in vegetables of the *Brassicaceae* family (cauliflower, broccoli) and in avocado fruit, have hypocholesterolemic effects when ingested in amounts of 1 to 3 g/day, and they are therefore seen as valuable allies in the prevention of cardiovascular diseases. The hypocholesterolemic effect of phytosterols and their reduced forms, phytostanols, is attributed to three metabolic actions: inhibition of intestinal absorption of cholesterol by competition, reduction of esterification, and stimulation of its excretion into the intestine. Phytosterols cause reduction in total plasma cholesterol and low-density lipoprotein (LDL) cholesterol without affecting levels of high-density lipoprotein (HDL) cholesterol (Piironen et al., 2000).

6.3.5 ORGANOSULFUR COMPOUNDS

Vegetables of the *Liliaceae* (onions and garlic) and *Brassicaceae* (broccoli, cabbage, Brussels sprouts, cauliflower) families are considered to have health-beneficial properties deriving from the presence of substances whose chemical structure contains sulfur atoms. Allicin (2-propene-1-sulfinothioic acid S-2-propenyl ester $CH_2=CH-CH_2-S(0)-S-CH_2-CH=CH_2$) is believed to be the biologically active molecule in garlic when this is macerated or homogenized. Allicin is not present in garlic until its precursor, alliin or (+)-S-allyl-L-cysteine sulfoxide ($CH_2=CH-CH_2-S(0)$ $CH_2-CH(NH_2)-COOH$), comes into contact with the enzyme alliinase when the cell membrane is ruptured by chopping or maceration of the garlic. Allicin is highly sensitive to heat, to low pH, and to organic solvents and produces various different degradation compounds when the garlic is processed, such as allyl sulfide, disulfide, and trisulfide (the latter the odoriferous factor), and allyl thiosulfinate, which is the molecule considered responsible for the health benefits. Also, allicin is responsible for garlic's bactericide action. In general, the following beneficial effects have been ascribed to garlic: antioxidant and free radical scavenging (a powerful inhibitor of lipid peroxidation), anti-inflammatory, anticoagulant, fungicide, antiviral (influenza, AIDS), interferon and immune system enhancement, bactericide, reduction of cholesterol levels, anticarcinogenic, and antimutagenic (Lawson, 1998; Seki et al., 2000; Benkeblia, 2004).

Numerous epidemiological studies have shown that a diet rich in vegetables of the *Brassicaceae* family (broccoli, cabbage, Brussels sprouts, cauliflower) reduces the incidence of some kinds of cancer (Wattenber, 1993; Michaud et al., 1999; Lund, 2003). This anticarcinogenic effect is attributed to the presence in its composition of a group of organosulfur minority constituents known as glucosinolates and to the capacity of some of its metabolites, isothiocyanates and indoles, to intervene in biotransformations catalyzed by enzymes associated with the antioxidant systems of the human organism, such as glutathione-S-transferase (Lampe and Peterson, 2002).

The nature of all the substances with health-beneficial activity present in fruits and vegetables is not yet fully understood, and also not fully understood are the chemical and biochemical mechanisms whereby these phytochemical substances affect certain physiological functions of the organism. It must be remembered that the protective effect of a diet rich in fruits and vegetables depends on substances of different chemical natures and different mechanisms, which may produce additive, synergistic, or even antagonistic effects. Some phytochemical compounds have been clearly associated with healthy effects on the organism, such as vitamin C or lycopene; in other cases, the beneficial effect has not been fully demonstrated by means of epidemiological studies. At the present time, the biological activity of fruit and vegetable constituents is still the object of research, in particular, substances originally known for their toxic action, such as glucosinolates and their hydrolysis products, isothiocyanates, which possess anticarcinogenic properties in normal doses of the vegetable. One example of this is the health-protective effect associated with ingestion of carrot, which has been attributed to the presence of β-carotene and α-carotene. However, other studies have shown that the polyacetylene falcarinol, which is present in carrots and whose physiological function in the plant is to protect against fungal attacks, presents considerable anticarcinogenic activity in *in vivo* experiments in which rats are treated with the carcinogen azoxymethane. The anticarcinogenic effect of the intake of carrot can therefore no longer be attributed only to carotenoid compounds (Brandt et al., 2004).

6.4   EFFECT OF MINIMAL PROCESSING ON PHYTOCHEMICAL COMPOUNDS AND ANTIOXIDANT CAPACITY IN PLANT PRODUCTS

There is growing interest in the nutritional value of foods in order to understand what the contribution of each individual food constituent is to daily nutritional needs and human well-being and how the food processing and preservation technologies affect its nutritional composition and health-promoting capacity. The fresh-cut fruit and vegetable industry is constantly growing due to consumer demand for plant food-derived products as the main carriers of the phytochemical compounds responsible for recognized beneficial health effects due to the intake of such types of food. In addition, the fresh-cut fruit and vegetable industry reflects the needs of the consumer with regard to faster preparation and more convenient products with nutritional value preserved, and natural and fresh color, flavor, and texture retained without the use of preservatives.

"Minimal processing" includes various nonthermal processing technologies that ensure product safety while maintaining the fresh appearance of fruit and vegetable products. The different stages in minimal processing of vegetables (peeling, cutting, washing, sanitizing) not only affect safety and sensory quality but can also cause changes in the nutritional quality and health-promoting properties of final products. The initial stages (peeling, cutting, shredding) promote a faster microbial degradation, physiological alterations (increasing respiration rate and ethylene synthesis), and biochemical changes induced by the rupture of plant cells and the leakage of intercellular products, facilitating reactions between oxidative enzymes and their substrates. This situation leads to various degradation reactions, the most important of which is enzymatic browning, which may promote color, texture, and flavor changes. For instance, phytochemical compounds can be oxidized in contact with oxygen or by light. Also, growth of spoilage or pathogenic microorganisms is facilitated at cutting sites, and different technologies are therefore needed to mitigate the effects of these degenerative processes. Storage temperature is the main factor affecting spoilage of minimally processed plant products. Other technologies can be employed such as the use of antioxidants, antimicrobial agents such as chlorines, and MAP to inhibit the decay of fresh-cut fruit and vegetables. Antimicrobial treatments with oxidizing agents like sodium hypochlorite can oxidize certain phytochemical compounds (Gil et al., 2006).

A global processing and storage design to achieve high-quality minimally processed foods requires a combination of different strategies and technologies to help reduce degradative processes in fresh-cut vegetables. Moreover, the content of bioactive compounds in minimally processed plant products can vary with genotype, environmental stress, growth conditions, and storage and processing conditions (Vallejo et al., 2002, 2003; West et al., 2004). In conclusion, the selection of cultivar and agricultural practices in combination with specific technologies to reduce the negative effects of processing (cutting, peeling, shredding) are necessary to obtain minimally processed plant products enriched in bioactive compounds. Special attention must be paid to some phases of the processing of minimally processed plant-derived products, such as the sanitation process. In these stages of the process, it is necessary to avoid enzymatic- and microbiological-induced alterations that may affect the stability of the phytochemical compounds responsible for the nutritional quality and health-promoting properties of fresh-cut vegetables. In the design of new processes to obtain precut fruits and vegetables with improved nutritional and health-promoting characteristics, processing and storage technology are selected on the basis of how they improved the nutritional constituents and antioxidant characteristics of the plant products.

Below are some results of the effect of minimal processing of fruits and vegetables on their phytochemical compounds and antioxidant properties.

6.4.1 Vitamin C

The first step in obtaining more nutritious and healthy minimally processed products is to use the starting material with a high content of bioactive compounds. The content of vitamin C in vegetables and fruits and their relationship with genetic, ripeness stage, agronomical practices, or climatic factors has been studied for different products and by diverse authors (Lee and Kader, 2000; Vallejo et al., 2002, 2003; Cano et al., 2008).

Importantly, there are not only differences between varieties in vitamin C content but also among different cultivars. Thus, a study of varieties of *Citrus* fruit (Cano et al., 2008) showed a higher vitamin C content in orange varieties (Navel, Sweet, and Sanguine) than in mandarin varieties (Clementines and Satsumes). Among orange varieties, Newhall and Navelate (Navel group) showed the highest vitamin C concentrations (64.2 and 47.8 mg/100 g, respectively), whereas Lanelate and Fukumoto presented the lowest values (37.5 and 39.5 mg/100 g, respectively). Among mandarin varieties, Oronules and Hernandine (Clementine group) showed the highest content (47.4 and 46 mg/100 g, respectively), whereas the Satsuma group (Owari, Oktisu, Avasa Pri-10, and Avasa Pri-19) presented the lower values (20, 21.3, 24.9, and 25.5 mg/100 g, respectively). Significant genotypic differences have been found between fruits of *Actinidia* family. Kiwifruit is well known today as a fruit with high vitamin C content. A study of eight *Actinidia* genotypes (Du et al., 2009) showed that vitamin C content varied from 41.7 to 1322.91 mg ascorbic acid/100 g f.w. Wild species such as *Actinidia eriantha* and *Actinidia latifolia* have higher vitamin C content (1284.87 and 1322.91 mg ascorbic acid/100 g f.w., respectively) than cultivars of *Actinidia chinensis* and *Actinidia deliciosa*. Thus, *Actinidia deliciosa* cv. Hayward and *Actinidia chinensis* cv. Xixuan showed lower vitamin C content with 63.41

TABLE 6.3
Effect of Minimal Processing on Vitamin C
Content

Tomato Cultivars	Vitamin C (mg/kg)
Rambo	
Whole	69.6 ± 2.4[a]
Fresh-cut	73.7 ± 7.7[a]
Durinta	
Whole	212.3 ± 1.5[b]
Fresh-cut	204.8 ± 9.0[b]
Bodar	
Whole	151.1 ± 3.5[c]
Fresh-cut	139.9 ± 9.3[c]
Pitenza	
Whole	94.5 ± 0.5[d]
Fresh-cut	108.1 ± 5.3[d]
Cencara	
Whole	81.3 ± 3.0[a]
Fresh-cut	74.3 ± 14.4[a]
Bola	
Whole	129.2 ± 5.1[e]
Fresh-cut	143.8 ± 2.6[e]

Note: Different letters indicate significant differences ($P < 0.05$).

Source: Odriozola-Serrano, I.; Soliva-Fortuny, R.; Martín-Belloso, O. 2008. Effect of minimal processing on bioactive compounds and color attributes of fresh-cut tomatoes. *LWT—Food Sci Technol.* 41: 217–226.

and 42.27 mg ascorbic acid/100 g f.w., respectively. Also, significant differences in vitamin C concentrations have been found in tomato cultivars (Rambo, Durinta, Bodar, Pitenza, Cencara, Bola). Vitamin C ranged from 69.6 mg ascorbic acid/100 g f.w. in Rambo to 212.3 mg ascorbic acid/100 g f.w. in Durinta tomatoes (Table 6.3) (Odriozola-Serrano et al., 2008).

In general, ascorbic acid content of vegetable products declines as a consequence of peeling and cutting operations, although this effect depends on the type of vegetable concerned and temperature and storage time (Lee and Kader, 2000; Gil et al., 2006). For instance, vitamin C content is significantly reduced in carrot, celery, potato, courgette, and white cabbage ($P < 0.05$) but is not significantly affected in red cabbage, radishes, and sweet potatoes after 2 days of storage at 15°C (Reyes et al., 2007) (Table 6.4).

Sometimes vitamin C showed no significant changes between whole vegetables and just-processed fresh-cut products. Thus, no effect of slicing and packaging

TABLE 6.4

Changes in Ascorbic Acid Content of Different Vegetables after Chopping and Storage for 2 Days at 15°C

	Ascorbic Acid (mg/kg f.w.)	
Products	Whole	Chopped
Potato	50.0 ± 11.0	33.8 ± 5.6
Sweet potato	162 ± 28	143 ± 21
Celery	33.6 ± 4.3	15.6 ± 4.2
White cabbage	350 ± 28	310 ± 23
Courgette	125 ± 20	59 ± 4.0
Carrot	16.7 ± 1.5	3.0 ± 0.2
Radish	225 ± 13	210 ± 14
Red cabbage	599 ± 24	633 ± 23

Source: Reyes, L.F.; Villareal, J.E.; Cisneros-Zevallos, L. 2007. The increase in antioxidant capacity after wounding depends on the type of fruit or vegetable tissue. *Food Chem.* 101:1254–1262. With permission.

was observed on vitamin C content in tomatoes regardless of cultivar studied (Table 6.3).

Another important factor influencing vitamin C retention in fresh-cut vegetables is the cutting mechanism used. For instance, in iceberg lettuce, 18% more ascorbic acid is retained by using a smooth knife instead of an undulated one. Also, 25% more vitamin C is retained when a mechanical cutter is used rather than cutting the lettuce with a knife by hand (Barry-Ryan and O'Beirne, 1999).

Cutting the product may promote the loss of vitamin C as a result of the increase of the surface area in contact with oxidizing agents such as oxygen that favors the action of ascorbate peroxidase (APX), the main enzyme responsible for degradation of vitamin C. Also, this vitamin loss could be attributed to the effect of the ethylene generated as a consequence of cutting.

Another important effect of the minimal process is related to the loss of vitamin C into the washing water (stage following cutting) through the cut site. These losses may also contribute to the differences in vitamin C content between cut and whole samples. Also, it is important to highlight the effect of sanitation treatments on the bioactive compounds in the plant tissues.

In the case of broccoli, the cutting stage reduced vitamin C content of fresh-cut product by 27% with respect to the uncut product (Figure 6.4). The vitamin C content of broccoli differs according to the part that is analyzed (stalk or floret); therefore, because the whole samples had the complete stalk, their vitamin C content was between 30% and 40% higher than that of cut samples. However, the different sanitation treatments used (water and a solution of 150 ppm sodium hypochlorite) did not

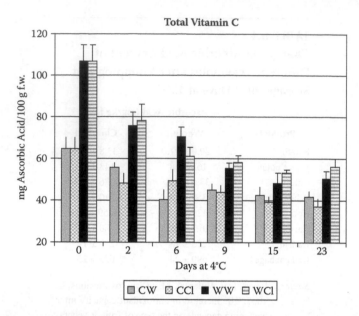

FIGURE 6.4 Effect of minimal processing on vitamin C in broccoli florets packaged under slightly modified atmosphere (16% O_2 + 4% CO_2) at 4°C. (CW, cut/washed in water at 4°C; CCL, cut/washed in 150 ppm sodium hypochlorite at 4°C; WW, whole/washed in water at 4°C; WCL, whole/washed in 150 ppm sodium hypochlorite at 4°C.) (From Martínez, J.A.; Sgroppo, S.; Sánchez-Moreno, C.; De Ancos, B.; Cano, M.P. 2005. Effects of processing and storage of fresh-cut onion on quercetin. *Acta Horticulturae.* 682:1889–1895.)

affect vitamin C content (Martínez et al., 2007) (Figure 6.4). When the equilibrium atmosphere was obtained on the second day of storage, there was a 28% reduction in total vitamin C concentration. Thereafter, storage under modified atmosphere effectively maintained the vitamin C concentration of the fresh-cut broccoli for 23 days at 4°C, demonstrating the protective effect of the modified atmosphere. These levels remained practically unchanged even when the cold chain was broken, after 24 hours of permanence of product at 20°C (Martínez et al., 2007).

Another example of vitamin C stability versus different treatments was reported for precut iceberg lettuce. Treatments with solutions of chloride (100 mg/L), citric acid (5 g/L), lactic acid (5 mL/L), and ozonated water (4 mg/L) maintained the initial levels of vitamin C in iceberg lettuce practically unchanged over 8 days at 4°C, with no differences deriving from the particular treatment used (Akbas and Ölmez, 2007).

In general, the vitamin C content of vegetables declines during storage. Low-oxygen atmospheres combined with low temperature help maintain the initial levels of vitamin C, although ascorbic acid (AA) is converted to dehydroascorbic acid (DAA) during storage (Figure 6.1). In these conditions, 87% of the initial vitamin C content was retained in green pepper rings after 15 days of chilled storage in a modified atmosphere (5% O_2 + 4% CO_2), while only 57% was retained in air (Howard and Hernandez-Brenes, 1998; Pilon et al., 2006). However, in some cases, better results are achieved when the cut vegetable is stored in air. For instance, vitamin C levels were maintained or even increased (16% to 18%) in different potato varieties after cutting and storing in air for up to 6 days at 4°C, whereas the vitamin C content fell by up to 36% when the potatoes were stored in a modified atmosphere (8.2 to 9.8 kPa CO_2 + 3.1 to 3.8 kPa O_2) (Tudela et al., 2002).

In some cases, vitamin C is reduced in an atmosphere containing a high concentration of carbon dioxide (10% to 30% CO_2). This is due to the prejudicial effect of the CO_2 which causes cytoplasmic acidification and, consequently, oxidative damage with increasing ascorbate oxidase activity. From a nutritional standpoint, CO_2-rich atmospheres may not be suitable for maintaining initial vitamin C levels in some vegetables. An increasing level of carbon dioxide in the atmosphere is accompanied by an increase in the concentration of dehydroascorbic acid (DAA). It has been reported that ascorbic acid (AA) is more susceptible than dehydroascorbic acid to degradation by CO_2-rich atmospheres, as the AA content tends to decline in atmospheres of this kind, whereas DAA tends to increase. It has been suggested that AA oxidation can be enhanced by CO_2-rich atmospheres (Agar et al., 1997). The increase in the percentage of DAA accompanying loss of total vitamin C has therefore been associated with oxidative stress, a situation in which AA is required to neutralize the reactive oxygenated species (ROS) generated, causing the conversion of AA to DAA followed by degradation of the latter to products lacking biological activity. For instance, it was reported that more vitamin C was retained in spinach stored in a modified atmosphere (6% O_2 + 14% CO_2) than stored in air, although the concentration of dehydroascorbic acid was greater in the modified atmosphere (Gil et al., 1999). On the other hand, in cut Swiss chard, which contains only DAA, total vitamin C was retained better in air than in the modified atmosphere (Gil et al., 1998b).

From a nutritional standpoint, modified or controlled atmospheres containing more than 10% of CO_2 may not be suitable for maintaining initial vitamin C levels in some fruits, such as minimally processed kiwi (Agar et al., 1999). For instance, the vitamin C content in kiwi slices stored in a controlled atmosphere with 5%, 10%, and 20% CO_2 declined by 14%, 22%, and 34%, respectively, after 12 days at 0°C. On the other hand, other minimally processed fruits such as strawberries or persimmons maintained their initial vitamin C levels virtually unchanged when stored at 0°C for 8 days in controlled atmospheres with a CO_2 concentration of 12% (Wright and Kader, 1997).

A number of different researchers investigated the effect of minimal processing on the vitamin C content of fruits (Wright and Kader, 1997; Agar et al., 1999; Gil et al., 2006). The effect of peeling and cutting operations on the vitamin C content of different fruits stored in air for 9 days at 5°C was investigated by Gil et al. (2006) (Figure 6.5). They found that the response to the stress caused by the preparation process (peeling and cutting) in terms of vitamin C content depended on the type of fruit. For instance, fresh-cut pineapple chunks stored for 6 days at 5°C presented a considerable decrease in ascorbic acid and a significant increase in dehydroascorbic acid; the final outcome was a significant (10%) increase in total vitamin C content of fresh-cut pineapple with respect to whole unprocessed pineapple stored in the same conditions (Figure 6.5). In the case of strawberries, there were no significant differences between fresh-cut and whole fruits; in both products, there was a gradual increase in vitamin C content, so at the end of storage at 5°C, the final concentration was 17% higher than at the outset. On the other hand, diced mango maintained its

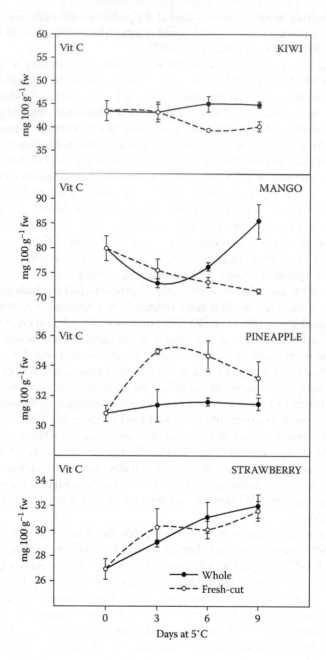

FIGURE 6.5 Total vitamin C content of whole and fresh-cut fruits stored in air at 5°C. (From Gil, M.I.; Aguayo, E.; Kader, A.A. 2006. Quality changes and nutrient retention in fresh-cut versus whole fruits during storage. *J Agric Food Chem.* 54: 4284–4296. With permission.)

initial vitamin C content over 6 days in storage at 5°C, with no significant differences from the whole fruit. Finally, the vitamin C content of fresh-cut kiwi was 7.5% lower than in the whole product at the end than after being in storage (Figure 6.5).

Another factor that may influence the nutritional quality of fresh-cut fruits and vegetables is the kind of cut applied. Argañosa et al. (2008) evaluated the effect of different cut types (cube, parallelepiped, cylinder, and sphere) on the quality of papaya cv. 'Sunrise' during storage at 4°C up to 10 days. In this study, cube-shaped fresh-cut papaya tended to retain higher ascorbic acid than the other cut types being cut in the cylinder, resulting in a loss of 18% in the ascorbic acid concentration during the 10-day shelf life at 4°C. In contrast, Artés-Hernandez et al. (2007) reported that wedges, slices, and half slices of fresh-cut lemon products ('Lisbon') preserved during 10 days at 0, 2, and 5°C, had good ascorbic acid retention (85%), and no differences were observed due to the type of cut.

Consumption of oranges is high throughout the world, and as such, they are considered to be the principal dietary source of vitamin C. Minimally processed orange segments are a new product that has been designed to facilitate acceptance by certain consumers (children, the elderly and infirm, etc.) or to serve the institutional catering sector (schools, hospitals, work canteens, hotels, etc.). Minimal processing of oranges entailing manual peeling and segmenting, packaging in a slightly modified atmosphere (19% O_2+ 3% CO_2), and low-temperature storage (4°C) produced no significant changes in vitamin C concentrations in orange segments over 10 days in chilled storage, while a decline in AA concentration of 24% was noted (Crespo et al., 2005). Piga et al. (2002) reported that minimal processing resulted in a 13% loss in AA concentration in segments of mandarin varieties after 12 days of storage at 4°C under MAP conditions. Del Caro et al. (2004) found an ascorbic acid decrease ranging from 1.63 to 5.10 mg/g dry matter in 'Minneola' tangelo and 'Salustiana' orange segments stored 12 days at 4°C, but no differences were observed in 'Shamouti' orange segments. A good vitamin C retention (about 85% of ascorbic acid and 15% of dehydroascorbic acid) was reported for other citrus fruits, such as fresh-cut lemons prepared as wedges, slices, and half slices and stored for 10 days at 0, 2, and 5°C (Artés-Hernández et al., 2007).

Vitamin C concentration remains relatively stable in minimally processed citrus products and does not differ significantly from whole oranges, lemons, or mandarins. The vitamin C concentration supplied by consumption of fresh-cut citrus fruits up to 12 days of refrigerated storage in MAP is similar to that supplied by an equivalent amount of freshly prepared citrus fruit (Del Caro et al., 2004; Crespo et al., 2005; Artés-Hernández et al., 2007).

Minimally processed fruits require a combination of new preservation techniques capable of keeping the safety and the quality of the fruit products for a longer storage time to meet consumer demands. In this sense, edible coatings could be applied to improve the shelf life of fresh-cut fruit packaged under MAP, thus reducing moisture and solute losses, gas exchange, and respiration and oxidative reaction rates, as well as reducing physiological disorders (Rojas-Graü et al., 2008). The effect of polysaccharide-based edible coatings such as pectin (2% w/v), gellano (0.5% w/v), and alginate (2% w/v) containing antibrowning substances such as N-acetylcysteine and glutathione on antioxidant constituents of fruits such as pears was studied by

Oms-Oliu et al. (2008a). The use of polysaccharide-based edible coatings with N-acetylcysteine and glutathione significantly reduced vitamin C loss of fresh-cut pears up to 11 days at 4°C (loss up to 18%) compared with uncoated or coated pieces without the incorporation of antioxidants (losses between 32% and 48%). The presence of antioxidants reduced the diffusion of oxygen protecting the vitamin C content of fresh-cut pears.

Initial processing operations do not seem to have a significant effect on the loss of vitamin C in fresh-cut fruits and vegetables, although the type of packaging and the storage temperature are determinant factors rather than initial processing stages.

6.4.2 PHENOLIC COMPOUNDS

Mechanical damage is sustained by plant tissue as a result of peeling and cutting increasing the synthesis of phenolics, which is associated with an increase in activity of the enzyme phenylalanine ammonio-lyase (PAL; EC 4.3.1.5), a physiological response of the plant tissue to the mechanical damage in order to reduce water loss and attack by pathogenic microorganisms (Cantos et al., 2001a; Reyes et al., 2007). Table 6.5 shows the increase in phenylalanine ammonio-lyase activity in some vegetables (carrot, lettuce, courgette, potato, green cabbage, red cabbage, radish, celery, etc.) as a result of cutting. The increase of phenylalanine ammonio-lyase activity indicates activation of phenol-synthesizing mechanisms in response to certain signals, such as the presence of ROS caused by stress, as shown in Figure 6.6.

TABLE 6.5
Changes in Activity of the Enzyme Phenylalanine Ammonio-Lyase (PAL) in Different Vegetables after Chopping and Storage for 2 Days at 15°C

	PAL Activity (μmol/(h g))	
Product	Whole	Chopped
Potato	0.02 ± 0.01	0.65 ± 0.1
Sweet potato	0.02 ± 0.01	0.12 ± 0.04
Celery	0.02 ± 0.01	0.17 ± 0.04
White cabbage	0.05 ± 0.01	2.81 ± 0.4
Courgette	0.06 ± 0.01	0.56 ± 0.2
Carrot	0.06 ± 0.01	4.20 ± 0.2
Lettuce	0.07 ± 0.01	0.70 ± 0.1
Radish	0.07 ± 0.01	1.56 ± 0.1
Red cabbage	0.08 ± 0.01	3.05 ± 0.5

Source: Reyes, L.F.; Villareal, J.E.; Cisneros-Zevallos, L. 2007. The increase in antioxidant capacity after wounding depends on the type of fruit or vegetable tissue. *Food Chem.* 101:1254–1262. With permission.

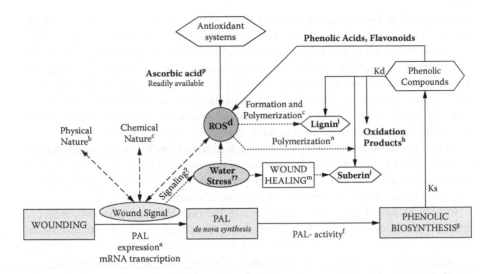

FIGURE 6.6 Mechanisms induced inside the plant cell following mechanical damage to the plant tissue. Involvement of reactive oxidative species (ROS) and antioxidant compounds. Ks and Kd are constants of the rates of synthesis and phenolic degradation, respectively. (From Reyes, L.F.; Villareal, J.E.; Cisneros-Zevallos, L. 2007. The increase in antioxidant capacity after wounding depends on the type of fruit or vegetable tissue. *Food Chem.* 101:1254–1262. With permission.)

The extent of the increase in total phenol concentration as a physiological response to mechanical damage depends on the plant tissue. For instance, in some products such as lettuce, carrot, celery, or sweet potato, there was a significant increase (81%, 191%, 30%, and 17%, respectively), while in others such as courgette, radish, or red cabbage, there was a decline in total phenols (26%, 15%, and 9%, respectively). This is due to a balance between the formation of phenols and their utilization in the synthesis of insoluble phenols (lignin and suberin) or in polymerization products resulting from oxidation (enzymatic browning) (Reyes et al., 2007) (Figure 6.6).

However, the buildup of phenolic compounds and increasing phenylalanine ammonio-lyase activity are inhibited when the cut vegetable is stored in a modified or controlled atmosphere. For instance, significant increases have been observed in soluble phenolic acids such as chlorogenic acid in carrots or caffeic acid in Lollo Rosso lettuce after cutting and during the early days of storage in air at 4 to 5°C, associated with an increase in the enzyme phenylalanine ammonio-lyase (Babic et al., 1993; Gil et al., 1998a). On the other hand, in a CO_2-rich (12% to 30%) modified atmosphere with or without O_2 (2%), or in an atmosphere containing 99% N_2, phenolic acid synthesis was halted, and concentrations remained constant throughout chilled storage or even declined by the end of the storage period (Gil et al., 1998a; Reyes et al., 2007). Similar behavior was reported for flavonoids in some vegetables such as red leaf lettuce, although the mechanical damage at the cutting stage affects flavonoids less than phenolic acids (Ferreres et al., 1997; Gil et al., 1998a). A different effect was reported in other vegetables, such as Swiss chard (*Beta vulgaris cycla*), where the concentration of flavonoids increased in the course of 8 days of storage at 6°C in both air and a modified atmosphere (7% O_2 + 10% CO_2) (Gil et al., 1998b), or spinach, where the flavonoid compounds remained constant over 7 days of storage at 8°C in air and in a modified atmosphere (Gil et al., 1999). A study on cut celery reports an increase to twice the initial concentrations of luteolin and apigenin, flavones accounting for 44% of the flavonoids present in this vegetable, 6 h after processing, and a recovery of the initial concentrations after 24 h in storage at 0°C (Viña and Chaves, 2007). This indicates that flavonoid synthesis is also affected by stress, but equilibrium is recovered more quickly than in the case of phenolic acids.

CO_2-rich atmospheres would therefore seem to be suitable for maintaining concentrations of flavonoid compounds, but how useful they are for augmenting concentrations of antioxidant phenolic acids in order to improve the nutritional and health-beneficial properties of cut vegetables depends on the type of vegetable concerned.

In general, the effect of cutting on phenolics has been thoroughly investigated, but there is little information on the influence of hygienizing treatments. It was found that while broccoli florets contained 11% more total phenols than whole broccoli, the hygienizing treatment used (a solution of 150 ppm sodium hypochlorite) did not affect the total phenol content (Figure 6.7). A modified atmosphere was found to protect phenolics during chilled storage, where levels remained constant for 21 days

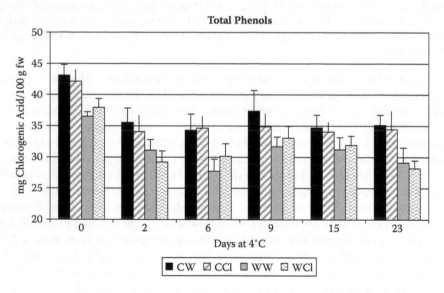

FIGURE 6.7 Effect of minimal processing on total phenolic content in broccoli florets packaged under slightly modified atmosphere (16 % O_2 + 4% CO_2) at 4°C. (CW, cut/ washed in water at 4°C; CCL, cut/washed in 150 ppm sodium hypochlorite at 4°C; WW, whole/ washed in water at 4°C; WCL, whole/washed in 150 ppm sodium hypochlorite at 4°C.) (From Martínez, J.A.; De Ancos, B.; Sánchez-Moreno, C.; Cano, M.P. 2007. Protective effect of minimally processing on bioactive compounds of broccoli (*Brassica oleracea* L. var. *italica*) during refrigerated storage and after cold chain rupture. In: *Abstracts of V Iberoamerican Congress in Postharvest and Agricultural Exports*. Postharvest Group of Polytechnic University of Cartagena and Iberoamerican Postharvest Association (Eds). pp. 694–706. Cartagena, Murcia, España.)

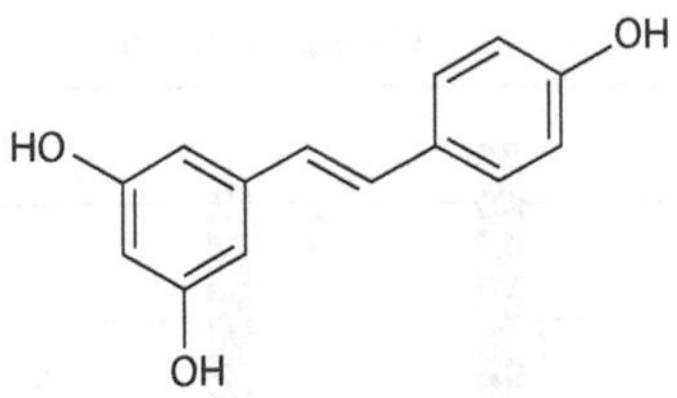

FIGURE 6.8 Structure of *trans*-resveratrol.

at 4°C despite a statistically significant drop (23.4%) on the second day of storage (Martínez et al., 2007).

Other hygienizing systems such as ionizing radiation can improve total phenol contents and concentrations of potential antioxidant constituents in some products like Chinese cabbage (*Brassica rapa* L.) through selection of the appropriate treatment conditions. For instance, while treatment with 0.5 kGy increases total phenol content, 1 kGy radiation causes a significant decrease in the same parameter. This increase in phenolic synthesis at 0.5 kGy is presumably connected with the formation of free radicals during irradiation, these serving as a stress signal that activates the antioxidant mechanisms in the plant tissue (Ahn et al., 2005).

Recent studies have shown that new hygienizing systems such as ultraviolet light (UV-C) with germicide activity (λ 265 nm) are capable of raising the concentrations of certain biologically active polyphenolics, such as anthocyanins and the stilbene known as *trans*-resveratrol, in red grape (Figure 6.8) (Cantos et al., 2001b).

Most studies on the effect of the initial minimal processing operations (peeling and cutting) on the anthocyanins in vegetables such as sweet potato, red cabbage, or 'Lollo Rosso' lettuce have found no significant changes, and the small decline observed has been associated with oxidation mechanisms catalyzed by the enzymes polyphenol oxidase and lipoxygenase (Gil et al., 1998a; Reyes et al., 2007). The effect of storage in air or in a modified atmosphere on anthocyanins also depends on the type of vegetable. For instance, the concentration of anthocyanins was not altered in green lettuce leaves stored in air but was slightly reduced in a modified atmosphere (14% CO_2 + 2.5% O_2). On the other hand, in red-leaf lettuce ('Lollo Rosso'), whose initial anthocyanin content is higher, the degradation was greater in the modified atmosphere than in air, indicating that modified atmosphere technology would not be a suitable tool for maintaining anthocyanin levels in red tissues (Gil et al., 1998a). Recently, however, it was reported that hygienizing treatments with ultraviolet light (UV-C, λ 254 nm, 30 to 510 W) effectively increased anthocyanin concentrations in products with high anthocyanin contents, such as red grapes (Cantos et al., 2001b).

Onion is characterized by high concentrations of the flavonol quercetin ($\approx$557 mg/kg f.w.), which vary depending on the variety. Neither the stress caused by cutting nor the stress caused by washing in a solution of 150 ppm sodium hypochlorite significantly affected the quercetin content of cut onion in a modified atmosphere during the first few days of storage, although after 30 days at 4°C, the samples treated with the sodium hypochlorite solution presented approximately 20% more quercetin than samples washed with water (Figure 6.9) (Martínez et al., 2005). In general, packag-

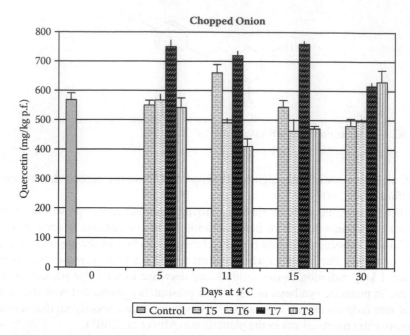

FIGURE 6.9   Effect of minimal processing (washing, packaging, chilled storage) on quercetin concentrations (mg/kg f.w.) in onion. (T5, washed in water at 4°C/vacuum-packaged; T6,
washed in water at 4°C/packaged in modified atmosphere; T7, washed in 150 ppm hypochlorite at 4°C/vacuum-packaged; T8, washed in 150 ppm hypochlorite at 4°C/packaged in modified atmosphere.) (From Martínez, J.A.; Sgroppo, S.; Sánchez-Moreno, C.; De Ancos, B.;
Cano, M.P. 2005. Effects of processing and storage of fresh-cut onion on quercetin. *Acta
Horticulturae.* 682:1889–1895.)

ing in a modified atmosphere had a protective effect on quercetin concentrations,
which remained practically unchanged over 30 days in storage (Figure 6.9).

Fruits are also good sources of phenolics. Red grapes, blackberries, whortleberries, raspberries, and strawberries are all rich in hydroxybenzoic acids, notably
ellagic acid, which has a protective effect against cancer (Maas and Galleta, 1991).
Citrus fruits are rich in glycosidated flavanones (hesperidin and narirutin), and most
fruits (apples, pears, peaches, strawberries, plums, melons, and others) contain flavonols such as quercetin and kaempferol. Finally, anthocyanins are found in berry
fruits such as strawberries, raspberries, blackberries, and whortleberries, or in other
fruits such as cherries.

The scale of the increase in concentration of total phenols as a physiological
response to mechanical damage caused to the plant tissue by peeling and cutting
depends on the type of fruit concerned. For instance, the total phenol content in
fresh-cut pineapple remained relatively stable and did not differ significantly
from that of whole pineapple stored in the same conditions (9 days in air at 5°C)
(Figure 6.10). Similar behavior was observed in mango and melon, where the total
phenols remained relatively stable for 9 days at 5°C, although in the case of cut
melon, the level was significantly lower by the end of storage (Figure 6.10).

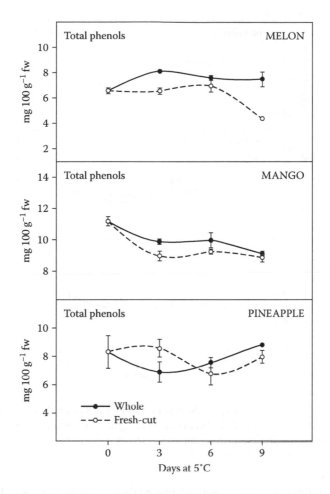

FIGURE 6.10 Total phenols in whole and fresh-cut fruits stored in air at 5°C. (From Gil, M.I.; Aguayo, E.; Kader, A.A. 2006. Quality changes and nutrient retention in fresh-cut versus whole fruits during storage. *J Agric Food Chem.* 54: 4284–4296. With permission.)

In strawberries, which contain high concentrations of phenolics (total phenols 60 mg/100 g f.w.), mostly anthocyanins (pelargonidin-3-glucoside) and flavonols (quercetin glycosides and kaempferol derivatives), there was no observable effect attributable to cutting; total phenols, flavonoids, and anthocyanins remained stable throughout storage at 5°C (Figure 6.11).

High-CO_2 (12% to 30%) modified atmospheres significantly influence the concentrations of phenolics. For instance, buildup of phenolic compounds and increasing phenylalanine ammonio-lyase activity are inhibited when the cut fruit is stored in a modified or controlled atmosphere (Reyes et al., 2007). Flavonoids behave in a similar way. For instance, a high-CO_2 (>20%) atmosphere caused degradation of strawberry anthocyanins, especially in the inner tissue. Color degradation was observed during chilled storage of strawberries in a controlled atmosphere, a development

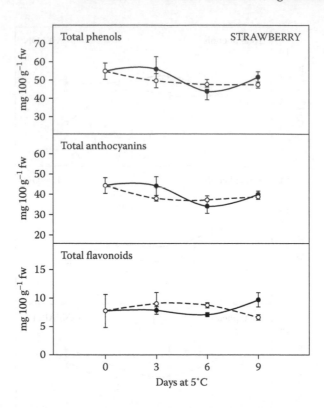

FIGURE 6.11 Total phenols, total anthocyanins, and total flavonoids in whole and fresh-cut strawberries stored in air at 5°C. (From Gil, M.I.; Aguayo, E.; Kader, A.A. 2006. Quality changes and nutrient retention in fresh-cut versus whole fruits during storage. *J Agric Food Chem.* 54: 4284–4296. With permission.)

related to a reduction in the activity of phenylalanine ammonio-lyase, the enzyme associated with synthesis of anthocyanins (Holcroft and Kader, 1999).

In general, high-O_2 atmospheres in the package headspace of fresh-cut fruits and vegetables induced a significant depletion in total phenolic content. The phenolic content in just-processed fresh-cut 'Piel de Sapo' melon was about 16 to 20 mg gallic acid 100 g f.w. packaged under 2.5 kPa O_2 + 7 kPa CO_2 and 10 kPa O_2 + 7 kPa CO_2 atmospheres, whereas concentration was 13 mg gallic acid 100 g f.w. under 70 kPa O_2. In these conditions, total phenolic content was best maintained or indeed increased in fresh-cut melon stored under 2.5 kPa O_2 + 7 kPa CO_2 up to 11 days at 4°C (Oms-Oliu et al., 2008c). The 2.5 kPa O_2 + 7 kPa CO_2 atmosphere induced a higher production of phenolic compounds compared with higher oxygen content atmospheres (10, 21, 30, and 70 kPa O_2), which may be related to an enhanced oxidative stress induced by too-low O_2 and high CO_2 concentration inside the package (Figure 6.6). Wounding may stimulate phenylalanine ammonio-lyase (PAL, E.C. 4.3.1.5) activity during minimal processing with consequent further production of the phenolic compounds. The PAL activation could be elicited through induced ROS (Figure 6.6). This atmosphere (2.5 kPa O_2 + 7 kPa CO_2) preserved fresh-cut pear

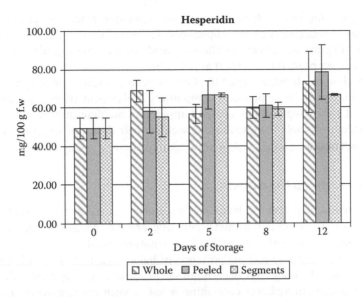

FIGURE 6.12 Effect of minimal processing on hesperidin in minimally processed orange products stored in a modified atmosphere (19% O_2 + 2.6% CO_2) at 4°C and whole orange stored at 8°C. (From Crespo, I.; Magalhaes, J.; Muñoz-Cuevas, M.; De Pascual-Teresa, S.; De Ancos, B.; Muñoz, M.; Cano, M.P. 2005. Changes of bioactive compounds and antioxidant capacity in minimally processed oranges. In: *Intrafood 2005, Innovations in Traditional Foods*. Fito, P., and Toldrá, F. (Eds). Elsevier (Ed), London, UK.)

phenol content better than 70 kPa O_2 atmospheric conditions during 14 storage days at 4°C (Oms-Oliu et al., 2008b). Cocci et al. (2006) reported that fresh-cut apples stored under air packaging (21 kPa O_2) could have led to a stronger degradation of phenolic compounds than if under low oxygen atmospheres (5 kPa O_2 + 5 kPa CO_2). This phenomenon could be due to the fast oxidation of phenolic compounds on the cut surface in contact with the O_2 in the package headspace.

Increased activity of the enzyme phenylalanine ammonio-lyase (PAL) in fresh-cut vegetables as a consequence of the stress produced by minimal processing steps was also observed in the flavonol compounds of fresh-cut citrus fruits such as oranges. The initial peeling and cutting steps did not significantly alter the concentration of total glycosidated flavanones, but storage in a slightly modified atmosphere combined with low temperatures (4°C) caused an increase in the concentration of flavanones. This increase was significant in the case of hesperidin in minimally processed orange segments after 12 days at 4°C (Crespo et al., 2005) (Figure 6.12). Quercetin, another flavonol compound, increased more than twofold the initial content in fresh-cut pear during 14 days of storage at 4°C under atmospheres with high (70 kPa) and low oxygen content (2.5 kPa O_2 + 7 kPa CO_2) (Oms-Oliu et al., 2008b).

Browning of cut surfaces is one of the main challenges in the development of fresh-cut fruits. The destruction of fruit cellular compartmentation allows the oxidation of phenolic compounds by polyphenol oxidase (PPO). Substances with antioxidant properties are used to prevent enzymatic browning such us *N*-acetylcysteine and glutathione. These natural thiol-containing compounds react with *o*-quinones formed during the initial phase of enzymatic browning reactions, yielding color-less adducts or reducing back to o-diphenols. The use of these types of antioxidants reduces the degradation of certain phenolic acids such as chlorogenic acid, as happens in fresh-cut pears (Oms-Oliu et al., 2008b).

The mechanical damage caused to plant tissue as a result of peeling and cutting increases the synthesis of phenolic compounds in fresh-cut fruits and vegetables. The type of packaging and the storage temperature are determining factors in the maintenance of phenolic compounds, as well as their chemical structure (phenolic acids or flavonoids).

6.4.3 CAROTENOID COMPOUNDS

The carotenoid compounds in fruits and vegetables do not generally undergo any major changes as a result of minimal processing, and the changes that do occur have been associated with oxidation mechanisms catalyzed by the enzymes polyphenol oxidase and lipoxygenase and the presence of light, oxygen, and low pH (Reyes et al., 2007). Chilled storage has been reported to cause a sustained time-dependent decline, although atmospheres containing a low oxygen concentration and a high percentage of nitrogen or carbon dioxide improve carotenoid retention in minimally processed vegetables. Packaging in a modified atmosphere maintains initial carotenoid concentrations in cut vegetables, whereas packaging in air results in some degradation in the same storage conditions. For instance, approximately 90% of the carotenoid compounds were retained in green pepper rings after 15 days of chilled storage in a modified atmosphere (5% O_2 + 4% CO_2), while only 52% were retained in air (Howard and Hernandez-Brenes, 1998; Pilon et al., 2006). The same is true with fresh-cut carrot, where the initial carotenoid concentration was kept constant during storage in a nitrogen-enriched atmosphere (90% N_2 + 5% CO_2) (Alasalvar et al., 2005).

Minimally processed fruits generally reach the end of their shelf life before there is any significant degradation of their carotenoid compounds (Wright and Kader, 1997). The effect of peeling and cutting on concentrations of carotenoid compounds will depend on the type and variety of fruit concerned (Gil et al., 2006) (Figure 6.13). The concentrations of carotenoid compounds in whole pineapple (cv. 'Tropical Gold') (250 µg/100 g f.w.) and in whole mango (cv. 'Ataulfo') (3,000 µg/100 g f.w.) presented no significant change over 9 days of storage in air at 5°C; in the case of chunks, on the other hand, levels dropped significantly (25%) after 3 and 9 days in the cases of pineapple and mango chunks, respectively. In melon (cv. 'Joaquin Gold'), the carotenoid content declined as a result of peeling and cutting (Figure 6.13).

Chilled storage causes a sustained time-dependent decline in concentrations of carotenoid compounds, although atmospheres containing a low oxygen concentration and a high percentage of nitrogen or carbon dioxide favor their retention in fresh-cut vegetables. For instance, the total antioxidant carotenes in Navelina oranges (142.78 µg/100 g f.w.)—relative contents 40% lutein, 27.5% β-cryptoxanthin, 20% zeaxanthin, 5% β-carotene, 4.6% α-cryptoxanthin, and 3% α-carotene—were not significantly altered in the early stages of minimal processing (peeling and cutting) to produce orange segments. During 12 days of chilled storage in a modified

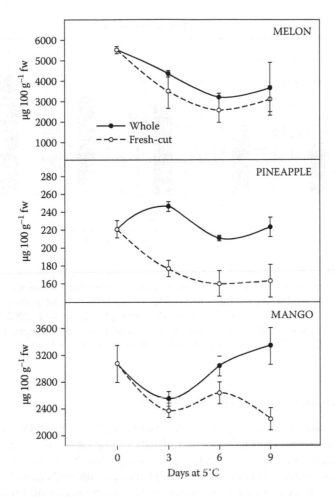

FIGURE 6.13 Total carotenoids in whole and fresh-cut fruits stored in air at 5°C. (From Gil, M.I.; Aguayo, E.; Kader, A.A. 2006. Quality changes and nutrient retention in fresh-cut versus whole fruits during storage. *J Agric Food Chem.* 54: 4284–4296. With permission.)

atmosphere, there was a significant increase in carotene concentrations in both whole unprocessed oranges (stored at 8°C) (89%) and in fresh-cut orange in segments (stored at 4°C) (23%). This fact is associated with biosynthesis of carotenoid compounds during chilled storage, as widely reported in the literature. This means that consumption of orange that is minimally processed in segments and stored at 4°C for up to 12 days supplies a similar concentration of antioxidant carotenoids to that provided by freshly prepared orange (Crespo et al., 2005) (Figure 6.14). Fresh-cut tomatoes retained their initial lycopene content for a period of 21 days at 4°C as a consequence of the synthesis of lycopene induced by ripening and the low oxidation of this carotenoid as a result of low availability of oxygen in the package headspace (Odriozola-Serrano et al., 2008).

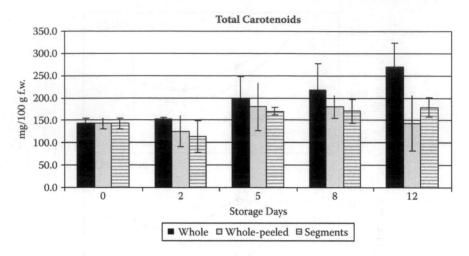

FIGURE 6.14 Effect of minimal processing on total carotenoids in minimally processed orange products (whole-peeled and segments) stored in a modified atmosphere (19% O_2 + 2.6% CO_2) at 4°C and whole orange stored at 8°C. (From Crespo, I.; Magalhaes, J.; Muñoz-Cuevas, M.; De Pascual-Teresa, S.; De Ancos, B.; Muñoz, M.; Cano, M.P. 2005. Changes of bioactive compounds and antioxidant capacity in minimally processed oranges. In: *Intrafood 2005, Innovations in Traditional Foods.* Fito, P., and Toldrá, F. (Eds.). Elsevier (Ed), London, UK.)

6.4.4 ANTIOXIDANT ACTIVITY

Augmented antioxidant activity (measured using the ABTS[•+] or DPPH radical scavenging methods) was reported in tissues of various vegetables in response to stress caused by mechanical damage (peeling and cutting); this has been associated with an increase or decrease in concentrations of phenolic compounds more than vitamin C concentration, as reported in numerous publications on minimal processing of vegetables (Reyes et al., 2007). Numerous studies have shown that its phenolic compounds confer antioxidant capacity (Proteggente et al., 2002), and that this capacity is one of the principal reasons why increased consumption of fruits and vegetables has been recommended as beneficial to health (Prior and Cao, 2000). We may therefore conclude that the healthy properties of certain vegetables can be enhanced by the mechanical damage caused by the peeling and cutting processes.

It is important to note that results also depend on the methodology used to measure antioxidant activity. For instance, in cut tomato packaged under a highly permeable film suitable for microwaving (Magnetron), it was found that the antioxidant capacity, determined as the ability to scavenge the radical ABTS[•+] and measured in lipophilic extracts, remained unchanged during storage, whereas the antioxidant capacity measured in hydrophilic extracts declined with time in storage at 5°C with respect to the antioxidant activity of whole tomato. Minimal processing, then, appears to reduce the capacity of tomato to contribute hydrophilic antioxidants to the diet. Once more, this confirms the influence of processing on a health-beneficial characteristic of vegetable products, in this case antioxidant capacity (Lana and Tijskens, 2006).

Minimal processing generally influences the antioxidant activity of vegetable products to the extent that it alters the bioactive compound responsible for that activity—in most cases, phenolic compounds and vitamin C. For instance, the antioxidant activity of celery after 24 hours at 0°C in a modified atmosphere presented no significant changes with respect to the control, just as the concentrations of luteolin and apigenin, flavones accounting for 44% of the flavonoids in celery, retained their initial concentration (Viña and Chaves, 2007). A similar trend was observed in fresh-cut tomato where no changes in antioxidant capacity might be due to the maintenance of the phytochemical compounds (lycopene, vitamin C, and phenolic compounds) during minimal processing operations and subsequent packaging and refrigerated storage (Odriozola-Serrano et al., 2008).

On the other hand, in fresh-cut spinach, in which ascorbic acid was oxidized to dehydroascorbic acid during chilled storage in a modified atmosphere, the antioxidant capacity was reduced. In this case, the loss of antioxidant capacity was associated with a decline in ascorbic acid concentration (Gil et al., 1999). The decrease of antioxidant capacity in fresh-cut mandarin, fresh-cut pears, and fresh-cut melon during the storage period has been related with the depletion of bioactive compounds (Piga et al., 2002; Oms-Oliu et al., 2008a, 2008b, 2008c).

The hygienizing systems used to obtain a safe fresh-cut vegetable product can also influence the capacity of these products to inhibit free radicals. For instance, the combined effect of ionizing radiation and modified atmospheres can improve antioxidant activity in some products such as Chinese cabbage (*Brassica rapa* L.) if appropriate treatment conditions are selected. The antioxidant capacity of cut Chinese cabbage was increased by treatment with 0.5 kGy radiations, whereas treatments exceeding 1 kGy produced the opposite effect. Irradiation, then, may be considered suitable for augmenting antioxidant activity in cut vegetable products (Ahn et al., 2005).

6.4.5 GLUCOSINOLATES

Epidemiological studies have shown that isothiocyanates, which are produced by hydrolysis of glucosinolates in vegetables of the *Brassicaceae* family (broccoli, cabbage, Brussels sprouts, cauliflower), reduce the incidence of some kinds of cancer (colon and lung) (Lund, 2003). This anticarcinogenic action is linked to the concentration of glucosinolates in the plant tissue, their conversion to isothiocyanates, and the bioavailability of the isothiocyanate metabolites. In a first stage, variation in the content of glucosinolates is closely linked with the genetic and environmental factors (cultivars, temperature, light, growth system, fertilization practices, postharvest storage and processing). Thus, the growing of broccoli in hydroponic conditions and the application of stress factor (i.e., NaCl) at head induction and during development could be beneficial for enrichment in glucosinolates (Moreno et al., 2008). In general, the action of cutting vegetables of the *Brassicaceae* family causes a reduction in the concentration of glucosinolates through the action of the enzyme myrosinase (Leoni et al., 1997), giving rise to various hydrolysis products, the most important in terms of biological activity being isothiocyanates. For instance, the initial concentration of glucosinolates (62 µmol/100 g f.w.) in broccoli florets dropped by 75% after 6 hours

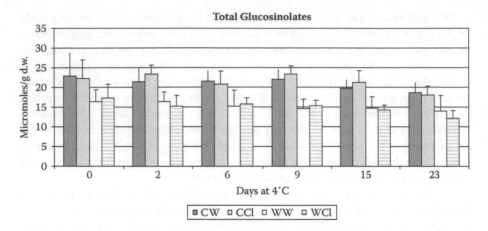

FIGURE 6.15 Effect of minimal processing on total glucosinolates in broccoli florets in a slightly modified atmosphere (16% O_2 + 4% CO_2) at 4°C. (CW, cut/water at 4°C; CCL, cut/150 ppm hypochlorite at 4°C; WW, whole/washed in water at 4°C; WCL, whole/washed in 150 ppm hypochlorite at 4°C.) (From Martínez, J.A.; De Ancos, B.; Sánchez-Moreno, C.; Cano, M.P. 2007. Protective effect of minimally processing on bioactive compounds of broccoli (*Brassica oleracea* L. var. *italica*) during refrigerated storage and after cold chain rupture. In: *Abstracts of V Iberoamerican Congress in Postharvest and Agricultural Exports*. Postharvest Group of Polytechnic University of Cartagena and Iberoamerican Postharvest Association (Eds.). pp. 694–706. Cartagena, Murcia, España.)

at 25°C, and only 50% of these were hydrolyzed to isothiocyanates. However, there were no significant differences between storage at ambient temperature and chilled storage (4 to 8°C); in both cases, the initial concentration of glucosinolates fell by 27% in 7 days (Song and Thornalley, 2007).

Other authors reported that removal of part of the stalk during preparation of the fresh-cut broccoli product caused a statistically significant increase (30% to 40%) in the concentration of glucosinolates (Figure 6.15) (Martínez et al., 2007). This increase in extracted glucosinolates could be due to the induction of glucosinolate synthesis in response to cutting (Verkerk et al., 2001), or again to an increase in the percentage of inflorescences in the cut product, where the glucosinolates, although distributed all over the plant, are present in significant greater concentration (Rangkadilok et al., 2002).

There has been little research into the effect of hygienizing treatments on glucosinolate integrity, but these do not appear to greatly affect storage. For instance, a hygienizing treatment with or without sodium hypochlorite did not significantly influence glucosinolate levels in cut broccoli florets stored for 23 days at 4°C (Martínez et al., 2007).

In general, controlled atmospheres (6% CO_2 + 2% O_2) combined with chilling at 4°C have been found to prolong the shelf life of cut broccoli as compared to storage in air, due essentially to a reduction in respiration rate. In these conditions, it is possible to control chlorophyll degradation and, hence, the appearance of yellowing as a sign of senescence in products of this kind. Some authors recommend CO_2

concentrations less than 8% and O_2 concentrations greater than 2% to prevent the development of unpleasant odors inside the packaging (Bastrash et al., 1993). During storage in a modified atmosphere, there was a nonsignificant downward trend in total glucosinolates (15%) (Figure 6.15) (Rangkadilok et al., 2002; Martínez et al., 2007). However, other authors reported a 71% drop in total glucosinolates when broccoli was stored for 7 days at 4°C in a similar atmosphere (17% O_2 + 3% CO_2) (Vallejo et al., 2003).

6.5 HEALTHY PROPERTIES OF MINIMALLY PROCESSED VEGETABLES

Studies published on the effect of some processing technologies coming under the heading of "minimal processing," such as high hydrostatic pressure, show that the healthy properties of fruit and vegetable derivatives are unchanged after processing. These entailed human intervention trials to examine parameters associated with the effects of intake of these products, such as the bioavailability of phytochemical compounds (vitamin C, carotenoids) and biomarkers for oxidative stress (isoprostanes) and inflammation (C-reactive protein [CRP] and prostaglandin E_2 [PGE_2]) (Sánchez-Moreno et al., 2003a, 2003b, 2005, 2006c). These studies confirmed that processing with high hydrostatic pressure keeps the bioactive compounds and healthy properties associated with consumption of fruit and vegetables virtually unchanged.

There are also human intervention studies on the effect of minimal processing of vegetables (cutting, hygienizing, modified-atmosphere packaging) on the bioavailability of phytochemical compounds (carotenoids, tocopherols, flavanones). For instance, the different steps followed to produce fresh-cut broccoli (cutting, washing with a 150 ppm sodium hypochlorite solution, modified-atmosphere packaging, storage for 9 days at 4°C) did not alter the bioavailability of carotenoid compounds or tocopherol in the product (Granado-Lorencio et al., 2008). Another study, of minimally processed orange segments stored in a modified atmosphere at 4°C, reported that with an intake of 200 g of product for 14 days, levels of flavanones, hesperidin, and naringenin had significantly increased after 7 days, remaining constant thereafter until the end of the study (De Pascual-Teresa et al., 2007). This indicates, then, that a daily intake of 200 g of minimally processed orange segments is sufficient to raise plasma levels of flavanones and their metabolites, thus accounting for the beneficial effect of consuming flavonoids of this family as evidenced in various epidemiological studies (Erlund, 2004).

In conclusion, we should stress that it is necessary to select specific processing conditions (cutting, hygienizing method, modified atmosphere, storage temperature) for each fruit or vegetable in order to achieve a safe fresh-cut product offering high nutritional quality with health benefits intact. The data on the beneficial effect of modified-atmosphere packaging combined with chilled storage on certain phytochemical compounds are generally contradictory; however, in some fresh-cut vegetable products, the concentrations of certain phytochemical compounds (carotenoids, vitamin C, polyphenolics) and the antioxidant activity are retained or even augmented. In general, it is safe to say that the concentrations of phytochemical compounds, the nutritional value, and the healthy properties of fresh-cut (minimally processed) fruits or vegetables or mixtures are similar to those of the whole vegetables from which they are derived.

6.6 REGULATIONS GOVERNING NUTRITIONAL DECLARATIONS AND HEALTHY PROPERTIES OF FOODS

In response to the new regulations governing nutritional labeling and application to minimally processed foods, there is a need for more scientific studies (*in vitro* and *in vivo*) to reliably determine the effect of minimal processing on the nutritional and bioactive substances in vegetable products and the healthy effects produced by their intake. This will facilitate the preparation of "nutritional profiles" so that health professionals and food technologists can inform consumers by means of reliable nutritional and health claims based on real scientific studies.

The Codex Alimentarius set up a Commission to implement the joint Food and Agriculture Organization (FAO)/World Health Organization (WHO) program of food standards. The guidelines on nutrition labeling drawn up by the Codex are founded on the principle that no food may be described or presented falsely or misleadingly (Codex Alimentarius Commission, 1992). The Codex Guidelines on Nutrition Labeling contain provisions for voluntary nutrient declarations and for the calculation and presentation of information on nutrients (CAC 2-1985, Rev. 1-1993: http://www.fao.org/docrep/005/Y2770S/y2770s06.htm); that is, it lays down general principles to be observed while leaving the definition of specific declarations up to national regulators.

With this mandate, various countries have drawn up specific regulations regarding nutrition and health claims for foods. The U.S. Food and Drug Administration (FDA), the body responsible for food legislation, indicates that nutritional labeling is mandatory for processed products (http://www.cfsa.fda.gov/label.html), while application of the regulation to fresh-consumed products is voluntary (71 FR42031 of August 2006). Since 1990, the FDA has approved a major set of rules regulating nutrition and health claims for foods, which are updated periodically. There are basically three categories of claims in labeling: health, nutrient contents, and structure/function. Each of these types of claim is authorized under different procedures designated by the FDA through the competent body (Office of Nutritional Products, Labeling, and Dietary Supplements) of the Center for Food Safety and Applied Nutrition (CFSAN) (http://www.cfsan.fda.gov/~dms/hclaims.html). By way of example, we would cite the *Food Labeling Guide*, which contains a summary of health claims that have been approved for use in the labeling of foods and dietary supplements in the United States (http://www.cdsan.fda.gov//~dms/flg-6c.html).

European regulation in this area lags far behind that of the United States. The new Community Regulation on nutritional and health claims made on foods (Reg. (EC) 1924/2006 of the European Parliament and of the Council of 20 December 2006), in force since 19 January 2007 and mandatory since July 2007, is based on the requirement that the nutrition and health claims made by the food industry be true and state the scientific basis on which they are founded. Consumer protection is an essential element of the new regulation, and this is assured by making it mandatory to market only safe foods with proper labeling to guarantee that nutrition and health claims are based on real scientific evidence and not on misleading, exaggerated, or scientifically baseless messages. For instance, in order to assure the veracity of claims made by the food industry, the new regulation requires that substances or products for which such claims are made must have been shown to possess a beneficial nutritional or physiological effect by means of internationally accepted scientific tests, which in Spain are to be regulated by the Spanish Food Safety and Nutrition Agency (*Agencia Española de Seguridad Alimentaria y Nutrición*/AESAN). Moreover, the bioactive substance about which the claim is made must be present in the end product in sufficient concentrations, or absent, or present in small enough quantities to produce the claimed nutritional or physiological effect. This bioactive substance must further be present in high enough concentrations to produce the claimed effect in an amount of food that can reasonably be consumed as part of a daily diet and can also be assimilated by the body once the food is ingested. Therefore, messages such as "light," "low-fat," "reduces cholesterol," or "high fiber content" may only be used in the labeling of products if the manufacturers/distributors comply with the requirements laid down in the regulation. For example, to be able to claim that a food is "low-sugar," it must be guaranteed not to contain more than 5 grams of sugar for every hundred grams of fresh product.

On the basis of these criteria, the regulation provides that nutrition and health claims may only be made in three categories:

1. A "nutrition claim" or "content claim" is one that states, suggests, or implies that a food has particular beneficial nutritional properties due to the energy it provides or the nutrients or other substances that it contains or does not contain (e.g., "low calorie/salt/sugar" or "high vitamin/fiber/protein").
2. A "health claim" is one that states, suggests, or implies that a relationship exists between a food category, a food, or one of its constituents and health (e.g., "reduces cholesterol").
3. A "reduction of disease risk claim" is one that states, suggests, or implies that the consumption of a food category, a food, or one of its constituents significantly reduces a risk factor in the development of a human disease (e.g., "reduces the risk of cerebral ischemia or strokes").

The regulation introduces the concept of "nutrient profiles" to generally establish and limit the composition of a food as the context in which claims may be made.

For purposes of implementation and monitoring, the regulation involves the health authorities of all the member states of the European Union. The European Food Safety Authority (EFSA) has a particularly important role in assessing the scientific bases on which claims are founded and in establishing "nutrient profiles." In Spain, the EFSA works with the AESAN (Ministry of Health and Consumer Affairs), which will be responsible for implementing the regulation in cooperation with the autonomous communities and will be the body called upon to examine claims by Spanish manufacturers.

ACKNOWLEDGMENTS

This work was made possible by research funding from the Spanish Ministry of Science and Innovation, AGL2002-04059-C02-02, AGL2005-03849/ALI, AGL2006-27824-E/ALI, 2006701081, the Programme Consolider-Ingenio 2010, FUN-C-FOOD, CSD2007-00063; and the National Institute for Agricultural Research -INIA, RTA04-171-C2-2. Special thanks are due to Project CYTED XI-22.

REFERENCES

Agar, I.T.; Massantini, R.; Hess-Pierce, B.; Kader, A.A. 1999. Postharvest CO_2 and ethylene production and quality maintenance of fresh-cut kiwifruit slices. *J Food Sci.* 64:433–440.

Agar, I.T.; Streif, J.; Bangerth, F. 1997. Effect of high CO_2 and controlled atmosphere (CA) on the ascorbic and dehydroascorbic acid content of some berry fruits. *Postharvest Biol Technol.* 11:47–55.

Ahn, H.-J.; Kin, J.-H.; Kim, J.-K.; Kim, D.-H.; Yook, H.-S.; Byun, M.-W. 2005. Combined effects of irradiation and modified atmosphere packaging on minimally processed Chinese cabbage (*Brassica rapa* L.). *Food Chem.* 89:589–597.

Akbas, M.Y.; Ölmez, H. 2007. Effectiveness of organic acid, ozonated water and chlorine dippings on microbial reduction and storage quality of fresh-cut iceberg lettuce. *J Sci Food Agric.* 87:2609–2616.

Alasalvar, C.; Al-Farsi, M.; Quantick, P.C.; Shahidi, F.; Wiktorowicz, R. 2005. Effect of chill storage and modified atmosphere packaging (MAP) on antioxidant activity, anthocyanins, carotenoids, phenolics and sensory quality of ready-to-eat shredded orange and purple carrots. *Food Chem.* 80:69–76.

Argañosa, A.C.S.J.; Raposo, M.F.J.; Teixeira, P.C.M.; Morais, A.M.M.B. 2008. Effect of cut-type on quality of minimally processed papaya. *J Food Sci Agric.* 88:2050–2060.

Artés-Hernández, F.; Rivera-Cabrera, F.; Kader, A.A. 2007. Quality retention and potential shelf-life of fresh-cut lemons affected by cut type and temperature. *Postharvest Biol Tech.* 43:245–254.

Aviram, M.; Dornfeld, L.; Rosenblat, M.; Volkova, N.; Kaplan, M.; Coleman, R.; Hayek, T.; Presser, D.; Fuhrman, B. 2000. Pomegranate juice consumption reduces oxidative stress, atherogenic modifications to LDL, and platelet aggregation: studies in humans and in atherosclerotic apoliproprotein E-deficient mice. *Am J Clin Nutr.* 71:1062–1076.

Babic, I.; Amiot, M.J.; Nguyen-The, C.; Aubert, S. 1993. Changes in phenolic content in fresh ready-to-use shredded carrots during storage. *J Food Sci.* 58:351–356.

Balla, C.S.; Farkas, J. 2006. Minimally processed fruits and fruit products and their microbiological safety. In: *Handbook of Fruits and Fruit Processing.* Chapter 7, pp. 115–128. Hui, Y.H., Barta, J., Cano, M.P., Gusek, T., Sidhu, J., Sinha, N. (Eds.), Blackwell, Oxford, UK.

Barry-Ryan, C.; O'Beirne, D. 1999. Ascorbic acid retention shredded iceberg lettuce as affected by minimal processing. *J Food Sci.* 64:498–500.

Bastrash, S.; Makhlouf, J.; Castaigne, F.; Willemot, C. 1993. Optimal controlled atmosphere conditions for storage of broccoli florets. *J Food Sci.* 58:338–341.

Benkeblia, L. 2004. Antimicrobial activity of essential oil extracts of various onions (*Allium cepa*) and garlic (*Allium sativum*). *LWT—Food Sci Technol.* 37:263–268.

Block, G.; Dietrich, M.; Norkus, E.P.; Morrow, J.D.; Hudes, M.; Caan, B.; Packer, L. 2002. Factors associated with oxidative stress in human populations. *Am J Epidemiol.* 156:274–285.

Boyer, J.; Liu, R.H. 2004. Apple phytochemicals and their health benefits. *Nutr J.* 3:5.

Brandt, K.; Christensen, L.P.; Hansen-Møller, J.; Hansen, S.L., Haraldsdottir, J.; Jespersen, L.; Purup, S.; Karazmi, A.; Barkholt, V.; Frøkiaer, H.; Købaek-Larsen, M. 2004. Health promoting compounds in vegetables and fruits: a systematic approach for identifying plant components with impact on human health. *Trends Food Sci Technol.* 15:384–393.

Cano, A.; Medina, A.; Bermejo, A. 2008. Bioactive compounds in different citrus varieties. Discrimination among cultivars. *J Food Compos Anal.* 21:377–381.

Cantos, E.; Espín, J.C.; Tomás-Barberán, F.A. 2001a. Effect of wounding on phenolic enzymes in six minimally processed lettuce cultivars upon storage. *J Agric Food Chem.* 49:322–330.

Cantos, E.; Espín, J.C.; Tomás-Barberán, F.A. 2001b. Postharvest induction modeling method using UV irradiation pulses for obtaining resveratrol-enriched table grapes: a new "functional fruit? *J. Agric Food Chem.* 49:5052–5058.

Castenmiller, J.J.M.; West, C.E. 1998. Bioavailability and bioconversion of carotenoids. *Annu Rev Nutr.* 18:19–38.

Cocci, E.; Rocculi, P; Romani, S.; Dalla Rosa, M. 2006. Changes in nutritional properties of minimally processed apples during storage. *Postharvest Biol Technol.* 39:265–271.

Crespo, I.; Magalhaes, J.; Muñoz-Cuevas, M.; De Pascual-Teresa, S.; De Ancos, B.; Muñoz, M.; Cano, M.P. 2005. Changes of bioactive compounds and antioxidant capacity in minimally processed oranges. In: *Intrafood 2005, Innovations in Traditional Foods.* Fito, P., and Toldrá, F. (Eds.). Elsevier (Ed), London, UK.

Davey, M.W.; Van Montagu, M.; Inzé, D.; Sanmartin, M.; Kanellis, A.; Smirnoff, N.; Benzie, I.J.J.; Strain, J.J.; Favell, D.; Fletcher, J. 2000. Plant L-ascorbic acid: chemistry, function, metabolism, bioavailability and effects of processing. *J Sci Food Agric.* 80:825–860.

Derbyshire, E.; Kemp, R.; Xingmin, M.; Formica, J.V.; Regelson, W. 1995. Review of the biology of quercetin and related bioflavonoids. *Food Chem Toxicol.* 33:1061–1080.

De Pascual-Teresa, S.; Sánchez-Moreno, C.; Granado, F.; Olmedilla, B.; De Ancos, B.; Cano, M.P. 2007. Short and mid-term bioavailability of flavanones from oranges in humans. *Curr Top Nutraceutical Res.* 2007. 5:129–134.

Del Caro, A.; Piga, A.; Vacca, V.; Agabbio, M., 2004. Changes of flavonoids, vitamin C, and antioxidant capacity in minimally processed citrus segments and juices during storage. *Food Chem* 84(1):99–105.

Du, G.; Li, M.; Ma, F.; Liang, D. 2009. Antioxidant capacity and the relationship with polyphenol and vitamin C in *Actinidia* fruits. *Food Chem.* 113:557–562.

Duthie, G.G.; Duthie, S.J.; Kyle, J.A.M. 2000. Plant polyphenols in cancer and heart disease: implications as nutritional antioxidants. *Nutr Res Rev.* 13:79–106.

Eastwood, M.A.; Morris, E.R. 1992. Physical properties of dietary fibre that influence physiological function. *Am J Clin Nutr.* 55:436–442.

Erlund, I. 2004. Review of the flavonoids quercetin, hesperetin, and naringenin. Dietary sources, bioactivities, bioavailability, and epidemiology. *Nutr Res.* 24:851–874.

FAO/WHO Food Standards Programme Codex Alimentarius Commissin, 2001. ISSN 0259–2916.

Ferreres, F.; Gil, M.I.; Castañer, M.; Tomás-Barberán, F.A. 1997. Phenolic metabolites in red pigmented lettuce (*Lactuca sativa*). Changes with minimal processing and cold storage. *J Agric Food Chem.* 45:4249–4254.

Gartner, C.; Stahl, W.; Sies, H. 1997. Lycopene is more bioavailable from tomato paste than from fresh tomatoes. *Am J Clin Nutr.* 66:116–122.

Gil, M.I.; Aguayo, E.; Kader, A.A. 2006. Quality changes and nutrient retention in fresh-cut versus whole fruits during storage. *J Agric Food Chem.* 54: 4284–4296.

Gil, M.I.; Castañer, M.; Ferreres, F.; Artés, F.; Tomás-Barberán, F.A. 1998a. Modified-atmosphere packaging of minimally processed 'Lollo Rosso' (*Lactuca sativa*): phenolic metabolites and quality changes. *Z Lebensm Unters Forsch A.* 206:350–354.

Gil, M.I.; Ferreres, F.; Tomás-Barberán, F.A. 1998b. Effect of modified atmosphere packaging on the flavonoids and vitamin C content of minimally processed Swiss chard (*Beta vulgaris* subspecies cycla). *J Agric Food Chem.* 46: 2007–2012.

Gil, M.I.; Ferreres, F.; Tomás-Barberán, F.A. 1999. Effect of postharvest storage and processing on the antioxidant constituents (flavonoids and vitamin C) of fresh-cut spinach. *J Agric Food Chem.* 47:2213–2217.

Giovannucci, E.; Ascherio, A.; Rimm, E.B.; Stampfer, M.J.; Colditz, G.A.; Willett, W.C. 1995. Intake of carotenoids and retinol in relation to risk of prostate cancer. *J Natl Cancer Inst.* 87:1767–1776.

González-Aguilar, G.; Gardea, A.A.; Cuamea-Navarro, F. (Eds). 2005. Nuevas Tecnologías de Conservación de Productos Vegetales Frescos Cortados. Centro de Investigación y Desarrollo, A.C. (CIAD, A.C.) (Editorial), Hermosillo, Mexico.

Granado-Lorencio, F.; Olmedilla-Alonso, B.; Herrero-Barbudo, C.; Sánchez-Moreno, C.; De Ancos, B.; Martínez, J.A.; Perez-Sacristán, B.; Blanco-Navarro, I. 2008. Modified-atmosphere packaging (MAP) does not affect the bioavailability of tocopherols and carotenoids from broccoli in humans: a cross-over study. *Food Chem.* 106:1070–1076.

Griffiths, G.; Trueman, L.; Crowther, T.; Thomas, B.; Smith, B. 2002. Onions: a global benefit to health. *Physiology Res.* 16:603–615.

Halliwell, B. 1987. Oxidants and human disease: some new concepts. *FASEB J.* 1:358.

Hasler, C.M.; Bloch, A.S.; Thomson, C.A. 2004. Position of the American Dietetic Association: functional foods. *J Am Diet Assoc.* 104:814–826.

Heard, G.M. 2005. Microbiology of fresh-cut produce. In: *Fresh-Cut Fruits and Vegetables: Science, Technology and Market.* Chapter 7, pp. 187–248. Lamikanra, O. (Ed.), CRC Press, Boca Raton, FL, EEUU.

Heinonen, O.P.; Albanes, D. 1994. The effect of vitamin E and beta-carotene on the incidence of lung cancer and other cancers in male smokers. *N Engl J Med.* 330:1029–1035.

Hennekens, C.H.; Buring, J.E.; Manson, J.E.; Stampfer, M.; Rosner, B.; Cook, N.R.; Belanger, C.; LaMotte, F.; Gaziano, J.M.; Ridker, P.M.; Willett, W.; Peto, R. 1996. Lack of effect of long-term supplementation with beta-carotene on the incidence of malignant neoplasms and cardiovascular disease. *N Engl J Med.* 334:1145–1149.

Holcroft, D.M.; Kader, A.A. 1999. Carbon dioxide-induced changes in color and anthocyanins synthesis of stored strawberry fruit. *Hortscience.* 34:1244–1248.

Howard, L.R.; Hernandez-Brenes, C. 1998. Antioxidant content and market quality of Jalapeño pepper rings as affected by minimal processing and modified atmosphere packaging. *J Food Qual.* 21:317–327.

Hurst, C.H. 2005. Safety aspects of fresh-cut fruits and vegetables. In: *Fresh-Cut Fruits and Vegetables: Science, Technology and Market.* Chapter 4, pp. 45–90. Lamikanra, O. (Ed.), CRC Press, Boca Raton, FL.

Jongen, W. (Ed.). 2005. *Improving the Safety of Fresh Fruit and Vegetables.* CRC Press, Boca Raton, FL.

Knekt, P.; Isotupa, I.; Rissanen, H.; Heliövaara, M.; Järvinen, R.; Häkkinen, S.; Aromaa, A.; Reunanen, A. 2000. Quercetin intake and the incidence of cerebrovascular disease. *Eur J Clin Nutr.* 54:415–417.

Knekt, P.; Järvinen, R.; Seppänen, R.; Heliövaara, M.; Teppo, L.; Pukkala, E.; Aromaa, A. 1997. Dietary flavonoids and the risk of lung cancer and other malignant neoplasms. *Am J Epidemiol.* 146:223–230.

Kohlmeier, L.; Su, L. 1997. Cruciferous vegetables consumption and colorectal cancer risk: meta-analysis of the epidemiological evidence. *FASEB J.* 11:A369.

Kris-Etherton, P.M.K.; Hecker, K.D.; Bonanome, A.; Coval, S.M.; Binkosdi, A.E.; Hilpert, K.F.; Griel, A.E.; Etherton, T.D. 2002. Bioactive compounds in foods: their role in the prevention of cardiovascular disease and cancer. *Am J Med.* 113: 71S–88S.

Lampe, J.W. 1999. Health effects of vegetables and fruit: assessing mechanisms of action in human experimental studies. *Am J Clin Nutr.* 70:475S–490S.

Lampe, J.W.; Peterson, S. 2002. Brassica, biotransformation and cancer risk: genetic polymorphism alters the preventive effects of cruciferous vegetables. *J Nutr.* 132:2991–2994.

Lana, M.M.; Tijskens, L.M.M. 2006. Effects of cutting and maturity on antioxidant activity of fresh-cut tomatoes. *Food Chem.* 97:203–211.

Lawson, L.D. 1998. Garlic: a review of its medicinal effects and indicated active compounds. In: Phytomedicines of Europe: Chemistry and Biological Activity. pp. 176–209. Lawson, L.D., and Baurer, R. (Eds.). ACS Symposium Series 691: American Chemical Society. Washington, DC.

Le Marchand, L; Murphy, S.P.; Hankin, J.H.; Wilkens, L.R.; Lolonel, L.N. 2000. Intake of flavonoids and lung cancer. *J Natl Cancer Inst.* 92:154–160.

Lee, I-Min; Cook, N.R.; Manson, J.A.; Buring, J.E.; Hennekens, C.H. 1999. β-carotene supplementation and incidence of cancer and cardiovascular disease: the Women's Health Study. *J Natl Cancer Inst.* 91:2102–2106.

Lee, S.K.; Kader, A.A. 2000. Preharvest and postharvest factors influencing vitamin C content in horticultural crops. *Postharvest Biol Technol.* 20:207–220.

Leoni, O.; Iori, R.; Palmieri, S.; Esposito, E.; Menegatti, E.; Cortesi, R.; Nastruzzi, C. 1997. Myrosinase-generated isothiocyanate from glucosinolate: isolation, characterization and *in vivo* antiproliferative studies. *Bioorg Med Chem.* 5:1799–1806.

Ling, W.H.; Jones, P.J. 1995. Dietary phytosterols: a review of metabolism, benefits and side effects. *Life Sci.* 57:195–206.

Liu, B.; Edgerton, S.; Yang, X.; Kim, A.; Ordoñez-Ercan, D.; Mason, T.; Alvarez, K.; McKimmey, C.; Liu, N.; Thor, A. 2005. Low dose dietary phytoestrogen abrogates tamoxigen-associated mammary tumor prevention. *Cancer Res.* 65:879–886.

Lund, E. 2003. Non-nutritive bioactive constituents of plants: dietary sources and health benefits of glucosinolates. *Int J Vitamin Nutr Res.* 73:135–143.

Maas, J.L.; Galleta, G.J. 1991. Ellagic acid, an anticarcinogen in fruits, especially in strawberries: a review. *HortScience.* 26:10–14.

Martínez, J.A.; De Ancos, B.; Sánchez-Moreno, C.; Cano, M.P. 2007. Protective effect of minimally processing on bioactive compounds of broccoli (*Brassica oleracea* L. var. *italica*) during refrigerated storage and after cold chain rupture. In: *Abstracts of V Iberoamerican Congress in Postharvest and Agricultural Exports.* Postharvest Group of Polytechnic University of Cartagena and Iberoamerican Postharvest Association (Eds.). pp. 694–706. Cartagena, Murcia, España.

Martínez, J.A.; Sgroppo, S.; Sánchez-Moreno, C.; De Ancos, B.; Cano, M.P. 2005. Effects of processing and storage of fresh-cut onion on quercetin. *Acta Horticulturae.* 682:1889–1895.

Maynard, M.; Gunnell, D.; Emmett, P.; Frankel, S.; Smith, G.D. 2003. Fruit, vegetables, and antioxidants in childhood and risk of adult cancer: the Boyd Orr cohort. *J Epidem Community Health.* 57:218–225.

Mayne, S.T. 1996. Beta-carotene, carotenoids and disease prevention in humans. *FASEB J.* 10:690–701.

Michaud, D.S.; Spiegelman, D.; Clinto, S.K.; Rimm, E.B.; Willett, W.C.; Giovannucci, E. 1999. Fruit and vegetable intake and incidence of bladder cancer in a male prospective cohort. *J Natl Cancer Inst.* 91:605–613.

Middleton, Jr. M.D.E. 1996. Biological properties of plant flavonoids: an overview. *Int J Pharmacognos.* 34:344–348.

Moreno, D.A.; López-Berenguer, C.; Martínez-Ballesta, M.C.; Carvajal, M.; García-Viguera, C. 2008. Basis for the new challenges of growing broccoli for health in hydroponics. *J Sci Food Agric.* 88:1472–1481.

Ness, A.R.; Khaw, K.T.; Bingham, S.; Day, N.E. 1996. Vitamin C status and serum lipids. *Eur J Clin Nutr.* 50:724–729.

Ness, A.R.; Powles, J.W. 1997. Fruit and vegetables and cardiovascular disease: a review. *Int J Epidemiol.* 26:1–13.

Odriozola-Serrano, I.; Soliva-Fortuny, R.; Martín-Belloso, O. 2008. Effect of minimal processing on bioactive compounds and color attributes of fresh-cut tomatoes. *LWT—Food Sci Technol.* 41:217–226.

Olmedilla, B.; Granado, F.; Blanco, I. 2001. Carotenoides y salud humana. In: *Serie Informes.* Fundación Española de Nutrición. Madrid, Spain.

Oms-Oliu, G.; Soliva-Fortuny, R.; Martín-Belloso, O. 2008a. Edible coatings with antibrowning agents to maintain sensory quality and antioxidant properties of fresh-cut pears. *Postharvest Biol Technol.* 50:87–94.

Oms-Oliu, G.; Odriozola-Serrano, I.; Soliva-Fortuny, R.; Martín-Belloso, O. 2008b. Antioxidant content of fresh-cut pears stored in high-O_2 active packages compared with conventional low-O_2 active and passive modified atmosphere packaging. *J Agric Food Chem.* 56:932–940.

Oms-Oliu, G.; Odriozola-Serrano, I.; Soliva-Fortuny, R.; Martín-Belloso, O. 2008c. The role of peroxidase on the antioxidant potential of fresh-cut 'Piel de Sapo' melon packaged under different modified atmospheres. *Food Chem.* 106:1085–1092.

O'Neill, M.E.; Carroll, Y.; Corridan, B.; Olmedilla, B.; Granado, F.; Blanco, I.; Van den Berg, H.; Hininger, I.; Rousell, A.-M.; Chopra, M.; Southon, S.; Thurnham, D.I. 2001. A European carotenoid database to assess carotenoid intake and its use in a five-country comparative study. *Br J Nutr.* 85:499–507.

Piga, A.; Agabbio, M.; Gambella, F.; Nicoli, M.C. 2002. Retention of antioxidant activity in minimally processed mandarin and satsuma fruits. *LWT—Food Sci Technol.* 35:344–347.

Piironen, V.; Lindsay, D.G.; Miettinen, T.A.; Toivo, J.; Lampi, A.-M. 2000. Plant sterols: biosynthesis, biological function and their importance to human nutrition. *J Sci Food Agric.* 80:939–966.

Pilon, L.; Oetterer, M.; Gallo, C.R.; Spoto, M.H.F. 2006. Shelf life of minimally processed carrot and green pepper. *Ciênc e Tecnol Aliment.* 26:150–158.

Plaza, L.; Sánchez-Moreno, C.; De Ancos, B.; Cano, M.P. 2006a. Carotenoid content and antioxidant capacity of Mediterranean vegetable soup (*gazpacho*) treated by high-pressure/temperature during refrigerated storage. *Eur Food Res Technol.* 223:210–215.

Plaza, L.; Sánchez-Moreno, C.; Elez,-Martínez, P.; De Ancos, B.; Martín-Belloso, O.; Cano, M.P. 2006b. Effect of refrigerated storage on vitamin C and antioxidant activity of orange juice processed by high-pressure or pulsed electric field with regard to low pasteurization. *Eur Food Res Technol.* 223:487–493.

Prior, R.L.; Cao, G. 2000. Antioxidant phytochemicals in fruits and vegetables: diet and health implications. *Hortscience.* 35:588–592.

Proteggente, A.R.; Pannala, A.S.; Paganga, G.; Van Buren, L., Wagner, E.; Wiseman, S.; Van De Put, F.; Dacombe, C.; Rice-Evans, C.A. 2002. The antioxidant activity of regularly consumed fruit and vegetables reflects their phenolic and vitamin C composition. *Free Radic Res.* 36:217–233.

Rangkadilok, N.; Tomkins, B.; Nicolas, M.E.; Premier, R.R.; Bennett, R.N.; Eagling, D.R.; Taylor, P.W.J. 2002. The effect of post-harvest and packaging treatments on glucoraphanin concentration in Broccoli (*Brassica oleracea* var. Italica). *J Agric Food Chem.* 50:7386–7391.

Reyes, L.F.; Villareal, J.E.; Cisneros-Zevallos, L. 2007. The increase in antioxidant capacity after wounding depends on the type of fruit or vegetable tissue. *Food Chem.* 101:1254–1262.

Rice-Evans, C.A.; Miller, N.J.; Paganga, G. 1996. Structure-antioxidant activity relationship of flavonoids and phenolic acids. *Free Radic Biol Med.* 20:933–956.

Rojas-Graü, M.A.; Tapia, M.S.; Martín-Belloso, O. 2008. Using polysaccharide-based edible coatings to maintain quality of fresh-cut Fuji apples. *LWT—Food Sci Technol.* 41:139–147.

Sánchez-Moreno, C.; Cano, M.P.; De Ancos, B.; Plaza, L.; Olmedilla, B.; Granado, F.; Martín, A. 2003a. High-pressurized orange juice consumption affects plasma vitamin C, antioxidative status and inflammatory markers in healthy humans. *J Nutr.* 133:2204–2209.

Sánchez-Moreno, C.; Cano, M.P.; De Ancos, B.; Plaza, L.; Olmedilla, B.; Granado, F.; Martín, A. 2003b. Effect of orange juice intake on vitamin C concentrations and biomarkers of antioxidant status in humans. *Am J Clin Nutr.* 78:454–460.

Sánchez-Moreno, C.; Plaza, L.; Elez-Martínez, P.; De Ancos, B.; Martín-Belloso, O.; Cano, M.P. 2005. Impact of high pressure and pulsed electric fields on bioactive compounds and antioxidant activity of orange juice in comparison with traditional thermal processing. *J Agric Food Chem.* 53:4403–4409.

Sánchez-Moreno, C.; Plaza, L.; De Ancos, B.; Cano, M.P. 2006a. Impact of high-pressure and traditional thermal processing of tomato purée on carotenoids, vitamin C and antioxidant activity. *J Sci Food Agric.* 86:171–179.

Sánchez-Moreno, C.; Plaza, L.; De Ancos, B.; Cano, M.P. 2006b. Nutritional characterisation of commercial traditional pasteurised tomato juices: carotenoids, vitamin C and radical-scavenging capacity. *Food Chem.* 98:749–756.

Sánchez-Moreno, C.; Cano, M.P.; De Ancos, B.; Plaza, L.; Olmedilla, B.; Granado, F.; Martín, A. 2006c. Mediterranean vegetable soup consumption increases plasma vitamin C, and decreases F_2-isoprostanes, prostaglandin E_2, and monocyte chemotactic protein-1 in healthy humans. *J Nutr Biochem.* 17:183–189.

Scalbert, A.; Manach, C.; Morand, C.; Rémésy, C.; Jiménez, L. 2005. Dietary polyphenols and the prevention of diseases. *Crit Rev Food Sci Nutr.* 45:287–306.

Seki, T.; Tsuji, K.; Hayato, Y.; Moritomo, T.; Ariga, T. 2000. Garlic and onion oils inhibit proliferations and induce differentiation of HL60 cells. *Cancer Lett.* 160:29–35.

Simon, J.A.; Hudes, E.S.; Tice, J.A. 2001. Relation of serum ascorbic acid to mortality among US adults. *J Am Coll Nutr.* 20:255–263.

Skibola, C.F.; Smith, M.T. 2000. Potential health impacts of excessive flavonoid intake. *Free Radical Biol Med.* 29:375–383.

Slattery, M.L.; Benson, J.; Curtin, K.; Ma, K.-N.; Schaeffer, D.; Potter, J.D. 2000. Carotenoids and colon cancer. *Am J Clin Nutr.* 71:575–582.

Song, L.; Thornalley, P.J. 2007. Effect of storage, processing and cooking on glucosinolate content of Brassica vegetables. *Food Chem Toxicol.* 45:216–224.

Steinmetz, K.A.; Potter, J.D. 1996. Vegetables, fruit and cancer prevention: a review. *J Am Diet Assoc.* 96:1027–1039.

Taylor, C.A.; Hampl, J.S.; Johnston, C.S. 2000. Low intakes of vegetables and fruits, especially citrus fruits, lead to inadequate vitamin C intakes among adults. *Eur J Clin Nutr.* 54:573–578.

Temple, N.J.; Gladwin, K.K. 2003. Fruit, vegetables, and the prevention of cancer: research challenges. *Nutrition.* 19:467–470.

Tribble, D.L. 1998. Further evidence of the cardiovascular benefits of diet enriched in carotenoids. *Am J Clin Nutr.* 68:521–522.

Trichopoulou, A.; Naska, A.; Antoniou, A.; Friel, S.; Trygg, K.; Turrini, A. 2003. Vegetable and fruit: the evidence in their favour and the public health perspective. *Int J Vitamin Nutr. Res.* 73:63–69.

Trumbo, P.R.; Yates, A.A.; Schlicker-Renfro, S.; Suitor, C. 2003. Dietary reference intakes: revised nutritional equivalents for folate, vitamin E and provitamin A carotenoids. *J Food Compos Anal.* 16:379–382.

Tudela, J.A.; Espín, J.C.; Gil, M.I. 2002. Vitamin C retention in fresh-cut potatoes. *Postharvest Biol Technol.* 26:75–84.

Vallejo, F.; García-Viguera, C.; Tomás-Barberán, F.A. 2003. Changes in broccoli (*Brassica oleracea* L. var. *Italica*) health-promoting compounds with inflorescence development. *J Agric Food Chem.* 51:3776–3782.

Vallejo, F.; Tomás-Barberán, F.A.; García-Viguera, C. 2002. Glucosinolates and vitamin C content in edible parts of broccoli inflorescences alter domestic cooking. *Eur Food Res Technol.* 215:310–316.

Van den Berg, H.; Faulks, R.; Granado, F.; Hirschberg, J.; Olmedilla, B.; Sandmann, G.; Southon, S.; Stahl, W. 2000. The potential for the improvement of carotenoid levels in foods and the likely systematic effects. *J Sci Food Agric.* 80:880–912.

Verkerk, R.; Dekker, M.; Jongen, W.M.F. 2001. Post-harvest increase of indolyl glucosinolates in response to chopping and storage of Brassica vegetables. *J Sci Food Agric.* 81:953–958.

Viña, S.Z.; Chaves, A.R. 2007. Respiratory activity and phenolics in pre-cut celery. *Food Chem.* 100:1654–1660.

Wang, J.; Mazza, G. 2002. Inhibitory effects of anthocyanins and other phenolics and nitric oxide production in LPS/IFN-gamma-activated RAW 264.7 macrophages. *J Agric Food Chem.* 50:850–857.

Wattenber, L.W. 1993. Inhibition of carcinogenesis by nonnutrient constituents of the diet. In: *Food and Cancer Prevention: Chemical and Biological Aspects.* pp. 12–24, Waldron, K.W., Johnson, I.T., Fenwick, G.R. (Eds.). Royal Society of Chemistry, London.

West, L.G.; Meyer, K.A.; Balch, B.A.; Rossi, F.J.; Shultz, M.R.; Hass, G.W. 2004. Glucoraphanin and 4-hydroxyglucobrassicin contents in seeds of 59 cultivars of broccoli, raab, kohlrabi, radish, cauliflower, Brussels sprouts, kale, and cabbage. *J Agric Food Chem.* 52:916–926.

Williamson, G.; Manach, C. 2005. Bioavailability and bioefficacy of polyphenols in humans. II. Review of 93 intervention studies. *Am J Clin Nutr.* 81:243S–255S.

Wright, K.P.; Kader, A.A. 1997. Effect of slicing and controlled-atmosphere storage on the ascorbate content and quality of strawberries and persimmons. *Postharvest Biol Technol.* 10:39–48.

7 Fruits and Vegetables for the Fresh-Cut Processing Industry

Marta Montero-Calderón and
María del Milagro Cerdas-Araya

CONTENTS

7.1 INTRODUCTION

Fruits and vegetables are a gift of nature and delight people of all ages. They provide a wide range of flavors and textures and are loaded with most of the nutrients required for good health and wellness.

Their consumption as fresh produce is largely recommended all over the world; however, nowadays, consumers have limited time for food preparation. This has led the fresh-cut fruits and vegetables market to grow, as they offer good and convenient produce, ready to eat, with the attributes of fresh produce.

Fresh-cut fruits and vegetables are composed of living cells that naturally spoil and deteriorate over time and are affected by all preparation operations and surrounding conditions during pre- and postharvest handling, processing operations, and storage. Respiration and ethylene production rates, color, aroma, and texture change as a response to physical damages produced during cutting operations, as well as to the activity of undesirable microorganisms (Wiley 1994; Aguayo et al. 2004; Artés-Hernández et al. 2007; Artes-Calero et al. 2009).

Quality attributes of the fruit and vegetables used as raw materials for processing have a great influence on the final quality and shelf-life of fresh-cut products; hence it is very important to identify and understand relevant changes for specific products and how they are affected by handling, processing, and storage.

Produce to be used for the fresh-cut industry should resist processing and maintain their attributes with minimum variations for as long as possible. This chapter focuses on the fruits and vegetables requirements for handling and conditioning for fresh-cut produce processing.

7.2 QUALITY OF INTACT FRESH FRUITS AND VEGETABLES

Fresh fruits and vegetables are expected to preserve quality during handling, storage, processing, and distribution. Even though quality criteria vary among products, it is generally associated with intrinsic properties of the food, such as visual appearance, texture, flavor, nutritive value, and safety issues during field production, handling, and processing. Appearance has a great influence on product selection for processing; shape, size, color, gloss, uniformity, and lack of wilting, browning, and decay symptoms give clues about stage of maturity, freshness, and expected process yield.

Texture attributes gather structural properties of the product and related sensorial attributes perceived as the product is bitten. Structures of fresh fruits and vegetables cells and tissues are complex in shape, chemical composition, adhesive and cohesive forces between cells, and how they are affected by turgidity, maturity stage, and other variables, resulting in a wide range of responses to force stresses during handling and processing (Schouten and Kooten 2004). Texture can be described by a series of parameters for specific characteristics, such as firmness or hardness, fracturability, adhesiveness, gumminess, crispiness, fibrousness, juiciness, flexibility, and others, with their relative importance dependent on the product and its final use. Mechanical response of intact produce is affected when fresh-cut products are prepared, because of injured cells and tissues, size reduction, elimination of protective skin, increased water losses, and wilt and decay.

Flavor embraces taste and aroma attributes, with a very wide range of combinations of sweetness, sourness, bitterness, and astringency, along with the characteristic aromas of each product, and the absence of undesirable off-flavors and off-odors. Kader (2008) highlights the importance of nutritive and better-flavored fruits and vegetables as a key factor in selecting cultivars, as a key to increase sales and consumption.

Nutrient content and biochemical composition vary with the products as they might come from different parts of the plant. Storage organs such as roots and tubers have high starch content; stems are rich in fibers and skeleton-type tissues with high lignin and cellulose content; and fruits are rich in sugars, organic acids, mineral salts, pectic substances, and enzymes (Maestrelli and Chourot 2002). Soluble solids content, total or titratable acidity, pH, water content, density, and the ratio of soluble solids content to acidity are commonly used as quality attributes. Fruits and vegetables are very good sources of vitamins A and C, minerals, carbohydrates, dietetic fiber, proteins, and antioxidant compounds such us carotenoids, flavonoids, and other phenolic compounds (Kader and Barret 2004); their composition and concentration vary among cultivars and are affected by pre- and postharvest practices.

Safety requirements related to fresh-cut produce include good agricultural practices (GAP) and good processing practices (GMF), freedom of plagues, mycotoxins, pesticide residues, and any other chemical or physical contamination that might risk consumer health.

Finally, it is important to consider that required quality attributes of fruits and vegetables for the fresh-cut industry may differ from those for the intact fruit market, because no alterations are done to the products in the latter case. Processors need intact fruits and vegetables that can withstand processing and maintain quality attributes of the fresh-cut product as long as possible, with high production yields, with very good and consistent quality, free from defects, with the right maturation stage; thus, the use of grocery store surplus or low quality and unmarketable products for processing should be avoided. The right intact fruits and vegetables used as raw matter must allow for the preparation of high-quality fresh-cut products, with uniform and consistent quality, suitable postcutting shelf-life, and consumer satisfaction.

7.3 DYNAMIC BEHAVIOR OF FRESH FRUITS AND VEGETABLES

Fruits and vegetables are composed of living tissues and have to withstand two major hurdles: harvest and processing. When they are harvested, water and nutrient supply from the mother plant ends, and tissues are injured at the incision point. Processing causes further physical damages as the products are cut, and the product skin is removed, causing an increase in the respiration and other metabolic activities continue all the way up to the consumer table.

For such a dynamic system, changes are continually occurring, even though they may not be obvious for the first hours or days after harvesting or cutting. Appearance and texture alterations are the first to be noticed, although many more physical, physiological, and biochemical changes are ongoing at the same time at different rates, and are influenced by internal and external factors.

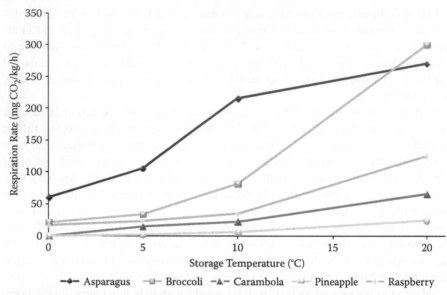

FIGURE 7.1 Effect of storage temperature on respiration rate of several intact fruits and vegetables. (From Barth, M.M., Zhuang, H., and Saltveit, M.E. 2004. Fresh-cut vegetables. In *The Commercial Storage of Fruits, Vegetables, and Florist and Nursery Stocks. USDA-ARS. Agriculture Handbook Number 66* (draft-revised 2004). Ed. K. Gross. http://www.ba.ars.usda.gov/ hb66/147freshcutvegetables.pdf (accessed August 25, 2009); Beaulieu, J.C. and Gorny, J.R. 2004; Fresh-cut fruits. In *The Commercial Storage of Fruits, Vegetables, and Florist and Nursery Stocks. USDA-ARS. Agriculture Handbook Number 66* (draft-revised 2004). Ed. K. Gross. http://www. ba.ars.usda.gov/hb66/146freshcutfruits.pdf (accessed August 25, 2009); Gross, K. 2004. Summary of respiration and ethylene production rates. In *The Commercial Storage of Fruits, Vegetables, and Florist and Nursery Stocks. USDA-ARS. Agriculture Handbook Number 66* (draft-revised 2004). Ed. K. Gross. http://www.ba.ars.usda.gov/hb66/010respiration.pdf (accessed August 25, 2009).)

The main intrinsic changes of fruits and vegetables are directly related to external factors, such as temperature, relative humidity, handling, microbiological, and other stresses occurring during pre- and postharvest operations.

7.3.1 RESPIRATION RATE

The rate of respiration is sensible to internal factors, such as product type and maturity stage, and also to external factors like temperature (Figure 7.1), ethylene concentration, stress caused during harvest, postharvest, and processing operations, pathogens, and physical injuries (Fonseca et al. 2002; Kader et al. 2002a; Varoquaux 2002). The faster the respiration rate is, the shorter will be the shelf life of the intact and fresh-cut products. Respiration rate is a good indicator of ongoing processes inside a product and how fast they are happening; products harvested during active growth (vegetables and immature fruits) usually have high respiration rates, while mature fruits and storage organs have relatively low rates (Saltveit 2004a).

7.3.2 ETHYLENE PRODUCTION

Ethylene has an active participation in growth, development, maturation, healing, and senescence of fresh produce (Kader 2002a; Saltveit 2004b). Its production rate varies among products; it increases with temperature, disease incidence, mechanical injuries, water stress, and maturity stage at harvest; and it can be partially controlled by low temperature storage, reduced oxygen composition, or elevated carbon dioxide levels surrounding the commodity. Very small concentrations of ethylene can damage sensible products (Table 7.1); hence, they should be handled separately from those that produce it.

7.3.3 WATER LOSS

Fruit and vegetable water content is very high and can be easily lost through the skin lenticels, stomatas, cuticule, and other structures when the surrounding atmosphere has low relative humidity. The skin of fruits and vegetables acts as a natural barrier that helps to partially control water loss, but its effectiveness depends on product morphology, surface characteristics, size, ratio of product surface area to weight and volume, maturity stage, physical injuries, as well as environmental temperature, relative humidity, and air movement around the product. Some products are more susceptible to lose water, such as lettuce and other leafy vegetables, which wilt and shrivel and generally deteriorate rapidly; others are more resistant to water loss, like apples and pears. The water loss for other products increases for those with rough, uneven, or extended surface area or with high stomata or lenticels density, like rambutan, with hair-like structures that favor water loss and desiccation, with the consequent external darkening occurring in a few days at 25°C and 60% relative humidity (Yingsanga et al. 2006), with losses of 7% to 11% weight during the first storage day.

On the other hand, fresh-cut products can lose water even more rapidly than whole products because of their increased surface area to volume ratio, as skin removal and damaged tissues resulting from cutting operations favor cellular content leakage. For such products, the use of sharp knives and proper packaging materials are key elements to reduce water migration.

7.3.4 INTERNAL AND EXTERNAL COLOR

Color is one of the main quality attributes of fresh commodities commonly used as selection criterion by the consumers and the fresh-cut industry, as well as an indicator of the overall quality and maturity stage of the product. Color varies among products, cultivars, and stages of maturity or development; it can be affected by preharvest factors such as plant nutrition, seasonality, climate conditions, production lot, temperature, relative humidity, storage time, and postharvest handling conditions. Color changes can result from the degradation or formation of pigment compounds such as chlorophylls (green), anthocyanins (red, blue, and purple), and carotenoids and flavonoids (yellow and orange) (Maestrelli and Chourot 2002; Kader and Barrett 2005). They can occur as part of the ripening process, but they also can be caused by mechanical stresses on cell wall, membranes, and tissues during produce handling and fresh-cut processing. Such damages favor fluids leakage, tissue softening, and enzymatic browning reactions, as enzymes and their substrates come into contact (Singh and Cadwaller 2004). Enzymatic activity of PPO (polyphenol oxidase) and POD (peroxidase) is also associated with color changes in fruits and vegetables. Mangoes, avocados, peaches, apples, pears, bananas, olives, potatoes, mushrooms, lettuce, grapes, and other fruits and vegetables are very sensitive to enzymatic browning due to PPO activity.

TABLE 7.1

Optimum Storage Temperature and Relative Humidity for Selected Fruits and Vegetables

Product	Storage Temperature °C	Relative Humidity %	Sensitivity to Chilling[a]	Freezing Temperature °C	Sensitivity to Ethylene[b]
Apple (summer)	0–2	85–95			
Fuji, Gala	0	90–95	√	−1.5	H
Golden, McIntosh	4	90–95	√	−1.5	H
Asparagus	2,5	95–100	√	−0.6	M
Avocado (ripe)	0–2	85–95	√	−1.6	H
Avocado (unripe cv. Fuerte, Hass)	3–7	85–90	√	−1.6	H
Banana	13–18	85–90	√	−0.8	H
Beets (topped)	0–2	98–100	√	−0.9	L
Blackberry	−0.5–0.0	90–95		−1.3	L
Broccoli	0	95–100	√	−0.6	H
Brussels sprouts	0	95–100	√	−0.8	H
Cabbage	0	98–100	√	−0.9	H
Cantaloupe	2–5	95	√	−1.2	M
Carambola	7–10	85–90	√	−1.2	H
Carrots	0	98–100	√	−1.4	H
Lettuce head					
Butterhead	0	98–100		−0.2	H
Iceberg	0	98–100		−0.2	H
Lime	9–10	85–90	√	−1.6	M
Mango	13	85–90	√	−1.4	H
Mushrooms	0	90	√	−0.9	M
Nectarine	−0.5–0.0	90–95	√	−0.9	M
Onion (mature bulbs, dried)	0	65–70	√	−0.8	L
Orange	7–10	85–90	√	−0.8	M
Oregano	0–5	90–95	√		H
Papaya (ripe)	13–18	85–90	√	−0.9	M
Passion fruit	7–10	85–90	√		M
Peach (ripe)	2–5	90–95	√	−1.2	M
Pear	7–10	85–90	√	−1.7	H
Pepper (Bell)	0	95–98	√	−0.7	L

Commodity	Temperature (°C)	RH (%)	Fresh-cut[a]	Freezing point (°C)	Rating[b]
Cassava	13–18	85–90	√		L
Cauliflower	0	95–98	√	−0.8	H
Celery	0	98–100	√	−0.5	M
Chayote	7	85–90	√		
Cherimoya (custard apple)	13–18	85–90	√	−2.2	H
Coconut	0–2	85–95	√		
Cucumber	10–12	85–90	√	−0.9	H
Eggplant	10–12	90–95	√	−0.5	L
Garlic	0	65–70	√	−0.8	
Granadilla	7–10	85–90	√	−0.8	
Guava	7–10	85–90	√	M	
Honeydew melon	13–18	85–90	√	−1.1	H
Kiwifruit (ripe)	0	90–95	√	−0.9	H
Lemon	10–13	85–90	√	−1.4	
Pineapple	7–13	85–90	√	−1.1	L
Plantain	13–15	90–95	√	−0.8	H
Pomegranate (arils)	5–7	90–95	√	−3.0	L
Potato (cured)	13–18	85–90	√	−0.8	M
Rambutan	12	90–95	√		H
Raspberry	−0.5–0.0	90–95	√	−0.9	L
Rhubarb	0	95–100	√	−0.9	L
Soursop	13	85–90	√		
Spinach	0	95–100	√	−0.3	L
Summer squash	7–10	95	√	−0.5	M
Sweet potato	13–18	85–90	√	−1.3	L
Strawberry	0	90–95	√	−0.8	L
Tomato (mature green)	10–13	90–95	√	−0.5	H
Tomato (firm ripe)	8–10	85–90	√	−0.5	L
Watermelon	10–15	90	√	−0.4	H

[a] o, no; √, yes.

[b] L, low; M, medium; H, high.

Sources: Beaulieu, J.C., and Gorny, J.R. 2004. Fresh-cut fruits. In *The Commercial Storage of Fruits, Vegetables, and Florist and Nursery Stocks. USDA-ARS. Agriculture Handbook Number 66* (draft-revised 2004). Ed. K. Gross. http://www.ba.ars.usda.gov/hb66/146freshcutfruits.pdf (accessed August 25, 2009) Thompson, J.F., and Kader, A.A. 2004. Wholesale distribution center storage. In *The Commercial Storage of Fruits, Vegetables, and Florist and Nursery Stocks. USDA-ARS. Agriculture Handbook Number 66* (draft-revised 2004). Ed. K. Gross. http://www.ba.ars.usda.gov/hb66/016wholesale.pdf (accessed August 25, 2009); Cantwell, M. 2002. Optimal handling conditions for fresh produce. In *Postharvest Technology of Horticultural Crops.* Ed. A.A. Kader. University of California Agriculture and Natural Resources. Publication 3311, pp. 511–518.

Color changes can be accelerated by external factors such as high temperature, low relative humidity environment, physical damages, and other stress conditions. Odriozola-Serrano et al. (2009) found that the degradation of anthocyanins responsible for the appealing bright red color of strawberries was significantly larger as the temperature rises and storage time increases.

Color requirements for fresh intact product consumption varies with those for fresh-cut processing. For the whole produce market, external color is one of the main quality attributes, while for the industry, both external and internal colors are important. In general, color should be bright, even, and preserve the natural appearance of the fresh product. Care must be taken with new cultivars with improved processing characteristics but with major differences in color, because they could be rejected by the consumer.

7.3.5 TEXTURE

Texture attributes vary during pre- and postharvest handling, as they are affected by stage of maturity, plant nutrition, water stress, storage temperature and relative humidity, rough handling, and ripening processes. Product softening, loss of turgidity, and increased elasticity or toughness are some of the changes occurring during product handling, which may reduce the product's value and utility for the fresh-cut industry. Changes can be due to water losses, as mentioned earlier, or to the activity of several enzymes or pathological breakdown in combination with handling conditions. Enzymes such as β-galactosidase, polygalacturonase, pectin methyl esterase, cellulose, phenilalanineammonia lyase, peroxidase, and cellulase participate in cell-wall modification; degradation of pectin compounds in tomatoes, melons, avocados, and peaches; pectin solubilization in strawberries; ripening initiation processes; tissue weakening and softening in raspberries, avocado, blackberries, mangoes, cherimoya, and tomatoes; and toughness development in asparagus (Bhowmik and Dris 2004).

Physical damages of fresh fruits and vegetables should be minimized throughout harvest, transportation, and postharvest and processing operations, because they have a negative effect on quality attributes and shelf life. Symptoms could show immediately, during processing, or after several days or weeks of storage; they are generally described as tissue darkening, loss of firmness, bruises, cracks, cuts, and perforations which lead to faster deterioration reactions than with intact products. Damaged tissues increase respiration rates, water loss, and other metabolic reactions; allow better contact between enzymes and their substrates; and favor microbial spoilage and other undesirable reactions (Singh and Anderson 2004).

Maestrelli and Chourot (2002) classified fruits as very fragile, fragile, resistant, or very resistant to handling. They found cultivar, stage of maturity, and handling practices influenced the response to mechanical injuries of peaches, pears, prunes, and apricots. Damages are caused by impact, compression, penetration, vibration, and shear forces against the product; hence, they can be controlled by protecting and immobilizing the product, which can be achieved by careful handling, reducing unnecessary movements, drop heights, and any cut edges or rough surfaces that could threaten product integrity.

Bruising response to processing varies among apple and peach cultivars, but the fruit susceptibility can also be influenced by preharvest practices, weather conditions, maturity stage, and other factors that might affect phenolic compound content and enzymatic activity (Varoquaux, 2002).

Impact bruising damages susceptibility is larger for sweet cherries handled below 10°C as compared with temperatures up to 20°C. Mechanical damages due to vibration are not affected by temperature (Crisosto et al. 1993). On the other hand, physiological disorders such as chilling and freezing injuries also contribute to tissue softening, water-soaked areas, ripening problems, surface and flesh discoloration, off-flavors and off-odors production, and increased susceptibility to microbial spoilage and breakdown (Kader 2002a; Kader and Barret 2004; Wang 2004).

7.3.6 COMPOSITIONAL CHANGES

Internal composition of fruits and vegetables keeps changing during growth and development and after harvest. Changes could be desirable or not, depending on the product and how and when it is going to be processed. Soluble solids content and acidity are two important parameters for fruits, frequently related to stage of maturity, flavor, and consumer preferences, and used as quality parameters for product selection for processing. Little changes occur in nonclimacteric fruits after harvest (pineapple, citrus fruits), while climacteric fruits suffer important changes as they continue to ripen.

Aroma compounds losses or the production of off-flavors and off-odors directly affect fruit and vegetable flavor. They can be associated to maturity stage, unfavorable storage conditions, or enzymatic activity of peroxidases and lypoxygenases in chile peppers, broccoli, asparagus, carrots, and green beans.

Antioxidant and other nutritional attributes can be lost during handling and processing. Kader and Barret (2004) pointed out the high sensibility of ascorbic acid to high temperatures, light, low humidity environments, physical damages, and chilling injuries.

7.3.7 GROWTH AND DEVELOPMENT

Some fresh produce continues to grow and develop after harvest. Rooting can occur on root crops and onions, seed germination on tomatoes and peppers, and elongation in asparagus (Kader 2002a). Proper selection of the harvesting index, handling and storage conditions, and special treatments are necessary to diminish these undesirable changes. Most of these changes are undesirable for fresh-cut processing, because metabolic activity increases and product appearance may rapidly deteriorate.

7.3.8 MICROBIAL SPOILAGE

Fruits and vegetables are good sources of nutrients and water for humans, but also for microorganisms, which can grow easily under ambient conditions. Microorganisms can get into the fruits at different stages of their field production, at the earlier stages of development, or during postharvest handling and processing operations. Some microbes can rapidly deteriorate fruit and vegetables' appearance and other quality attributes, but do not represent a risk for consumers (phytopathogens), while others can be harmful for consumer health (food safety related microorganisms). They can be separated into two different groups:

- Fruit and vegetables microorganisms: Include molds, yeast, bacteria and viruses which affect product quality, but do not represent a risk for consumer health.
- Harmful microorganisms: They can cause illness and even death to consumers, but in general they affect neither the product appearance nor other quality attributes of the fruit or vegetable. Since no deterioration symptoms are observed in the product, care must be taken to avoid product contamination. Sanitation programs in the fields and processing area and good hygienic practices of the operators are needed to avoid product contamination.

7.4 TECHNOLOGICAL TOOLS TO PRESERVE FRESH PRODUCE FOR PROCESSING

Quality is a key factor for processed foods, but it is particularly true and important for fresh-cut fruits and vegetables, which must preserve the quality attributes of the intact produce after cutting stresses, without being subjected to any strong temperature-stabilizing treatment. Hodges and Toivonen (2008) highlighted the importance of recognizing that all processes applied to a fruit or vegetable cause stress-induced changes in the tissue physiology and metabolism. They also pointed out the need to understand how these changes occur to set effective strategies to preserve the product quality and extend its shelf life.

The fresh-cut processing industry requires high-grade fruits and vegetables as raw materials. They should have good appearance, texture, taste, odor, and nutritive attributes, and they must be safe for the consumer. They should be free from mechanical injuries, decay, insects, and other damages, and they must resist process operations and further handling and storage procedures.

In addition, effective technological tools must be used in each of the slabs of the chain, from the production fields to the processing industry, to assure proper quality produce supply and minimize product losses.

The success of a fresh-cut product will depend on the fulfillment of the target consumer needs and expectations, the use of the right fruits and vegetables at their optimum maturity stage for processing, and the application of the right operation procedures and packages, as well as the utilization of appropriate market tactics.

Strategies for better intact fruits and vegetables for processing involving preharvest, harvest, and postharvest handling will be discussed later in this chapter.

7.4.1 Cultivar Evaluation and Selection

For each type of produce, there could be a few or many cultivars in the market, each with particular benefits and limitations for processing. Studies comparing cultivars usually show considerable variation in respiration rates, color, texture, bruising susceptibility, size, shape, appearance, nutritional value, sensory, and other characteristics of many fruits and vegetables (Crisosto et al. 1993; Gorny et al. 2000; Crisosto and Mitchell 2002; Maestrelli and Chourot 2002; Schouten and Kooten 2004; Deepa et al. 2007).

Cultivar selection for the fresh-cut industry looks for intact fresh fruits or vegetables that can meet the desired quality attributes preestablished for their fresh-cut product, resist transportation and handling before processing, tolerate processing operations with minor quality alterations, and have a prolonged after-cutting shelf life. These four elements are interrelated because of the dynamic behavior of intact fresh fruits and vegetables, because their characteristics and quality are continually changing and influenced by environmental conditions, production technology during preharvest and postharvest handling, fresh-cut processing and packaging, and distribution to the final market.

Industry needs reliable agricultural products with high quality throughout the year, when possible, in order to produce uniform fresh-cut products without interruption. Cultivar selection is the first step, and it has to be followed by proper preharvest practices, harvesting indicators, postharvest handling operations and storage before processing, and food safety programs to avoid consumer risks (Chiesa et al. 2003).

Ideal fruits and vegetables for fresh-cut processing are those with the best and homogeneous quality attributes: right stage of development or maturity; high field production and processing yields; availability year-round; free from physical, physiological, or pathological disorders; easy to handle; highly resistant to handling and all processing operations and stabilizing treatments; little susceptible to external conditions; with a prolonged shelf life after processing to maintain quality attributes all the way to the consumer and safe consumption. It should also meet consumer likes and preferences and market requirements.

For specific products, particular requirements are needed. Processors need to clearly define the attributes of their final products to evaluate cultivar response to handling, processing, and after-cutting handling and to determine limiting factors, such as juice leaking, discoloration, browning, wilting, microbial spoilage, or others that might restrict fresh-cut product shelf life. The effects of harvesting indicators, handling, and preparation for processing practices on final product quality should also be evaluated.

Some examples of studies conducted to select the best cultivars for fresh-cut processing include one of Gorny et al. (1999), who found that the critical factors that impact fresh-cut peach and nectarine slice quality were ruled by product response to cutting, which was better when the flesh firmness was between 13 and 27 N, and the fruits were stored at 0°C and 90% to 95% relative humidity. In another study, the same authors (2000) compared the suitability of Anjou, Bartlett, Bosc, and Red Anjou pear cultivars for fresh-cut slices production and found significant differences in respiration and ethylene production rates, flesh firmness, color, and susceptibility to cut-surface browning, which was very intense for Anjou and Red Anjou cultivars. Apple sensibility to browning of five apple cultivars was also studied by Milani and Hamedi (2005), who found differences in browning rate; the Red Delicious cultivar exhibited the highest browning rate, followed by Golden Delicious with a medium rate, and Granny Smith and Golden Smoty had a weak browning rate. Sweet cherry cultivars were evaluated for fresh-cut processing by Toivonen et al. (2006), who concluded that most of them were adequate for fresh-cut processing, as they maintain firmness, even though they showed differences in postcutting bleed, weight loss, and decay throughout storage.

Following are some hints for fruits and vegetables cultivar selection for fresh-cut processed products:

- Define desirable quality parameters and tolerance ranges for the fresh-cut product (color, shape, size, flavor, soluble solids content, acidity, texture, juiciness, nutritional content, other).
- Identify available cultivars with consistent and reliable quality that can meet product concept.
- Look for limiting factors to fresh-cut product quality and shelf life, based on visual and eating quality and microbial safety; consider deteriorative changes that reduce their marketability, such as browning, water-soaked tissues, translucency, softening, composition changes, microbial growth, decay, or others.
- Evaluate cultivar aptitude for processing by studying product response to handling and processing, and susceptibility to deteriorative changes before, during, and after cutting.
- Determine required harvest maturity stage and ripening treatments when needed.
- Evaluate processing yields (usable product per kilogram of intact fruit or vegetable, processing time per kilogram of prepared fresh-cut product, etc.).
- For those cases in which a cultivar is preferred though it may have some limitations, evaluate alternative treatments to control undesired browning, softening, or other changes on the fresh-cut product.
- Determine expected postcutting shelf life of the final product at handling temperatures.
- Study product compatibility among products for mixed fresh-cut fruits or vegetables.
- Seek possible suppliers, production sites, agricultural practices, traceability possibility, product quality, and availability throughout the year.
- When possible and convenient, evaluate product response to mechanization to reduce processing time, reduce contamination risks, and improve yields.

7.4.2 PREHARVEST PRACTICES TO IMPROVE INTACT PRODUCE QUALITY

Genetic material, sowing, growing conditions, light intensity along production period, pruning, product thinning, harvest maturity, nutrients and water supply, soil quality, fertilization, weeds control, and pest management affect product quality together with climate conditions (Hodges and Toivonen 2008; Kader 2008). Thus, careful production plans must be implemented, looking to strengthen the preferred characteristic of a particular crop and, consequently, the final quality and shelf life of fresh-cut products. The effects of some preharvest practices are provided in advance.

Crop rotation is recommended for its positive effect on product quality contrast with the decay inoculum of soilborne fungi, bacteria, and nematodes which would build up with repeated cropping of the same vegetable in the production fields (Crisosto and Mitchell 2002). Fruit size and yield are affected by fruitlet thinning, position inside the tree, pruning, and other cultural practices according to the same authors.

Irrigation is very important for all crops, because plants tissues need water to live. Low water supply might stress product and increase its sensibility to sunburns, alter maturation processes in pears, and provoke a leather-like texture on peaches, while moderate water stress can reduce fruit size and increase soluble solids content, acidity, and ascorbic acid content (Kader 2002a). Gelly et al. (2003) also reported that deficit irrigation on peaches (*Prunus pérsica* L.) increased soluble solids content and helped to maintain fruit color longer. On the other hand, excess water stress could lead to cracking failures in cherries, apricots, tomatoes, and other products; reduce firmness and soluble solids content; and cause an increased susceptibility to mechanical injuries due to an excess of turgidity (Kader 2002b). Plants can also be stressed because of salt presence in irrigation water. Kim et al. (2008b) evaluated stress due to water salinity on romaine lettuce (*Lactuca sativa* cultivar Clemente). Sodium chloride concentrations above 100 mM resulted in 1.5- to 3-fold reduction of lettuce height and weight, as compared with control treatment without salt. Color losses increased with sodium chloride concentrations.

Carotenoids and phenolic compounds contents are also affected by irrigation with salt water. The number of days before harvest at which irrigation is stopped also influenced product quality in Iceberg lettuce, as observed by Fonseca (2006), who found out that when irrigation was stopped 4 days before harvesting instead of 16 days, the product weight and diameter were larger, but the aerobic bacteria counts increased, resulting in faster deterioration of the quality of the product. For intact tomatoes, Kim et al. (2008a) found a 30% increase in lycopene and vitamin C content when they were irrigated with salt water, though phenolic compounds content was not affected. Type of substrate also affects quality attributes.

Fertilization affects both quality at harvest and postharvest shelf life of fruits and vegetables. Nutrients should be balanced, because deficiencies or excesses can favor physiological disorders and reduce product quality and shelf life. High nitrogen fertilization is used to increase product size, but it can reduce volatile compounds production and promote changes in product flavor; other elements also show opposite response, such as high levels of potassium, which can reduce color disorders, while high levels of magnesium can increase them (Crisosto and Mitchell 2002). Plant nutrition differences can affect product size, firmness, and weight loss susceptibility. Calcium has been associated with a reduction of respiration rate and ethylene production, firmness increase, and ripening and deteriorative reactions slowdown (Kader 2002b).

Hodges and Toivonen (2008) compared quality attributes of tomato slices grown in hairy vetch and black polyethylene mulch and found that those grown in the hairy vetch mulch were firmer and had less water-soaked areas and less increase in electrical conductivity, stresses associated with chilling injuries and membrane damages, respectively.

Calcium chloride sprays have been successfully used to reduce browning core, cork spots, superficial scalds disorders, and external and internal rots on Anjou pears, with an overall enhancement of fruit appearance and an improvement of fruit juiciness and fruit color (Alcarez-Lopez et al. 2009; Raese and Drake 2000).

Climatic conditions (temperature, rain, wind, light) affect internal quality attributes of the products as well as their susceptibility to handling and processing. The effect of climatic conditions can be partially controlled in the growing areas by shades, drainage, and wind stoppers; they can also be precisely controlled in greenhouse plantations. Jolliffe (1996) found an important reduction on skin chlorophyll content and shelf life for low light intensity on greenhouse-grown English cucumbers; such a reduction can be avoided by the use of supplemental light during growing which increases product yields, external and internal quality of many vegetables, including dry matter content and skin chlorophyll content on cucumber, higher ascorbic acid and sugar content in tomato, and better head firmness on lettuce (Hovi-Pekkanen and Tahvonen 2008).

There is a trend in Europe and Latin America to start moving from traditional growing to protected areas cultivation for better control during produce growing. A wide range of simple and complex technologies are used to control temperature, relative humidity, and irrigation control, and more recently, floating trays with a nutrients solution supply were incorporated, and small leaves grow with a significant reduction in nitrates accumulation and microbial load, two characteristics appreciated for fresh-cut processing (Rodríguez-Hidalgo et al. 2006).

7.4.3 Harvest and Maturity Indices

Stage of maturity at harvest is critical not only to assure product quality and shelf life of intact fruits and vegetables for the fresh market but also for the fresh-cut industry; it affects product composition, postharvest tolerance to handling and processing operations, and their postcutting life (Kader 2002b; Toivonen and DeEll 2002; Kader and Barret 2004; Martín-Belloso and Rojas-Graü 2005; Kader 2008).

Maturity at harvest influences fruits and vegetables response to processing and deterioration reactions. Toivonen (2008) studied the effect of maturity at harvest on the susceptibility of antibrowning treated apple slices to cut-edge browning. He found that the cutting surface of slices from Granny Smith apples picked prior to proper harvest maturity are more susceptible to browning even after commercial antibrowning treatments. Fruit ripeness of pear slices at cutting affected their shelf life (Gorny et al. 2000); it varied from 2 days at 0°C for ripe fruit to more than 8 days for partially ripe and mature-green pears; and it also affected surface darkening at 0°C, which was significantly reduced for partially ripe and mature-green fruit. However, the eating quality of mature-green pear slices exhibited lack of juiciness and aroma.

Harvesting criteria vary among products: how they are to be consumed or processed, distance to marketplaces, intended storage time and temperature, industry requirements, consumer preferences, and many other parameters. Some produce may be consumed in several stages of maturity, such as mangoes, papayas, and plantains, which have different uses for green-mature and fully ripe products; vegetables are obtained from different parts of the plant, such as leaves, flowers, sprouts, roots, and tubers that reach their best quality attributes at various stages of growth and development of the plant, so there is a wide range of possibilities for harvesting, depending on the final destination of the produce, the desired quality attributes, and their resistance or tolerance to withstand handling and processing.

Maturation or harvesting indices have been set to describe the right time to harvest for better quality and shelf life. Harvest indices must be simple, easy to understand and apply, reliable, the product of an objective measurement, and nondestructive if possible (Reid 2002). Table 7.2 shows maturity indices commonly used for fruits and vegetables harvesting. Production yields and quality parameters of the product are

TABLE 7.2
Maturity Indices Commonly Used for Fruits and Vegetables

Index	Examples
Elapsed days from full bloom to harvest	Apples, pears
Mean heat units during development	Peas, apples, sweet corn
Development of abscission layer	Some melons, apples, feijoas
Surface morphology and structure	Cuticle formation on grapes, tomatoes; netting of some melons; gloss of some fruits (development of wax)
Size	All fruits and many vegetables
Specific gravity	Cherries, watermelons, potatoes
Shape	Angularity of banana fingers; full cheeks of mangoes; compactnes of broccoli and cauliflower
Solidity	Lettuce, cabbage, Brussels sprouts
Textural properties:	
Firmness	Apples, pears, stone fruits
Tenderness	Peas, apples, sweet corn
External color	All fruits and most vegetables
Internal color and structure	Formation of jellylike material in tomato fruits; flesh color of some fruits
Compositional factors:	
Starch content	Apples, pears
Sugar content	Apples, pears, stone fruits, grapes
Acid content, sugar:acid ratio	Pomegranates, citrus, papaya, melons, kiwifruit
Juice content	Citrus fruits
Oil content	Avocados
Astringency (tannin content)	Persimmons, dates
Internal ethylene concentration	Apples, pears

Source: Reid, M.S. 2002. Maturation and maturity indices. In *Postharvest Technology of Horticultural Crops.* Ed. A.A. Kader. University of California. Agriculture and Natural Resources Publication 3311, pp. 55–62. With permission.

taken into consideration, as well as market prices and buyers' requirements. Early harvesting results in low production yields and underdeveloped quality attributes, whereas late harvesting leads to overmature products and excessive postharvest losses.

To develop a maturity index, Reid (2002) suggested the following steps:

1. Determine changes in the commodity throughout its development.
2. Look for features whose changes correlate well with the stages of the commodity's development.
3. Carry out storage trials and taste panels to determine maturity indices that fulfill minimum acceptability and required shelf life.
4. Select maturity indices and assigned minimally acceptable values.
5. Test harvesting indices over several years in several growing locations to ensure that they reflect the quality of the harvested product.

The chosen index or indices should be adjusted for fruits and vegetables to be used for fresh-cut processing, after evaluating their response to process, handling, and storage operations.

Nonfruit vegetables generally include diameter, length, shape, color, firmness and compactness, and other appearance parameters as the main harvesting indices; some examples are asparagus, celery, rhubarb, and okra length; bud size of artichokes; compactness of broccoli, cauliflower, Brussels sprouts, cabbage, and some lettuce cultivars; and color in lima beans, broccoli, collards, and peas.

Color, size, firmness, appearance, and internal quality attributes are used for fruit vegetables harvested, along with observations about natural incision of the fruit to the plant. For instance, for cantaloupes and other melons, stem appearance and how the stem naturally breaks are good indicators for harvesting, together with soluble solids content, aroma, fruit and rind color changes, or even days from bloom. Tomato harvesting indices selection will depend on the use given to the product: They can be harvested mature-green, which are very firm tomatoes, with color changing from green to light green; ripe with full red color and soft, but still firm; or in the middle, known as breaker tomatoes, which are firmer than ripe tomatoes but softer than mature-green and exhibit a pink to red color on the blossom end. Color and firmness are very important parameters for cucumber, eggplant, bell pepper, watermelon, and other fruit vegetables. Asghary et al. (2005) found 'Semsory' muskmelon (*Cucumis melo* L. var. reticulates) harvested at the first stages of yellow color development had higher sugar content, better color, taste, aroma, and market value than those harvested at the mature green stage.

Beaulieu et al. (2004) highlighted tissue softening as a serious problem and limiting factor for fresh-cut products, and listed softening enzymes, decreased turgidity due to water loss, and stage of maturity as the main causes of texture changes. They studied the effect of product firmness on the postcutting sensory attributes of fresh-cut cantaloupe stored at 4°C, prepared from melons harvested at four distinct maturity stages (one-quarter to full-slip), and found that those from three-quarters mature cantaloupes exhibit less firmness loss than those from full-slip maturity fruits. Antioxidant characteristics also vary with cultivars and maturity stage; total and individual phenolic content, antioxidant capacity, carotenoids, ascorbic acid, and capsaicin content varied among sweet pepper genotypes and maturity stage (Marin et al. 2004; Deepa et al. 2007). Kader (2008) pointed out that nonfruit vegetables have better quality taste when they are harvested immature, while fruit vegetables and fruits get better when they are harvested fully ripe.

Harvesting criteria for fruits also include shape, size, and appearance parameters, but flavor and aroma take an important role. They have great influence on product quality, because aroma-related volatile and nonvolatile compound synthesis increases as the product matures and ripens (Kader 2008). Optimum levels of such compounds do not always match the harvesting criteria, because other parameters have to be considered, such as the type of product, resistance to handling and processing, time required to reach the final market, produce prices, how it is processed or consumed, and others. For apples, harvest date is determined by several parameters, including days from the full bloom as a rough idea of fruit maturity, background color, ease of separation of the fruit from the spur, soluble solids content, starch conversion into sugars, flesh firmness, and internal ethylene concentration (Gast 1994; Toivonen 2008).

For climacteric fruits, proper selection of harvesting indicators is very important, because if fruits are picked prior to physiological maturation, in an early period of preclimateric stage, fruits quality attributes would not reach desired levels. Robles et al. (2006) studied changes in Ataulfo mangoes as the fruit ripened and found an important increase in total soluble solids and ethylene production rate, accompanied by a gradual reduction of the respiration rate, acidity, and firmness. Maradol papaya, Keitt mangoes, and Red Spanish pineapple give better results for fresh-cut processing, when processed before the full ripeness stage (Hernández et al. 2007), explained by less firmness and color alterations during postcutting storage.

In summary, maturity at harvest affects quality attributes and after-cutting shelf life of intact and fresh-cut produce, and harvesting indices should be adjusted to produce response to handling and processing. Under- or overmature produce results in deficient quality attributes and low yields, while overmature products diminish postcutting shelf life of fresh-cut products and increase susceptibility to deterioration, mechanical damages, microbial spoilage, and other damages.

7.4.4 Postharvest Strategies to Reduce Undesirable Changes

7.4.4.1 Use Optimum Storage Temperatures to Reduce Metabolic Activity

Temperature is the most important external factor to control during postharvest handling and storage previous to processing, because it rules most of the changes occurring inside an intact or fresh-cut fruit or vegetable. As the temperature drops, most reactions slow down; hence quality attributes can withstand for longer periods. Optimum storage temperature should always be chosen for intact and fresh-cut fruit and vegetables (Table 7.1). As a general rule, fresh-cut products should be stored at 5°C or below, but the optimum temperature for the intact products could be higher for chilling sensitive fruits and vegetables, and it must be considered for produce storage before processing. Some produce can withstand temperatures near freezing (0°C and below), some need temperatures near 0°C, and those sensitive to chilling injury disorders cannot be stored at temperatures below 7 to 13°C, depending on the product. Storage at lower temperatures than those tolerated by the intact produce will result in uneven ripening, flavor, and color, and aroma losses, texture changes, and other undesirable changes.

For every 10°C rise on the produce handling temperature in the range from 0 to 30°C, the rates of respiration and deterioration increase two to three times for non-chilling-sensitive commodities (Kader 2002a; Saltveit 2004a). Crisosto et al. (1993) observed that sweet cherry respiration rates of four cultivars rapidly increased from nearly 10 mg CO_2/kg/h at 0°C to 45 to 50 mg CO_2/kg/h at 20°C, though response to temperature varied among cultivars exhibiting differences in fruit sensibility to temperature changes (Crisosto et al. 1993).

Exposure to high temperatures is also detrimental, though some product tissues can tolerate them for short periods; it causes phytotoxic symptoms that lead to accelerated deterioration (Saltveit 2004a). Prolonged exposure to sun in the fields, transportation trucks, or during storage should be avoided to reduce quality losses.

Once the produces are processed, temperature must be held at 5°C to minimize changes on the quality attributes of the fresh-cut products as well as microbial spoilage.

7.4.4.2 Relative Humidity and Water Loss Control

Following temperature, relative humidity is the second factor in importance for quality maintenance. Shelf life and value of fruits and vegetables decreases with water loss because it causes appearance deterioration, tissue softening, wilting, shriveling, and weight loss. Such changes also affect product suitability for the fresh market and the fresh-cut industry, because commodity resistance and yield during processing and handling deteriorate, and it is more sensible to have a shorter shelf life for the product.

Fresh produce are not solid-pack but porous materials filled with their own internal atmosphere, which has a high relative humidity. They lose water through the skin or abscission cuts, because of relative humidity differences between the internal atmosphere and that surrounding the product. Because of this, fresh produce should be stored under high relative humidity environments, as a complement to optimum storage temperature. However, storage requirements vary because water losses also depend on skin and other product characteristics, which make some products more susceptible to losses than others (Table 7.2). Díaz-Pérez et al. (2007) determined water loss relations with intrinsic characteristics of bell pepper, such as fruit size, maturity stage, cuticle thickness, natural wax over the product surface, and reported larger water loss through the calyx or stem scar than from the product skin, as previously reported for eggplants and tomato.

Water loss cannot be completely stopped, but it can be reduced by careful handling and proper storage temperature and relative humidity conditions. Temperature should be as low as the product can tolerate without chilling injury symptoms (Table 7.1), and relative humidity should be higher than 80% for most products, and up to 95% to 100% for products very sensitive to water loss, such as leafy vegetables and strawberries. Packaging materials, produce waxing, and reduced exposure to air movement could also help reduce water losses.

7.4.4.3 Air Movement

Cold air is normally used for produce cooling and storage; it removes heat from the produce and delivers it to the evaporator of the refrigeration system. The faster the air passes through the produce, the quicker the product cools. However, once the product is cold, excess air movement favors water losses, and thus, it should be kept as low as possible to allow proper ventilation without major losses. Adequate package sizes and ventilations and design proper product layout in the storage rooms can help to control excessive exposure to air.

7.4.4.4 Light

Exposure of potatoes to light favors greening during storage because of the production of solanine and chlorophyll. Such changes are undesirable and can be avoided by storing in darkness. Prolonged storage of green vegetables without light could also discolor them. The light effect starts at the fields or greenhouses; light intensity also affects flavonoids, thiamine, riboflavin, carotenoids, ascorbic acid, and other compounds found in fruits and vegetables during growing, thus affecting their composition and nutritional quality (Kader 2002b).

7.4.4.5 Atmosphere Composition

Cells and tissues require oxygen and produce carbon dioxide during respiration. Low oxygen and high carbon dioxide concentration in the atmosphere surrounding the fruit or vegetable can be used to delay deterioration and extend shelf life. Table 7.1 shows recommended atmosphere composition for several commodities; however, product benefits and shelf-life extension can significantly vary among products and cultivars.

7.4.4.6 Ethylene

Ethylene is a plant regulator that affects growth, development, ripening, and senescence processes and postharvest quality (Watkins 2006). Very low concentrations of ethylene in the atmosphere surrounding the product can trigger ripening processes of climacteric fruits and undesirable reactions on some fruits and vegetables, such as color loss and senescence reactions. Sensibility to ethylene varies among products, and changes can be desirable or not, but as a general rule, very low concentrations of ethylene are needed to affect product quality. Controlled application of ethylene can be used for uniform maturation and degreening but should be avoided for long-term storage. As a general rule, ethylene producers must always be separated from ethylene-sensitive products.

7.4.4.7 Handling and Processing

Mechanical damages on fruits and vegetables are caused by impact, compression, and shear and puncture forces applied to the product while harvesting and handling it all the way to the consumer or processing industry. Such damages accelerate metabolic processes and favor microbiological spoilage. Some of these damages symptoms can be detected only after several days of storage, during processing, or during subsequent storage, but they greatly affect product quality, stability, and shelf life. Stress caused by physical efforts during handling should be minimized in order to supply raw materials suitable to resist further stress processes during fresh-cut processing.

Varoquaux (2002) suggested that peeling and cutting damage product cells and cause an increase of the membrane permeability and, probably, a reduction of phospholipids biosynthesis. These events trigger the reactions of restoration of cellular microstructures and membrane integrity, entail the production of aldehydes of long carbonated chains, cause an increase in respiration rate, and lead to rapid consumption of cellular metabolites and consequent deterioration. However, produce response to stress also depends on the type of fruit or vegetable, its stage of development or maturity, and environmental conditions, and thus, it can vary for a single product, such as kiwi for which ethylene production rise due to cutting stress can be very rapid, or it might take several hours (Varoquaux et al. 2002).

7.4.4.8 Field and Storage Packages

The main function of packaging is to protect fresh fruits and vegetables against mechanical injuries, contamination, or any other damages throughout postharvest handling. Packages should have smooth surfaces and edges, be resistant to staking, have a suitable size and shape for the product, be easy to handle, allow proper ventilation for cooling, and be readily available. Some packages should have water loss barriers or some other special requirement. Packages for fresh fruits and vegetables used for fresh-cut processing are used only to protect the product as it is carried from the fields to the processing plant, and for short-term storage prior to processing.

7.4.5 CONDITIONING AND STORAGE BEFORE PROCESSING

Fresh-cut products quality starts with fresh fruits and vegetables, properly handled during preharvest, harvest, and postharvest. The fresher the prime matter, the better is the final product.

Temporary storage between harvesting and processing also affects the quality and shelf life of fresh-cut products; in general, the longer the delay before processing, the shorter the shelf life is going to be. However, because fresh-cut products are very perishable, lasting between 1 and 2 weeks, temporary storage of intact fruits and vegetables will be convenient. Tropical or template fruits, roots, or tubers, or other vegetables brought from distant markets could be processed at the final market, though some quality attributes could be partially compromised.

As for the cultivar selection, the effect of storage prior to processing should be studied for specific intact fruits and vegetables and their fresh-cut products.

Products to be stored prior to processing should be conditioned before storage or transportation to the market where they are going to be prepared, as a means to preserve their quality. Some common preparation operations include product selection (separation of culls), washing, classification based on quality criteria, stabilizing treatments such as application of growth regulators, antifungal treatments, curing, packaging, and cooling.

Induced fruit ripening could be useful to obtain uniform product characteristics prior to processing. It is generally carried out under controlled conditions of temperature, humidity, and air circulation. Ethylene generator devices yield very good results with banana, tomato, avocado, plantain, and other fruits.

In contrast, application of 1-methylcyclopropene (1-MCP) has been widely used to reduce the action of ethylene in fruits and vegetables. Several authors reported that color changes, tissue softening, and other changes occurring during ripening are substantially delayed (Schouten and Kooten 2004). 1-MCP (1-methylcyclopropene) inhibits ethylene action and ripening reactions. It is applied at 20 to 25°C in low concentrations (2.5 nL/L to 1 µL/L) for 12 to 24 hours, but results depend on cultivar, development stage, time from harvest to treatment, and multiple applications. Effects vary among fruits and vegetables, including delay in respiration rate, ethylene production, volatile production, color changes, chlorophyll degradation, membrane changes, softening, acidity and sugars variation, and development of disorders and diseases. It protects products from endogenous and exogenous sources of ethylene. Chlorophyll degradation and color changes are prevented or delayed in oranges, broccoli, tomato, avocado, and other green vegetables, and volatile development is inhibited in several apple cultivars, apricots, melons, bananas, and other fruits. Product softening is also delayed in fruits such as avocado, custard apple, mango, papaya, apple, apricots, pears, mature plums, peaches, nectarines, and tomato (Blankenship and Dole 2004; Watkins 2006; Manganaris et al. 2008). As product ripening and senescence processes are delayed or inhibited, the product shelf life is increased.

McArtney et al. (2008) studied preharvest applications of 1-MCP in Golden Delicious apples, and they were able to reduce the rate of softening of the fruit during storage.

Storage of intact fruits and vegetables should be carried out at their optimal temperature and relative humidity levels.

7.5 CONCLUSIONS

Intact and fresh-cut fruits and vegetable characteristics have an intrinsic dynamic behavior because they are composed of living tissues that keep changing over time and are influenced by environmental conditions and handling practices.

Initial quality of the intact fruits and vegetables used for fresh-cut processing will fix the maximum attainable quality and after-cutting shelf life of the processed product.

Temperature control and reduction of mechanical injuries of the intact produce before processing are key factors in maintaining their quality and suitability for processing.

REFERENCES

Aguayo, E., Escalona, V.H., and Artes, F. 2004. Metabolic behavior and quality changes of whole and fresh processed melon. *Journal of Food Science* 69: 148–155.

Alcarez-Lopez, D., Botia, M., Alcarez, C.F., and Riquelme, F. 2003. Effects of foliar sprays containing calcium, magnesium and titanium on plum (*Prunus domestica* L.) fruit quality. *Journal of Plant Physiology* 160(12): 1441–1446.

Artes-Calero, F., Aguayo, E., Gómez, P., and Artes-Hernández, F. 2009. Productos vegetales mínimamente procesados o de la cuarta gama. *Horticultura Internacional* 69: 52–59.

Artés-Hernández, F., Rivera-Cabrera, F., and Kader, A.A. 2007. Quality retention and potential shelf-life of fresh-cut lemons as affected by cut type and temperature. *Postharvest Biology and Technology* 43(2): 245–254.

Asghary, M., Babalar, M., Talaei, A., and Kashi, A. 2005. The influence of harvest maturity and storage temperature on quality and postharvest life of 'Semsory' muskmelon fruit. In *Proceedings Fifth International Postharvest Symposium*. Eds. F. Menacarelli and P. Tonutti. *Acta Horticulturae* 682: 107–110.

Barth, M.M., Zhuang, H., and Saltveit, M.E. 2004. Fresh-cut vegetables. In *The Commercial Storage of Fruits, Vegetables, and Florist and Nursery Stocks. USDA-ARS. Agriculture Handbook Number 66* (draft-revised 2004). Ed. K. Gross. http://www.ba.ars.usda.gov/hb66/147freshcutvegetables.pdf (accessed August 25, 2009).

Beaulieu, J.C., and Gorny, J.R. 2004. Fresh-cut fruits. In *The Commercial Storage of Fruits, Vegetables, and Florist and Nursery Stocks. USDA-ARS. Agriculture Handbook Number 66* (draft-revised 2004). Ed. K. Gross. http://www.ba.ars.usda.gov/hb66/146freshcutfruits.pdf (accessed August 25, 2009).

Beaulieu, J.C., Ingram, D.A., Lea, J.M., and Bett-Garber, K.L. 2004. Effect of harvest maturity on the sensory characteristics of fresh-cut cantaloupe. *Journal of Food Science* 69(7): 250–258.

Bhowmik, P.K., and Dris, R. 2004. Enzymes and quality factors of fruits and vegetables. In *Production Practices and Quality Assessment of Food Crops, Volume 3, Quality Handling and Evaluation*. Eds. R. Dris and S.M. Jain. Dordrecht, The Netherlands: Kluwer Academic, pp. 1–25.

Blankenship, S.M., and Dole, J.M. 2004. 1-Methylcyclopropene: a review. *Postharvest Biology and Technology* 28: 1–25.

Cantwell, M. 2002. Optimal handling conditions for fresh produce. In *Postharvest Technology of Horticultural Crops*. Ed. A.A. Kader. University of California Agriculture and Natural Resources. Publication 3311, pp. 511–518.

Chiesa, A., Frezza, D., Fraschina, A., Trinchero, G., Moccia, S., and Leon, A. 2003. Pre-harvest factors and fresh-cut vegetables quality. In *Proceedings International Conference on Quality in Chains. E. An Integrated View on Fruit and Vegetable Quality*. Eds. L.M.M. Tijskens and H.M. Vollebregt. *Acta Horticulturae* (ISHS) 604: 153–159.

Crisosto, C., Garner, D., Doyle, J., and Day, K.R. 1993. Relationship between fruit respiration, bruising susceptibility and temperature in sweet cherries. *Hortscience* 28(2): 132–135.

Crisosto, C., and Mitchell, J. 2002. Pre-harvest factors affecting fruit and vegetable quality. In *Postharvest Technology of Horticultural Crops*. Ed. A.A. Kader. University of California Agriculture and Natural Resources. Publication 3311, pp. 39–54.

Deepa, N., Kaur, C., George, B., Singh, B., and Kapoor, H.C. 2007. Antioxidant constituents in some sweet pepper (*Capsicum annuum* L.). *LWT—Food Science and Technology* 40: 121–129.

Díaz-Pérez, J.C., Muy-Rangel, M.D., and Mascorro, A.G. 2007. Fruit size and stage of ripeness affect postharvest water loss in bell pepper fruit (*Capsicum annuum* L.). *Journal of the Science of Food and Agriculture* 87:68–73.

Fonseca, J.M. 2006. Postharvest quality and microbial population of head lettuce as affected by moisture at harvest. *Journal of Food Science* 71(2):45–49.

Fonseca, S.C., Oliveira, F.A.R., and Brecht, J.K. 2002. Modeling respiration rate of fresh fruits and vegetables for modified atmosphere packages: a review. *Journal of Food Engineering* 52: 99–119.

Gast, K.L.B. 1994. *Harvest Maturity Indicators for Fruits and Vegetables. Postharvest Management of Commercial Horticultural Crops*. Kansas State University Agricultural Experiment Station and Cooperative Service, 7pp.

Gelly, M., Recasens, L., Girona, J., Mata, M., Arbones, A., Rufat, J., and Marsal, J. 2003. Effects of stage II and postharvest deficit irrigation on peach quality during maturation and alter cold storage. *Journal of the Science of Food and Agriculture* 84:563–570.

Gorny, J.R., Cifuentes, R.A., Hess-Pierce, B., and Kader, A.A. 2000. Quality changes in fresh-cut slices as affected by cultivar, ripeness stage, fruit size, and storage regime. *Journal of Food Science* 65(3): 541–544.

Gorny, J.R., Hess-Pierce, B., and Kader, A.A. 1999. Postharvest physiology and quality maintenance of fresh-cut nectarines and peaches. In *Proceedings International Symposium on Effect of Pre- and Post Harvest Factors on Storage of Fruit*. Ed. L. Michalczuk. *Acta Horticulturae* 485, ISHS: 173–179.

Gross, K. 2004. Summary of respiration and ethylene production rates. In *The Commercial Storage of Fruits, Vegetables, and Florist and Nursery Stocks. USDA-ARS. Agriculture Handbook Number 66* (draft-revised 2004). Ed. K. Gross. http://www.ba.ars.usda.gov/hb66/010respiration.pdf (accessed August 25, 2009).

Hernández, Y., González, M., and Lobo, M.G. 2007. Importancia del grado de madurez en el procesado mínimo de frutas. In *V Congreso Iberoamericano de Tecnología Postcosecha y Agroexportaciones* (S6-O203). Universidad Politécnica de Cartagena, pp. 837–847.

Hodges, D.M., and Toivonen, P.M.A. 2008. Quality of fresh-cut fruits and vegetables as affected by exposure to abiotic stress. *Postharvest Biology and Technology* 48(2): 155–162.

Hovi-Pekkanen, T., and Tahvonen, R. 2008. Effects of interlighting on yield and external fruit quality in year-round cultivated cucumber. *Scientia Horticulturae* 116(2): 152–161.

Kader, A.A. 2002a. Postharvest biology and technology: an overview. In *Postharvest Technology of Horticultural Crops*. Ed. A.A. Kader. University of California, Agriculture and Natural Resources Publication 3311, pp. 39–47.

Kader, A.A. 2002b. Quality parameters of fresh-cut fruit and vegetable products. In *Fresh-Cut Fruits and Vegetables: Science, Technology, and Market*. Ed. O. Lamikanra. Boca Raton, FL: CRC Press, pp. 12–19.

Kader, A.A. 2008. Perspective flavor quality of fruits and vegetables. *Journal of the Science of Food and Agriculture* 88: 1863–1868.

Kader, A.A., and Barrett, D.M. 2004. Classification, composition of fruits, and postharvest maintenance of quality. In *Processing Fruits*. Eds. D.M. Barrett, L. Somogyi, and H. Ramaswamy. Boca Raton, FL: CRC Press, pp. 3–22.

Kim, H.J., Fonseca, J.M., Kubota, CH., Kroggel, M., and Choi, J. 2008a. Quality of fresh-cut tomatoes as affected by salt content in irrigation water and post-processing ultraviolet-C treatment. *Journal of the Science of Food and Agriculture* 88:1969–1974.

Kim, H.J., Fonseca, J.M., Kubota, C.H., Choi, J., and Kwon, D. 2008b. Salt in irrigation water affects the nutritional and visual properties of romaine lettuce (*Lactica sativa* L.). *Journal of Agricultural and Food Chemistry* 56(10): 3772–3776.

Lin, W.C., and Jolliffe, P.A. 1996. Light intensity and spectral quality affect fruit growth and shelf life of greenhouse-grown long English cucumber. *Journal of the American Society for Horticultural Science* 121: 1168–1173.

Maestrelli, A., and Chourot, J.M. 2002. Sélection des cultivars en relation avec la transformation. In: *Technologies de transformation des fruits*. Eds. G. Albagnac, J.C. Varoquaux, and J.C. Montigaud. Collection Sciences et Techniques Agroalimentaries, Lavoiser, Paris, pp. 41–75.

Manganaris, G.A., Crisosto, C.H., Bremr, V., and Holcroft, D. 2008. Novel 1-methylcyclopropene immersion formulation extends shelf life of advanced maturity 'Joanna Red' plums (*Prunus salicina* Lindell). Research Note. *Postharvest Biology and Technology* 47: 429–433.

Marín, A., Ferreres, F., Tomás-Barberán, F.A., and Gil, M.I. 2004. Characterization and quantification of antioxidant constituents of sweet pepper (*Capsicum annuum* L.). *Journal of Agriculture and Food Chemistry* 52(12): 3861–3869.

Martin-Belloso, O., and Rojas-Graü, A. 2005. Características de calidad de los vegetales frescos cortados. In *Nuevas tecnologías de conservación de productos vegetales frescos cortados*. Eds. G. González-Aguilar, A. Gardea, and F. Cuamea-Navarro. Mexico, Logiprint Digital S. de R.L. de C.V., pp. 77–93.

McArtney, S.J., Obermiller, J.D., Schupp, J.R., Parker, M.L., and Edgington, T.B. 2008. Preharvest 1-methylcyclopropene delays fruit maturity and reduces softening and superficial scald of apples during long-term storage. *Hortscience* 43(2): 366–371.

Milani, J., and Hamedi, M. 2005. Susceptibility of five apple cultivars to enzymatic browning. In *Proceedings V International Postharvest Symposium*. Eds. F. Mencarelli and P. Tonutti. *Acta Horticulturae* (ISHS) 682: 2221–2226.

Odriozola-Serrano, I., Soliva-Fortuny, R., and Martín-Belloso, O. 2009. Influence of storage temperature on the kinetics of the changes in Anthocyanins, Vitamin C, and antioxidant capacity in fresh-cut strawberries stored under high-oxygen atmospheres. *Journal of Food Science*, 2: 184–191.

Raese, J.T., and Drake, S.R. 2000. Effect of calcium sprays, time of harvest, cold storage, and ripeness on fruit quality of 'Anjou' pears. *Journal of Plant Nutrition* 23(6): 843–853.

Reid, M.S. 2002. Maturation and maturity indices. In *Postharvest Technology of Horticultural Crops*. Ed. A.A. Kader. University of California. Agriculture and Natural Resources Publication 3311, pp. 55–62.

Robles-Sánchez, R., Villegas-Ochoa, M., Cruz-Valenzuela, M., Vázquez-Ortiz, F., Castelo, A., Zavala-Ayala, F., and González-Aguilar, G. 2006. Determinación del estado de madurez óptimo de mango 'Ataulfo' destinado a procesamiento mínimo. CIAD, México. *Simposio Poscosecha 2006*. Orihuela. http://www.horticom.com/pd/imagenes/66/149/66149.pdf (accessed August 25, 2009), pp. 21–25.

Rodríguez-Hidalgo, S., Silveira, A.C., Artés-Hernández, F., and Artes, F. 2006. Evolución de la calidad de cuatro variedades de espinaca "baby" cultivadas en bandejas flotantes y mínimamente procesadas en fresco. Universidad Politécnica de Cartagena, España. 2006. *Simposio Poscosecha 2006*, Orihuela, pp. 227–231.

Rojas-Graü, M.A., Soliva-Fortuny, R., and Martín-Belloso, O. 2008. Effect of natural antibrowning agents on color and related enzymes in fresh-cut Fuji apples as an alternative to the use of ascorbic acid. *Journal of Food Science* 73: 267–272.

Salteit, M.E. 2004a. Respiratory metabolism. In *The Commercial Storage of Fruits, Vegetables, and Florist and Nursery Stocks. USDA-ARS. Agriculture Handbook Number 66* (draft-revised 2004). Ed. K. Gross. http://www.ba.ars.usda.gov/hb66/019respiration.pdf (accessed August 25, 2009).

Salteit, M.E. 2004b. Ethylene effects. In *The Commercial Storage of Fruits, Vegetables, and Florist and Nursery Stocks. USDA-ARS. Agriculture Handbook Number 66* (draft-revised 2004). Ed. K. Gross. http://www.ba.ars.usda.gov/hb66/020ethylene.pdf (accessed August 25, 2009).

Schouten, R.E., and Kooten, O. 2004. Genetic and physiological factors affecting colour and firmness. In *Understanding and Measuring the Shelf-Life of Food*. Ed. R. Steele. Cambridge: Woodhead, pp. 69–90.

Singh, R.P., and Anderson, B.A. 2004. The major types of food spoilage: an overview. In *Understanding and Measuring the Shelf-Life of Food*. Ed. R. Steele. Cambridge: Woodhead, pp. 3–23.

Soliva-Fortuny, R.C., Oms-Oliu, G., and Martín-Belloso, O. 2002. Effect of ripeness stages on the storage atmosphere, color, and textural properties of minimally processed apple slices. *Journal of Food Science* 67(5): 1958–1963.

Thompson, J.F., and Kader, A.A. 2004. Wholesale distribution center storage. In *The Commercial Storage of Fruits, Vegetables, and Florist and Nursery Stocks. USDA-ARS. Agriculture Handbook Number 66* (draft-revised 2004). Ed. K. Gross. http://www.ba.ars.usda.gov/hb66/016wholesale.pdf (accessed August 25, 2009).

Toivonen, P.M.A. 2008. Influence of harvest maturity on cut-edge browning of 'Granny Smith' fresh apple slices treated with anti-browning solution after cutting. *LWT—Food Science and Technology* 41: 1607–1609.

Toivonen, P.M.A., and DeEll, J. 2002. Physiology of fresh-cut fruits and vegetables. In *Fresh-Cut Fruits and Vegetables: Science, Technology, and Market.* Ed. O. Lamikanra. Boca Raton, FL: CRC Press, pp. 91–123.

Toivonen, P.M.A., Kappel, F., Stan, S., McKenzie, D.L., and Hocking, R. 2006. Factors affecting the quality of a novel fresh-cut sweet cherry product. *LWT—Food Science and Technology* 39: 240–246.

Varoquaux, P. 2002. Fruits frais prêts à l'emploi dits de 4ᵉ gamme. In *Technologies de Transformation des Fruits.* Eds. G. Albagnac, P. Varoquaux, and J.C. Montigaud. Collection Sciences and Techniques Agroalimentaries. Lavoiser, Paris.

Wang, C.Y. 2004. Chilling and freezing injury. In *The Commercial Storage of Fruits, Vegetables, and Florist and Nursery Stocks. USDA-ARS. Agriculture Handbook Number 66* (draft-revised 2004). Ed. K. Gross. http://www.ba.ars.usda.gov/hb66/ 018chilling.pdf (accessed August 25, 2009).

Watkins, C.B. 2006. The use of 1-methylcyclopropene (1-MCP) on fruits and vegetables. Research review paper. *Biotechnology Advances* 24: 389–409.

Wiley, C.R. 1994. *Minimally Processed Refrigerated Fruits and Vegetables.* New York: Chapman and Hall.

Yingsanga, P., Srilaong, V., and Kanlayanarat, S. 2006. Morphological differences associated with water loss in rambutan fruit cv. 'Rongrien' and 'See-Chompoo'. In *Proceedings IV International Conference on Managing Quality in Chains.* Eds. A.C. Purvis et al. *Acta Horticulture* (ISHS) 712: 453–459.

8 Treatments to Ensure Safety of Fresh-Cut Fruits and Vegetables

Maria Isabel Gil, Ana Allende,
and Maria Victoria Selma

CONTENTS

8.1 INTRODUCTION

The highest priority of fresh-cut processors is the safety of their products. Fresh-cut products are considered "ready to eat" owing to the wash process used in their preparation. It is possible to reduce the numbers of pathogens on produce by washing produce in water, but it is not possible to eliminate human pathogens through any technology other than thorough cooking or irradiation. Thus it is essential to prevent the presence of pathogens with produce food safety programs including Good Agricultural Practices (GAPs), Good Manufacturing Practices (GMPs), and Hazard Analysis Critical Control Point (HACCP) programs (IFPA, 2001).

8.2 WASHING OF FRESH-CUT PRODUCE

The proper washing of fresh-cut produce immediately after cutting is one of the most important steps in fresh-cut processing (Gil and Selma, 2006). Food safety guidelines for the fresh-cut produce industry generally specify a washing or sanitizing step to remove dirt, pesticide residues, and microorganisms responsible for quality loss and decay. The washing is based on conveying the product under water across an upward flow of air bubbles or pressurized water into the water tank, which might contain a sanitizing agent. This step is also used to pre-cool cut produce and remove cell exudates that adhere to product cut surfaces supporting microbial growth or resulting in discoloration.

Washing with water alone removes only the free cellular contents that are released by cutting; it is ineffective in assuring fresh-cut produce safety (Allende et al., 2008a). Pathogens can survive for relatively long times in water, or in plant residue entrapped in process line equipment or biofilms, and can subsequently contaminate clean product that passes through that water (Allende et al., 2008b; López-Gálvez et al., 2008). The large operational costs of water use have resulted in the industry-wide common practice of reusing or recirculating washwater (Luo, 2007). This causes the efficacy of the washing to be affected by the presence of organic matter in the water and allows microorganisms to survive for relatively long times in water or attached to plant tissue (Allende et al., 2008b). The effectiveness of the washing depends, therefore, on the quality of the washwater. The product should be washed and rinsed with processing water visually free of dust, dirt, and other debris and sanitized (Sapers, 2003). The U.S. Food and Drug Administration (FDA) guide to minimize microbial food safety hazards of fresh-cut fruits and vegetables points out that adequate water quality is critical in a fresh-cut processing facility (FDA/CFSAN, 2008).

It is well known that fresh-cut processors usually rely on wash-water sanitizers to reduce initial bacterial populations in their fresh-cut products as a strategy to maintain their quality and extend their shelf life (Zagory, 1999). There is a concern that reduction in surface populations reduces competition for space and nutrients and could promote the growth of potential pathogenic microorganisms (Brackett, 1992; Parish et al., 2003). Some of the researchers evaluated the sanitizer efficacy by determining microbial reductions at day 0 (Beuchat et al., 2004; Ukuku et al., 2005; Allende et al., 2008c), but these differences disappear after storage. This fact has been supported by several authors who suggested that microbial populations of fresh-cut fruits and vegetables could increase most rapidly and even reach equal or highest numbers after treatment with antimicrobial solutions compared to water controls (Beltrán et al., 2005b; Allende et al., 2007; Gómez-López et al., 2007; Stringer et al., 2007; Allende et al., 2008b). Some authors asserted that microbiology will affect overall product quality (Brackett, 1992; Marchetti et al., 1992; Lavelli et al., 2006). However, it has been demonstrated that neither the level of total count nor the level of specific spoilage microorganisms per se can directly predict the sensory quality of a product (Allende et al., 2008a). Many published studies support this view, as in many cases total bacterial numbers bear little or no relationship to product quality or shelf life (Bennik et al., 1998; Zagory, 1999; Gram et al., 2002). Furthermore, authors who initially correlated food quality and safety have demonstrated that given enough time, microorganisms can grow to high populations in packaged produce in the absence of obvious sensory defects (Brackett, 1999; Berrang et al., 1989). Therefore, the goal of washing is not necessarily to remove as many microorganisms as possible from the produce. Rather, the challenge is to ensure that the microorganisms present do not create human health risk and that if harmful microorganisms should inadvertently be present, that environmental conditions prevent their growth. Therefore, the best method to eliminate pathogens from produce is to prevent contamination in the first place (Parish et al., 2003). It is more difficult to decontaminate produce than it is to avoid contamination. However, this is not always possible, and the need to wash and sanitize many types of produce remains of vital importance to prevent disease outbreaks.

8.3 WHY WATER DISINFECTION IS NEEDED

Washwater in tanks, recirculated water, or water that is reused in a spray-wash system can become contaminated with pathogens if contaminated product coming in from the field is washed in that water (Allende et al., 2008b). Disinfection is the treatment of process water to inactivate or destroy pathogenic bacteria, fungi, viruses, and other microorganisms. Water is one of the key elements in maintaining the quality and safety of the fresh-cut products. The goal of disinfection is to prevent the transfer of microorganisms from process water to produce and from a contaminated produce to another produce over time (Suslow, 2001). Even though washing can remove some of the surface microorganisms, it cannot remove all of them. Microorganisms adhere to the surface of produce and may be present in nooks and crannies where water and washwater disinfectants cannot penetrate (Parish et al., 2003; Sapers, 2003). Commercial sanitizers can be very effective in eliminating free-floating or exposed microorganisms on produce surface. However, they are especially effective to disinfect, not to sanitize, produce. Clean, disinfected water is necessary to minimize the potential transmission of pathogens. The risk of cross-contamination is not eliminated by using large quantities of water. Pathogens can be rapidly acquired and taken up on plant surfaces, and natural openings or wounds can serve as points of entry (Burnett and Beuchat, 2001). Once fruits and vegetables have been contaminated with bacterial pathogens or parasites, none of these methods will ensure the safety of the product. Disinfection of water is therefore a critical step to minimize the potential transmission of pathogens from water to produce and among produce over time (Suslow, 2001; IFPA, 2002). In fact, the fresh-cut industry continues to use sanitizers, in spite of their limited direct microbiological benefits on produce, to extend the use of washwater or confer some improvement in quality during early to mid-storage.

8.4 CONSIDERATIONS DURING FRESH-CUT PRODUCE WASH

Three parameters have to be controlled when washing fresh-cut fruits and vegetables: quantity of water used (5 to 10 L/kg of product), temperature of water to cool the product, and concentration of the sanitizer (Yildiz, 1994). For the product being rinsed, water temperature must be as cold as possible, and 0°C is the optimal water temperature for most products. However, when the water temperature is much lower than the produce temperature, the internal gas contracts, thereby creating a partial vacuum that will draw in water through pores, channels, or punctures and be sufficient to draw water into the fruit (Bartz, 1999; Sapers, 2003). Infiltration of washwater into intact fruit has been demonstrated with several fruits and vegetables. Washwater contaminated with microorganisms, including pathogens, can infiltrate the intercellular spaces of produce through pores.

Water used in the washing of fresh-cut product may become a source of contamination if the washwater contains human pathogens and if there is insufficient wash-water disinfectant present (Suslow, 2001). When the fresh-cut produce is fully submerged in water, for washing or as a means of cooling, they are more likely to have washwater infiltration into the tissues. The reason is that microorganisms, including human pathogens, have a greater affinity to adhere to cut surfaces than uncut surfaces (Seo and Frank, 1999). Growing conditions, particularly conditions such as soil type (sand, muck, etc.), may have a profound effect on washwater disinfectant efficacy as well as the potential for removal of soil particles (e.g., difficulty in removing sand particles from crinkly leaf spinach products). Microbial reduction on produce and water by disinfectant is a concentration-by-time-dependent relationship, and it must be remembered that human pathogens, if present on the surface, may not be completely eliminated by washing (Suslow, 2001).

Water quality has traditionally been addressed separately in terms of public health impact of waterborne pathogens and outbreaks due to contaminated water. Recently, however, the role of water as a contributing factor to foodborne disease has been explored (Mena, 2006). Water has been considered the "forgotten food, particularly in terms of its impact as an ingredient of a manufactured food product, and now water source and quality are becoming more recognized as important and potentially impacting components during food production and processing, particularly for fresh-cut produce.

8.5 SANITIZING TREATMENTS FOR FRESH-CUT FRUITS AND VEGETABLES

Sanitizing treatments are recommended to minimize the potential for growth of microorganisms and contamination of fresh-cut produce. Based on scientific reports, the FDA elaborates guides of recommendations to minimize microbial food safety hazards for fresh fruits and vegetables. The guide published in 1998 defines "sanitize" as "to treat clean produce by a process that is effective in destroying or substantially reducing the numbers of microorganisms of public health concern" (FDA, 1998). This guide also referred to reducing other undesirable microorganisms, without adversely affecting the quality of the product or its safety for the consumer." However, in the last FDA guide published in February 2008, the washing of fresh produce has been approached in a different way. In this new version, the guideline states that "washing raw agricultural commodities before any processing of the produce occurs may reduce potential surface contamination. However, washing, even with disinfectants, can only reduce, not eliminate, pathogens if present, as washing has little or no effect on pathogens that have been internalized in the produce." Moreover, the importance of the maintenance of water quality during washing has attracted much attention as it is now specified that "antimicrobial chemicals, when used appropriately with adequate quality water, help minimize the potential for microbial contamination of processing water and subsequent cross contamination of the product." Many of the most recent findings about fresh-cut washing agree with this new approach (Allende et al., 2008b; López-Gálvez et al., 2008).

A variety of methods are used to reduce the potential for microbial contamination. Each method has distinct advantages and limitations depending upon the type of produce, mitigation protocol, and other variables. In the last few years, important information has been published concerning the efficacy of washing and sanitizing treatments in reducing microbial populations on fresh produce. A clear and well-documented comparison of different sanitation methods has been compiled in the Food Safety Guidelines for the Fresh-Cut Produce Industry review (IFPA, 2001). This topic is covered in the review by Parish et al. (2003) and that of Sapers (2003). The efficiency of numerous chemical and physical methods for assuring the microbiological safety of fresh-cut produce is covered in this chapter, which includes the latest and most significant research findings (Table 8.1). This chapter excludes other treatments with lethal effects on plant tissue, such as heat treatment, freezing, drying, fermentation, or those that can be unsafe for consumers.

Traditional methods of reducing microbial populations on produce involve chemical and physical treatments. The control of contamination requires that these treatments be applied to equipment and facilities as well as to produce. Methods of cleaning and sanitizing produce surfaces usually involve the application of water, cleaning chemicals (for example, detergent), and mechanical treatment of the surface by brush or spray washers, followed by rinsing with potable water (Artés and Allende, 2005). The efficacy of the method used to reduce microbial populations is usually dependent upon the type of treatment, type and physiology of the target microorganisms, characteristics of produce surfaces (cracks, crevices, hydrophobic tendency, and texture), exposure time, and concentration of cleaner/sanitizer, pH, and temperature (Sapers, 2001; Parish et al., 2003). Bacteria tend to concentrate where there are more binding sites. Attachment also might be in stomata (Seo and Frank, 1999), indentations, or other natural irregularities on the intact surface where bacteria could lodge. Bacteria also might attach at cut surfaces (Takeuchi and Frank, 2000; Liao and Cook, 2001) or in punctures or cracks in the external surface (Burnett et al., 2000). Sapers (2003) described that microbial resistance to sanitizing washes will also depend in part on the time interval between contamination and washing.

In each country, the regulatory status of washwater disinfectants depends on some particular agencies. The definition of the product used to disinfect washwater depends on the type of product to be washed and in some cases, the location where the disinfectant is used (IFPA, 2002). In the United States, for instance, if the product to be washed is a fresh-cut produce, the washwater disinfectant is regulated by the FDA as a secondary direct food additive, unless it may be considered Generally Recognized as Safe (GRAS). If the product is a raw agricultural commodity that is washed in a food-processing facility, such as a fresh-cut facility, both the U.S. Environmental Protection Agency (EPA) and FDA have regulatory jurisdiction and the disinfectant product must be registered as pesticides with EPA. A selected list of washwater disinfectants that have been approved by FDA (21 CFR 173.315; 21 CFR 178.1010) is reported in the Food Safety Guidelines for the Fresh-Cut Produce Industry (IFPA, 2001).

TABLE 8.1

Main Considerations of Washwater Sanitizers

Sanitizers	Considerations
Hypochlorite	The pH of the water should be kept between 6.0 and 7.5 to ensure the concentration of active chlorine (hypochlorous acid) is high enough. A rinse step of produce with potable water is necessary.
Chlorine dioxide	Produces fewer potentially carcinogenic chlorinated reaction products than chlorine, and it has greater activity at neutral pH.
Acidified sodium chlorite	More soluble than sodium hypochlorite (NaOCl) in water and has greater oxidizing capacity than hypochlorous acid.
Organic acid formulations	COD of water increases significantly after the addition of organic acid formulations such as Citrox and Purac. Tsunami is a good alternative as a sanitizing agent, but it is more expensive than chlorine.
Alkaline-based sanitizers	The high pH of alkaline washing solutions (11 to 12) and concerns about environmental discharge of phosphates may be limiting factors for use of certain alkaline compounds on produce.
Hydrogen peroxide	GRAS for some food applications but has not yet been approved as an antimicrobial wash-agent for produce.
Lactoperoxidase technology	The active molecule with disinfectant activity is hypothiocyanite, which does not remain in the finished product because it has very short life duration.
Ozone	A good option for washwater disinfection, reducing the need for water replacement, but it is not a substitute for the washing tank sanitizer.
UV-C illumination	Its efficacy as a washwater disinfectant is significantly impacted by turbidity.
Advanced oxidation processes	They are effective reducing microorganisms, chemical oxygen demand, and turbidity of water from the fresh-cut industry. Water could be reused for a longer time, but it is not a substitute for the washing tank sanitizer.
High pressure, pulsed electric field, oscillating magnetic fields	With the high capital expenditure together with the expensive process of optimization and water treatment, it is unlikely that the fresh produce industry would take up these technologies.

8.6 CHLORINE-BASED SANITIZERS

Chlorine-based sanitizers have been extensively used to sanitize produce and surfaces within produce processing facilities, as well as to reduce microbial populations in water for several decades and are perhaps the most widely used sanitizers in the food industry (Ogawa et al., 1980; Walker and LaGrange 1991; Cherry, 1999). Chlorine is obtained as a gas (Cl_2) or as liquid in the form of sodium (NaOCl) or calcium hypochlorites [$Ca(OCl)_2$]. Generally, it is used in the 50 to 200 ppm concentration range with a contact time of 1 to 2 min. The antimicrobial activity of chlorine compounds depends on the amount of hypochlorous acid (HOCl) present in the water which depends on the pH of the water, the concentration of organic material in the water, and, to some extent, the temperature of the water. Above pH 7.5, very little chlorine occurs as active hypochlorous acid (HOCl), but rather more as inactive hypochlorite (OCl⁻). If the concentration of active chlorine (hypochlorous acid) is not high enough, it was not effective as a disinfectant. Therefore, the pH of the water should be kept between 6.0 and 7.5 to ensure chlorine activity (Sapers, 2001). In contrast, if the water pH gets below 6.0, chlorine gas may be formed which is a health hazard for employees. Chlorine will not confer any benefits if the washwater is not kept relatively free of organic matter, or if the chlorine is not allowed to contact the product for a long enough time. The concentration of available chlorine determines the oxidizing potential of the solution and its disinfecting power. Although minimum concentrations of available chlorine (1 ppm) are required to inactivate microorganisms in clean water, higher concentrations are commonly used for most commodities.

Chlorination systems must be properly monitored and operated to ensure food safety (Suslow, 2001). The chlorine concentration and pH of chlorinated process water should be checked frequently using test paper strips, colorimetric kits, or electronic sensors. ORP or REDOX measured by electrical conductivity across a pair of electrodes is a method used to determine chlorine concentration, particularly applicable to water used in dump tanks or for cleaning or cooling purposes (Suslow, 2004). Some automated cooling systems monitor the oxidation reduction potential (ORP) of process water using probes that measure activity in millivolts (mV). The relationships between ORP, contact time, and microbial inactivation for chlorine-based oxidizers are used to establish the setting for the system. For example, an ORP set point of 600 to 650 mV is commonly used. There are readily available commercial systems for in-line monitoring and application of chlorine to maintain water cleanliness and to monitor periodically the concentration of hypochlorous acid (Adeskaveg, 1995). Colorimetric test kits are commonly used and are the most accurate for routine evaluations. For postharvest use, however, inaccurate measurements can result due to other salts dissolved in the water from organic and soil particles washed off of the commodity. Therefore, ORP is often used as a more accurate guide to detect sudden changes in conductivity that may influence free chlorine concentration.

Chlorine may incompletely oxidize organic materials to produce undesirable byproducts in process water, such as chloroform or other trihalomethanes, which are known or suspected of being carcinogenic (Chang et al., 1988; Parish et al., 2003). Furthermore, concerns about their impacts on human and environmental safety have also been raised in recent years. For this reason, its use for treatment of minimally processed vegetables is banned in several European countries, including Germany, The Netherlands, Switzerland, and Belgium (Artés and Allende, 2005). Also, some restrictions in the use of chlorine are implemented in other countries. Therefore, other alternatives to chlorine or improvements of chlorine-based sanitizers have been investigated. From the U.S. government regulatory perspective, the benefits of proper chlorination as a primary tool for sanitation outweigh concern for the potential presence of these by-products.

Chlorine dioxide (ClO_2) is more effective against many classes of microorganisms at lower concentrations than free chlorine. Its major advantages over HOCl include reduced reactivity with organic matter and greater activity at neutral pH. Therefore, it produces fewer potentially carcinogenic chlorinated reaction products than chlorine (Tsai et al., 1995; Rittman, 1997), although its oxidizing power is reported as 2.5

times higher than that of chlorine (Apel, 1993; White, 1992). However, stability of chlorine dioxide may be a problem (White, 1992; Parish et al., 2003). Also, chlorine dioxide generation systems are generally more expensive than that of hypochlorite, as they require on-site generation, specialized worker safety programs, as well as close injection systems for containment of concentrate leakage and volatilization fumes. However, stabilized liquid formulations, which are cheaper and easier to use, are now available (Tecsa®Clor, Protecsa S.A.; ProMinent Gugal S.A.).

Acidified sodium chlorite is a highly effective antimicrobial produced by lowering the pH (2.5 to 3.2) of a solution of sodium chlorite ($NaClO_2$) with any GRAS acid (Warf, 2001). Reactive intermediates of this compound are highly oxidative with broad-spectrum germicidal activity (Allende et al., 2008c, 2008d). Currently, acidified sodium chlorite is commercially supplied as a combination of citric acid and sodium chlorite and is known as Sanova (Ecolab®). This chemical combination produces active chlorine dioxide (ClO_2) that is more soluble than sodium hypochlorite (NaOCl) in water and has about 2.5 times greater oxidizing capacity than hypochlorous acid (HOCl) (Inatsu et al., 2005). Acidified sodium chlorite has been approved for use on certain raw fruits and vegetables as either a spray or dip in the range of 0.5 to 1.2 g L^{-1} followed by a potable water rinse (21CFR173.325). Recent studies have demonstrated its efficacy reducing total viable and *E. coli* counts of fresh-cut lettuce, escarole, cilantro, and carrots, which is similar to that of chlorine and even higher (González et al., 2005; Allende et al., 2008a, 2008c, 2008d).

Electrolyzed oxidizing (EO) water is a special case of chlorination where a dilute salt solution (1% NaCl) is passed through an electrochemical cell where the anode and cathode are separated by a diaphragm (Guentzel et al., 2008). In this case, the concentration of available chlorine is between 10 and 100 ppm, and it has a high oxidation potential between 1000 and 1150 mv (Sapers, 2003). During the process, there is a conversion of chloride ions and water molecules into chlorine oxidants such as Cl_2 and $HOCl/ClO^-$. Research studies have demonstrated the efficacy of EO water as an antimicrobial decontamination agent for use in food preparation and water purification (Park et al., 2004; Buck et al., 2003; Izumi, 1999).

8.7 OTHER CHEMICAL SANITIZERS

Acidification of low-acid products may act to prevent microbial proliferation because many pathogens generally cannot grow at pH values below 4.5 (Parish et al., 2003). Most of the fruits naturally possess significant concentrations of organic acids such as acetic, benzoic, citric, malic, sorbic, succinic, and tartaric acids which negatively affect the viability of contaminating bacteria. Other fruits such as melons and papayas as well as the majority of vegetables contain lower concentrations of organic acids, and pH values above 5.0 do not suppress the growth of pathogenic bacterial contaminants. Organic acids, in particular, vinegar and lemon juice have potential as inexpensive, simple household sanitizers; however, it is not clear whether organic treatments would produce off-flavors in treated produce (González et al., 2005).

Organic acid formulation such as Citrox®, containing phenolic compounds as the active ingredient (Citrox Limited, Middlesbrough, UK) and Purac® (PURAC Bioquímica, Spain), made of lactic acid, are commercially available for use as washwater and produce disinfectants. The manufacter's recommendations for produce treatment are concentrations of 5 ml L^{-1} applied for 5 min in the case of Citrox and 20 ml L^{-1} for 3 min for Purac. They are as effective as chlorine, catallix system, and Sanova for initial sanitation of leafy fresh-cut produce such as fresh-cut escarole and lettuce (Allende et al., 2008a). However, Citrox and Purac are not effective in reducing the *E. coli* population in washwater even at the highest manufacturer's recommended doses. Moreover, chemical oxygen demand (COD) of washwater increases significantly after the addition of both sanitizers and did not prevent transfer of *E. coli* cells between inoculated and uninoculated fresh-cut lettuce and, therefore, did not prevent cross-contamination (López-Gálvez et al., 2008).

Other organic acid formulations, such as peroxyacetic acid (CH_3CO_3H), which is actually an equilibrium mixture of the peroxy compound, hydrogen peroxide, and acetic acid (Ecolab, 1997; Sapers, 2003) are commercially available for use as wash-water disinfectants (Tsunami™ Ecolab, Mendota Heights, MN). It is a colorless liquid with an acrid odor. Peroxyacetic acid is active over a broad pH range and is less sensitive to organic matter than sodium hypochlorite. The high antimicrobial properties of peroxyacetic acid are well known (Block, 1991). This agent is recommended for use in treating process water, but one of the major suppliers is also claiming substantial reductions in microbial populations on produce surface (Sapers, 2003). However, after application of peroxyacetic acid for disinfection, produce must be rinsed with potable water. Peroxyacetic acid is approved for addition to washwater (21CFR173.315). It decomposes into acetic acid, water, and oxygen, all harmless residuals. It is a strong oxidizing agent and can be hazardous to handle at high concentrations but not at strengths marketed to the produce industry. Peroxyacetic acid is also recommended as water disinfection agents preventing *E. coli* cross-contamination of produce during processing (López-Gálvez et al., 2008).

Various high pH cleaners containing sodium hydroxide, potassium hydroxide, sodium bicarbonate, and sodium orthophenylphenate (with or without surfactants) have also been effective for microbial inactivation (Pao et al., 2000). High pH waxes used on fresh market citrus have shown antimicrobial activity when applied on orange fruit surfaces (Pao et al., 1999). Among alkaline-based sanitizers, trisodium phosphate (TSP) has been shown to be effective as a wash to decontaminate and reduce *Salmonella* risk in tomatoes (Zhuang and Beuchat, 1996) and apples (Sapers et al., 1999; Liao and Sapers, 2000). However, resistance of *Listeria monocytogenes* to TSP have also been reported (Zhang and Farber, 1996). Experimental applications for disinfection of fresh and fresh-cut produce have employed concentrations from 2% to 15% during 15 s to 5 min (Zhuang and Beuchat, 1996; Liao and Sapers, 2000; Sapers, 2003). Trisodium phosphate is classified as GRAS (21CFR182.1778) when used in accordance with GMPs (Sapers, 2003). However, the high pH of typical alkaline wash solutions (11 to 12) and concerns about environmental discharge of phosphates may be limiting factors for use of certain alkaline compounds on produce.

Hydrogen peroxide (H_2O_2) is GRAS for some food applications (21CFR184.1366) but has not yet been approved as an antimicrobial wash-agent for produce. It produces no residue because it is broken down to water and oxygen by catalase (Sapers, 2003). Solutions of 5% H_2O_2 alone or combined with commercial surfactants can achieve substantially higher log reductions for inoculated apples than 200 ppm chlorine (Sapers et al., 1999). Treatment by dipping in H_2O_2 solution reduced microbial populations on fresh-cut produce without altering sensory characteristics. H_2O_2 vapor treatments have been used to inhibit postharvest decay in some fruits and vegetables (Ukuku and Sapers, 2001; Ukuku et al., 2001). However, H_2O_2 is phytotoxic to some commodities, causing browning in lettuce and mushrooms and anthocyanin bleaching in raspberries and strawberries (Sapers and Simmons, 1998).

The H_2O_2 plus the sodium thiocyanate generates hypothiocyanite ($OSCN^-$) in the presence of peroxidise. This technology is named the lactoperoxidase (LPS) technology, and its commercial name is Catallix®. The active molecule with disinfectant activity is $OSCN^-$. The $OSCN^-$ molecule does not remain in the finished product because it has very short life duration. It is only recently that Catallix has been targeted at food processing, and especially for fresh-cuts (Allende et al., 2008a). In the past, the use has been investigated but always proved to be too expensive compared to chlorine. Due to the innovative approach of the LPS technology, the overall costs are similar. Catallix is approved for use on fresh-cut products as a processing aid. It has been found that Catallix is as effective as chlorine, Tsunami, Purac, and Sanova for initial sanitation of leafy fresh-cut produce such as escarole and lettuce (Allende et al., 2008a). However, as observed for peroxyacetic acid, the use of Catallix increases the COD of washwater.

Ozonated water and gaseous ozone (O_3) are applied to fresh-cut vegetables for sanitation purposes, reducing microbial populations, preventing browning, and extending the shelf life of some of these products (Beltrán et al., 2005a, 2005b; Selma et al., 2006, 2007, 2008c, 2008d). Ozone is also effective for reducing microbial flora of fresh-cut onion, escarole, carrot, and spinach washwaters collected from the industry (Selma et al., 2008b). Ozone is reported to have 1.5 times the oxidizing potential of chlorine and 3,000 times the potential of hypochlorous acid. It decomposes spontaneously during water treatment via a series of complex reaction mechanisms that involve the generation of hydroxyl free radicals ($^•OH$) (Hoigné and Bader, 1983a, 1983b; Glaze, 1987). The $^•OH$ radicals are the principal reactive oxidizing agents in water and are highly active in the inactivation of bacteria and virus (Vorontsov et al., 1997; Kim et al., 2003; Selma et al., 2006, 2007). When O_3 breaks down, it forms oxygen, and it has not been identified as creating undesirable disinfection byproducts, such as trihalomethanes. However, O_3 may indirectly form them if halogens are present in the washwater. Ozone was approved in 2001 by the FDA for use under GMP, as an antimicrobial agent for the treatment of raw and fresh-cut fruits and vegetables, in gas and aqueous phases (Graham, 1997; Xu, 1999). Typical O_3 use rates for disinfection of postharvest water are 0.5 to 4 ppm and 0.1 ppm for flume water (Strasser, 1998; IFPA, 2002). Ozone is not generally affected by pH within a range of 6 to 8, but its decomposition increases with high pH especially above pH 8. Disinfection, however, may still occur at a high pH because the biocidal activity of the compound is relatively rapid (White, 1992). Ozone treatments have been shown to be effective, reducing COD and turbidity of water from the fresh-cut industry (Selma et al., 2008b). An often-cited disadvantage of using O_3 as a disinfectant is its high instability, making it difficult to predict how O_3 reacts in the presence of organic matter (Cho et al., 2003). As a consequence, the high antimicrobial efficacy of O_3 treatments demonstrated in potable tap water is lower in the case of vegetable

washwaters from the fresh-cut industry, as the influence of physicochemical characteristics (mainly turbidity and organic matter) of the water affects the effectiveness
of the treatments (Bryant et al., 1992; Selma et al., 2006, 2007, 2008b). Treatment
cost of an O_3 system could be critical for its acceptability and implementation in
the food industry. However, recent improvements in this technology, through the
generation of smaller bubbles by dissolved O_3 flotation technique, benefit the O_3
generation process by increasing the effectiveness of O_3, as well as reducing power
consumption and related operating costs (Lee et al., 2008). Therefore, O_3 technology
is a good option for the wash-water disinfection for the fresh-cut industry, because it
will reduce the need for water replacement and for high sanitizer concentration such
as chlorine during vegetable washing.

8.8   PHYSICAL TREATMENTS

Other alternatives to chemicals for disinfection of recycled or recirculating process
water and fruit and vegetable sanitization are physical treatments, such as ultraviolet (UV) illumination (Allende et al., 2006; Selma et al., 2008b). In fact, the most
energetic fraction of the UV spectra, corresponding to the UV-C range (200 to 290
nm), is used as an antibacterial agent in water and air treatments, allowing effective
disinfection rates using germicidal lamps (254 nm) (Chang et al., 1985). The inactivation of microorganisms as a result of UV radiation is almost entirely attributable to
photobiochemical reactions that are induced within the microorganisms. Microbial
inactivation is directly related to the UV energy dose received by the microorganism
and is measured in joules per square meter (J/m^2) (Lucht et al., 1998). UV technology has been FDA approved (21CFR179.39) for use as a disinfectant to treat food
as long as the proper wavelength of energy is maintained (200 to 300 nm). Many
UV light systems are available for water sterilization. UV light systems do not leave
any chemical residues and they are not affected by water chemistry, but their efficacy as washwater disinfectant is significantly impacted by turbidity and so requires
clear water to be effective. In fact, the effectiveness of UV-C in disinfection depends
on the presence of particle-associated microorganisms, which may have a negative
impact on the disinfection process, because UV cannot penetrate particles by transmission through solid material. Therefore, turbidity, suspended solids, and absorbing
compounds interfere in UV efficacy on water disinfection (Selma et al., 2008b). As
a consequence, filtration should be used as a pretreatment to reduce the total number
of particles. Moreover, in contrast with O_3, turbidity is not significantly reduced by
UV-C treatment (Selma et al., 2008b).

Other physical technologies such as high pressure, pulsed electric field, pulsed
light, oscillating magnetic fields, and ultrasound treatments, have been investigated
to reduce or eliminate microorganisms (FDA/CFSAN, 2000). The effects of moderate thermal and pulsed electric field treatments on textural properties of carrots,
potatoes, broccoli, and apples have been investigated (Lebovka et al., 2004; Stringer
et al., 2007), but there are no findings suggesting their use in controlling microorganisms on fresh products. Because of the high capital expenditure together with the
expensive process of optimization and water treatment, it is unlikely that the fresh
produce industry would take up these technologies. Furthermore, the additional

1 log reduction achieved by applying ultrasound to a chlorinated water wash does not eliminate the risk of pathogens on fresh produce (Seymour et al., 2002).

8.9 COMBINED METHODS AND HURDLES

The concept of using multiple intervention methods is analogous to hurdle technology where two or more preservation technologies are used to prevent growth of microorganisms in or on foods (Leistner, 1994). It would be expected that combinations of sanitizers and other intervention methods, such as heat or irradiation, would have additive, synergistic, or antagonistic interactions (Allende et al., 2006). For instance, chlorine–ozone combinations may have beneficial effects on the shelf life and quality of lettuce salads, as well as on the water used for rinsing or cleaning the lettuce (García et al., 2003; Baur et al., 2004). Adding approved surfactants to process water reduces water surface tension and may increase the effectiveness of sanitizers.

Advanced oxidation processes (AOPs) represent the newest development in sanitizing technology, and two or more oxidants are used simultaneously. The result is the on-site total destruction of even refractory organics without the generation of sludge or residues. This technology is being widely applied to treat contaminated groundwaters, to purify and disinfect drinking waters and process waters, and to destroy trace organics in industrial effluents. The most common process used to generate ·OH is through the use of combined catalytic oxidants such as ozone-ultraviolet (O_3-UV), hydrogen peroxide-ultraviolet (H_2O_2-UV), and hydrogen peroxide-ozone (H_2O_2-O_3) (Gottschalk et al., 2000). Although these processes can produce ·OH, the O_3-UV combination provides the maximum yield of ·OH per oxidant (Gottschalk et al., 2000). For this reason, the O_3-UV process has been attracting increasing research interest (Teo et al., 2002; Beltrán et al., 2005a; Selma et al., 2006, 2007). The O_3-UV process has also been applied to fresh-cut vegetables for sanitation purposes, reducing microbial populations, preventing browning, and extending the shelf life of some of these products (Beltrán et al., 2005a, 2005b; Selma et al., 2006, 2007). Furthermore, the O_3-UV combination is an effective disinfection treatment on vegetable washwater collected from the industry, achieving microbial reductions of 6.6 log CFU/ml (Selma et al., 2008b). Also, O_3-UV treatments have been shown to be effective at reducing COD and turbidity of water from the fresh-cut industry (Selma et al., 2008b). Therefore, water quality can remain constant for longer periods and could, therefore, be reused.

Heterogeneous photocatalysis is also considered an AOP that uses UV-C illumination and titanium dioxide (TiO_2-UV) (Fujishima and Honda, 1972). All the reactions occur at nanometric scale on the pure silica fiber surface that fixes the titanium dioxide activated by a special lamp. Electron-hole pairs are generated on the TiO_2 photocatalyst surface by UV illumination and photon absorption, promoting the formation of hydroxyl radicals (·OH) through a reductive pathway. The ·OH have far more oxidizing power than O_3, H_2O_2, hypochlorous acid, and chlorine (Suslow, 2001) and they have been described to inactivate bacteria and virus (Vorontsov et al., 1997; Srinivasan and Somasundaram, 2003). The heterogeneous photocatalytic technology has been proposed as one of the best disinfection tools for water applications, because contaminant disinfection by-products or malodorous halogenated compounds derived from the secondary chloramines reaction with the organic matter are not incorporated in the water. It acts on virus, bacteria, seaweed, and other species, standing out in its effectiveness against the *Legionella*. Heterogeneous photocatalysis is an effective disinfection technology on vegetable washwaters, reducing microbial counts, including bacteria, molds, and yeasts (Selma et al., 2008a). TiO_2-UV resulted in being as effective as the O_3-UV treatment method for microbial inactivation of vegetable washwater from the fresh-cut produce industry (Selma et al., 2008a). Heterogeneous photocatalytic systems are also effective in reducing water turbidity, although COD was unaffected after the treatments (Selma et al., 2008a). In contrast with O_3 and O_3-UV equipment, photocatalytic systems are easy to install and maintain and do not require special precautions because there is no risk for the users. However, commercial photocatalytic systems are able to treat water but not foodstuffs. Another disadvantage of using this technology as a disinfection technique is that water turbidity can negatively affect the photocatalytic inactivation (Kim et al., 2003). As a consequence, filtration should be used as a pretreatment to reduce the total number of particles.

8.10   CONCLUSIONS

Proper washing of fresh-cut produce immediately after cutting is one of the most important steps in fresh-cut processing, but its effectiveness depends on the quality of the washwater. Sanitizers, despite their limited direct microbiological benefits on produce, are necessary in vegetable washwater to minimize the risk of cross-contamination from water to produce and among produce over time.

8.11   REMARKS

- Ensure that water used to wash produce after cutting is of sufficient microbial and physicochemical quality.
- Ensure that sufficient but not excessive concentrations of approved water disinfectant are present in produce washwater to reduce the potential for cross-contamination. Thus, monitor the disinfectant level in the water at a frequency sufficient to ensure that it is of appropriate microbial quality for its intended use.
- Evaluate water quality variables such as pH, organic load, turbidity, soil, and sanitizer concentration to ensure that the wash-water disinfectant of choice is effective in reducing the potential for water-to-produce cross-contamination.
- Minimize use for fresh-cut production of fruits and vegetables that have visible signs of decay due to the possible increased risk of the presence of human pathogens associated with decay or damage.
- Evaluate process design to accommodate raw product variability (e.g., variations in soil and weather conditions) that may affect wash-water efficacy. For example, evaluate specific product wash-water disinfectant demand, monitor product-to-water volume ratio, assess use of filtration systems to remove sand or soil from water during processing, or assess when water should be changed or added.

8.12 FUTURE TRENDS

Among sanitizers, ease of use and relative low cost mean that chlorine-based agents are still the best alternatives as disinfection agents able to prevent pathogen cross-contamination of produce during washing. However, hyperchlorination (use of high levels of chlorine) of wastewater with high organic matter content may have potentially negative effects on product sensory quality, the environment, and human health. Thus, lower doses of chlorine-based agents could be used in fresh-cut produce washing without losing efficacy, reducing the risk of chlorine byproducts. Additionally, the implementation of a proper wastewater treatment able to remove undesirable physical, chemical, and microbiological components, such as AOPs, should be selected with the goal of minimizing the effective dose to reduce safety hazards, especially in the fresh and fresh-cut industry where water reuse practices are used. The use of a good wastewater treatment combined with the addition of enough sanitizer to maintain residual antimicrobial activity represents the best barrier to keep the quality of the water and to eliminate microorganisms before they attach to the produce.

REFERENCES

Adeskaveg, J.E., Postharvest sanitation to reduce decay of perishable commodities, in *Perisahables Handling Newsletter*, Issue No. 82. University of California, Division of Agriculture and Natural Resources, Oakland, CA. 1995.

Allende, A., Tomás-Barberán, F.A., and Gil, M.I., Minimal processing for healthy traditional foods. *Trends in Food Sci. Technol.*, 17, 513–519, 2006.

Allende, A., Martínez, B., Selma, M.V., Gil, M.I., Suárez, J.E., and Rodríguez, A., Growth and bacteriocin production by lactic acid bacteria in vegetable broth and their effectiveness at reducing *Listeria monocytogenes in vitro* and in fresh-cut lettuce. *Food Microbiol.*, 24, 759–766, 2007.

Allende, A., Selma, M.V., López-Gálvez, F., Villaescusa, R., and Gil, M.I., Role of commercial sanitizers and washing systems on epiphytic microorganisms and sensory quality of fresh-cut escarole and lettuce. *Postharvest Biol. Technol.*, 49, 155–163, 2008a.

Allende, A., Selma, M.V., López-Gálvez, F., Villaescusa, R., and Gil, M.I., Impact of washwater quality on sensory and microbial quality, including *Escherichia coli* cross-contamination, of fresh-cut escarole. *J. Food Protection*, 71, 2514–2518, 2008b.

Allende, A., McEvoy, J., Tao, Y., and Luo, Y., Antimicrobial effect of acidified sodium chlorite, sodium chlorite, sodium hypochlorite, and citric acid on *Escherichia coli* O157:H7 and natural microflora of fresh-cut cilantro. *Food Control*, 20, 230–234, 2008c.

Allende, A., Gonzalez, R.J., McEvoy, J., and Luo, Y., Assessment of sodium hypochlorite and acidified sodium chlorite as antimicrobial agents to inhibit growth of *Escherichia coli* O157:H7 and natural microflora on shredded carrots. *Int. J. Veg. Sci.*, 13, 51–63, 2008d.

Apel, G., Chlorine dioxide. *Tree Fruit Postharvest J.* 4, 12–13, 1993.

Artés, F., and Allende, A., Processing lines and alternative preservation techniques to prolong the shelf-life of minimally fresh processed leafy vegetables. *Europ. J. Hort. Sci.*, 70, 231–245, 2005.

Bartz, J.A., Washing fresh fruits and vegetables: lessons from treatment of tomatoes and potatoes with water. *Dairy Food Environ Sanit.* 19, 853–864, 1999.

Baur, S., Klaiber, R., Hammes, W.P., and Carle, R., Sensory and microbiological quality of shredded, packaged iceberg lettuce as affected by pre-washing procedures with chlorinated and ozonated water. *Innov. Food Sci. Emerg. Technol.* 5, 45–55, 2004.

Beltrán, D., Selma, M.V., Marín, A., and Gil, M. I., Ozonated water extends the shelf life of fresh-cut lettuce. *J. Agric. Food Chem.* 53, 5654–5663, 2005a.

Beltrán, D., Selma, M.V., Tudela, J.A., and Gil, M.I., Effect of different sanitizers on microbial and sensory quality of fresh-cut potato strips stored under modified atmosphere or vacuum packaging. *Postharvest. Biol. Technol.* 37, 37–46, 2005b.

Bennik, M.H.J., Vorstman, W., Smid, E.J., and Gorris, L.G.M., The influence of oxygen and carbon dioxide on the growth of prevalent Enterobacteriaceae and *Pseudomonas* species isolated from fresh and controlled-atmosphere stored vegetables. *Food Microbiol.* 15, 459–469, 1998.

Berrang, M.E., Brackett, R.E., and Beuchat, L.R., Growth of *Listeria monocytogenes* on fresh vegetables stored under controlled atmosphere. *J. Food Prot.* 52, 702–705, 1989.

Beuchat, L.R., Adler, B.B., and Lang, M.M., Efficacy of chlorine and a peroxyacetic acid sanitizer in killing *Listeria monocytogenes* on iceberg and romaine lettuce using simulated commercial processing conditions. *J. Food Prot.* 67, 1238–1242, 2004.

Block, S.S., Peroxygen compounds, in *Disinfection, sterilization, and preservation,* Block, S.S., Ed., Lea and Febiger, Philadelphia, 1991, pp. 182–190.

Brackett, R.E., Shelf stability and safety of fresh produce as influenced by sanitation and disinfection, *J. Food Prot.* 55, 808–814, 1992.

Brackett, R.E., Incidence, contributing factors, and control of bacterial pathogens in produce. *Postharvest Biol. Technol.* 15, 305–311, 1999.

Bryant, E.A., Fulton, G.P., and Budd, G.C., Disinfection alternatives for safe drinking water. VanNostrand Reinhold, New York, 1992, pp. 518.

Buck, J.W., van Iersel, M.W., Oetting, R.D., and Hung, Y.-C., Evaluation of acidic electrolyzed water for phytotoxic symptoms on foliage and flowers of bedding plants. *Crop. Prot.* 22, 73–77, 2003.

Burnett, S.L., Chen, J., and Beuchat, L.R., Attachment of *Escherichia coli* O157:H7 to the surfaces and internal structures of apples as detected by confocal scanning laser microscopy. *Appl. Environ. Microbiol.,* 66, 4679–4687, 2000.

Burnett, S.L., and Beuchat, L.R., Food-borne pathogens: Human pathogens associated with raw produce and unpasteurized juices, and difficulties in decontamination. *J. Ind. Microbiol. Biotechnol.,* 27, 104–110, 2001.

21CFR173.315, Chemicals used in washing or to assist in the peeling of fruits and vegetables. Code of Federal Regulations 21, Part 173, Section 173.315.

21CFR173.325, Secondary Direct. Food additives permitted in food for human consumption: acidified sodium chlorite solutions, Code of Federal Regulations. Title 21, Part 173.325.

21CFR178.1010, Sanitizing solutions. Code of Federal Regulations 21, Part 178, Section 178.1010.

21CFR179.39, Ultraviolet radiation for the processing and treatment of food. Code of Federal Regulations 21, Part 179, Section 179.39.

21CFR182.1778, Sodium phosphate. Code of Federal Regulations 21, Part 182, Section 182.1778.

21CFR184.1366, Hydrogen peroxide. Code of Federal Regulations 21, Part 170–199, Section 184.1366, 463.

Chang, J.C.H., Ossoff, S.F., Lobe, D.C., Dorfman, M.H., Dumais, C.M., Qualls, R.G., and Johnson, J.D., UV inactivation of pathogenic and indicator microorganisms. *Appl. Environ. Microbiol.,* 49, 1361–1365, 1985.

Chang, T.L., Streicher, R., and Zimmer, H., The interaction of aqueous solutions of chlorine with malic acid, tartaric acid, and various fruit juices. A source of mutagens. *Anal. Lett.* 21, 2049–2067, 1988.

Cherry, J.P., Improving the safety of fresh produce with antimicrobials. *Food Technol.* 53, 54–57, 1999.

Cho, M., Chung, H., and Yoon, J., Disinfection of water containing natural organic matter by using ozone-initiated radical reactions. *Appl. Environ. Microbiol.* 69, 2284–2291, 2003.

Ecolab and Inc. Catching the wave. *Food Qual.* 4, 51–52, 1997.

FDA, Food and Drug Administration. Guidance for Industry. Guide to minimize microbial food safety hazards for fresh fruits and vegetables, Food Safety Initiative Staff. HFS-32, U.S. Food and Drug Administration, Center for Food Safety and Applied Nutrition, Washington, DC. http://www.foodsafety.gov/~dms/ prodguid.html, 1998.

FDA/CFSAN, Food and Drug Administration, Center for Food Safety and Applied Nutrition. Kinetics of microbial inactivation for alternative food processing technologies. Food and Drug Administration, Center for Food Safety and Applied Nutrition. http://vm.cfsan.fda.gov/~comm/ift-toc.html, 2000.

FDA/CFSAN, Food and Drug Administration, Center for Food Safety and Applied Nutrition. Guidance for industry: Guide to minimize food safety hazards for fresh-cut fruits and vegetables. *Federal Register*, 72, 11364–11368, 2008.

Fujishima, A., and Honda, K., Electrochemical photocatalysis of water at semiconductor electrode. *Nature.* 238, 27–38, 1972.

García, A., Mount, J.R., and Davidson, P.M., Ozone and chlorine treatment of minimally processed lettuce. *J. Food Sci.* 68, 2747–2751, 2003.

Gil, M.I., and Selma, M.V., Overview of hazards in fresh-cut produce production. Control and management of food safety hazards, in *Microbial hazard identification in fresh fruits and vegetables*, James, J.A., Ed., John Wiley and Sons, New York, 2006, pp. 155–219.

Glaze, W.H., Drinking-water treatment with ozone. *Environ. Sci. Technol.* 21, 224–230, 1987.

Gómez-López, V.M., Devlieghere, F., Ragaert, P., and Debevere, J., Shelf-life extension of minimally processed carrots by gaseous chlorine dioxide. *Int. J. of Food Microbiol.* 116, 221–227, 2007.

González, R., Allende, A., Ruiz, S., and Luo, Y., Sanitizantes utilizados en productos vegetales cortados, in *Nuevas tecnologías de conservación de vegetales frescos cortados,* González-Aguilar, G.A., Gardea-Bejar, A., and Cuamea-Navarro, F., Eds., Guadalajara, Jalisco, México 2005, pp. 263–285.

Gottschalk, J., Libra, J.A., and Saupe, A., Ozonation of water and wastewater, in *A practical guide to understanding ozone and its application*, Wiley-VCH, New York, 2000.

Graham, D.M., Use of ozone for food processing. *Food Technol.* 51, 72–75, 1997.

Gram, L., Ravn, L., Rasch, M., Bruhn, J.B., Christensen, A.B., and Givskov, M., Food spoilage—interactions between food spoilage bacteria. *Int. J. Food Microbiol.* 78, 79–97, 2002.

Guentzel, J.L., Lam, K.L., Callan, M.A., Emmons, S.A., and Dunham, V.L, Reduction of bacteria on spinach, lettuce, and surfaces in food service areas using neutral electrolyzed oxidizing water. *Food Microbiol.* 25, 36–41, 2008.

Hoigné, J., and Bader, H., Rate constants of reactions of ozone with organic and inorganic compounds in water- I. Non-dissociating organic compounds. *Water Research.* 17, 173–183, 1983a.

Hoigné, J., and Bader, H., Rate constants of reactions of ozone with organic and inorganic compounds in water -II. Dissociating organic compounds. *Water Research* 17, 185–194, 1983b.

Inatsu, Y., Bari, M.L., Kawasaki, S., Isshiki, K., and Kawamoto, S., Efficacy of acidified sodium chlorite treatments in reducing *Escherichia coli* O157:H7 on Chinese cabbage. *J. Food Prot.,* 68, 251–255, 2005.

IFPA, International Fresh-Cut Produce Association, See http://www.fresh-cuts.org. Food Safety Guidelines for the Fresh-Cut Produce Industry, J.R. Gorny, Ed., 4th ed. Alexandria, VA, 2001.

Izumi, H., Electrolyzed water as a disinfectant for fresh-cut vegetables. *J. Food Sci.* 64, 536–539, 1999.

Kim, B., Kim, D., Cho, D., and Cho, S., Bactericidal effect of TiO_2 photocatalyst on selected food-borne pathogenic bacteria. *Chemosphere* 52, 277–281, 2003.

Lavelli, V., Pagliarini, E., Ambrosoli, R., Minati, J.L., and Zanoni, B., Physicochemical, microbial, and sensory parameters as indices to evaluate the quality of minimally-processed carrots. *Postharvest Biol. Technol.* 40, 34–40, 2006.

Lebovka, N.I., Praporscic, I., and Vorobiev, E. Effect of moderate thermal and pulsed electric field treatments on textural properties of carrots, potatoes and apples. *Inn. Food Sci. Emerging Technol.* 5, 9–16, 2004.

Lee, B.H., Song, W.C., Manna, B., and Ha, J.K., Dissolved ozone flotation (DOF)—a promising technology in municipal wastewater treatment. *Desalination.* 225, 260–273, 2008.

Leistner, L., Food design by hurdle technology and HACCP. Adalbert Raps Foundation, Kulmbach, Germany, 1994.

Liao, C.H., and Cook, P.H., Response to trisodium phosphate treatment of *Salmonella* Chester attached to fresh-cut green pepper slices. *Can. J. Microbiol.* 47, 25–32, 2001.

Liao, C.H., and Sapers, G.M., Attachment and growth of *Salmonella* chester on apple fruit and *in vivo* response of attached bacteria to sanitizer treatments. *J. Food Prot.* 63, 876–883, 2000.

López-Gálvez, F., Allende, A., Selma, M.V., and Gil, M.I., Prevention of *Escherichia coli* cross-contamination by different commercial sanitizers during washing of fresh-cut lettuce *Int. J. Food Microbiol.* 133, 167–171, 2009.

Lucht, L., Blank, G., and Borsa, J., Recovery of food-borne microorganisms from potentially lethal radiation damage. *J. Food Prot.* 61, 586–590, 1998.

Luo, Y., Fresh-cut produce washwater reuse affects water quality and packaged product quality and microbial growth in romaine lettuce. *HortScience.* 42, 1413–1419, 2007.

Marchetti, R., Casadei, M.A., and Guerzoni, M.E., Microbial population dynamics in ready-to-use vegetable salads. *Int. J. Food Sci.* 2, 97–108, 1992.

Mena, K., Produce quality and foodborne disease: assessing water's role, in *Microbial hazard identification in fresh fruits and vegetables.* James, J.A., Eds., John Wiley and Sons, New York, 2006, pp. 95–114.

Ogawa, J.M., Hoy, M.W., Manji, B.T., and Hall, D.H., Proper use of chlorine for postharvest decay control of fresh market tomatoes. California Tomato Rama, *Informational Bulletin* 27, 2, 1980.

Pao, S., Davis, C.L., Kelsey, D.F., and Petracek, P.D., Sanitizing effects of fruit waxes at high pH and temperature on orange surfaces inoculated with *Escherichia coli. J. Food Sci.* 64, 359–362, 1999.

Pao, S., Davis, C.L., and Kelsey, D.F., Efficacy of alkaline washing for the decontamination of orange fruit surfaces inoculated with *Escherichia coli. J. Food Prot.* 63, 961–964, 2000.

Parish, M.E., Beuchat, L.R., Suslow, T.V., Harris, L.J., Garret, E.H., Farber, J.N., and Busta, F.F., Methods to reduce/eliminate pathogens from fresh and fresh-cut produce, in *Comprehensive reviews in food science and food safety*, Kroger M., Ed., IFT/FDA, Chicago, 2003, Chapter V. http://www.cfsan.fda.gov/ ~comm/ift3-toc.html.

Park, H., Hung, Y.-C., and Chung, D., Effects of chlorine and pH on efficacy of electrolyzed water for inactivating *E. coli* O157:H7 and *Listeria monocytogenes. Int. J. Food Microbiol.* 91, 13–18, 2004.

Rittman, D.D., "Can you have your cake and eat it too" with chlorine dioxide? *Water/Eng. Mag.* April, 1997.

Sapers, G.M., Efficacy of washing and sanitizing methods for disinfection of fresh fruit and vegetable products. *Food Technol. Biotechnol.,* 39, 305–311, 2001.

Sapers, G.M., Washing and sanitizing raw materials for minimally processed fruit and vegetable products, in *Microbial safety of minimally processed foods.* Novak, J.S., Sapers, G.M., and Juneja, V.K., Eds., CRC Press, Boca Raton, FL. 2003, pp. 221–253.

Sapers, G.M., Miller, R.L., and Mattrazzo, A.M. Effectiveness of sanitizing agents in inactivating *Escherichia coli* in golden delicious apples. *J. Food Sci.* 64, 734–737, 1999.

Sapers, G.M., and Simmons, G.F., Hydrogen peroxide disinfection of minimally processed fruits and vegetables. *Food Technol.* 52, 48–52, 1998.

Selma, M.V., Beltrán, D., Chacon-Vera, E., and Gil, M.I., Effect of ozone on the inactivation of *Yersinia enterocolitica* and reduction of natural flora on potatoes. *J. Food Prot.* 69, 2357–2363, 2006.

Selma, M.V., Beltrán D., Allende, A., Chacon-Vera, E., and Gil, M.I., Elimination by ozone of *Shigella sonnei* in shredded lettuce and water. *Food Microbiol.* 24, 492–499, 2007.

Selma, M.V., Allende, A., López-Gálvez, F., Conesa, M.A., and Gil, M.I., Heterogeneous photocatalitic desinfection of washwaters from the fresh-cut industry. *J. Food Prot.* 71, 286–292, 2008a.

Selma, M.V., Allende, A., López-Gálvez, F., Conesa, M.A., and Gil, M.I., Disinfection potential of ozone, ultraviolet-C and their combination in washwater for the fresh-cut vegetable industry. *Food Microbiol.* 25, 809–814, 2008b.

Selma, M.V., Ibáñez A.M., Allende, A., Cantwell, M., and Suslow, T.V., Effect of gaseous ozone and hot water on microbial and sensory quality of cantaloupe and potential transference of *E. coli* O157:H7 during cutting. *Food Microbiol.* 25, 162–168, 2008c.

Selma, M.V., Ibáñez, A.M., Cantwell, M., and Suslow, T.V., Reduction by gaseous ozone of *Salmonella* and microbial flora associated with fresh-cut cantaloupe. *J. Food Microbiol.* 25, 558–565, 2008d.

Seo, K.H., and Frank, J.F., Attachment of *Escherichia coli* O157:H7 to lettuce leaf surface and bacterial viability in response to chlorine treatment as demonstrated by using confocal scanning laser microscopy. *J. Food Prot.* 62, 3–9, 1999.

Seymour, I.J., Burfoot, D., Smith, R.L., Cox, L.A., and Lockwood, A., Ultrasound decontamination of minimally processed fruits and vegetables. *Int. J. Food Sci. Technol.* 37, 547–557, 2002.

Srinivasan, C., and Somasundaram, N., Bactericidal and detoxification effects of irradiated semiconductor catalyst, TiO_2. *Curr. Sci.* 85, 1431–1438, 2003.

Strasser, J., Ozone applications in apple processing. Tech Application, Electric Power Research Institute, Inc., Palo Alto, CA, 1998.

Stringer, S.C., Plowman, J., and Peck, M.W, The microbiological quality of hot water-washed broccoli florets and cut green beans. *J. Appl. Microbiol.* 10, 241–250, 2007.

Suslow, T.V., Water disinfection: a practical approach to calculating dose values for preharvest and postharvest applications, University of California, Division of Agriculture and Natural Resources, 2001. Publication 7256 and http://vric.ucdavis.edu.

Suslow, T.V., Oxidation-reduction potential (ORP) for water disinfection monitoring, control, and documentation. University of California, Division of Agriculture and Natural Resources, 2004. Publication 8149 and http://anrcatalog.ucdavis.edu/pdf/8149.pdf.

Takeuchi, K., and Frank, J.F., Penetration of *Escherichia coli* O157:H7 into lettuce tissues as affected by inoculum size and temperature and the effect of chlorine treatment on cell viability. *J. Food Prot.* 63, 434–440, 2000.

Teo, K.C., Yang, C., Xie, R.J., Goh, N.K., and Chia L.S., Destruction of model organic pollutants in water using ozone, UV and their combination. *Water Sci. Technol.* 47, 191–196, 2002.

Tsai, L.H., Higby, R., and Schade, J., Disinfection of poultry chiller water with chlorine dioxide: consumption and byproduct formation. *J. Agric. Food Chem.* 43, 2768–2773, 1995.

Ukuku, D.O, Bari, M.L., Kawamoto, S., and Isshiki, K., Use of hydrogen peroxide in combination with nisin, sodium lactate and citric acid for reducing transfer of bacterial pathogens from whole melon surfaces to fresh-cut pieces. *Int. J. Food Microbiol.* 104, 225–233, 2005.

Ukuku, D.O., Pilizota, V., and Sapers, G.M., Influence of washing treatment on native micro-flora and *Escherichia coli* 25922 populations of inoculated cantaloupes. *J. Food Safety*, 21, 31–47, 2001.

Ukuku, D.O., and Sapers, G.M., Effect of sanitizer treatments on *Salmonella* Stanley attached to the surface of cantaloupe and cell transfer to fresh-cut tissues during cutting practices. *J. Food Prot.* 64, 1286–1291, 2001.

Vorontsov, A.V., Savinov, E.N., Barannik, G.B., Troitsky, V.N., and Parmon, V.N., Quantitative studies on the heterogeneous gas-phase photooxidation of CO and simple VOCs by air over TiO_2. *Catal. Today.* 39, 207–218, 1997.

Walker, H.W., and LaGrange, W.S., Sanitation in food manufacturing operations, in *Disinfection, sterilization, and preservation.* Block, S.E., Ed., 4th ed., Lea and Febiger, Philadelphia, 1991.

Warf, C.C., The chemistry and mode of action of acidified sodium chlorite, in *2001 IFT annual meeting*, June 23–June 27, New Orleans, LA, 2001, pp. 1–91.

White, G.C., *Handbook of chlorination and alternative disinfectants,* 3rd ed., Van Nostrand Reinhold, New York, 1992.

Xu, L., Use of ozone to improve the safety of fresh fruits and vegetables. *Food Technol.* 53, 58–61, 1999.

Yildiz, F., Initial preparation, handling, and distribution of minimally processed refrigerated fruits and vegetables, in *Minimally processed refrigerated fruits and vegetables,* Wiley, R.C., Ed., Chapman and Hall, New York, 1994, pp. 41–48.

Zagory, D., Effects of post-processing handling and packaging on microbial populations. *Postharvest Biol. Technol.* 14, 313–321, 1999.

Zhang, S., and Farber, J.M., The effects of various disinfectants against *Listeria monocytogenes* on fresh-cut vegetables. *Food Microbiol.* 13, 311–321, 1996.

Zhuang, R.Y., and Beuchat, L.R., Effectiveness of trisodium phosphate for killing *Salmonella montevideo* on tomatoes. *Lett. Appl. Microbiol.* 22, 97–100, 1996.

9 Use of Additives to Preserve the Quality of Fresh-Cut Fruits and Vegetables

J. Fernando Ayala-Zavala and
Gustavo A. González-Aguilar

CONTENTS

9.1   INTRODUCTION

Fresh-cut fruits and vegetables are highly perishable products due to their intrinsic characteristics and minimal processing. Microbial growth, sensorial attributes decay, and loss of nutrients are among the major causes that compromise quality and safety of fresh-cut produce (Ayala-Zavala et al., 2008a). These drawbacks are caused by the steps involved in minimal processing, like peeling and cutting, which promote an increment in the metabolic rate, enzymatic reactions, and released juice (Rapisarda et al., 2006). Chemical synthetic additives can reduce decay rate, but consumers are concerned about chemical residues in the product, which could affect their health and cause environmental pollution (White and McFadden, 2008), thereby giving rise to the need to develop alternative methods for controlling fresh-cut fruit decay. One of the major emerging technologies for the control of postharvest diseases is the application of natural additives.

In recent years, the interest in natural antimicrobial compounds has increased, and numerous studies on the antimicrobial activity of a wide range of natural compounds have been reported (Ayala-Zavala et al., 2008b). Many pathogenic microorganisms that can be the cause of foodborne diseases and fresh food decay can be inhibited using natural compounds (Fisher and Phillips, 2008). Among these, several essential oils, alcohols, organic acids, and aromatic compounds have found to be biologically active against postharvest diseases.

The main reason for promoting the application of natural products in fresh fruits and vegetables is the consumer's demand for natural and organic methods to preserve foods. There is an increasing number of consumers choosing convenient and ready-to-use fruits and vegetables with a fresh-like quality, containing only natural ingredients (Roller and Lusengo, 1997). Different studies have been focused on improving the efficiency of natural compounds as emerging technologies to preserve fresh-cut fruit safety and quality (Ayala-Zavala et al., 2008a, 2008b, 2008c). However, regulatory actions on the use of natural alternative additives are still being analyzed. Demands from increasingly mistrustful consumers have led to numerous legislation reviews, which are expected to result in well-planned laws regarding regulations on natural food additives.

9.2   LIMITING FACTORS THAT AFFECT QUALITY AND ENHANCE DECAY OF FRESH-CUT FRUITS AND VEGETABLES

Decay can seriously limit the shelf life of fresh-cut fruits and vegetables. During processing and storage, decay and loss of quality can be observed in most of the fresh produce (Busta et al., 2003). The main attributes that affect quality of fresh produce are loss of sensory acceptability and microbial load, which could be avoided using different additives (Bett-Garber et al., 2003). The next section covers the relevance of sensory and microbiological attributes limiting shelf life of fresh-cut produce.

9.2.1 SUSCEPTIBILITY OF FRESH-CUT FRUIT TO MICROBIAL ATTACK

On the basis of nutrient content, fruits and vegetables can be a good source of nutrients for bacteria, yeast, and mold growth (Ayala-Zavala et al., 2008a). The microecology of fresh produce is especially important to consider, because these produce can change the microenvironment through their metabolic activity. Natural microbial flora may have about 10^4 to 10^5 colony-forming units per gram (CFU/g) (Busta et al., 2003). The commonly encountered microflora of fruits and vegetables are *Pseudomonas* spp., *Erwinia herbicola*, *Enterobacter agglomerans*, lactic acid bacteria, and molds and yeasts (Busta et al., 2003). Although this microflora is largely responsible for the spoilage of fresh produce, it can vary greatly for each product, depending on the medium's pH, nutrient availability, water activity, storage conditions, among other factors. Temperature can play an important role in determining the outcome of the final microflora found in refrigerated fruits and vegetables (Gonzalez-Aguilar et al., 2004). The humidity at which fruits and vegetables are stored can also affect microbial development: Low humidity will discourage bacteria from growing on the surface of fruits and vegetables (Ayala-Zavala et al., 2008a).

9.2.2 FRUIT COMPOSITION IS RELATED TO THE RATE OF DETERIORATION PROCESSES

The fact that fresh-cut fruits are complex and active systems in which microbiological, enzymatic, and physicochemical reactions are simultaneously taking place makes the deterioration processes faster in them than in whole fruits (Artés et al., 2007). Fresh-cut fruit deterioration is dependent on the understanding of these reactions and their respective mechanisms. Fruit composition involves different compounds that will affect the shelf life of the product. Key factors include moisture (water activity), sugar, acid content, and pH (Brecht, 2006). Water activity (aw) is directly related to the relative humidity (RH) equilibrium of a given food and describes the degree at which water is "bound" in the system, controlling its availability to act as a solvent and participating in chemical/biochemical reactions and growth of microorganisms (Ayala-Zavala et al., 2008a). Fungi are the most important microorganisms causing postharvest wastage of fresh-cut fruit, where the relatively acid conditions tend to suppress bacterial growth (Frazier and Westhoff, 1993). However, in vegetables, bacterial infections are more common due to their high pH. Bacteria are also important as agents of both spoilage and foodborne diseases. These important properties can be used to predict the stability and safety of food with respect to microbial growth, rates of deteriorative reactions, and chemical and physical properties.

9.2.3 How Is Sensorial Appeal Affected by Decay?

Sensorial appeal of fresh-cut fruits and vegetables by consumers is a major factor in the purchase decision (Toivonen and Brummell, 2008). The most important sensory attributes of fresh-cut produce include appearance, color, texture, flavor, and aroma. Color affects consumer acceptance of fresh fruits and vegetables (Gonzalez-Aguilar et al., 2000). Another aspect is the deterioration of texture. The texture of fruits and vegetables is often interpreted in terms of firmness, crispness, juiciness, and toughness (attributed by the fibrousness of plant tissue), where firm or crispy tissues are generally desired in fresh produce (Toivonen and Brummell, 2008). Flavor and aroma are often the major indicators of shelf life from the consumer's point of view (Beaulieu and Lea, 2003). However, it has to be pointed out that after minimal processing and storage of fresh-cut fruit, deterioration occurs, compromising sensorial appeal.

Fresh-cut fruit and vegetables are living tissues constituted by metabolic active cells; however, during peeling and cutting, cells are disrupted, and the interactions between substrates and enzymes are favored, enhancing sensorial decay (Toivonen and Brummell, 2008). Browning reactions have generally been assumed to be a direct consequence of polyphenol oxidase (PPO) action on polyphenols (Gonzalez-Aguilar et al., 2005), although some have attributed at least a partial role to the action of phenol peroxidase (POD) on polyphenols (Toivonen and Brummell, 2008). Slicing operations also result in dramatic losses in firmness of fruit tissues. Pectinolytic and proteolytic enzymes exuding from bruised cells may diffuse into inner tissues (Karakurt and Huber, 2003). Following processing of fresh-cut produce, the two primary mechanisms of flavor loss are metabolic and diffusional (Forney, 2008). Metabolic changes in flavor are the result of the synthesis or catabolism of either flavor compounds or compounds responsible for off-flavors. These metabolic processes are dependent on product physiology, which is influenced by maturity and a variety of environmental factors.

Fresh-cut products are wounded tissues, and consequently they deteriorate more rapidly and their physiology differs from that of intact fruit and vegetables. The various processes of peeling, coring, and chopping, slicing, dicing, or shredding cause cell injury, releasing its content at the sites of wounding. Subcellular compartmentalization is disrupted at the cut surfaces, and the mixing of substrates and enzymes, which are normally separated, can initiate reactions that do not occur typically. Therefore, preservative technologies are needed to maintain quality and safety of fresh-cut produce.

9.3 CONVENTIONAL ADDITIVES APPLIED TO FRESH-CUT FRUITS AND VEGETABLES

Many of the treatments and storage conditions applied to fresh-cuts are designed to ameliorate the initial effects of wounding and wounding-induced responses. Several additive compounds are being applied to minimize or reduce fresh-cut produce decay, and others represent an alternative to use in the fresh-cut industry. They range from appropriate sanitizers, antioxidants, and texturizers.

9.3.1 SANITIZERS

Sanitizers are chemicals that may be used to reduce microorganisms from the surfaces of whole and cut produce, because fresh produce can be a vehicle of viruses, parasites, spoilage bacteria, molds, and yeast, as well as occasional pathogenic bacteria (Artés et al., 2007).

"Methods to reduce/eliminate pathogens from fresh and fresh-cut produce" summarizes the uses of several sanitizers: chlorine (hypochlorite), chlorine dioxide, and acidified sodium chlorite, bromine, iodine, quaternary ammonium compounds, acidic compounds, alkaline compounds (phosphates), and hydrogen peroxide. Some examples of the use of sanitizers for fresh-cut fruits are discussed here (Parish et al., 2003).

9.3.1.1 Chlorine

Chlorine has been used for sanitation purposes in food processing for several decades and is perhaps the most widely used sanitizer in the food industry (Artés et al., 2007). Chemicals that are chlorine based are often used to sanitize produce and surfaces within produce processing facilities, as well as to reduce microbial populations in water used during whole fruit sanitation, cleaning and packing operations, and for fresh-cut fruits and vegetables disinfection (Artés et al., 2007). At the foodservice, chlorine remains a convenient and inexpensive sanitizer for use against many foodborne pathogens. The most common forms of free chlorine include liquid chlorine and hypochlorite, although other chlorine forms are commercially available, like chlorine dioxide and acidified sodium chlorite. Liquid chlorine and hypochlorite are generally used in the 50 to 200 ppm concentration range with a contact time of 1 to 2 min to sanitize produce surfaces and processing equipment. Hypochlorous acid (HOCl) is the form of free available chlorine that has the highest bactericidal activity against a broad range of microorganisms. In aqueous solutions, the equilibrium between hypochlorous acid (HOCl) and the hypochlorite ion (OCl^-) is pH dependent with the concentration of HOCl increasing as pH decreases. Typically, pH values between 6.0 and 7.5 are used in sanitizer solutions to minimize corrosion of equipment while yielding acceptable chlorine efficacy. Maximum solubility in water is observed near 4°C; however, it has been suggested that the temperature of processing water should be maintained at least 10°C higher than that of produce items in order to reduce the possibility of microbial infiltration caused by a temperature-generated pressure differential. The opportunity for infiltration of microorganisms is also minimized when the sanitary condition of the water is maintained.

Various studies have reported the efficacy to a different extent of chlorine sanitation on fruit inoculated with different pathogens. Chlorine sanitation (200 to 250 ppm Cl_2) inhibited $1.74 \log_{10}$ CFU of *Escherichia coli* on inoculated apples (Sapers et al., 2002). Chlorine and nisin-EDTA (ethylene diamine tetra acetic acid) treatments inhibited $1.86 \log_{10}$ CFU of gram (–) bacteria on fresh-cut cantaloupes (Ukuku and Fett, 2002). Moreover, in another study, where chlorine (>200 ppm) was utilized to reduce *Salmonella* and *E. coli* O157:H7, the reduction was of $2.3 \log_{10}$ CFU/cm^2 in apples, tomatoes, and lettuces (Parish et al., 2003). A more effective sanitation process was achieved using chlorine to reduce *Salmonella* on inoculated apples, with an inhibition of $3.2 \log_{10}$ CFU/g (Parnell and Harris, 2003).

Because chlorine reacts with organic matter, components leaching from tissues of cut produce surfaces may neutralize some of the chlorine before it reaches microbial cells, thereby reducing its effectiveness. Additionally, crevices, cracks, and small fissures in produce, along with the hydrophobic nature of the waxy cuticle on the surface of many fruit and vegetables, may prevent chlorine and other sanitizers from reaching the microorganisms. Surfactants, detergents, and solvents, alone or coupled with physical manipulation such as brushing, may be used to reduce hydrophobicity or to remove part of the wax to increase exposure of microorganisms to sanitizers.

Safety concerns about the production of chlorinated organic compounds, such as trihalomethanes, and their impact on human and environmental safety have been raised in recent years, and alternatives to chlorine have been investigated. However, regulations and guidelines applicable to chlorine dioxide and chlorite are reported by the U.S. Food and Drug Administration (FDA). Chlorine, as acidified sodium chlorite, is applied as a direct food additive permitted in food for human consumption. Its use as an antimicrobial agent in water to sanitize fruits and vegetables is permitted at levels from 50 to 1500 ppm (CFSAN/FDA, 2006).

9.3.1.2 Iodine

Iodophores have a broad spectrum of antimicrobial activity, are less corrosive than chlorine at low temperatures, and are less volatile and irritating to skin than other types of iodine solutions (Parish et al., 2003). However, iodine-containing sanitizer solutions may be corrosive (upon vaporization above 50°C), have reduced efficacy at low temperature, and may stain equipment, clothes, and skin. The use of iodine-containing solutions as direct-contact sanitizers for produce is further limited due to a reaction between iodine and starch resulting in a blue-purple color. Despite these limitations, iodine solutions such as iodophores (combinations of elemental iodine and nonionic surfactants or carriers) are commonly used as sanitizers for food contact surfaces and equipment in the food-processing industry (Parish et al., 2003).

As with most sanitizers, iodophores are more active against vegetative cells than against bacterial spores. Decimal reduction values for vegetative bacterial cells are between 3 and 15 seconds at 6 to 13 ppm available iodine at neutral pH (Hays et al., 1967). D values for spores of *Bacillus cereus, Bacillus subtilis*, and *C. botulinum* Type A treated with 10 to 100 ppm of iodophore are 10- to 1000-fold greater than for vegetative cells (Odlaug, 1981). Although iodophores are not approved for direct food contact, they might have some usefulness for treatment of produce items that are peeled before consumption. This type of use would require regulatory approval and a demonstration that produce treated with these compounds is safe for consumption.

The primary residue of iodine-potassium iodide is the inorganic halide, iodide (I). In the presence of organic matrices, such as food items, the iodine in the iodine-potassium iodide complex is very rapidly reduced to iodide, with a reaction rate in the order of seconds. Due to the natural occurrence in all fruits and vegetables of antioxidants, which are the likely agents in this reduction, there is little likelihood of iodine remaining intact in any crop matrix. Iodine has been approved by the FDA for use in drugs and has been deemed generally recognized as safe (GRAS) to be used as a food additive (CFSAN/FDA, 2006). There are also a number of antimicrobial uses already approved by the EPA for iodine and iodophores, including sanitization of food handling equipment.

9.3.1.3 Quaternary Ammonium Compounds

Commonly called "quats," quaternary ammonium compounds are cationic surfactants that are odorless, colorless, stable at high temperatures, noncorrosive to equipment, nonirritating to skin, and able to penetrate food contact surfaces more readily than other sanitizers (Parish et al., 2003). The antimicrobial activity of quats is greater against fungi and Gram-positive bacteria than against Gram-negative bacteria. Thus, *Listeria monocytogenes* is more sensitive to quats than coliforms, *Salmonella* spp., pathogenic *E. coli*, or *Pseudomonas*. Due to their high surface-active capability, the mechanism of activity for quats possibly involves a breakdown of the cell membrane/wall complex (Marriott, 1999). Exposure of orange fruit during 30 s to a 500 ppm quat solution reduced *Xanthomonas campestris* as effectively as 150 to 250 ppm chlorine for 2 min (Brown and Schubert, 1987).

Quat sanitizers form a residual antimicrobial film when applied to most hard surfaces and are relatively stable to organic compounds. They are most effective when used at pH 6 to 10 and are not compatible with acidic environments, soaps, or anionic detergents. Although they are not approved for direct food contact, quats may have some limited usefulness with whole produce that must be peeled prior to consumption. As with iodine compounds, direct food contact would require regulatory approval and a demonstration that produce treated with quats is safe for consumption.

The food additive, quaternary ammonium chloride combination, may be safely used in food in accordance with the following conditions: The additive contains the following compounds: *n*-dodecyl dimethyl benzyl ammonium chloride (CAS Reg. No. 139–07–1), *n* dodecyl dimethyl ethylbenzyl ammonium chloride (CAS Reg. No. 27479–28–3), *n*-hexadecyl dimethyl benzyl ammonium chloride (CAS Reg. No. 122–18–9), *n*-octadecyl dimethyl benzyl ammonium chloride (CAS Reg. No. 122–19–0), *n*-tetradecyl dimethyl benzyl ammonium chloride (CAS Reg. No. 139–08–2), *n*-tetradecyl dimethyl ethylbenzyl ammonium chloride (CAS Reg. No. 27479–29–4). The additive meets the following specifications: pH (5% active solution) 7 to 8; total amines, maximum 1% as combined free amines and amine hydrochlorides (CFSAN/FDA, 2006).

9.3.1.4 Acidic Compounds

Organic acids are commonly used as antimicrobial acidulants and antibrowning agents in fresh-cut produce (Table 9.1) (Ruiz-Cruz et al., 2007). Because many pathogens cannot grow at pH values much below 4.5, acidification may act to prevent microbial proliferation. Organic acids may also possess bactericidal capabilities. The antimicrobial action of organic acids is due to the pH reduction in the environment, disruption of membrane transport and permeability, anion accumulation, or a reduction in internal cellular pH by the dissociation of hydrogen ions from the acid. Many types of produce, especially fruit, naturally possess significant concentrations of organic acids, such as acetic, benzoic, citric, malic, sorbic, succinic, and tartaric acids, which negatively affect the viability of contaminating bacteria.

TABLE 9.1

Organic Acids Used as Sanitizers in Fresh-Cut Fruits and Vegetables

Organic Acid	Dose	Produce	Effect	Reference
Citric acid	As lemon juice	Fresh-cut papaya	Reduced *Salmonella typhi*	Fernandez-Escartin et al. (1989)
Sodium lactate	2%	Cabbage, broccoli, and mung bean sprouts	Reduced *Listeria monocytogenes*	Bari et al. (2005)
Potassium sorbate	0.02%			
Phytic acid	0.02%			
Citric acid	10 mM			
Tartaric acid	1,500 ppm	Vegetable salad	Reduced total bacteria	Shapiro and Holder (1960)
Lactic acid	1%	Vegetable salad	Reduced coliform bacteria	Torriani et al. (1997)

Organic acids, including lactic acid, acetic acid, citric acid, ascorbic acid, and other organic acids are approved or listed in FDA regulations for various technical purposes (e.g., as acidulants, antioxidants, flavoring agents, pH adjusters, nutrients, and preservatives). The FDA's decision about this use of organic acids was based on industry requests that were supported by data that showed that this application of organic acids meets FDA's definition (CFSAN/FDA, 2006). Therefore, products made from organic acid–treated produce do not have to declare the organic acids in the ingredients statement on the product label.

9.3.1.5 Alkaline Compounds

In contrast to acidulants, alkaline compounds work like antimicrobial agents raising the pH. *E. coli* O157:H7 populations were reduced 5 and 6 log after a 30-s treatment with 1% trisodium phosphate (TSP) at 10°C and room temperature, respectively (Somers et al., 1994). *Campylobacter jejuni* was almost as sensitive as *E. coli* O157:H7 to TSP. Treatment with 8% TSP decreased populations of *L. monocytogenes* only 1 log cycle. A 5-min treatment with 2% TSP produced a 1 log reduction of *Salmonella chester* attached to the surface of apple disks (Liao and Sapers, 2000). *Salmonella montevideo* populations on the surface of tomatoes were reduced from 5.2 log CFU/cm^2 to nondetectable levels after 15 s in 15% TSP (Zhuang and Beuchat, 1996). A significant reduction in population was observed after 15 s in 1% TSP.

Various high pH cleaners containing sodium hydroxide, potassium hydroxide, sodium bicarbonate, and sodium orthophenylphenate (with or without surfactants) reduced populations of *E. coli* on orange surfaces (Pao et al., 2000). These same researchers determined that high pH waxes used on fresh market citrus provided substantial inactivation of *E. coli* on orange fruit surfaces (Pao et al., 1999). The high pH of typical alkaline wash solutions (11 to 12) and concerns about environmental discharge of phosphates may be limiting factors for the use of certain alkaline compounds on produce.

9.3.1.6 Ozone

The use of ozone as an antimicrobial agent in food processing has been reviewed (Kim et al., 1999); however, little has been reported about the inactivation of pathogens on produce. *Salmonellae* and *E. coli* populations were reduced 3 to 4 log/g in ground black pepper after 60-min treatment with ozonated air (6.7 mg/L at a flow rate of 6 L/min); however, significant changes in the volatile oil profiles were also noted (Zhao and Cranston, 1995). Volatile oils in whole black peppercorns treated in ozonated water were not significantly affected.

Ozone is an effective treatment for drinking water and will inactivate bacteria, fungi, viruses, and protozoa (Restaino et al., 1995). According to Restaino et al. (1995), bacterial pathogens, such as *Salmonella typhimurium, Yersinia enterocolitica, Staphylococcus aureus*, and *L. monocytogenes*, are sensitive to treatment with 20 ppm ozone in water. Treatment of *Cryptosporidium parvum* oocysts with 1 ppm ozone for 5 min resulted in < 1 log inactivation. In the same study, *Giardia* spp. cysts were more sensitive than *C. parvum* to ozone treatment (Finch and Fairbairn, 1991).

Treatment with ozonated water can extend the shelf life of apples, grapes, oranges, pears, raspberries, and strawberries by reducing microbial populations and by oxidation of ethylene to retard ripening (Beuchat et al., 1998). Microbial populations on berries and oranges were reduced by treatment with 2 to 3 ppm and 40 ppm, respectively. Kim et al. reported a 2 log/g reduction in total counts for shredded lettuce suspended in water ozonated with 1.3 mM ozone at a flow rate of 0.5 L/min (Kim et al., 1999). Fungal growth during storage of blackberries was inhibited by 0.1 to 0.3 ppm ozone (Barth et al., 1995). Treatment of grapes by ozone increased shelf life and reduced fungal growth (Sarig et al., 1996). Spoilage of vegetables, such as onions, potatoes, and sugar beets, was reduced upon storage in an ozone-containing atmosphere (Kim et al., 1999).

Due to its strong oxidizing activity, ozone may cause physiological injury in produce (Parish et al., 2003). Bananas treated with ozone developed black spots after 8 days of exposure to 25 to 30 ppm gaseous ozone. Carrots exposed to ozone gas during storage had a lighter, less intense color than untreated carrots (Liew and Prange, 1994). Ozone can also cause corrosion of metals and other materials in processing equipment. It is capital intensive and may be difficult to monitor and control in situations where highly variable organic loads are likely to occur. As with other sanitizers, employee safety and health issues must be addressed, and appropriate safeguards must be in place when using ozone as a sanitizing agent. Because ozone produces toxic vapors, adequate ventilation is necessary for employee safety. However, because it has excellent ability to penetrate and does not leave a residue, ozone may be useful for treatment of process water, food contact surfaces, or whole produce. Industry representatives indicate that the postharvest use of ozone for treatment of produce is increasing.

9.3.1.7 Hydrogen Peroxide

Hydrogen peroxide (H_2O_2) possesses bactericidal and inhibitory activity due to its properties as an oxidant, and due to its capacity to generate other cytotoxic oxidizing species such as hydroxyl radicals. The sporicidal activity of H_2O_2 coupled with rapid breakdown makes it a desirable sterilant for use on some food contact surfaces and for packaging materials in aseptic filling operations. The residual H_2O_2 level may vary dependent on the presence or absence of peroxidase in the produce item.

Use of H_2O_2 on whole and fresh-cut produce has been investigated in recent years. *Salmonella* populations on alfalfa sprouts were reduced approximately 2 log CFU/g after treatment for 2 min with 2% H_2O_2 or 200 ppm chlorine (Beuchat and Ryu, 1997). Less than 1 log CFU/g reduction was observed on cantaloupe cubes under similar test conditions. Treatment with 5% H_2O_2 bleached sprouts and cantaloupe cubes. Treatment of whole cantaloupes, honeydew melons, and asparagus spears with 1% H_2O_2 was less effective at reducing levels of inoculated salmonellae and *E. coli* O157:H7 than hypochlorite, acidified sodium chlorite, or a peracetic acid-containing sanitizer (Park and Beuchat, 1999). H_2O_2 (3%), alone or in combination with 2% or 5% acetic acid sprayed onto green peppers reduced *Shigella* populations approximately 5 log cycles, compared to less than a 1-log reduction by water alone (Peters, 1995). In the same study, *Shigella* inoculated onto lettuce was reduced approximately 4 log after dipping in H_2O_2 combined with either 2% or 5% acetic acid; however, obvious visual defects were noted on the treated lettuce. The same treatment gave similar results for *E. coli* O157:H7 inoculated onto broccoli florets or tomatoes with minimal visual defects.

Microbial populations on whole cantaloupes, grapes, prunes, raisins, walnuts, and pistachios were significantly reduced upon treatment with H_2O_2 vapor (Sapers and Simmons, 1998). Treatment by dipping in a H_2O_2 solution reduced microbial populations on fresh-cut bell peppers, cucumber, zucchini, cantaloupe, and honeydew melon, but did not alter sensory characteristics. Treatment of other produce was not as successful. H_2O_2 vapor concentrations necessary to control *Pseudomonas tolaasii* caused mushrooms to turn brown, while anthocyanin bleaching occurred in strawberries and raspberries. Shredded lettuce was severely browned upon dipping in a solution of H_2O_2. Combinations of 5% H_2O_2 with acidic surfactants at 50°C produced a 3 to 4 log reduction of nonpathogenic *E. coli* inoculated onto the surfaces of unwaxed Golden Delicious apples (Sapers et al., 1999). Further research is necessary to determine the usefulness of H_2O_2 treatment on other fruits and vegetables.

9.3.2 ANTIOXIDANTS

As previously described, browning is a major detrimental factor of the quality of white-fleshed fresh-cut fruits and vegetables. To avoid this problem, several additives are applied mainly by dipping, spraying, or vacuum impregnation. Antioxidants are grouped in accordance to their mode of action (i.e., as acidulants, reducing, chelating, complexing, enzyme, and inhibitors).

9.3.2.1   Acidulants

The optimum pH for polyphenoloxidase has been reported to be from acid to neutral in most fruits and vegetables, and the optimum activity is observed at pH 6 to 6.5, and minimum activity is detected below pH 4.5. This is the reason behind the use of chemicals that lower the product's pH or acidulants to help control enzymatic browning. Acidulants are used in combination with other treatments, because reducing browning by controlling only the pH is difficult. Acidulants, such as citric, malic, and phosphoric acids, are capable of lowering the pH of a system, thus reducing the polyphenol oxidase activity (Rojas-Graü et al., 2007).

Citric acid is widely used as an acidulant and is typically applied at levels ranging between 0.5% and 2% (w/v) for the prevention of browning in fruits and vegetables. Citric acid can be used in combination with other antibrowning agents, such as ascorbic or erythorbic acids and their neutral salts, for the chelation of prooxidants and for the inactivation of PPO. In addition to lowering the pH, citric acid acts by chelating the copper at the active site of the enzyme (Marshall et al., 2000). De Souza et al. (2006) used treatments of citric acid, calcium chloride, and reduced oxygen (2.5%), or high carbon dioxide (5% to 40%) atmospheres for mango (Kensington) stored at 3°C. They concluded that the use of citric acid had little positive effect and appeared to promote softening. The best treatment was low oxygen and calcium chloride, which allowed for a shelf life of 15 days (De Souza et al., 2006).

9.3.2.2   Reducing Agents

Reducing agents react with quinones, reducing them to phenols, and act on the enzyme PPO by linking irreversibly the copper of the enzyme. Reducing compounds are very effective in the control of browning (Lamikanra, 2002). One of the most widely used antibrowning agents is ascorbic acid. Ascorbic acid is a moderate reducing compound. It is acidic in nature, forms neutral salts with bases, and is water soluble. Erythorbic acid, which is the D isomer of ascorbic acid but without the vitamin C activity, is cheaper than vitamin C and is believed to have the same antioxidant properties (Gonzalez-Aguilar et al., 2005).

Sulfites are inhibitors of enzymatic browning. These compounds include sulfur dioxide (SO_2) and several forms of inorganic sulfites that release SO_2. Although they are very effective, the FDA has restricted their use due to potential allergic reactions (FDA, 2000).

Ascorbic acid reduces PPO browning by reducing *o*-quinones back to phenol compounds before they form brown pigments. However, ascorbic acid is consumed in the process, providing only temporary protection unless used at higher concentrations. Gorny et al. determined that 2% ascorbic acid with 1% calcium lactate reduced the browning of fresh-cut peaches initially, but after 8 days at 0°C, the difference was minimal (Gorny et al., 2002). Gil et al. determined that 2% ascorbic acid was effective in reducing the browning of fresh-cut Fuji apple slices, but in combination with low oxygen atmosphere storage (Gil et al., 1998).

Another reducing agent is cysteine, but for complete browning control, the amount of cysteine required is often incompatible with product taste (Lamikanra, 2002). The thiol-containing compounds, such as *N*-acetyl *L*-cysteine and reduced glutathione, are natural chemicals that react with quinones formed during the initial phase of enzymatic browning reactions. Oms-Oliu et al. (2006) used combinations of N-acetyl-L-cysteine, reduced glutathione, ascorbic acid, and 4-hexylresorcinol and concluded that 0.75% of N-acetyl-L-cysteine was effective to prevent browning of fresh-cut pears up to 28 days at 4°C, and 0.7% glutathione was effective up to 21 days at 4°C. There was also an enhanced effect combining N-acetyl-L-cysteine with reduced glutathione. Ascorbic acid and 4-hexylresorcinol were not effective (Oms-Oliu et al., 2006). Rojas-Graü et al. (2006) compared the browning inhibition of N-acetyl-cysteine, glutathione, ascorbic acid, and 4-hexylresorcinol with 'Fuji' apples stored for 14 days at 4°C. They determined that the best treatment concentrations were at least 0.75% of N-acetyl-cysteine, 0.60% of N-acetyl-cysteine, with 0.60% of glutathione. The sensory effects of the treatments were not determined (Rojas-Graü et al., 2006). Gonzalez-Aguilar et al. (2005) compared N-acetyl cysteine with ascorbic acid and isoascorbic acid as antibrowning agents for fresh-cut pineapple stored for 14 days at 10°C. Although the treatment with N-acetyl-cysteine (0.05 M) was the most effective in reducing browning and better appearance, higher levels of sugars and vitamin C (0.05 M) resulted from isoasorbic acid (0.1 M) and ascorbic acid. The level of antibrowning used did not affect other sensory characteristics (Gonzalez-Aguilar et al., 2005).

9.3.2.3 Chelating Agents

Chelating agents prevent enzymatic browning through the formation of a complex between these inhibitors and copper through an unshared pair of electrons in their molecular structures. Some of the chelating agents used on fruits and vegetables are citric acid and EDTA (Alzamora et al., 2000). EDTA is used with other antibrowning chemicals in concentrations up to 500 ppm (Lamikanra, 2002). Some tests using EDTA as an inhibitor of peach PPO were not totally effective (Marshall et al., 2000).

Phosphates have been used as components of commercial browning inhibitors. Pilizota and Sapers used combinations of sodium hexametaphosphate, ascorbic acid, calcium chloride, sodium chloride, and sodium erythorbate, with different levels of citric acid to adjust the pH, to develop an acidic browning inhibitor targeted at the core browning of fresh-cut apple slices but without affecting the tissue by the lower pH. The best treatments were 3% ascorbic acid + 1% citric acid + 1% sodium hexametaphosphate that had a pH of 2.9, but the problem was that in some cases, sodium hexamethaphosphate induced tissue breakdown with both varieties tested but only at 10°C. Although no formal sensory evaluation was done, some sour flavor was detected (Pilizota and Sapers, 2004).

Kojic acid, 5-hydroxy-2-hydroxymethyl-4H-pyran-4-one, is a γ-pyrone derivative and a fungal metabolite produced by many species of *Aspergillus* and *Penicillium* and a good metal ion chelator (Marshall et al., 2000). Son et al. (2001) used kojic acid among 36 other antibrowning compounds to compare the inhibitory effect on apple slices. Kojic acid, oxalic acid, oxalacetic acid, ascorbic acid, cysteine, glutathione, N-acetyl-cysteine, and 4-hexyl resorcinol were grouped as the ones to show the highest inhibitory activity on apple browning. The minimal concentrations for an effective antibrowning activity were 0.25% oxalacetic acid, 0.05% oxalic acid, 0.05% cysteine, and 0.05% kojic acid (Son et al., 2001).

9.3.2.4   Enzyme Inhibitors

4-Hexylresorcinol is an antibrowning agent with potential for application to fresh-cut products. Dong et al. (2000) used 4-hexylresorcinol with a combination of other compounds to extend the shelf life of fresh-cut Anjou pears. They determined that 4-hexylresorcinol (0.005% and 0.01%) was effective to prevent browning in combination with 0.5% ascorbic acid, but there was no effect without ascorbic acid. Sensory evaluation indicated that 0.01% of 4-hexylresorcinol was detected by panelists (Dong et al., 2000).

Salts of calcium, zinc, and sodium have been tested as anti-browning agents that act by inhibiting the enzyme polyphenol oxidase. However, chloride is a weak inhibitor; some authors report that the chloride levels required to inhibit the enzyme are too high and compromise taste (Lamikanra, 2002). Other studies tested a mix of ascorbic acid–sodium chloride, which inhibited 90% to 100% of the PPO activity (Alzamora et al., 2000).

Lu et al. (2007) used sodium chlorite in fresh-cut apple slices in dipping treatment solutions for 1 min. The treatments were sodium chlorite, sodium chlorite acidified with organic acids, and other salts. Apple slices treated in acidified sodium chlorite or sodium chlorite alone had a significantly smaller decrease in lightness value ($L*$), indicating less browning than those treated in citric acid or water control at 4 h. After 2 weeks of storage, only sodium chlorite (0.5 to 1 g/L), sodium bisulfite (0.5 g/L), and calcium l-ascorbate (10 g/L) continued to inhibit browning. Treatment with 0.5 g/L sodium chlorite and pH adjusted in the range from 3.9 to 6.2 using citric acid reduced browning more effectively than 0.5 g/L sodium chlorite without pH adjustment. Two organic acids, salicylic acid and cinnamic acid, when added to a sodium chlorite solution, were found to achieve even better inhibition of browning than citric acid at the same pH value (Lu et al., 2007).

9.3.2.5   Other Antibrownings

Enzymatic treatments with proteases that attack PPO have been suggested as alternative prevention treatments for enzymatic browning. Some preliminary tests used small pieces of apples and potatoes dipped for 5 min in a 2% enzyme solution; results showed that papain worked best on apples, whereas ficin (enzyme from figs) worked better on potatoes (Lamikanra, 2002). Forget et al. (1998) studied the antibrowning efficiency of papain extracts by studying their activity through two mechanisms: PPO inactivation and presence of quinone trapping substances.

9.3.3   Texturizers

There are two factors that most influence the mouth feel of a fruit or vegetable: firmness and juiciness. Firmness is determined largely by the physical anatomy of the tissue, particularly cell size, shape, and packing, cell wall thickness and strength, and the extent of cell-to-cell adhesion, together with turgor status. The most common additives to preserve fresh-cut fruit texture are calcium and ethylene blockers.

9.3.3.1 Calcium

Calcium forms used in the fresh-cut industry are calcium lactate, calcium chloride, calcium phosphate, calcium propionate, and calcium gluconate, which are used more when the objective is preservation or enhancement of the product firmness (Luna-Guzman and Barrett, 2000). Selection of the appropriate source depends on several factors: bioavailability and solubility are the most significant, followed by flavor change and interaction with the food matrix.

Calcium chloride has been widely used as a preservative and firming agent in the fruits and vegetables industry for fresh-cut commodities. Luna-Guzman and Barrett (2000) compared the effect of calcium chloride and calcium lactate dips on fresh-cut cantaloupe firmness. The use of calcium chloride is associated with bitterness and off-flavors (Ohlsson, 1994), mainly due to the residual chlorine remaining on the surface of the product.

Calcium lactate, calcium propionate, and calcium gluconate have shown some of the benefits of the use of calcium chloride, such as product firmness improvement, and avoid some of the disadvantages, such as bitterness and residual flavor (Yang and Lawless, 2003). The use of calcium salts other than calcium chloride could inhibit the formation of carcinogenic compounds (chloramines and trihalomethanes) linked to the use of chlorine. Calcium incorporation by impregnation with two calcium sources, calcium lactate and calcium gluconate, has been studied in fresh-cut apples (Anino et al., 2006).

9.3.3.2 Ethylene Blockers

1-Methylcyclopropene (1-MCP) has been considered one of the best options for extending quality and shelf-life of fresh-cut fruits and vegetables (Blankenship and Dole, 2003). 1-MCP exerts its action by tightly binding ethylene receptors, preventing ethylene binding, and consequently, inhibiting its action (Sisler and Serek, 1997). Treatment can be applied to either whole or sliced fruit (Laurie, 2005), and therefore it has a potential use in modified atmosphere packaging systems. Some studies have focused on the effects of 1-MCP treatment of whole fruits on the quality and shelf-life of fresh-cut produce. In some fruits, like banana and persimmon, whole fruit showed better results than fresh-cut fruit in terms of texture preservation (Vilas-Boas and Kader, 2001). Several reports showed excellent results in fresh-cut apples (Jiang and Joyce, 2002; Bai et al., 2004; Calderón-López et al., 2005). Weight loss of 1-MCP-treated apple slices was less than 2% after 21 days, with improved firmness preservation (Calderón-López et al., 2005). Furthermore, the firmness of sliced tomatoes was improved, and water soaking reduced by treating with 1-MCP (Jeong et al., 2004).

9.4 EMERGING ADDITIVES APPLIED TO FRESH-CUT FRUITS AND VEGETABLES

Research actions on the effect of several natural additive compounds have been undertaken, evaluating their mode of action, activity, toxicology, and effect on sensorial, biochemical, and physiological properties of the treated fresh-cut produce, all aimed at fulfilling consumer demands of healthy and safe produce with excellent quality. Therefore, considerable research has been recently directed toward the development of effective natural food preservatives. However, regulatory legislation of these additives is still in progress.

9.4.1 Natural Additives with Antimicrobial and Flavoring Potential

Food safety and quality have always been important to consumers and continue to be a basic requirement of a modern food system. Chemical control of fresh-cut fruit decay (synthetic additives) has been used since the initiation of this industry as a reliable preservative factor to control the amount of deteriorative factors in fresh-cut fruits and vegetables. However, most of these compounds do not satisfy the concepts of "natural" and "healthy" that consumers prefer and that the food industry, consequently, needs to provide (IFT, 2008). This necessity is underlined by many in agro-industries, legislatures, and consumer organizations around the world.

Natural antimicrobial compounds are a re-emerging alternative to fresh-cut produce preservation. The antimicrobial power of plant and herb extracts has been recognized for centuries and mainly used as natural medicine. Plants produce a wide range of volatile compounds, some of which are important flavor quality factors in fruits, vegetables, spices, and herbs (Lanciotti et al., 2004). A number of volatile compounds inhibit the growth of microorganisms (Burt, 2004). The ability of plant volatiles to inhibit microorganism growth is one of the reasons why there is an increased interest in using them to control post-harvest and post-processing decay of fruits and vegetables (Yao and Tian, 2005). Plant volatiles have been widely used as food flavoring agents, and many are GRAS.

Essential oils (EOs) represent the most important aromatic fraction of plants and plant produce, constituted by a complex mixture of terpenes, alcohols, cetones, aldehydes, esters, and sulfur compounds, depending on the source of the plants. Their antibacterial mode of action has been related to their individual active compounds. Tripathi and Dubey (2003) reported that the exact mode of action of antimicrobial compounds, such as thymol, eugenol, and carvacrol, has not been well determined, although it seems that they may inactivate essential enzymes, react with the cell membrane, or disturb genetic material functionality.

Plant EOs have shown a wide antimicrobial range of action against several bacteria and their toxins produced in foods, yeast, and molds (Tripathi and Dubey, 2003; Burt, 2004). Therefore, EOs present huge potential as food preservatives, especially because most are classified as GRAS. Citrus EO preserved the quality of fresh-cut fruit salads without affecting consumer acceptance of the product (Lanciotti et al., 2004). Garlic oil preserved overall quality and antioxidant capacity of fresh-cut tomatoes (Ayala-Zavala et al., 2008b). Cinnamon leaf and garlic oils showed antifungal activity against *Alternaria alternata* (Ayala-Zavala et al., 2008c).

The antimicrobial activities of a variety of naturally occurring phenolics from different plant sources have been studied in detail (Burt, 2004). Phenolics from spices, such as gingeron, zingerone, and capsaicin, have been found to inhibit germination of bacterial spores. Natural plant phenolic compounds are important food preservative factors and have, as a group, a remarkable antimicrobial range (Burt, 2004).

Methyl jasmonate (MJ) is a natural compound widely distributed in plants. It was first detected as a sweet fragrant compound in Jasminum essential oil and other plant species (González-Aguilar et al., 2006). MJ is known to regulate plant development and response to environmental stress (Demo et al., 2005; Yao and Tian, 2005), affecting many biochemical and physiological reactions in the tissue of whole and fresh-cut fruits and vegetables and extending shelf-life of whole and fresh-cut tomatoes, mangoes, guavas, and strawberries (González-Aguilar et al., 2006).

Ethanol, a GRAS compound, has shown to be effective for controlling decay of whole fruits and vegetables, inhibiting microbial growth (Karabulut et al., 2004). The mode of action of ethanol is by interaction with the membrane of microorganisms. Several devices have been designed to control ethanol release in the headspace of packed fruit (Kalathenos and Russell, 2003b). Ayala-Zavala et al. (2005, 2008) reported that ethanol treatment in conjunction with MJ increased antioxidant capacity, volatile compounds, and post-harvest life of strawberry fruit, as well as extended shelf-life of fresh-cut tomatoes. Plotto et al. (2006) concluded that ethanol vapor applied for 20 h, prior to processing whole mangoes, did not delay ripening; however, shorter time of exposure (10 h) suppressed fruit ripening.

Appropriate or compatible use of natural antimicrobial agents would involve using these compounds to add positive sensory characteristics in addition to improve food safety and extend shelf-life of fresh fruits and vegetables (Tripathi and Dubey, 2003; Ayala-Zavala et al., 2008a). Essential oils are effective antimicrobials; however, their aromatic volatile constituents can be absorbed by the food product. By choosing the right combination in aromas between the antimicrobial essential oil and the fresh-cut produce, quality involving safety and flavor can be improved (Ayala-Zavala et al., 2008a).

9.4.2 Natural Additives with Antibrowning and Texturizer Potential

Complexing agents entrap or form complexes with the substrates of the enzyme PPO or with reaction products. Some of the complexing agents are cyclodextrins of cyclic nonreducing oligosaccharides of six or more D-glucose residues. The problem that some researchers have observed is that β-cyclodextrin has low water solubility and, in some experiments with apples, was not effective or not consistent in controlling browning (Lamikanra, 2002).

Lopez-Nicolas et al. (2007) used different types of cyclodextrins as secondary antibrowning agents in apple juice and determined that maltosyl-β-cyclodextrix can enhance the ability of ascorbic acid to prevent the enzymatic browning due to the protective effect against ascorbic acid oxidation. Alvarez-Parilla et al. (2007) compared the polyphenol oxidase inhibitory effect of β-cyclodextrix, 4-hexylresorcinol, and methyl jasmonate in red delicious apple; the inhibitory strength was higher for 4-hexylresorcinol followed by β-cyclodextrin, and then methyl jasmonate. There was also a dual synergistic effect between β-cyclodextrix and 4-hexylresorcinol.

Honey has been studied for its antioxidant capacity and is believed to contain a small peptide that inhibits the activity of PPO (Marshall et al., 2000). Jeon and Zhao (2005) evaluated ten different honeys from floral sources and their antibrowning effects on fresh cut apples. The apples were vacuum impregnated in 10% honey solutions and the color was monitored for 10 days during storage at 4°C and 80% RH.

The Wildflower and Alfalfa honeys significantly inhibited browning discoloration, although there was an initial reduction of lightness as a result of the color from honey. A honey with light color may be preferred to be used as an anti-browning agent for fresh-cut apples.

Lozano de Gonzales et al. (1993) used pineapple juice for anti-browning, considering that pineapple contains the enzyme bromelain, which is also capable of inhibiting enzymatic browning just as is ascorbic acid. Pineapple juice was an effective browning inhibitor in both fresh and dried apples. All fractions of pineapple juice separated by different extraction methods inhibited enzymatic browning at least by 26%, as measured colorimetrically and by visual examination. Fractioning identified that the inhibitor is a neutral compound of low molecular weight.

Song et al. (2007) used rhubarb juice as a natural anti-browning agent for fresh-cut apple slices. They found that juices at 20% concentration containing 67 mg/100 g of oxalic acid inhibited browning. Yoruk and Marshall (2003) investigated the mode of inhibition of oxalic acid on PPO and determined that by binding with copper to form an inactive complex, it reduces catechol-quinone product formation. Oxalic acid was a more potent inhibitor of PPO as compared with other structurally related acids. Other compounds, such as benzoic and cinnamic acids, are PPO inhibitors but have been found not to give prolonged protection along storage time (Lamikanra, 2002).

9.4.3 FORTIFICANTS

Vitamins and minerals, called micronutrients, play a very important role in our health even though they make up only a very small part of the foods that we eat each day. Diets that do not contain adequate amounts of vital micronutrients often result in deficiency diseases, including blindness, mental retardation, and reduced resistance to infectious diseases, depending on the particular micronutrient. In this context, several additives with possibility to be applied as fresh-cut fruit and vegetable preservatives could be contemplated.

Most fruit and vegetables contain low amounts of some important vitamins and minerals, such as vitamin E, calcium, and zinc. For example, 100 g of raw apple (with skin) contains only 0.18 ± 0 mg of vitamin E, 6 ± 0.34 mg of calcium, and 0.04 ± 0.004 mg of zinc, according to the U.S. Department of Agriculture (USDA) National Nutrient Database for Standard Reference. Only 8% of men and 2.4% of women in the United States met current recommendations for vitamin E intake from food sources (Maras et al., 2004). Instead, vitamin E supplements are consumed daily by more than 35 million people in the United States (Traber, 2004). Leonard et al. (2004) showed that the bioavailability of vitamin E from vitamin E-enriched foods is much greater than that of an encapsulated supplement. Along with increased consumption of fresh-cut apples due to their convenience, fresh-like taste, and consumers' awareness of their health benefit, fresh-cut apples enriched with vitamin E or other bioactive compounds would be a good choice to develop functional foods and to provide opportunities for increasing intake of these nutraceuticals.

Xie and Zhao (2003) reported that use of diluted (20%) high fructose corn syrup (HFCS) containing calcium and zinc as solutions prevents browning discoloration of fresh-cut apples while significantly increasing calcium and zinc contents in apples. Park et al. (2005) further evaluated the use of 20% HFCS and other polymers, such as 1% pectin, 15 hydroxypropyl methylcellulose (HPMC), and 1% calcium casein-ate (CC) as VI solutions for incorporating vitamin E and minerals. It was found that HPMC and CC, large-molecular-weight polymers commonly used as edible coating materials, can be used to make VI solutions for fresh-cut apples, resulting in mini-mal impact on their physicochemical properties.

Food fortification has been applied with success in both developed and devel-oping countries to address and prevent micronutrient deficiencies. Due to varying eating habits and type of deficiency disease, the food vehicle, and the micronutrient added, the amounts that are added would not be the same for each country.

9.4.4 PROBIOTICS

Probiotics are live microorganisms that transit the gastrointestinal tract and, in doing so, benefit the health of the consumer (Tannock et al., 2000). Several members of the genera *Lactobacillus* and *Bifidobacterium* have gained recognition as probiotic bacteria because of their various therapeutic health benefits, mainly resistance to enteric pathogens, anti-colon cancer effect, immune system modulation, and relief of allergies.

Food industries, especially dairy industries, have been quick to realize the huge market potential created by the numerous positive health benefits of these probiotic bacteria. Both *Lactobacillus* spp. and *Bifidobacterium* spp. have been accorded the GRAS status because of their long history of safe use in foods (Salminen et al., 1998). Among the food industry and due to the advances in probiotic research, these additives are being contemplated in fresh-cut fruits and vegetables.

The addition of probiotics to obtain functional edible films to treat fresh-cut fruits has been reported (Tapia et al., 2007). Fresh-cut apple and papaya cylinders were successfully coated with 2% (w/v) alginate or gellan-film-forming solutions contain-ing viable bifidobacteria. Upon culture, bifidobacteria added to the film-forming solutions yielded viable populations of *B. lactis* Bb-12 in the order of 9.93 and 9.67 $\log_{10}$ CFU/g for the gellan or alginate coatings, respectively. Immediately after coat-ing (d 0), viable counts of *B. lactis* Bb-12 on the coated papaya pieces were 6.89 and 7.52 $\log_{10}$ CFU/g for alginate or gellan, respectively, while for coated apple pieces the values were 7.91 and 7.78 $\log_{10}$ CFU/g for alginate or gellan, respectively. This represents approximately a 2-log-cycle decrease compared to the concentration of the original film-forming solution; this drop was caused by dilution effects. *Bifidus* population remained viable and constant during the 10-d storage period at 2°C. The survival and maintenance of *B. lactis* Bb-12 in the alginate- or gellan-based edible coatings on both fresh-cut papaya and apples may be regarded as satisfactory, as their values remained between 6 and 7 $\log_{10}$ CFU/g. In order to confer health benefits to humans, the viable count of bifidobacteria at the time of consumption should be 10^6 CFU/g (Samona and Robinson, 1991). Ingestion in the number of $\geq 10^6$ cells per gram has been recommended for a classic probiotic food such as yogurt (Kurman and Rasic, 1991). It is also important for manufacturers and retailers to be able to confirm the viable count of these organisms in bifidus-containing products.

9.5   FUTURE TRENDS

Considering the increasing demands of consumers and the food industry, as well as environmental preservation, the use of additives based on natural compounds could be an alternative to chemical synthesized compounds in the preservation of fresh-cut fruits and vegetables. Future research should be focused on optimization of the industrial application, the influence of temperature and food matrix on the functional activity, finding optimal doses of these compounds, and evaluating toxicological activity, as well as assessing sensorial and overall quality of fresh-cut fruit. All these studies will be useful to understand the mode of action of the additives and will allow offering producers a practical method to preserve fresher, more natural foods containing less artificial preservatives, maintaining an ever increasing quality of fresh-cut fruits and vegetables.

REFERENCES

Alvarez-Parilla, E., de la Rosa, L., García, J., Escobedo, R., Mercado, G., Moyer, E., Vásquez, A., Gonzalez-Aguilar, G. A. 2007. Dual effect of β-cyclodextrin on the inhibition of apple polyphenol oxidase by 4-hexylresorcino and methyl jasmonate. *Food Chemistry* 101(4): 1346–1356.

Alzamora, S. M., Tapia, M. S., López-Malo, A. 2000. *Minimal Processing of Fruit and Vegetables. Fundamental Aspects and Applications.* Gaithersburg, MD, Aspen.

Anino, S. V., Salvatori, D. M., Alzamora, S. M. 2006. Changes in calcium level and mechanical properties of apple tissue due to impregnation with calcium salts. *Food Research International* 39: 154–164.

Artés, F., Gómez, P., Artés-Hernández, F., Aguayo, E., Escalona, V. 2007. Improved strategies for keeping overall quality of fresh-cut produce. *Acta Horticulturae (ISHS)* 746: 245–258.

Ayala-Zavala, J. F., del-Toro-Sanchez, L., Alvarez-Parrilla, E., Gonzalez-Aguilar, G. A. 2008a. High relative humidity in-package of fresh-cut fruits and vegetables: advantage or disadvantage considering microbiological problems and antimicrobial delivering systems? *Journal of Food Science* 73(4): R41–R47.

Ayala-Zavala, J. F., Oms-Oliu, G., Odriozola-Serrano, I., Gonzalez-Aguilar, G. A., Alvarez-Parrilla, E., Martin-Belloso, O. 2008b. Bio-preservation of fresh-cut tomatoes using natural antimicrobials. *European Food Research and Technology* 226(5): 1047–1055.

Ayala-Zavala, J. F., Soto-Valdez, H., González-Leon, A., Alvarez-Parrilla, E., Martin-Belloso, O., Gonzalez-Aguilar, G. A. 2008c. Microencapsulation of cinnamon leaf (*Cinnamomum zeylanicum*) and garlic (*Allium sativum*) oils in beta-cyclodextrin. *Journal of Inclusion Phenomena and Macrocyclic Chemistry* 60(3–4): 359–368.

Ayala-Zavala, J. F., Wang, S. Y., Wang, C. Y., Gonzalez-Aguilar, G. A. 2005. Methyl jasmonate in conjunction with ethanol treatment increases antioxidant capacity, volatile compounds and postharvest life of strawberry fruit. *European Food Research and Technology* 221(6): 731–738.

Bai, J., Baldwin, E. A., Soliva-Fortuny, R. C., Mattheis, J. P., Stanley, R., Perera, C., Brencht, J. K 2004. Effect of pretreatment of intact 'Gala' apple with ethanol vapor, heat, or 1-methylcyclopropene on quality and shelf life of fresh-cut slices. *Journal of the American Society for Horticultural Science* 1: 583–593.

Bari, M. L., Ukuku, D. O., Kawasaki, T., Inatsu, Y., Isshiki, Y., Kawamoto, S. 2005. Combined efficacy of nisin and pediocin with sodium lactate, citric acid, phytic acid, and potassium sorbate and EDTA in reducing the *Listeria monocytogenes* population of inoculated fresh-cut produce. *Journal of Food Protection* 68(7): 1381–1387.

Barth, M. M., Zhou, C., Mercier, M., Payne, F. A. 1995. Ozone storage effects on anthocyanin content and fungal growth in blackberries. *Journal of Food Science* 60: 1286–1287.

Beaulieu, J. C., Lea, J. M. 2003. Volatile and quality changes in fresh-cut mangos prepared from firm-ripe and soft-ripe fruit, stored in clamshell containers and passive MAP. *Postharvest Biology and Technology* 30(1): 15–28.

Bett-Garber, K. L., Beaulieu, J. C., Ingram, D. A. 2003. Effect of storage on sensory properties of fresh-cut cantaloupe varieties. *Journal of Food Quality* 26(4): 323–335.

Beuchat, L. R., Nail, B. V., Adler, B. B., Clavero, M. R. S. 1998. Efficacy of spray application of chlorinated water in killing pathogenic bacteria on raw apples, tomatoes, and lettuce. *Journal of Food Protection* 61: 1305–1311.

Beuchat, L. R., Ryu, J. H. 1997. Produce handling and processing practices. *Emerging Infectious Diseases* 3(4): 459–465.

Blankenship, S. M., Dole, J. M. 2003. 1-Methylcyclopropene: a review. *Postharvest Biology and Technology* 28: 1–25.

Brecht, J. K. 2006. Shelf-life limiting quality factors in fresh-cut (sliced) tomatoes: anti-ethylene treatment and maturity and variety selection to ensure quality retention. *Oral Presentation on the 2006 Tomato Breeders Round Table and Tomato Quality Workshop.*

Brown, G. E., Schubert, T. S. 1987. Use of *Xanthamonas campestris* pv. vesicatoria to evaluate surface disinfectants for canker quarantine treatment of citrus fruit. *Plant Diseases* 4: 319–323.

Burt, S. 2004. Essential oils: their antibacterial properties and potential applications in foods—a review. *International Journal of Food Microbiology* 94: 223–253.

Busta, F. F., Suslow, T. V., Parish, M. E., Beuchat, L. R., Farber, J. N., Garrett, E. H., Harris, L. J. 2003. The use of indicators and surrogate microorganisms for the evaluation of pathogens in fresh and fresh-cut produce. *Comprehensive Reviews in Food Science and Food Safety* 2(s1).

Calderón-López, B., Bartsch, B. J. A., Lee, C. Y., Watkins, C. B. 2005. Cultivar effects on quality of fresh cut apple slices from 1-methylcyclopropene (1-MCP)-treated apple fruit. *Journal of Food Science* 70: 221–227.

CFSAN/FDA, Center for Food Safety and Applied Nutrition/U.S. Food and Drug Administration. 2006, 03/10/2008. Food additives list. Retrieved June 7, 2008, from http://www.foodsafety.gov/~dms/opa-appa.html.

De Souza, B., O'Hare, T., Durigan, J., De Souza, P. 2006. Impact of atmosphere, organic acids, and calcium on quality of fresh cut Kensington mango. *Postharvest Biology and Technology* 42(2): 161–167.

Demo, M., Oliva, M. M., López, M. L., Zunino, M. P., Zygadio, J. A. 2005. Antimicrobial activity of essential oils obtained from aromatic plants of Argentina. *Pharmaceutical Biology* 43: 129–134.

Dong, X., Wrolstad, R., Sugar, D. 2000. Extending shelf life of fresh-cut pears. *Journal of Food Science* 65(1): 181–186.

FDA. 2000, April, 19, 2001. Sulfites: an important food safety issue. An update on regulatory status and methodologies. Retrieved June 10, 2008, from http://www.cfsan.fda.gov/~dms/fssulfit.html.

Fernandez-Escartin, E. F., Castillo-Ayala, A., Saldana-Lozano, J. 1989. Survival and growth of *Salmonella* and *Shigella* on sliced fresh fruit. *Journal of Food Protection* 52(7): 471–472.

Finch, G. R., Fairbairn, N. 1991. Comparative inactivation of poliovirus type 3 and MS2 coliphage in demand-free phosphate buffer by using ozone. *Applied Environmental Microbiology* 57(11): 3121–3126.

Fisher, K., Phillips, C. 2008. Potential antimicrobial uses of essential oils in food: is citrus the answer? *Trends in Food Science and Technology* 19(3): 156–164.

Forget, R., Cerny, M., Rigald, D., Fayad, E., Dahouk, N., Varoquaux, P. 1998. *Antibrowning efficiency of papaine extracts*. Second International Electronic Conference on Synthetic Organic Chemistry.

Forney, C. F. 2008. Flavour loss during postharvest handling and marketing of fresh-cut produce. *Stewart Postharvest Review* 4: 1–10.

Frazier, W. C., Westhoff, D. C. 1993. *Microbiología de los alimentos*. Zaragoza, Spain, Acribia.

Gil, M., Gorny, J., Kader, A. 1998. Responses of Fuji apples slices to ascorbic acid treatments and low oxygen atmospheres. *Hortscience* 33(2): 305–309.

Gonzalez-Aguilar, G. A., Ayala-Zavala, J. F., Ruiz-Cruz, S., Acedo-Felix, E., Diaz-Cinco, M. E. 2004. Effect of temperature and modified atmosphere packaging on overall quality of fresh-cut bell peppers. *Lebensmittel-Wissenschaft und-Technologie-Food Science and Technology* 37(8): 817–826.

González-Aguilar, G. A., Ruiz-Cruz, S., Soto-Valdez, H., Vazquez-Ortiz, F., Pacheco-Aguilar, R., Wang, C. Y. 2005. Biochemical changes of fresh-cut pineapple slices treated with antibrowning agents. *International Journal of Food Science and Technology* 40(4): 377–383.

González-Aguilar, G. A., Tiznado-Hernández, M., Wang, C. Y. 2006. Physiological and biochemical responses of horticultural products to methyl jasmonate. *Stewart Postharvest Review* 2: 1–9.

González-Aguilar, G. A., Wang, C. Y., Buta, J. G. 2000. Maintaining quality of fresh-cut mangoes using antibrowning agents and modified atmosphere packaging. *Journal of Agricultural and Food Chemistry* 48(9): 4204–4208.

Gorny, J., Hess-Pierce, B., Cifuentes, R., Kader, A. 2002. Quality changes in fresh-cut pear slices as affected by controlled atmospheres and chemical preservatives. *Postharvest Biology and Technology* 24: 271–278.

Hays, H., Elliker, P. R., Sandine, W. E. 1967. Microbial destruction by low concentrations of hypochlorite and iodophor germicides in alkaline and acidified water. *Applied Microbiology* 15: 575–581.

IFT, Institute of Food Technologists. 2008. Food production is getting greener; are you ready? Retrieved June 10, 2008, from http://members.ift.org/IFT/Pubs/Newsletters/weekly/nl_070908.htm#meetings1.

Jeon, M., Zhao, Y. 2005. Honey in combination with vacuum impregnation to prevent enzymatic browning of fresh-cut apples. *International Journal of Food Sciences and Nutrition* 56(3): 165–176.

Jeong, J., Brecht, J. K., Huber, D. J., S. A., S. 2004. 1-Methylcyclopropene (1-MCP) for Maintaining Textural Quality of Fresh-Cut Tomato. *Journal of the American Society for Horticultural Science* 39: 1359-1362.

Jiang, Y. M., Joyce, D. C. 2002. 1-Methylcyclopropene treatment effects on intact and fresh-cut apple. *Journal of Horticultural Science and Biotechnology* 77: 19–21.

Kalathenos, P., Russell, N. J. 2003. Ethanol as a food preservative. *Food Preservatives*. N. J. Russell and G. W. Gould. New York, Kluwer Academic/Plenum: 196–217.

Karabulut, O. A., Gabler, F. M., Mansour, M., Smilanick, J. L. 2004. Postharvest ethanol and hot water treatments of table grapes to control gray mold. *Postharvest Biology and Technology* 34: 169–177.

Karakurt, Y., Huber, D. J. 2003. Activities of several membrane and cell-wall hydrolases, ethylene biosynthetic enzymes, and cell wall polyuronide degradation during low-temperature storage of intact and fresh-cut papaya (*Carica papaya*) fruit. *Postharvest Biology and Technology* 28(2): 219–229.

Kim, J. G., Yousef, A. E., Dave, S. 1999. Application of ozone for enhancing the microbiological safety and quality of foods: a review. *Journal of Food Protection* 62(9): 1071–1087.

Kurman, J. A., Rasic, J. L. 1991. The health potential of products containing bifidobacteria. *Therapeutic Properties of Fermented Milks*. R. K. Robinson. London, Elsevier Applied Food Sciences: 117–158.

Lamikanra, O. 2002. *Fresh-Cut Fruit and Vegetables: Science, Technology, and Market*. Boca Raton, FL, CRC Press.

Lanciotti, R., Gianotti, A., Patrignani, F., Belletti, N., Guerzoni, M. E., Gardini, F. 2004. Use of natural aroma compounds to improve shelf-life and safety of minimally processed fruits. *Trends in Food Science and Technology* 15(3–4): 201–208.

Laurie, S. 2005. Application of 1-methylcyclopropene to prevent spoilage. *Stewart Postharvest Review* 1(4): 1–9.

Leonard, S. W., Good, C. K., Gugger, E. T., Traber, M. G. 2004. Vitamin E bioavailability from fortified breakfast cereal is greater than that from encapsulated supplements. *American Journal of Clinical Nutrition* 79: 86–92.

Liao, C., Sapers, G. M. 2000. Attachment and growth of Salmonella Chester on apple fruits and in vivo response of attached bacteria to sanitizer treatments. *Journal of Food Protection* 63: 876–883.

Liew, C. L., Prange, R. K. 1994. Effect of ozone and storage temperature on postharvest diseases and physiology of carrots (*Caucus carota* L.). *Journal of the American Society for Horticultural Science* 119: 563–567.

Lopez-Nicolas, J., Nunez, E., Sanches, A., Garcia, F. 2007. Kinetic model of apple juice enzymatic browning in the presence of cyclodextrins: the use of maltosyl-β-cyclodextrin as secondary antioxidant. *Food Chemistry* 101(3): 1164.

Lozano de Gonzales, P., Barret, D., Wrolstad, R., Durst, R. 1993. Enzymatic browning inhibited in fresh and dried apple rings by pineapple juice. *Journal of Food Science* 58(2): 399–404.

Lu, S., Luo, Y., Turner, E., Feng, H. 2007. Efficacy of sodium chlorite as an inhibitor of enzymatic browning in apple slices. *Food Chemistry* 104(2): 824–829.

Luna-Guzman, I., Barrett, D. 2000. Comparison of calcium chloride and calcium lactate effectiveness in maintaining shelf stability and quality of fresh-cut cantaloupes. *Postharvest Biology and Technology* 19: 61–72.

Maras, J. E., Bermudez, O. I., Qiao, N., Bakun, P. J., Boody-Alter, E. L., Tucker, K. L. 2004. Intake of alpha-tocopherol is limited among US adults. *Journal of the American Dietitian Association* 104: 567–575.

Marriott, N. G. 1999. *Principles of Food Sanitation*. 4th ed. Gaithersburg, MD, Aspen.

Marshall, M., Kim, J., Wei, C. 2000. *Enzymatic Browning in Fruits, Vegetables and Seafood*. Rome, Food and Agriculture Organization of the United Nations.

Odlaug, T. E. 1981. Antimicrobial activity of halogens. *Journal of Food Protection* 44(8): 608–613.

Ohlsson, T. 1994. Minimal processing—preservation methods of the future—an overview. *Trends in Food Science and Technology* 5: 341–344.

Oms-Oliu, G., Aguilo-Aguayo, I., Martin-Belloso, O. 2006. Inhibition of browning on fresh-cut pear wedges by natural compounds. *Journal of Food Science* 71(3): S216–S224.

Pao, S., Davis, C. L., Kelsey, D. F. 2000. Efficacy of alkaline washing for the decontamination of orange fruit surfaces inoculated with *Escherichia coli*. *Journal of Food Protection* 63: 961–964.

Pao, S., Davis, C. L., Kelsey, D. F., Petracek, P. D. 1999. Sanitizing effects of fruit waxes at high pH and temperature on orange surfaces inoculated with *Escherichia coli*. *Journal of Food Science* 64(2): 359–362.

Parish, M. E., Beuchat, L. R., Suslow, T. V., Harris, L. J., Garrett, E. H., Farber, J. N., Busta, F. F. 2003. Methods to reduce/eliminate pathogens from fresh and fresh-cut produce. *Comprehensive Reviews in Food Science and Food Safety* 2(s1).

Park, C. M., Beuchat, L. R. 1999. Evaluation of sanitizers for killing *Escherichia coli* O157:H7, Salmonella and naturally occurring microorganisms on cantaloupes, honeydew melons, and asparagus. *Dairy Food and Environmental Sanitation* 19: 842–847.

Park, S., Kodihalli, I., Zhao, Y. Y. 2005. Nutritional, sensory, and physicochemical properties of vitamin E- and mineral-fortified fresh-cut apples by use of vacuum impregnation. *Journal of Food Science* 70(9): S593–S599.

Parnell, T. L., Harris, L. J. 2003. Reducing Salmonella on apples with wash practices commonly used by consumers. *Journal of Food Protection* 66: 741–747.

Peters, D. L. 1995. Control of enteric pathogenic bacteria on fresh produce. Lincoln, University of Nebraska Graduate College.

Pilizota, V., Sapers, G. M. 2004. Novel browning inhibitor formulation for fresh cut apples. *Journal of Food Science* 69(4): 140–143.

Plotto, A., Bai, J., Narciso, J. A., Brecht, J. K., Baldwin, E. A. 2006. Ethanol vapor prior to processing extends fresh-cut mango storage by decreasing spoilage, but does not always delay ripening. *Postharvest Biology and Technology* 39: 134–145.

Rapisarda, P., Caggia, C., Lanza, C. M., Bellomo, S. E., Pannuzzo, P., Restuccia, C. 2006. Physicochemical, microbiological, and sensory evaluation of minimally processed Tarocco clone oranges packaged with 3 different permeability films. *Journal of Food Science* 71(3): S299–S306.

Restaino, L., Frampton, E. W., Hemphill, J. B., Palnikar, P. 1995. Efficacy of ozonated water against various food-related microorganisms. *Applied Environmental Microbiology* 61(9): 3471–3475.

Rojas-Graü, M. A., Grasa-Guillem, R., Martin-Belloso, O. 2007. Quality changes in fresh-cut Fuji apple as affected by ripeness stage, antibrowning agents, and storage atmosphere. *Journal of Food Science* 72(1): S36–S43.

Rojas-Graü, M. A., Sobrino-López, A., Soledad Tapia, M., Martin-Belloso, O. 2006. Browning inhibition in fresh-cut 'Fuji' apple slices by natural antibrowning agents. *Journal of Food Science* 71(1): S59–S65.

Roller, S., Lusengo, J. 1997. Developments in natural food preservatives. *Agro Food Industry Hi-Tech* 8(4): 22–25.

Ruiz-Cruz, S., Islas-Osuna, M. A., Sotelo-Mundo, R. R., Vazquez-Ortiz, F., Gonzalez-Aguilar, G. A. 2007. Sanitation procedure affects biochemical and nutritional changes of shredded carrots. *Journal of Food Science* 72(2): S146–S152.

Salminen, S., Vonwright, A., Morelli, L., Marteau, P., Brassart, D., Devos, W. M., Fonden, R., Saxelin, M., Collins, K., Mogensen, G., Birkeland, S. E., Mattilasandholm, T. 1998. Demonstration of safety of probiotics—a review. *International Journal of Food Microbiology* 44(1/2): 93–106.

Samona, A., Robinson, R. K. (1991). Enumeration of bifidobacteria in dairy products. *Journal of the Society of Dairy Technology* 44: 64–66.

Sapers, G. M., Miller, R. L., Annous B.A., Burke, A. M. 2002. Improved antimicrobial wash treatments for decontamination of apples. *Journal of Food Science* 67: 1886–1891.

Sapers, G. M., Miller, R. L., Mattrazzo, A. M. 1999. Effectiveness of sanitizing agents in inactivating *Escherichia coli* in golden delicious apples. *Journal of Food Science* 64(4): 734–737.

Sapers, G. M., Simmons, G. F. 1998. Hydrogen peroxide disinfection of minimally processed fruits and vegetables. *Food Technology* 52(2): 48–52.

Sarig, P., Zahavi, T., Zutkhi, Y., Yannai, S., Lisker, N., Ben-Arie, R. 1996. Ozone for control of postharvest decay of table grapes caused by Rhizopus stolonifer. *Physiology and Molecular Plant Pathology* 48: 403–415.

Shapiro, J. E., Holder, I. A. 1960. Effect of antibiotic and chemical dips on the microflora of packaged salad mix. *Applied Microbiology* 8: 341.

Sisler, E. C., Serek, M. 1997. Inhibitors of ethylene responses in plants at the receptor level. Recent developments. *Physiology Plantarum* 100: 577–582.

Somers, E. B., Schoeni, J. L., Wong, A. C. L. 1994. Effect of trisodium phosphate on biofilm and planktonic cells of *Campylobacter jejuni, Escherichia coli* O157:H7, *Listeria monocytogenes* and *Salmonella typhimurium. International Journal of Food Microbiology* 22: 269–276.

Son, S., Moon, K., Lee, C. 2001. Inhibitory effects on various antibrowning agents on apple slices. *Food Chemistry* 73(1): 23–30.

Song, Y., Yao, Y., Zhai, H., Du, Y., Che, F., Shu-wei, W. 2007. Polyphenolic compounds and the degree of browning in processing apple varieties. *Agricultural Science in China* 6(5): 607–612.

Tannock, G. W., Munro, K., Harmsen, H. J. M., Welling, G. W., Smart, J., Gopal, P. K. 2000. Analysis of the fecal microflora of human subjects consuming a probiotic product containing *Lactobacillus rhamnosus* DR 20. *Applied Environmental Microbiology* 66: 2578–2588.

Tapia, M. S., Rojas-Graü, M. A., Rodriguez, F. J., Ramirez, J., Carmona, A., Martin-Belloso, O. 2007. Alginate- and gellan-based edible films for probiotic coatings on fresh-cut fruits. *Journal of Food Science* 72(4): E190–E196.

Toivonen, P. M. A., Brummell, D. A. 2008. Biochemical bases of appearance and texture changes in fresh-cut fruit and vegetables. *Postharvest Biology and Technology* 48(1): 1–14.

Torriani, S., Orsi, C., Vescovo, M. 1997. Potential of *Lactobacillus casei*, culture permeate, and lactic acid to control microorganisms in ready-to-use vegetables. *Journal of Food Protection* 60: 1564–1567.

Traber, M. G. 2004. Vitamin E, nuclear receptors and xenobiotic metabolism. *Archives of Biochemistry and Biophysic* 423: 6–11.

Tripathi, P., Dubey, N. K. 2003. Exploitation of natural products as an alternative strategy to control postharvest fungal rotting of fruit and vegetables. *Postharvest Biology and Technology* 32: 235–245.

Ukuku, D. O., Fett, W. F. 2002. Effectiveness of chlorine and nisin-EDTA treatments of whole melons and fresh-cut pieces for reducing native microflora and extending shelf-life. *Journal of Food Safety* 22: 231–257.

Vilas-Boas, E. V., Kader, A. A. 2001. Effect of 1-MCP on fresh-cut fruits. *Perishables Handling Quarterly* 108: 25.

White, J. M. L., McFadden, J. P. 2008. Contact allergens in food ingredients and additives: atopy and the hapten-atopy hypothesis. *Contact Dermatitis* 58(4): 245–246.

Xie, J., Zhao, Y. 2003. Nutritional enrichment of fresh apple (Royal Gala) by vacuum impregnation. *International Journal of Food Sciences and Nutrition* 54: 338–397.

Yang, H. H., Lawless, H. T. 2003. Descriptive analysis of divalent salts. *Journal of Sensory Studies* 20: 97–113.

Yao, H. J., Tian, S. P. 2005. Effects of biocontrol agent and methyl jasmonate on postharvest diseases of peach fruit and the possible mechanisms involved. *Journal of Applied Microbiology* 98: 941–950.

Yoruk, R., Marshall, M. 2003. A survey on the potential mode of inhibition for oxalic acid on polyphenol oxidase. *Journal of Food Science* 68(8): 2479–2485.

Zhao, J., Cranston, P. M. 1995. Microbial decontamination of black pepper by ozone and the effect of the treatment on volatile oil constituents of the spice. *Journal of the Science of Food and Agriculture* 68: 11–18.

Zhuang, R. Y., Beuchat, L. R. 1996. Effectiveness of trisodium phosphate for killing *Salmonella montevideo* on tomatoes. *Letters Applied Microbiology* 22: 97–100.

10 Modified Atmosphere Packaging of Fruits and Vegetables
Modeling Approach

*Carole Guillaume, Valérie Guillard,
and Nathalie Gontard*

CONTENTS

10.1 INTRODUCTION

Consumer demand and market trends for more convenient, safer, and longer storage products with low preservative contents are increasing. In this context, preservation of raw or fresh-cut fruits and vegetables constitutes one of the most challenging applications: They are respiring produce and must remain living all along the distribution chain. Lowering respiration is the key element to delay physiological and biochemical changes and consequently increase shelf life. These changes go also with microbial spoilage and are emphasized by minimal processing such as cutting, stoning, or peeling. Though sanitation methods (as chlorine washes) are commonly employed, they are not totally effective because microorganisms can locate in subsurface structures of produces and then survive (Takeuchi and Frank 2000; Burnett and Beuchat 2001). Common preserving processes (such as thermal treatments) or additions of preservative agents (except ascorbic or citric acid that might be incorporated into washing water) are not adapted to these fragile products. Then there is an obvious opportunity for the food packaging industry to develop innovative solutions that fulfilled this type of food requirements to prolong its shelf life while maintaining its quality and safety. Modified atmosphere packaging (MAP) of fresh food is the fastest-growing sector, with an average annual growth rate of 13.6% over the last 5 years, and will continue to grow due to research and development (R&D) and legislative efforts on modern concepts of active packaging. These packaging concepts deliberately incorporate active agents intended to release or to absorb substances into, onto, or from the packaged food or the environment surrounding the food. Active system leaders for MAP applications include oxygen scavengers, ethylene absorbers/removers, and moisture controllers (Brody et al. 2001). From pad or sachet that was associated to package, new trends are now to directly incorporate active agent onto or into the material and mainly concern antimicrobial packaging that can release volatile antimicrobial agents for MAP applications (Appendini and Hotchkiss 2002; Suppakul et al. 2003).

Technological aspects of modified atmosphere and its potential to extend shelf life for many foods, including fresh fruits and vegetables, have already been well documented (Jayas and Jeyamkondan 2002; Brecht et al. 2003; Hertog 2003; Saltveit 2003; Varoquaux and Ozdemir 2005; Rico et al. 2007). A combination of low oxygen (O_2) levels and medium carbon dioxide (CO_2) content is commonly recommended to slow respiration of raw or fresh-cut products and delay spoilage. Nevertheless, there is no unique recommended atmosphere because optimal environmental conditions vary according to specie, variety, and processing. These optimal conditions can be easily obtained in controlled atmosphere storage, but it is more hazardous in MAP. At this time, MAP in the food industry still relies on an empirical approach that was previously called the pack-and-pray approach (Hertog and Banks 2003). Then, to support the development of and achieve breakthroughs in the design of innovative MAP solutions in a rational way, there is still a need to formulate strategic options for reducing time-consuming step-by-step trials. This could be done through prediction tools based on modeling and transdisciplinary approaches that should integrate the knowledge of mass transfer properties of packaging materials and their incidence on food quality and safety evolution.

In this chapter, we set up a critical review of existing mathematical models for developing structure/transfer and transfer/reaction couplings dedicated to MAP of fresh fruits and vegetables.

10.2 MODIFIED ATMOSPHERE PACKAGING SYSTEMS FOR FRESH FRUITS AND VEGETABLES

Technically, MAP of fresh or minimally processed fruits and vegetables can be achieved in either a passive or an active way. In the first case, gas or vapor exchanges within the package are due to both physiological responses (respiration, maturation, transpiration) of the vegetal and gas or vapor diffusions through the packaging, as indicated in Figure 10.1. After a transient period, gas partial pressure of packaging headspace reaches a steady state when diffusive exchanges through the film exactly compensate gas or vapor production or consumption. This is called the equilibrium MAP (EMAP). EMAP should occur as quickly as possible after packaging the products and should be close to optimal recommended atmosphere to preserve quality and safety of the packaged product. Thereafter, generally for a long storage period, physiological disorders marked by a respiratory crisis, as well as important microbiological development, might occur and lead to a deviation period no longer consistent with previous mass transfer considerations. Active MAP incorporates gas flushing or use of gas or vapor scavenging or emitting systems integrated into the packaging material or added in a separate sachet or label. Principles for gas exchanges within

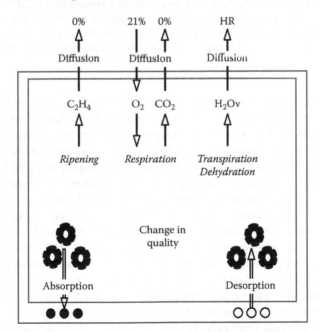

FIGURE 10.1 Gas or vapor exchange in passive or active modified atmosphere packaging of fresh fruits and vegetables (physiological reactions and mass transports in packaging material).

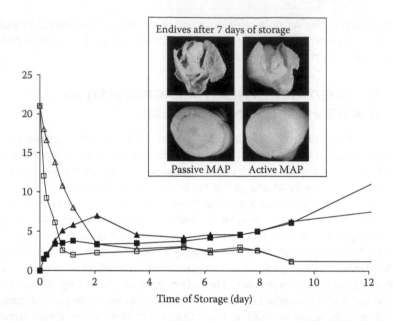

FIGURE 10.2 Change in O_2 (open marks) and CO_2 (full marks) partial pressures within passive modified atmosphere packaging (MAP) ($\triangle$, $\blacktriangle$) and active MAP ($\square$, $\blacksquare$) of 550 g of endives stored at 20°C and its effect on head opening and browning of basal part after 7 days of storage. Passive MAP corresponds to low-density polyethylene (LDPE) pouches and active MAP to LDPE pouches containing an O_2 scavenger sachet. (Adapted from Charles, F., Guillaume, C., and Gontard, N. 2008. Effect of passive and active modified atmosphere packaging on quality changes of fresh endives. *Postharvest Biology and Technology* 48: 22–29.)

the package are the same as for passive MAP, but initial atmosphere might be different from that of normal air (gas flushing), or passive transfers might be combined to active absorption or desorption (scavengers or emitters, respectively). Therefore, kinetics of absorption or desorption have to be taken into consideration, as illustrated in Figure 10.1. The main goal of active MAP is to shorten or avoid the transient period that could be detrimental when products are sensitive to enzymatic browning, such as fresh-cuts. An example of differences in modified atmosphere between passive and active MAP is given in Figure 10.2. Endives have been packed in low-density polyethylene (LDPE) pouches without (passive MAP) or with an O_2 scavenger sachet (active MAP) and stored at 20°C for 12 days. Active MAP did not modify the equilibrium partial pressures in O_2 and CO_2, which was the result of gas diffusion through the LDPE film and endive respiration rate. However, it considerably reduced the time at which the EMAP is reached: 2 days, against 5 days with passive MAP. This EMAP reduction induced a delay in endives head greening and opening as well as browning of the basal part. Then, active MAP maintained acceptable visual aspects of endives longer than passive MAP. Active packaging should be designed to match product requirements as for passive packaging and constitutes an alternative when passive packaging solutions result in a fair level of quality or shelf life of food product. Table 10.1 presents several systems that are, or could be, employed in MAP of fresh fruits and vegetables, with related research papers.

10.2.1 MICROPERFORATED VERSUS PERMSELECTIVE MATERIALS

Microperforated films are commonly used in MAP of high respiration fresh fruits and vegetables to enhance gas diffusion through the film compared to conventional dense polymers. They enable a large range of O_2 permeability, depending on the density of perforations. Biaxial-oriented polypropylene (BOPP) film with two microperforations leads to a suitable EMAP for preserving visual appearance and nutritional compounds in a mix of broccoli and cauliflower florets (Schreiner et al. 2007) for up to 7 days at 8°C. Polyethylene terephthalate (PET)/polypropylene (PP) films perforated with one or three micro-holes appeared well adapted to maintaining the quality of wild strawberries up to 6 days at 10°C (Almenar et al. 2007). Even though they are already commercially used for packing ready-to-cook fresh vegetables, microperforated films do not meet all the requirements of fresh produce, especially CO_2-sensitive ones with high respiration rate such as mushrooms (Exama et al. 1993; Barron et al. 2002). With a permselectivity value (ratio between CO_2 permeability and O_2 permeability) close to 1, CO_2 diffuses through the film at a similar rate to O_2. Then EMAP with both low O_2 and low CO_2 levels can never be reached (Fishman et al. 1995; Varoquaux and Ozdemir 2005) with such materials.

High permselective and hydrophilic films such as Pebax® or Saran film® exhibit a permselective ratio of 8.6 at 23°C (Barron et al. 2002) or 10.2 at 4°C (Rogers 1975), respectively—for comparison, it ranges from 3 to 6 in other synthetic material (Exama et al. 1993). Pebax was commonly used for overwrapping fresh fruits and vegetables on the Japanese market, and its efficacy was previously demonstrated: It prevented chilling injury symptoms in MAP of papaya (Singh and Rao 2005) and maintained the same EMAP of mushrooms stored at 10 or 20°C (Barron et al. 2002). In addition, hydrophilic materials can modify their gas transport properties as a function of temperature. Temperature effect can be expressed by the gas permeability activation energy of materials and by the gas permeability Q_{10}, mostly used by biologist for describing the degree by which a biological or chemical process is accelerated by a rise of 10°C. For instance, the permeability activation energy of Saran film is around 66.5 KJ.mol^{-1} for O_2 and 51.5 KJ.mol^{-1} for CO_2, and its permeability Q_{10} reaches 2.82 for O_2 and 2.23 for CO_2 (Rogers 1975). In other plastic materials, the Q_{10} for O_2 or CO_2 permeability remains lower than 2, whereas the Q_{10} for the respiration rate ranges from 1.8 to 3 for most raw fresh fruits and vegetables (Exama et al. 1993). Then only hydrophilic materials can adapt their gas permeabilities to the respiration of fresh products as a function of temperature. This seems worthy of investigation for overcoming detrimental EMAP changes that occur in the case of chill chain disruption with other synthetic materials (perforated or not). Landec Corp. (USA) developed a liner (Intellipac™) able to self-adjust permselectivity, from 1.5 to 18, when increasing temperature (Brody et al. 2001). Other hydrophilic materials such as protein-based materials also appeared to be of great interest for future packaging material development; the permselectivity ratio increases up to 38 in wheat gluten-based film stored at 45°C and 95% relative humidity (RH)

TABLE 10.1

Some Available Systems Used or That Could Be Used in Commercial Applications for Modified Atmosphere Packaging of Fresh Fruits and Vegetables

Systems	Formats	Active Agents	Trade Names	References
Gas-permeable and nonselective materials	Microperforated petrochemical material		All film manufacturers	Almenar et al. (2007), Schreiner et al. (2007)
Gas-permeable and permselective materials	Copolymer block or chemical modification of side chains in petrochemical material		Pebax® (Arkema, France), Saran Film® (Dow Chemical Co., USA), Intellipac™ (Landec Corp., USA)	Rogers (1975), Barron et al. (2002), Singh and Rao (2005)
Oxygen scavengers	Sachet, label	Iron: moisture-activated oxidation; Glucose oxydase: self-activated enzyme-mediated oxidation	ATCO®LH (Laboratoire Standa, France), Ageless® (Mitsubishi Gas Chemical Company Inc., Japan) Bioka (Bioka LTD, Finland)	Tewari et al. (2002), Charles et al. (2003), Charles et al. (2008)
Carbon dioxide scavengers	Sachet, label	Calcium hydroxide: moisture-activated carbonation	ATCO®CO (Laboratoire Standa, France)	Charles et al. (2005), Charles et al. (2006)
Ethylene scavenger/remover	Sachet, multilayered paper or cardboard; Petrochemical films as low-density polyethylene	Silica gel to adsorb ethylene and potassium permanganate to oxidize it; Mineral as zeolite; Silica gel	Ethylene Control Inc. (USA), Frisspack® (Dunapack Ltd., Hungary) Green bag (Evert-Fresh Corp., USA)	Brody et al. (2001), Correa et al. (2005)
Moisture controllers	Sachet	Silica gel	Condensationguard™ (Grace division, USA)	Paull (1999), Brody et al. (2001)
Antimicrobial	Multilayered paper or cardboard; Sachet; Label, coated petrochemical, or paper-based material	Sodium metabisulfite: moisture-activated sulfite emission; Ethanol microcapsules: moisture-activated emission; AIT microcapsules	Kontroll® (Kontek SRL, Italy), Ethicap® (Freund Industrial Co., Japan) Wasaouro® (Mitsubishi Kagaku Foods Co., Japan)	Valverde et al. (2005), Lurie et al. (2006), Martinez-Romero et al. (2007), Ayala-Zavala et al. (2008), Utto et al. (2008)

(Gontard et al. 1996) and has been successfully tested for maintaining the same EMAP whatever the temperature (Barron et al. 2002). In addition, and comparing to all kind of synthetic films, protein-based materials exhibit the unique ability to adapt their mass transfer properties to RH. At 25°C, the O_2 permeability value was 1.24 $amol.m^{-1}.s^{-1}.Pa^{-1}$ in dry conditions and increased to 1,290 $amol.m^{-1}.s^{-1}.Pa^{-1}$ at 95% RH (Gontard et al. 1996). The effect of RH was more pronounced on CO_2 permeability than O_2 permeability, and consequently, permselectivity increased from 5 to 30 when RH increased from 0% to 95%. It should be stressed that a ratio of 30 cannot be reached with synthetic material, and this could be interesting for CO_2-sensitive produce such as mushrooms (Barron et al. 2002).

10.2.2 ACTIVE SYSTEMS

Active packaging engineered to scavenge O_2 found a variety of uses in food packaging because it can react with and eliminate O_2 from the headspace or that is dissolved into food product. Major O_2 scavengers systems are sachets or labels made of a porous material (paper coated with perforated PP in ATCO®-LH, for instance) and containing active agents. The most significant O_2 scavengers are those based on the oxidation of ferrous ion, which is moisture activated, but other active agents such as unsaturated fatty acids and hydrocarbons, enzymes, and ascorbic acid, already exist (Brody et al. 2001). More and more, O_2 scavengers are directly incorporated into multilayered plastic materials such as Cryovac®OS1000 (Cryovac Sealed Air Corp., USA). However, these materials are currently barriers to O_2 and cannot be used for MAP of fresh fruits and vegetables, because it would sharply lead to anoxia involving a turn into fermentative catabolism. O_2 scavenger sachets containing iron or enzyme have already been studied and could be used in active MAP of fresh products such as for tomatoes (Charles et al. 2003) or endives (Charles et al. 2005, 2008) as discussed in paragraph 1 (Figure 10.2).

Whether beneficial effects of CO_2 have been previously reported, high levels of this gas in MAP can be detrimental to fresh CO_2-sensitive product quality: CO_2 injuries might occur when a fresh product is exposed to a level above its tolerance limit, such as formation of brown spots on lettuce or yellowing of mushrooms, which are common visual degradations caused by high CO_2 content (Zagory and Kader 1988; Lopez Briones et al. 1992). CO_2 scavenger sachets already exist and are composed of a physical absorbent such as zeolite or a chemical one such as calcium hydroxide, sodium carbonate, or magnesium hydroxide. Their use is limited to fermented or roasted foods, but it could be of great interest for preventing the formation of a CO_2 peak, during the transient period, in MAP of high-respiring products. Even when using hydrophilic materials, this peak appeared in MAP of fresh mushrooms, leading to an irreversible yellowing of the commodity (Barron et al. 2002). Charles et al. (2005) demonstrated its complete disappearance in active MAP of endives with CO_2 scavengers (ATCO®CO).

Ethylene scavengers are already commercialized to delay ripening of climacteric fruits and are mainly based on ethylene oxidation by potassium permanganate (Frisspack®, Dunpack Ltd., Hungary) or the incorporation of mineral clay such as zeolite within packaging materials (Green bag, Evert-Fresh Corp., USA). The effect of mineral clay was attributed to its adsorption of ethylene and also to the formation of pores within the packaging material that could modify ethylene permeability of the film (Brody et al. 2001).

In MAP, moisture vapor is produced by transpiration, and a dehydration phenomenon occurs spontaneously with the diffusion of moisture vapor from the high moisture fresh product to the lower moisture surrounding environment. Water loss is an important marketing factor of fresh fruits and vegetables. Because fruits and vegetables are mainly constituted by over 90% of water, an average 5% water loss (ranging from 3% to 10% depending on the commodity) is visually noticeable, lowering the grade of the product and resulting in a depression of its commercial value. Major effects of water loss are a reduction in weight basis and a wilted appearance. A reduction of the nutritional value also is seen because the amount of water-soluble components decreases while water is released, leading to aroma and flavor loss, and then sensitivity to chilling injuries is emphasized (Paull 1999; Maguire et al. 2006). If RH should be higher than 90% to delay product dehydration, 100% RH is not recommended because it favors microbial growth and spoilage. Several moisture control systems already exist, and CondensationGard™ (Grace Division, USA) appears to be the most adapted for fresh fruits and vegetables preservation. It only adsorbs excessive moisture to avoid condensation while retaining suitable moisture balance and is currently available at the 90% RH grade. This active system should be tested for active MAP of fresh commodities.

Another kind of active MAP is antimicrobial packaging that contains active agents releasing volatile substances, with antimicrobial activity to control undesirable growth of microorganisms at the surface of food. Such systems are commercially available for fresh fruits and vegetables such as Wasaouro® (Mitsubishi Kagaku Foods CO., Japan) releasing allylisothiocyanate (AITC) or Kontroll® films (Kontek SRL, Italy) releasing sulfite for table grape preservation, in high moisture conditions. Ethanol emitters, mainly used in MAP of pastry, could be adapted to fresh products because their beneficial effects on quality were previously reported on table grapes (Lurie et al. 2006). More commonly, volatile aroma compounds (including AITC) constitute alternative strategies to control postharvest spoilage of fresh fruits and vegetables (Valverde et al. 2005; Martinez-Romero et al. 2007; Ayala-Zavala et al. 2008; Utto et al. 2008).

10.3   MODELING MASS TRANSPORT IN PACKAGING MATERIALS

10.3.1   Transport of Gas and Vapors in Polymers

Gas transport through dense polymer obeys the solution-diffusion mechanism, where gases dissolve into the polymer matrix at the face of the sample exposed to higher gas pressure, diffuse across the polymer film to the face of the sample exposed to lower pressure, and then desorb from that surface (Matteuci et al. 2006). This transport is usually modeled using first Fick's law describing gradient-driven fluxes through a medium with a certain resistance. If the gradient-driven flux is a partial pressure difference, this resistance is represented by the coefficient of permeability or permeability of gas A (P_A) as follows:

$$J_A = P_A S \frac{P_2 - P_1}{e} \tag{10.1}$$

where J_A is the steady-state flux of gas A through the film, S is the film area, p_1 and p_2 are the gas partial pressures across the film, and e is the film thickness.

As regards Equation 10.1, for a given permeability value, gas permeates into (or out of) the package faster with increased film area, with thinner films, or with larger partial pressure differences. The gas permeability of a film typically depends on the material used (i.e., intrinsic parameters such as chemical nature of the monomer, percentage of polymerization, and crystallinity). With the current range of polymers available (synthetic or biosourced ones), a wide range in permeability can be realized.

If Fick's law is obeyed, then permeability can be expressed as follows:

$$P_A = D_A \times S_A \tag{10.2}$$

where D_A is the effective, concentration-averaged diffusion coefficient of gas A in the film. The solubility coefficient of gas A in the polymer, S_A, is given by Henry's law by

$$S_A = C/p \tag{10.3}$$

where C is the gas concentration in the polymer at the film surface in contact with gas pressure p.

The ability of a polymer to separate two gases A and B is characterized by the ideal selectivity, $\beta_{A/B}$, which is defined as the ratio of permeabilities:

$$\beta_{A/B} = \frac{P_A}{P_B} \tag{10.4}$$

Substituting Equation 10.2 into Equation 10.4 gives a relationship between the ideal selectivity, the diffusion selectivity, D_A/D_B, and the solubility selectivity, S_A/S_B:

$$\beta_{A/B} = \frac{D_A}{D_B} \times \frac{S_A}{S_B} \tag{10.5}$$

Most films are selective barriers with different permeability values for the different gases (e.g., O_2 permeability is usually lower than CO_2 permeability, and high values of selectivity could be achieved by using biodegradable materials based on proteins, for example) (MujicaPaz and Gontard 1997; Irissin-Mangata et al. 1999; Cuq et al. 1998). Diffusivity selectivity depends essentially on the relative penetrant sizes and the size-sieving ability of the polymer, whereas solubility selectivity is controlled primarily by the relative condensability of the penetrants and the relative affinity of the penetrants for the matrix (Matteucci et al. 2008). Consequently, solubility is more influenced by the composition of the film, whereas diffusivity (and thus, permeability) is more influenced by the structure of the material: multilayer, composite, or nanocomposite materials.

10.3.1.1 Multi-Layer Films

In order to combine the respective performance of several materials, multi-layers could be processed. In this case, the overall permeability (P_T) of gas A through the material is given by

$$\frac{e_T}{P_T} = \sum_{i=1}^{n} \frac{e_i}{P_i}$$

(10.6)

where e_T is the overall film thickness and e_i and P_i are, respectively, the thickness and permeability of the layer i, and n the number of layers in the material. The permeability of the highest barrier layer governs the overall permeability of the multi-layers.

10.3.1.2 Composite and Nanocomposite Films

In the case of composite or nanocomposite materials, many models have been derived to describe permeability as a function of composition or structural organization. For example, Bruggeman's model has been used to describe permeability in heterogeneous materials over a wide range of particle loading (Matteucci et al. 2008):

$$\frac{(P_C/P_M)-(P_D/P_M)}{1-(P_D/P_M)}\left(\frac{P_C}{P_M}\right)^{1/3} = 1-\phi_D$$

(10.7)

where P_C, P_M, and P_D are the permeabilities of the composite, polymer matrix, and disperse phase, respectively. φ_D is the dispersed phase volume fraction. In Equation 10.7, the dispersed phase is composed of spherical particles. In heterogeneous films containing an impermeable dispersed phase (i.e., $P_D = 0$), Equation 10.7 reduces to (Bouma et al. 1997)

$$\frac{P_C}{P_M} = (1-\phi_D)^{3/2}$$

(10.8)

If the dispersed phase is much more permeable than the matrix (i.e., $P_D > P_M$), Equation 10.7 becomes (Bouma et al. 1997)

$$\frac{P_C}{P_M} = \frac{1}{(1-\phi_D)^3}$$

(10.9)

Barrer et al. (1963) found that CH_4 permeability of natural rubber filled with impermeable, micron-sized ZnO particles obeys Equation 10.8 to a good approximation, and Matteucci et al. (2008) found that a good estimation of CO_2, CH_4, N_2, and H_2 permeability in TiO_2 = filled 1,2-polybutadiene could be obtained from the permeability of each gas in the unfilled polymer by using Bruggeman's model. However, Bruggeman's model as presented in Equation 10.7 assumes that the dispersed phase is spherical. For nonspherical geometry, Equation 10.7 is less accurate. That is why alternative models for predicting the permeability or diffusivity of a composite or nanocomposite have been developed for various geometry and pellets orientation. For example, Cussler et al. (1988) gave the following expression for the effective diffusivity for a suspension of impermeable flakes oriented perpendicularly to the diffusion flux:

$$\frac{D_C}{D_0} = \left(1 + \frac{\alpha^2 \phi_D^2}{1 - \phi_D}\right)^{-1} \tag{10.10}$$

where ϕ_D is the dispersed phase volume fraction, α is the particles aspect ratio, and D_C and D_0 are the diffusion coefficients with and without pellets, respectively.

To take into account the resistance to diffusion in the slits between adjacent pellets in the same horizontal plane, Aris (1986) proposed the following model:

$$\frac{D_C}{D_0} = \left(1 + \frac{\alpha^2 \phi_D^2}{1 - \phi_D} + \frac{\alpha \phi_D}{\sigma} + \frac{4\alpha \phi_D}{\pi(1 - \phi_D)} \ln\left(\frac{\pi \alpha^2 \phi}{\sigma(1 - \phi)}\right)\right)^{-1} \tag{10.11}$$

where σ is the slit shape: the ratio between slit width (ε) and slit thickness (t) (Figure 10.3). However, Equation 10.10 and Equation 10.11 require that the impermeable domains be aligned either parallel or perpendicular to the polymer surface and no interface regions be present between the polymer bulk and pellets. Diffusion through heterogeneous media containing impermeable anisotropic domains is strongly influenced by domain orientation relative to the diffusion direction. To take into account this orientation, Sorrentino et al. (2006) proposed a modified form of Equation 10.11 to predict the effective diffusivity through polymer clay nanocomposites as a function of clay sheets orientation, volume fraction, polymer clay orientation, and aspect ratio:

$$\frac{D_C}{D_0} = \frac{(1 + \beta \phi_D)}{\left[(1 - \phi_D) + \phi_D \left(\frac{L + 2t}{L \sin\theta + 2t \cos\theta}\right)^2\right]} \tag{10.12}$$

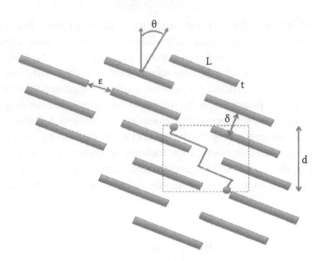

FIGURE 10.3 Idealized model for hindered factor calculation for plane pellets exfoliated in a polymer matrix. (Adapted from Sorrentino, M., Tortora, M., and V. Vittoria. 2006. Diffusion behavior in polymer-clay nanocomposites. *Journal of Polymer Science Part B: Polymer Physics* 44: 265–274.)

with

$$\beta = \left(\frac{V_s}{V_f} \frac{D_s}{D_0} - \frac{V_s + V_f}{V_f} \right) \qquad (10.13)$$

where V_S and D_S are the volume and diffusion coefficient of the interface region, and V_P is the volume of the unmixed polymer. L is the length of a pellet, and θ is an angle characterizing the pellet orientation (Figure 10.3). The parameters V_S and D_S are no longer as easily calculated. They depend on many factors as the type of permeant, the type of the surface modifier utilized, the presence of micro-voids, or the morphology of the sample. However, if t and θ are known, the parameter β can be used as a fitting parameter. If the presence of the interface region can be neglected, $\beta = -1$. Sorrentino et al. (2006) successfully validated Equation 10.12 on polycaprolactone, polyurethanes, and polypropylene nanocomposites in which modified montmorillonite clays were introduced.

10.3.1.3 Microperforated Films

The diffusion of O_2 and CO_2 through air is 8.5 and 1.5 million times greater, respectively, than through low-density polyethylene films (Mannapperuma et al. 1989). This difference in gas diffusion means that the gas exchange through a microperforated material occurs almost entirely through the microperforations. In this case, from a theoretical point of view, Equation 10.1 does not apply. Various models have been thus proposed which attempt to describe the exchange of gases through perforation. Some authors have used the flow equation according to the Stephan–Maxwell law. However, if the Stefan–Maxwell law is more accurate, it is less versatile due to its high complexity; therefore, some authors prefer to apply Fick's law or derivations of Fick's law, or even empirical models (Table 10.2). Gonzalez et al. (2008) have measured O_2 and CO_2 transmission rates through microperforated films and have compared the relative performance of various models (Becker 1979; Fishman et al. 1996; Fonseca et al. 1996; Heiss et al. 1954) and their own empirical one to predict the gas transmission rates. They found that their empirical equation that relates the area of microperforation with the transmission rate of oxygen and carbon dioxide (Table 10.2) can be used for a wide range of conditions which distinguishes it from the other models and equations applicable with a more limited range.

The gas permeability of a film depends on extrinsic parameters such as temperature and relative humidity.

10.3.1.4 Temperature Effect

The effect of temperature on gas permeability (diffusivity or solubility) is usually represented using the Arrhenius' relation:

$$P = P_0 \exp\left(-\frac{E_a}{RT}\right) \qquad (10.14)$$

where P_0 is a temperature-dependent factor, E_a the energy of activation for permeation, R the gas constant, and T the absolute temperature.

Another accepted relationship is the Q_{10}, the ratio of permeabilities at temperature T to that at $T + 10$:

$$Q_{10} = \frac{P(T+10)}{P(T)} \qquad (10.15)$$

In MAP science and technology, Equation 10.15 is often preferred to Equation 10.14 to describe the effect of temperature on gas and vapor permeability values. The effect of temperature on respiration rate is always described using the Q_{10} ratio (Equation 10.29). However, knowing the activation energy for permeation of a gas A in a polymer, the corresponding Q_{10} could be found using the following relation:

$$E_a = R\frac{T_1 T_2}{10}\ln Q_{10} \qquad (10.16)$$

When perforated films are considered, the effect of temperature on permeation through pores (air) is much less as compared to its effect on permeation through the polymer. As permeation through the pores accounts for most of the total permeation through a perforated film, the activation energies for perforated films are close to zero. Perforated films are not able to compensate for the effect of temperature variations during storage on fruit and vegetable respiration.

TABLE 10.2

Models for Predicting Gas Exchanges through Perforations

Reference	Model	Equation	Thickness (L) and Diameter (d) of the Perforations	Number of Perforations (n)
Becker (1979)	Fick's law	Flow of gas A through the holes from compartment 1 to 2: $$J_{hA} = -\frac{D_A S_h (C_{A,1} - C_{A,2})}{L}$$ where S_h is the perforation surface; L and S the film thickness and surface, respectively; and D_A the diffusivity of gas A. Gas A transmission rate: $TR_A = \dfrac{D_A S}{L}$	$L = 270$ μm, 40 μm $\leq d \leq 600$ μm	$N = 1$
Edmond et al. (1991)	Fick's law, empirical TR	$TR_A = a_1 + a_2 d^2 + a_3 L + a_4 T + a_5 d^4 + a_6 L^2 + a_7 T^2 + a_8 d^2 L + a_{10} LT$ T is expressed in K, L, and d in m; and a_i are constants	$L = 1.59, 7.14,$ and 12.7 mm $d = 6, 8.85,$ and 11 mm	
Fonseca et al. (1996)		$TR_{O_2} = (9.12 \pm 3.53) \times 10^{-6} d^{1.47 \pm 0.08} L^{-0.55 \pm 0.03}$ $TR_{CO_2} = (0.82 \pm 0.04) TR_{O_2}$ d and L expressed in m	5 mm $\leq L \leq 30$ mm, 8 mm $\leq d \leq 17$ mm	
Heiss (1954), Lange et al. (2000)	Modification of Fick's law	$J_{hA} = -\dfrac{D_A S (C_{A,1} - C_{A,2})}{L + \varepsilon}$ $\dfrac{5}{12} < \varepsilon < \dfrac{10}{12} d$ depending on air speed outside the leak	$L \geq 10$ μm, $d \geq 100$ μm 40 μm $\leq L \leq 600$ μm, 200 μm $\leq d \leq 1{,}000$ μm	
Fishman et al. (1996), Chung et al. (2003)		$TR_A = \dfrac{D_A S}{L + \varepsilon}$ $\varepsilon = 0.5 d$	$L = 20$ μm, $d = 2$ mm Theoretical analysis	$N = 4$
Gonzalez et al. (2008)		$TR_A = a_1 S_h^{a_2}$ with S_h the perforation surface	All L values up to $L > 1.6$ mm All d values up to $d < 6$ mm	

10.3.1.5   Relative Humidity Effect

Contrary to the impact of temperature which could be well modeled using the Arrhenius or Q_{10} relation, the effect of relative humidity on gas permeability is still subject to study. No generalized relationship could have been found related to variations of gas or vapor permeability with relative humidity. Some attempts have been made, for example, to relate C_2H_4 permeability in wheat gluten films to the moisture content of the matrix and then the relative humidity of the external atmosphere (MujicaPaz et al. 2005), but the polynomial relationship proposed by the authors is of value only for wheat gluten films and in the conditions of the study. The general statement is that an increase in relative humidity increases the gas or vapor permeability value. At high humidities, water can be absorbed and interact with polar groups in the polymeric films, leading to a swelling and plasticizing effect (increase in free volume and in macromolecules chain mobility). As a consequence, the permeability for other gases and vapors changes as well. There is a real need for a theoretical approach describing the impact of RH on parameters such as diffusion or permeability.

10.3.2   Absorption or Release of Gas or Vapor

10.3.2.1   Absorption of Gas and Vapor

As previously mentioned, it could be necessary to introduce active agents for absorbing unsuitable gases or vapors and reducing the transient period. This has to be carefully controlled to avoid detrimental effects on the product, for instance, a total depletion of O_2 would induce a turn into fermentative catabolism. In this condition, absorption kinetics must be known. A typical saturation experimental curve was obtained with several O_2 scavenger sachets whatever the initial O_2 content, from 5 to 21 kPa (Tewari et al. 2002; Charles et al. 2006). Although O_2 absorption rate did not depend on the initial O_2 concentration at high partial pressures, it was slightly affected at low partial pressures (less than 5 kPa). This effect was attributed to a diffusion phenomenon that limited absorption at a low O_2 level. A significant parasite CO_2 absorption kinetic was identified in ATCO®LH sachet (see Table 10.1) and cannot be neglected for MAP application (Charles et al. 2006). First-order (Equation 10.17) and second-order (Equation 10.18) relations were proposed to calculate the absorption rate:

$$\ln A_t = kt + \ln A_0 \tag{10.17}$$

$$\frac{1}{A_t} = kt + \frac{1}{A_0} \tag{10.18}$$

where A is the amount of reactant at any time A_t or at intial time A_0, and k is the rate constant. The first-order equation accurately explained the data and was chosen to express the absorbed O_2 at any time:

$$N_{O_2t} = N_{O_20} \times (1 - \exp^{-kt})$$ (10.19)

where N_{O_20} is the maximal absorption capacity.

An Arrhenius-type equation was validated to model the effect of temperature on the absorption rate constant k.

$$k = X \times \exp^{\left(\frac{Ea}{RT}\right)}$$ (10.20)

where X is a temperature-dependent factor, E_a the energy of activation for permeation, R the gas constant, and T the absolute temperature.

The first-order equation and the Arrhenius law were also appropriate to describe the parasite CO_2 absorption by the O_2 scavenger and by CO_2 scavengers (Charles et al. 2006). For the latter, no parasite O_2 absorption was found in controlled atmosphere experiments.

It should be stressed that experimental values of absorption rates had high standard deviations for the same type of scavenger, and the knowledge of this variation should be taken into account in modeling for better food control and prevention in food safety. Another important point is that most scavengers are moisture activated, and at this time, no model relates the O_2 absorption rate to the surrounding RH.

10.3.2.2   Release of Active Agents

If in the MAP system the packaging material is an active film, in which, for example, antimicrobial compounds are entrapped in the polymer matrix and, then, are slowly released as a function of time, an additionnal mass transfer to the O_2/CO_2 and H_2O/C_2H_4 fluxes occurs (Figure 10.4). In this case, the concentration in migrant varies

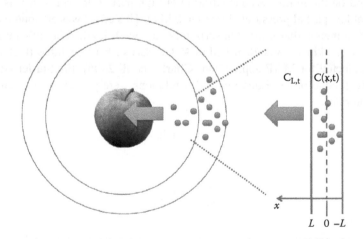

FIGURE 10.4 Scheme of the package of thickness 2 L in the food/internal atmosphere/ active packaging system: controlled release of active agents.

with time and with the position within the matrix (transient state), and obeys the second Fick's law as follows:

$$t > 0, \ -L < 0 < L, \ \frac{\partial C}{\partial x} = D\frac{\partial^2 C}{\partial x^2} \tag{10.21}$$

where x is the distance (m), C the packaging concentration in migrant (kg additive/kg of polymer), and D the migrant diffusivity within the packaging (m²/s). To solve this partial differential equation, two boundary conditions and one initial condition are required.

At the packaging–internal atmosphere interface, the equilibrium is assumed to be reached instantaneously. Moreover, the rate at which the substance is transferred into the atmosphere is constantly equal to the rate at which this substance is brought to the surface by internal diffusion through the polymer packaging, leading to the following relationship:

$$t > 0, \ x = L, \ -D\frac{\partial C}{\partial x} = k\left(C_{L,t} - C_{L,\infty}\right) \tag{10.22}$$

where k is the coefficient of mass transfer at the interface packaging/atmosphere (m/s), $C_{L,t}$ is the concentration of the diffusing substance on the surface of the polymer, and $C_{L,\infty}$ is the concentration of the diffusing substance on the surface of the polymer required to maintain equilibrium with the concentration of this substance in the internal atmosphere at time t.

Chalier et al. (2008) succeeded in predicting the release of a volatile compound (carvacrol) from a soya proteins matrix into atmosphere using Equation 10.15 and appropriate boundary conditions. These authors have put in evidence the triggering effect of relative humidity and temperature on the carvacrol release from the matrix. This particular behavior of soya proteins (related to the glass transition phenomena of the soya protein network) would be very interesting throughout the life cycle of the material. As shown in Figure 10.5, one may imagine, for example, a storage of the film prior to use at a low relative humidity (60%, e.g., below the glass transition of the polymer) to favor the retention of the antimicrobial compound. At this relative humidity, the remaining quantity of antimicrobial compound in the film would be higher than 50% of the initial quantity after 1 month of storage. Then, once in contact with the food product (relative humidity close to 100%), the diffusion of the volatile compound into the internal atmosphere would suddenly increase to maintain a high concentration in preservative at the food surface where microorganisms grow. In case of temperature abuse during storage of the packed food, the response of the active film would be once more in favor of the preservation of the fresh product: The quantity of antimicrobial compound release in the internal headspace would increase exponentially with the temperature increase (Figure 10.3).

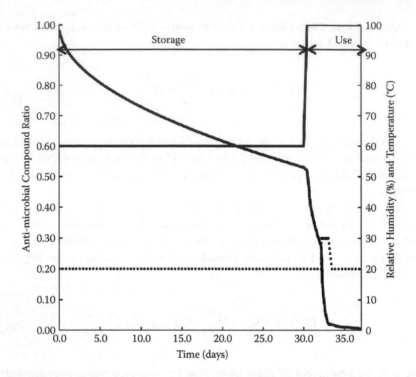

FIGURE 10.5 Predicted controlled release of an antimicrobial compound from a bio-sourced polymer during the life cycle of the polymer (stored at 60% of relative humidity during 1 month before being used during 5 days as packaging of a fresh food product—100% RH): effect of a temperature abuse of 1 day at 35°C during storage of the packed product at 20°C. (Calculated from data in Chalier, P., Ben-Arfa, A., Guillard, V., and Gontard, N. 2008. Moisture and temperature triggered release of a volatile active agent from soy protein coated paper: effect of glass transition phenomena on carvacrol diffusion coefficient. *Journal of Agricultural and Food Chemistry* 57:658–665.)

Some authors use a simplified derivation of the previous Fickian model:

$$\frac{M_t}{M_0} = kt^{\frac{1}{2}}$$

(10.23)

where

$$\frac{M_t}{M_0}$$

is the fraction of migrant released, t is the release time, and k is a constant characteristic of each sample. The value of the diffusion coefficient, D, can be calculated according to the following relation:

$$k = 4\sqrt{\frac{D}{\pi L^2}} \qquad (10.24)$$

where L corresponds to the thickness of the film. This model is valid for release of less than 60% of initial load. Equation 10.23 was successfully used by Lin et al. (2007) to model the controlled release of aspirin from poly(methyl) methacrylate (PMMA)/silica composite material. In a general manner, this model is often used for the study of a drug delivery system when the active system is immersed in a liquid (Reinhard et al. 1991; Cypes et al. 2003; Frank et al. 2005).

10.4 MODELING PHYSIOLOGICAL REACTIONS

The complexity of biological systems, such as fruits and vegetables, contributes to the difficulty in modeling the biochemical reality of respiration and transpiration. However, empirical and simplified kinetics models can be used to describe overall physiological pathways.

10.4.1 MODELING RESPIRATION

In its simplest form, aerobic respiration can be considered as a single limiting enzymatic reaction in which the substrate is O_2. Then, the dependence of the O_2 respiratory or consumption rate $\left(R_{O_2}\right)$ on O_2 partial pressure $\left(p_{O_2}\right)$ can be expressed by a Michaelis-Menten-type equation as follows:

$$R_{O_2} = \frac{R_{max O_2} \times p_{O_2}}{Km_{app O_2} + p_{O_2}} \qquad (10.25)$$

in which, $R_{max O_2}$ is the maximum rate of O_2 consumption, and $Km_{app O_2}$ is the Michaelis constant for O_2 consumption (apparent dissociation for the enzyme–substrate complex). This simplification was first applied for respiration at the cell level in 1973 (Chevillote 1973) and tends to correctly fit experimental data. Nevertheless, Equation 10.25 should be completed when taking into account the role of carbon dioxide through inhibition mechanisms. Four models can explain the CO_2 inhibition of O_2 consumption. Competitive inhibition, in which O_2 and CO_2 compete for the same active site of the single enzyme, can be used when the O_2 consumption rate is lowered at high CO_2 concentrations (see Equation 10.26). Uncompetitive refers to a reaction between the inhibitor and only the enzyme–substrate complex; in this case, the O_2 respiratory rate is not influenced at high CO_2 contents (see Equation 10.27). In other cases, noncompetitive inhibition might be used when CO_2 can react equally with both the enzyme and the enzyme–substrate complex and the effect of high CO_2 concentration on the O_2 uptake rate comes in between the concentrations obtained with the two previous types of inhibition (see Equation 10.28). If this inhibition is widely used for its simplicity, it could be necessary to consider the reality of the respiration process. Because many enzymes reactions are involved, each can potentially be inhibited by CO_2 according to one of the previous inhibition models. Then, it could be necessary to use a more complex equation that combines competitive and uncompetitive inhibition, where each type differs in its relative activity (see Equation 10.29). Models of these inhibitions have been previously described in well-documented reviews (Peppelenbos and van't Leven 1996; Fonseca et al. 2002) and need the introduction of p_{CO_2}, the CO_2 partial pressure; and Ki_{CO_2}, $K'i_{CO_2}$, $K''i_{CO_2}$, the Michaelis constant for the CO_2 inhibition of O_2 consumption (apparent dissociation of the inhibitor–enzyme complex, Ki, or the inhibitor–enzyme–substrate complex, K'i, or both complexes, K''i, depending on the type of inhibition):

$$R_{O_2} = \frac{R_{\max O_2} \times P_{O_2}}{Km_{appO_2} \times \left(1 + \dfrac{P_{CO_2}}{Ki_{CO_2}}\right) + P_{O_2}} \tag{10.26}$$

$$R_{O_2} = \frac{R_{\max O_2} \times P_{O_2}}{Km_{appO_2} + P_{O_2} \times \left(1 + \dfrac{P_{CO_2}}{K'i_{CO_2}}\right)} \tag{10.27}$$

$$R_{O_2} = \frac{R_{\max O_2} \times P_{O_2}}{\left(Km_{appO_2} + P_{O_2}\right) \times \left(1 + \dfrac{P_{CO_2}}{K''i_{CO_2}}\right)} \tag{10.28}$$

$$R_{O_2} = \frac{R_{\max O_2} \times P_{O_2}}{Km_{appO_2} \times \left(1 + \dfrac{P_{CO_2}}{Ki_{CO_2}}\right) + P_{O_2} \times \left(1 + \dfrac{P_{CO_2}}{K'i_{CO_2}}\right)} \tag{10.29}$$

Although inhibition models are accurate enough for experimental data, they all consider that gas exchanges are constant in time. They do not describe the respiratory crisis occurring in postharvested climacteric organs when ripened, as well as the temporary increase in gas exchange rate noted in minimally processed fruits and vegetables. It appears necessary to enhance understanding and interpretation of gas exchanges by taking into account gas diffusion within the product. At this time, models describe the effect of external gas partial pressure on physiological reactions; but from a biological point of view, only internal gas partial pressure is involved in physiological processes. As diffusion properties of materials can affect gas composition in the headspace of the packaging, diffusion through wax layers, cracks, pores, or the cell membrane can alter the internal gas/vapor atmosphere within the fresh product. Mathematical models of biological diffusion should be the same as for packaging diffusion and should be based on Fick's law (Equation 10.1). This could be helpful to describe the effect of wounding on physiological reactions in fresh-cut products and would also be applied to transpiration phenomenon.

10.4.1.1 CO_2 and Fermentative Respiration

In the oxidative metabolism, the CO_2 respiratory or production rate might also be modeled by taking into account the respiratory quotient (RQ), which is the ratio between CO_2 production and O_2 consumption:

$$R_{CO_2} = RQ \times R_{O_2} \tag{10.30}$$

Whereas respiration rates of different plant tissues vary, RQ value remains close to unity, from 0.7 to 1.3 depending on the substrate catabolized (carbohydrates, lipids, proteins, or organic acids) and then on the aging of the product. However, at low O_2 partial pressure, the RQ might reach higher values. Considering Equation 10.25 and Equation 10.30 at low O_2 partial pressure, few or no CO_2 would be produced. However, in most plant tissues, such conditions lead to a fermentative respiratory pathway in which additional CO_2 is formed. To overcome this drawback, several authors tentatively develop some models based on the CO_2 balance accounting for the oxidative (ox) and fermentative (ferm) respiratory pathway:

$$R_{CO_2} = R_{CO_2(ox)} + R_{CO_2(ferm)} \text{ with } R_{CO_2(ox)} = RQ_{(ox)} \times R_{O_2} \tag{10.31}$$

Models are compared in a previous review (Peppelenbos et al. 1996). Some are based on the O_2 inhibition of CO_2 production in fermentation, and others rely on the inhibition of fermentative CO_2 production by the ATP concentration, considering that carbohydrates are nonlimiting. At this time, there is no physiological relevance of the models for attributing the increase in fermentation rates to low O_2 content or to energy fluxes. In addition, none takes into account the effect of CO_2 partial pressure on the fermentative CO_2 production, which has been noted for some products such as mung bean sprouts, asparagus, or apples. There is still a need to develop accurate models for this respiratory pathway.

10.4.1.2 Temperature Dependence of Respiration

The effect of temperature on O_2 or CO_2 respiration rate might be independently described with the Arrhenius law or the Q_{10} relation. According to the Arrhenius equation, the O_2 respiration rate (same with CO_2) is expressed as follows:

$$R_{O_2} = R_{O_2 0} \exp\left(-\frac{E_a}{RT}\right) \tag{10.32}$$

where $R_{O_2 0}$ is a temperature-dependent factor, E_a is the energy of activation for O_2 consumption, R is the gas constant, and T is the absolute temperature. In such a nonactivated process, the activation energy loses its physical meaning and quantifies the temperature dependence.

Considering the Q_{10} relation, Equation 10.33 is used:

$$Q_{10} = \left(\frac{R_{O_2 1}}{R_{O_2 2}}\right)^{\frac{10}{(T_2 - T_1)}} \tag{10.33}$$

It should be stressed that the Q_{10} value is widely used by physiologists but more and more, it is replaced by E_a to simplify the global equation in coupling models where diffusion through gas material is also considered.

An example of modeled change in O_2 consumption and CO_2 production rate versus time at several temperatures of storage is given in Figure 10.6. Initial gas composition was chosen as air, the headspace volume was known, and the system was considered as closed. At low temperature (less than 8°C), O_2 uptake and CO_2 release remain low and constant. At higher temperature, initial O_2 and CO_2 respiratory rate are higher, meaning that O_2 depletion is accelerated. It leads to a rapid anoxia and a reduction

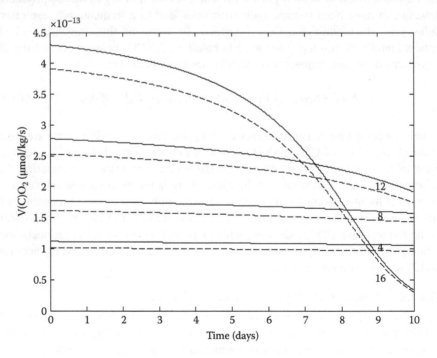

FIGURE 10.6 Example of predicted O_2 consumption (VO2 in μmol/kg/s; solid line) and CO_2 production (VCO2 in μmol/kg/s; dotted line) of 'Elsanta' strawberries packed in a tight-closed finite volume of air as a function of time and different storage temperature (from 4 to 16°C). (Calculated from data in Hertog, M. L. A. T. M., Boerrigter, H. A. M., Van den Boogaard, G. J. P. M., Tijskens, L. M. M., and Van Schaik, A. C. R. 1999. Predicting keeping quality of strawberries (cv. 'Elsanta') packed under modified atmospheres: an integrated model approach. *Postharvest Biology and Technology* 15: 1–12.)

of O_2 consumption, typical of a turn into a fermentative pathway. Because the mathematical model used does not take into account additional CO_2 production occurring in such a case, the reduction of O_2 consumption is concomitant to a reduction of CO_2 production.

10.4.2 MODELING TRANSPIRATION

Most models describe the moisture transfer through the skin of fresh products as a function of biophysical and thermophysical properties (thermal diffusivity, surface cellular structure, skin thickness, pores, geometry) (Fockens and Meffert 1972; Hayakawa and Succar 1982; Sastry and Buffington 1982; Gaffney et al. 1985; Veraverbeke et al. 2003). None of them took into consideration the influence of the respiration rate that depends on gas composition in the packaging headspace. Some authors tentatively developed a mathematical model for describing transpiration in link with respiration (Kang and Dong 1998; Song et al. 2002), which is based on heat and mass transfers. The hypothesis for such a model is that a large amount of energy produced during respiration is dissipated as heat (from 80% to 100%) and that carbohydrates are nonlimiting substrates. Then, the internal heat source can be expressed from the oxidation of glucose:

$$C_6H_{12}O_6 + 6O_2 \rightarrow 6CO_2 + 6H_2O + 2816kJ \tag{10.34}$$

The internal heat Q_i can be expressed by taking into account the weight of fresh commodity w and the heat produced during respiration according to Equation 10.35:

$$Q_i = w \times \left(\frac{2816}{6}\right) \times \left(\frac{R_{O_2} + R_{CO_2}}{2}\right) \times a \tag{10.35}$$

where α is the ratio of energy dissipated as heat (from 0.8 to 1). This relies on the assumption that the respiration rate is the average of O_2 and CO_2 respiratory rates.

The convective heat occurring at the surface of product (or external heat) Q_e is described as follows:

$$Q_e = h \times A \times (T_e - T) \tag{10.36}$$

where h is the convective heat transfer coefficient of the surface of the product, A is the surface area of the product, T is the Celsius temperature of the product surface or T_e of the surrounding.

By taking into account the latent heat of moisture vaporization and the sensible heat for increasing the temperature of the product, the heat balance can be written as follows:

$$Q_i = Q_e = (m \times \lambda) + \left(w \times C_s \times \frac{dT}{dt}\right) \tag{10.37}$$

where m is the rate of vaporization, λ is the latent heat of vaporization, and C_s is the specific heat of the product.

This equation can be simplified by considering that the temperature at the surface of the product is equal to that of the surroundings. Q_e is therefore neglected, and the rate of vaporization corresponds to

$$m = \frac{Q_i - \left(w \times C_s \times \dfrac{dT}{dt} \right)}{\lambda}$$ (10.38)

This model has been validated on apple at a fixed temperature and RH (Kang and Lee 1998) as well as on blueberry when coupling with water vapor permeation through packaging material (Song et al. 2002).

10.4.3 How to Assess Physiological Parameters

Accurate assessments of respiration rate and other physiological parameters are essential for optimizing physiological models. Respiration rate can be assessed at a required temperature according to usual methods that are the closed system, the flowing system, the permeable system, and the automatic respirometer. Whatever the technique used, it is necessary to know the weight and the apparent density of the product as well as the volume of the container to calculate the void volume. In the closed jar technique, fresh products are placed in a hermetic container. The initial gas composition inside jars can be set up by gas flushing or considered as air. Gas samples are withdrawn through a septum, and O_2 and CO_2 are monitored using gas chromatography at regular time intervals for a short period. It is fast and simple but does not allow measurements under stable atmospheric conditions, and modification of partial pressures quickly influences the respiration rate.

This system can be adapted to ensure permanent renewal of the atmosphere in the vessel at a constant rate: This is the flowing system. Comparison of gas composition between the inlet and the outlet permit assessment of the O_2 uptake and CO_2 production by the product. This technique is time consuming, requires considerable amounts of gases, and is not adapted to low respiring products.

In the permeable system, products are placed into a flexible pouch with known permeability and diffusion area. O_2 and CO_2 respiratory rates are then proportional to the steady-state concentrations within the packaging. This method is the least accurate because it involves the determination of more variables (packaging characterization) and time to achieve steady state might be a limitation (high respiring products, for instance).

The automatic respirometer consists of two vessels placed in a thermostated bath. Fresh products are introduced into vessel 1; the other vessel is used as a pressure reference. Vessels are closed under normal air or flushed with a preset gas composition through mass flow meters. Internal gas composition is continuously measured through probes and gas chromatograph. When CO_2 partial pressure is 0.1 kPa higher than the preset value, a valve is automatically activated to pump the gas contained in vessel 1 through a CO_2 trap until the initial preset CO_2 partial pressure is restored.

Removal of excessive CO_2 leads to a decrease in pressure in vessel 1 which is equilibrated by the injection of pure nitrogen. The run is stopped once the O_2 is completely consumed. A complete description of this system is given in Varoquaux et al. (1999). This technique relies on a whole arsenal of machinery compared to the closed jar technique, but it allows the measurements of accurate data without CO_2 inhibition.

Once these measurements are recorded, and whatever the technique, O_2 or CO_2 partial pressure are plotted against time, and a regression curve is calculated to fill through the data points (commonly a second-degree polynomial equation). The first derivative is applied to get the estimation of R_{O_2} and R_{CO_2} at any time and, therefore, the RQ. Then,

$$\frac{1}{R_{O_2}}$$

is expressed as a function of

$$\left(\frac{1}{P_{O_2}}\right) \text{ or } \left(\frac{1}{P_{CO_2}}\right)$$

according to the linear plot, and the linear part is used to estimate $R_{\max O_2}$ and $Km_{app O_2}$ or Ki_{CO_2}, $K'i_{CO_2}$, $K''i_{CO_2}$ (depending on the inhibition mechanism), respectively.

To assess the Ea or the Q_{10} of O_2 uptake and CO_2 production, R_{O_2} and R_{CO_2} are measured at several temperatures. Then $\ln R_{O_2}$ is plotted as a function of $(1/T)$ according to Equation 10.35 to estimate Ea. Otherwise, Equation 10.36 is used to get the Q_{10} value:

$$\ln R_{O_2} = \ln R_{O_2 0} - \frac{Ea}{RT} \tag{10.39}$$

$$\log R_{O_2} = AT_{oc} + \log R_{O_2 0} \quad Q_{10} = 10^{10A} \tag{10.40}$$

Respect of SI (International System of Units) units is really important when constituting such a database for modeling. For instance, O_2 or CO_2 respiratory rates should be expressed in $mol.kg^{-1}.s^{-1}$, but at this time, other units such as $mg.kg^{-1}.h^{-1}$ are encountered. The same applies to gas film permeability that should be expressed in $mol.m^{-1}.Pa^{-1}.s^{-1}$ and is found as, for instance, $mL.\mu m.m^{-2}.atm^{-1}.day^{-1}$. There is still a need to standardize units for easier use in databases.

10.5 CONCLUSION

This review clearly shows the interest in and also the complexity of active and passive MAP of fresh fruits and vegetables. It is demonstrated that understanding mass transfer phenomena is essential to promote rational and specific design of packaging materials for fresh fruits and vegetables for the sake of all stakeholders. Such an approach will allow the enhancement of the quality preservation of such fragile but nutritionally recommended products for consumer health. This will also enable the development of tailor-made packaging with a reduced quantity of raw material used and promote the development of renewable and biodegradable packaging solutions for environmental considerations. Future trends in scientific work should be to develop integrated models able to combine mass transfer models with both physiological and quality ones. This could be done by first using already existing mathematical models for predicting some fresh fruits and vegetables qualities (sensorial evaluation, browning, spoilage, etc.) and by developing more suitable ones with required associated databases. In addition, the behavior of ecofriendly packaging, such as agropolymer-based materials, could be really interesting when it is fully understood and thus controlled, especially the moisture sensitivity of the packaging.

REFERENCES

Almenar, E., Del-Valle, V., Hernandez-Munoz, P., Lagaron, M. J., Catala, R., and R. Gavara. 2007. Equilibrium modified atmosphere packaging of wild strawberries. *Journal of the Science of Food and Agriculture* 87: 1931–1939.

Appendini, P., and J. H. Hotchkiss. 2002. Review of antimicrobial food packaging. *Innovative Food Science and Emerging Technologies* 3: 113–126.

Aris, R. 1986. *Archive for Rational Mechanics and Analysis* 95: 83–91.

Ayala-Zavala, J. F., Oms-Oliu, G., Odriozola-Serrano, I., Gonzales-Aguilar, G. A., Alvarez-Parrilla, E., and O. Martin-Belloso. 2008. Bio-preservation of fresh-cut tomatoes using natural antimicrobials. *European Food Research and Technology* 226: 1047–1055.

Barrer, R. M., Barrier, J. A., and M. G. Rogers. 1963. Heterogeneous membranes: diffusion in filled rubber. *Journal of Polymer Science Part A: Polymer Chemistry* 1: 2565–2586.

Barron, C., Varoquaux, P., Guilbert, S., Gontard, N., and B. Gouble. 2002. Modified atmosphere packaging of cultivated mushroom (*Agaricus biscporus* L.) with hydrophilic films. *Journal of Food Science* 671: 251–255.

Becker, K. 1979. Water vapour permeability of open pores and other holes in packages. *Verpackungs Rundschau* 30: 87–90.

Bouma, R. H. B., Checchetti, A., Chidichimo, G., and E. Drioli. 1997. Permeation through a heterogeneous membrane: the effect of the dispersed phase. *Journal of Membrane Science* 128: 141–149.

Brecht, J. K., Chau, K. V., Foncesca, S. C., and F. A. R. Oliveira. 2003. CA transport of fresh produce in MAP: designing systems for optimal atmosphere conditions throughout the postharvest handling chain. *Acta Horticulturae* 600: 799–801.

Brody, A. L., Strupinsky, E. R., and L. R. Kline. 2001. *Active packaging for food applications*. Lancaster, PA: Technomic.

Burnett, A.B., and L. R. Beuchat. 2001. Comparison of sample preparation methods for recovering *Salmonella* from raw fruits, vegetables and herbs. *Journal of Food Protection* 64: 1459–1465.

Chalier, P., Ben-Arfa, A., Guillard, V., and N. Gontard. 2008. Moisture and temperature triggered release of a volatile active agent from soy protein coated paper: effect of glass transition phenomena on carvacrol diffusion coefficient. *Journal of Agricultural and Food Chemistry* 57:658–665.

Charles, F., Guillaume, C., and N. Gontard. 2008. Effect of passive and active modified atmosphere packaging on quality changes of fresh endives. *Postharvest Biology and Technology* 48: 22–29.

Charles, F., Sanchez, J., and N. Gontard. 2003. Active modified atmosphere packaging of fresh fruits and vegetables: modeling with tomatoes and oxygen absorber. *Journal of Food Science* 68: 1736–1742.

Charles, F., Sanchez, J., and N. Gontard. 2005. Modeling of active modified atmosphere packaging of endives exposed to several postharvest temperatures. *Journal of Food Science* 70: 443–449.

Charles, F., Sanchez, J., and N. Gontard. 2006. Absorption kinetics of oxygen and carbon dioxyde scavengers as part of active modified atmosphere packaging. *Journal of Food Engineering* 72: 1–7.

Chevillote, P. 1973. Relation between reaction cytochrome oxidase-oxygen and oxygen uptake in cells *in vivo*. The role of diffusion. *Journal of Theoretical Biology* 39: 277–295.

Chung, D. W., Papadakis, S. E., and K. L. Yam. 2003. Simple models for evaluating effects of small leaks on the gas barrier properties of food packages. *Packaging Technology and Science* 16:77–86.

Correa, S. F., Filho, M. B., Da Silva, M. G., Oliveira, J. G., Aroucha, E. M. M., Silva, R. F., Pereira, M. G., and H. Vargas. 2005. Effect of the potassium permanganate during papaya fruit ripening: ethylene production. *Journal de Physique IV* 125: 869–871.

Cuq, B., Gontard, N., and S. Guilbert. 1998. Proteins as agricultural polymers for packaging production. *Cereal Chemistry* 75: 1–9.

Cussler, E. L., Hughes, S. E., Ward, W. J., and R. Aris. 1988. Barrier membranes. *Journal of Membrane Science* 38: 161–174.

Cypes, S. H., Saltzmann, W. M., and E. P. Giannelis. 2003. Organosilicate-polymer drug delivery systems: controlled release and enhanced mechanical properties. *Journal of Controlled Release* 90: 163–169.

Edmond, J. P., Castainge, F., Toupin, C. J., and D. Desilets. 1991. Mathematical modelling of gas exchange in modified atmosphere packaging. *Transactions of the American Society of Agricultural Engineers* 34: 239–245.

Exama, A., Arul, J., Lencki, R. W., Lee, L. Z., and C. Toupin. 1993. Suitability of plastic films for modified atmosphere packaging of fruits and vegetables. *Journal of Food Science* 58: 1365–1370.

Fishman, S., Rodov, V., and S. Ben-Yehoshua. 1996. Mathematical model for perforation effect on oxygen and water vapour dynamics in modified atmosphere packages. *Journal of Food Science* 61: 956–961.

Fishman, S., Rodov, V., Peretz, J., and S. Ben-Yehoshua. 1995. Model for gas exchange dynamics in modified atmosphere packages of fruits and vegetables. *Journal of Food Science* 60: 1078–1083.

Fockens, F. H., and H. F. T. Meffert. 1972. Biophysical properties of horticultural products as related to loss of moisture during cooling down. *Journal of the Food Science and Agriculture* 23: 285–298.

Fonseca, S. C., Oliveira, F. A. R., and J. K. Brecht. 2002. Modelling respiration rate of fresh fruits and vegetables for modified atmosphere packages: a review. *Journal of Food Engineering* 52: 99–119.

Fonseca, S. C., Oliveira, F. A. R., and K. V. Chau. 1996. Modeling oxygen and carbon dioxide exchange through a perforation, for development of perforated modified atmosphere bulk packages Abstract and poster session at IFT annual meeting and food expo, June 22–26, New Orleans, LA.

Frank, A., Rath, S. K., and S. S. Venkatraman. 2005. Controlled release from bioerodible polymers: effect of drug type and polymer composition. *Journal of Controlled Release* 102: 333–344.

Gaffney, J. J., Baird, C. D., and K. V. Chau. 1985. Influence of airflow rate, respiration, evaporative cooling, and other factors affecting weight loss calculations for fruits and vegetables. *ASHRAE Transactions* 91: 690–707.

Gontard, N., Thibault, R., Cuq, B., and S. Guilbert. 1996. Influence of relative humidity and film composition on oxygen and carbon dioxide permeabilities of edible films. *Journal of Agricultural and Food Chemistry* 44: 1064–1069.

Gonzalez, J., Ferrer, A., Oria, R., and M. L. Salvador. 2008. Determination of O_2 and CO_2 transmission rates through microperforated films for modified atmosphere packaging of fresh fruits and vegetables. *Journal of Food Engineering* 86:194–201.

Hayakawa, K., and J. Succar. 1982. Heat transfer and moisture loss of spherical fresh produce. *Journal of Food Science* 47: 596–605.

Heiss, R. 1954. *Verpackung feuchtigkeitsempfindlicher Lebensmittel*. Berlin: Springer.

Hertog, M. L. A. T. M. 2003. MAP performance under dynamic temperature conditions. In *Novel food packaging techniques*, ed. R. Ahvenainen, pp. 563–575. Cambridge: Woodhead.

Hertog, M. L. A. T. M., and N. H. Banks. 2003. Improving MAP through conceptual models. In *Novel food packaging techniques*, ed. R. Ahvenainen, pp. 563–575. Cambridge: Woodhead.

Hertog, M. L. A. T. M., Boerrigter, H. A. M., Van den Boogaard, G. J. P. M., Tijskens, L. M. M., and A. C. R. Van Schaik. 1999. Predicting keeping quality of strawberries (cv. 'Elsanta') packed under modified atmospheres: an integrated model approach. *Postharvest Biology and Technology* 15: 1–12.

Irissin-Mangata, J., Boutevin, B., and G. Bauduin. 1999. Bilayer films composed of wheat gluten and functionalized polyethylene: permeability and other physical properties. *Polymer Bulletin* 43: 441–448.

Jayas, D. S., and S. Jeyamkondan. 2002. Modified atmosphere storage of grains, meats, fruits and vegetables. *Biosystems Engineering* 82: 235–251.

Kang, J. S., and S. L. Dong. 1998. A kinetic model for transpiration of fresh produce in a controlled atmosphere. *Journal of Food Engineering* 35: 65–73.

Lange, J., Büsing, B., Hertlein, J., and S. Hediger. 2000. Water vapour transport through large defects in flexible packaging. Modelling gravimetric measurement and magnetic resonance imaging. *Packaging Technology and Science* 13: 139–147.

Lin, M., Wang, H., Meng, S., Zhong, W., Li, Z., Cai, R., Chen, Z., Zhou, X., and Q. Du. 2007. Structure and release behaviour of PMMA/silica composite drug delivery system. *Journal of Pharmaceutical Sciences* 96: 1518–1526.

Lopez Briones, G., Varoquaux, P., Chambroy, Y., Bouquant, J., Bureau, G., and B. Pascat. 1992. Storage of common mushroom under controlled atmospheres. *International Journal of Food Science and Technology* 27: 493–505.

Lurie, S., Pesis, E., Gadiyeva, O., Feygenberg, O., Ben-Arie, R., Kaplunov, T., Zutahy, Y., and A. Lichter. 2006. Modified ethanol atmosphere to control decay of table grapes during storage. *Postharvest Biology and Technology* 42: 222–227.

Maguire, K. M., Sabarez, H. T., and D. J. Tanner. 2006. Chapter 3: Postharvest preservation and storage. In *Handbook of vegetable preservation and processing*, eds. Y. H. Hui, S. Ghazala, D. Graham, K. D. Murell, and W. K. Nip. New York: Marcel Dekker.

Mannapperuma, J. D., Zagory, D., Sigh, R. P., and A. A. Kader. 1989. Design of polymeric packages for modified atmosphere storage of fresh produce. In *Proceedings of the fifth international controlled atmosphere research conference*, vol. 2, ed. J. K. Fellman, pp. 225–233, June 14–16, Wenatchee, WA.

Martinez-Romero, D., Guillen, F., Valverde, J. M., Bailen, G., Zapata, P., Serrano, M., Castillo, S., and D. Valero. 2007. Influence of carvacrol on survival of *Botrytis cinerea* inoculated in table grapes. *International Journal of Food Microbiology* 115:144–148.

Matteucci, S., Kusuma, V. A., Swinnea, S., and B. D. Freeman. 2008. Gas permeability, solubility and diffusivity in 1,2-polybutadiene containing brookite nanoparticles. *Polymer* 49: 757–773.

Matteucci, S., Yampol'skii, Y. P., Freeman, B. D., and I. Pinneau. 2006. Transport of gases and vapors in glassy and rubbery polymers. In *Materials science of membranes for gas and vapor separations*, eds. Y. P. Yampol'skii, B. D. Freeman, and I. Pinneau, pp. 1–48. London: John Wiley and Sons.

MujicaPaz, H., and N. Gontard. 1997. Oxygen and carbon dioxide permeability of wheat gluten film: effect of relative humidity and temperature. *Journal of Agricultural and Food Chemistry* 45: 4101–4105.

MujicaPaz, H., Guillard, V., Reynes, M., and N. Gontard. 2005. Ethylene permeability of wheat gluten film as a function of temperature and relative humidity. *Journal of Membrane Science* 256: 108–115.

Paull, R. E. 1999. Effect of temperature and relative humidity on fresh commodity quality. *Postharvest Biology and Technology* 15: 263–277.

Peppelenbos, H. W., and J. van't Leven. 1996. Evaluation of four type of inhibition for modelling the influence of carbon dioxide on oxygen consumption of fruits and vegetables. *Postharvest Biology and Technology* 7: 27–40.

Peppelenbos, H. W., Tijskens, L. M. M., van't Leven, J., and E. C. Wilkinson. 1996. Modelling oxidative and fermentative carbon dioxide production of fruits and vegetables. *Postharvest Biology and Technology* 9: 283–295.

Reinhard, C. S., Radomsky, M. L., Saltzman, W. M., Hilton, J., and H. Brem. 1991. Polymeric controlled release of dexamethasone in normal rat brain. *Journal of Controlled Release* 16: 331–340.

Rico, D., Martin-Diana, A. B., Barat, J. M., and C. Barry-Ryan. 2007. Extending and measuring the quality of fresh-cut fruit and vegetables: a review. *Trends in Food Science and Technology* 18: 373–386.

Rogers, C. E. 1975. Permeability and chemical resistance. In *Engineering design for plastics*, ed. E.I. Baur, pp. 682–683. New York: Robert E. Krieger.

Saltveit, M. E. 2003. Is it possible to find an optimal controlled atmosphere. *Postharvest Biology and Technology* 27: 3–13.

Sastry, S. K., and D. E. Buffington. 1982. Transpiration rates of stored perishable commodities. A mathematical model and experiments on tomatoes. *ASHRAE Transactions* 88: 391–397.

Schreiner, M., Peters, P., and A. Krumbein. 2007. Changes of glucosinolates in mixed fresh-cut broccoli and cauliflower florets in modified atmosphere packaging. *Journal of Food Science* 72: 585–589.

Singh, S. P., and D. V. S. Rao. 2005. Effect of modified atmosphere packaging (MAP) on the alleviation of chilling injury and dietary antioxidants levels in 'Solo' papaya during low temperature storage. *European Journal of Horticultural Science* 243: 243–251.

Song, Y., Vorsa, N., and K. L. Yam. 2002. Modeling respiration-transpiration in a modified atmosphere packaging system containing blueberry. *Journal of Food Engineering* 53: 103–109.

Sorrentino, M., Tortora, M., and V. Vittoria. 2006. Diffusion behavior in polymer-clay nanocomposites. *Journal of Polymer Science Part B: Polymer Physics* 44: 265–274.

Suppakul, P., Miltz, J., Sonneveld, K., and S. W. Bigger. 2003. Active packaging technologies with an emphasis on antimicrobial packaging and its application. *Journal of Food Science* 68: 408–420.

Takeuchi, K., and J. F. Frank. 2000. Penetration of *Escherichia coli* O157: H7 into lettuce tissues as affected by innoculum size and temperature and the effect of chlorine treatment on cell viability. *Journal of Food Protection* 63: 434–440.

Tewari, G., Jayas, D. S., Jeremiah, L. E., and R. A. Holley. 2002. Absorption kinetics of oxygen scavengers. *International Journal of Food Science and Technology* 37: 209–217.

Utto, W., Mawson, A. J., and J. E. Bronlund. 2008. Hexanal reduces infection of tomatoes by *Botrytis cinerea* whilst maintaining quality. *Postharvest Biology and Technology* 47: 434–437.

Valverde, J. M., Guillen, F., Martinez-Tomero, D., Castillo, S., Serrano, M., and D. Valero. 2005. Improvement of table grapes quality and safety by the combination of modified atmosphere packaging (MAP) and eugenol, menthol, or thymol. *Journal of Agricultural and Food Chemistry* 53: 7458–7464.

Varoquaux, P., Gouble, B., Barron, C., and F. Yildiz. 1999. Respiratory parameters and sugar catabolism of mushroom (*Agaricus bisporus* Lange). *Postharvest Biology and Technology* 16: 51–61.

Varoquaux, P., and I. S. Ozdemir. 2005. Packaging and produce degradation. In *Produce degradation: pathways and prevention*, eds. O. Lamikanra, S. Imam, and D. Ukuku, pp. 117–153. Boca Raton, FL: Taylor and Francis.

Veraverbeke, E. A., Verboven, P., Oostveldt, V. O., and B. M. Nicolai. 2003. Prediction of moisture loss across the cuticle of apple (*Malus sylvestris* subsp. Mitis (Wallr.)) during storage: Part 2. Model simulations and practical applications. *Postharvest Biology and Technology* 30: 89–97.

Zagory, D., and J. Kader. 1988. Modified atmosphere packaging of fresh produces. *Food Technology* 42:70–74.

11 Use of Edible Coatings for Fresh-Cut Fruits and Vegetables

M. Alejandra Rojas-Graü, Robert Soliva-Fortuny, and Olga Martín-Belloso

CONTENTS

11.1 INTRODUCTION

Fresh-cut processing alters the integrity of fruits and vegetables, bringing about negative effects on product quality such as browning, off-flavor development, texture breakdown, and proliferation of microorganisms, thus reducing the shelf life of fresh-cut fruit commodities. Various techniques such as modified atmospheres, enzyme and browning inhibitors, texture stabilizers, and antimicrobial dips are employed to delay these negative effects. The application of each technique has advantages and drawbacks. For this reason, development of new preservation strategies is still needed to improve the quality and shelf-life of fresh-cut commodities. Edible coatings offer excellent prospects for extending the shelf life of fresh-cut produce by reducing the deleterious effects caused by minimal processing operations. Edible coatings from polysaccharides, proteins, and lipids can serve as barriers to moisture migration, prevent diffusion of gases and control microbial growth. They can also enhance quality and appearance of fresh produce by preventing flavor and aroma migration and by providing structural integrity. Edible coatings have also been studied as potential carriers of active ingredients such as antioxidants, antimicrobials, coloring agents, vitamins, probiotics, and nutraceuticals.

This chapter focuses on the use of edible coatings on fresh-cut fruits and vegetables, including a brief description of the main compounds used to form edible coatings and their most relevant properties, as well as an overview of the applications that have been researched for improving quality and extending the shelf life of fresh-cut fruits and vegetables. Some industrial considerations and regulatory status are also reviewed.

11.2 EDIBLE FILMS AND COATINGS

Edible films and coatings are generally defined as continuous matrices that can be prepared from edible materials such as polysaccharides, proteins, and lipids. Plasticizers and other additives are incorporated into formulations to modify their physical properties or functionality. Coatings are either applied to or formed directly on foods, while films are self-supporting structures that can be used to wrap food products. They are located either on the food surface or as thin layers between different components of a food product (Debeaufort and Voilley, 2009).

The use of edible coatings on fresh commodities consists of the application of a thin layer of any edible material on the surface of a fruit or vegetable with the purpose of providing it with a modified atmosphere, retarding gas transfer, reducing moisture and aroma losses, delaying color changes, and improving the general appearance of the product through storage. In the last years, edible coatings have been evaluated in order to improve quality and prolong the shelf-life of some fresh-cut fruits and vegetables by their ability to function as carriers of active compounds. However, the application of edible coatings to foods with the aim of prolonging their shelf life is not new. Wax coatings on whole fruits and vegetables have been used since the 1800s. In fact, coating of fresh citrus fruits (oranges and lemons) with wax to retard desiccation was practiced in China in the 12th and 13th centuries (Hardenburg, 1967) but was only commercially utilized on apples and pears as recently as the 1930s (Baldwin et al., 1995; Park, 1999). Currently there are several applications of edible films and coatings in the food industry, including shelf-life extension of oxygen-sensitive foods such as nuts, fresh whole fruits, and fresh-cut fruits and vegetables, reduction of oil migration into surrounding food components (e.g., nuts in chocolate); increase of integrity of fragile foods, such as breakfast cereals and freeze-dried foods, decrease of moisture loss of foods with high water activity, improvement of the moisture barrier of crisp inclusions such as nuts, cookies, and candies in ice cream, carriers of seasonings in snack foods, prevention of miscellaneous phenomena such as oxidations, moisture loss, aroma, or color migration in frozen foods, separation of heterogeneous foods or food components, and film pouches for dry food ingredients (Cuq, Gontard, and Guilbert, 1995; Han, 2000). Edible films and coatings have been used as a strategy to reduce the deleterious effects that minimal processing imposes on intact vegetable tissues.

On the other hand, edible films and coatings have a high potential to include active compounds due to their ability to carry ingredients such as antibrowning agents, nutrients, and antimicrobial compounds that can extend product shelf-life and reduce the risk of pathogen growth on cut surfaces (Martín-Belloso et al., 2009). The application of edible coatings to deliver active substances is one of the major advances reached so far in order to increase the shelf life of fresh-cut produce. In this way, a new generation of edible coatings is under development, with the aim of allowing the incorporation of or controlled release of active compounds using nanotechnological solutions such as nanoencapsulation and multilayered systems. These techniques could be used to satisfactorily coat highly hydrophilic food systems such as fresh-cut fruits and vegetables (Vargas et al., 2008).

11.3 FORMULATION OF EDIBLE COATINGS

A very wide range of compounds can be used in the formulation of edible coatings. Polysaccharides, proteins, and lipids are the common coating-forming materials that can be used individually or combined. Other minor components such as plasticizers, emulsifiers, and surfactants are usually included to enhance coating properties (Baldwin et al., 1995). Their presence and abundance determine the barrier properties of the coating layer with regard to water vapor, oxygen, carbon dioxide, and lipid transfer in food systems.

Some basic factors need to be taken into account when formulating edible coatings for fresh-cut commodities. Mechanical structure of the film and the affinity between the coating material and the food are important factors to be controlled. To get the full advantage of an edible coating, the coating must adhere to the food surface (Lin and Zhao, 2007). The degree of adhesion depends on the chemical and electrostatic affinity of the coating material with the food surface. Higher adhesion ensures longer durability of the film on the surface of the fruit or vegetable (Olivas and Barbosa-Cánovas, 2005). Nevertheless, when coatings are applied onto the surface of fresh-cut commodities, they need to be first dissolved and then adsorbed by the wet surfaces to form a unique layer. For improving surface adhesion of coatings, surfactants are typically added into coating formulations to improve wettability and adhesion (Choi et al., 2002; Lin and Krochta, 2005).

11.3.1 POLYSACCHARIDE-BASED COATINGS

Polysaccharides are the most widely used components found in edible coatings for fresh-cut commodities, as they are present in most commercially available formulations. Polysaccharides exhibit effective gas barrier properties, although they are highly hydrophilic and thus have a high water vapor permeability in comparison with commercial plastic films. Nevertheless, the oxygen and moisture barrier properties of polysaccharide films and coatings can contribute to protect fresh-cut fruit and vegetables from dehydration and, in some cases, deplete their respiration rate. These coatings may help to retard ripening and increase shelf life of coated produce, without creating severe anaerobic conditions (Baldwin et al. 1995). The method by which polysaccharide coatings reduce moisture loss is by acting as a sacrificial moisture barrier to the atmosphere, so that moisture content of the coated food can be maintained (Kester and Fennema, 1986). Polysaccharides that have been used for coating applications in fresh-cut fruits and vegetables include those derived from plants (cellulose and derivatives, pectin, starch), seaweed extracts (alginates, carrageenan), shell crustaceans extracts (chitosan), and some mucilage compounds.

Cellulose is the most abundant natural polymer on earth. It is highly crystalline, fibrous, and insoluble. Several water-soluble, composite coatings are made commercially from cellulose. Cellulose derivatives such as methylcellulose (MC), hydroxy-propylmethyl-cellulose (HPMC), and the ionic carboxymethyl-cellulose (CMC) are commonly found in the formulation of edible coatings, especially in commercial products, for extending shelf life of fresh fruits and vegetables. These coatings reduce oxygen uptake without causing an equivalent increase in the carbon dioxide level in internal atmospheres of fruit or vegetable tissues, preventing the occurrence of anaerobic respiration. As a result, a coating manufactured by Agricoat Industries Ltd. as Semperfresh™ is formulated with sucrose esters of fatty acids, mono- and diglycerides, and the sodium salt of CMC. It forms an invisible coating that is odorless and tasteless, and creates a barrier that is differentially permeable to oxygen and carbon dioxide. Semperfresh coating has been successfully applied to extend the shelf life and preserve important flavor compounds of fresh fruits and vegetables, such as apples (Drake et al., 1987; Bauchot et al., 1995; Sumnu and Bayindirli, 1995), bananas (Banks, 1984), mangoes (Dhalla and Hanson, 1988; Carrillo-Lopez et al., 2000), and tomatoes (Tasdelen and Bayindirli, 1998).

Pectin is a complex anionic polysaccharide composed of (1®4)-α-D-galacturonic acid units interrupted by single (1®2)-α-L-rhamnose residues (Ridley et al., 2001). Currently, apple pomace and citrus peel are the main sources of commercial pectin, while other potentially valuable sources remain unused because of certain undesirable structural properties. The carboxyl groups of the galacturonic acid units are partly esterified by methyl groups. By principle, if the degree of methyl esterification (DE) is greater than 50%, the pectin is called high methoxyl (HM) grade or high ester, and if it is less than 50%, it is called low methoxyl (LM) or low ester pectin (Nieto, 2009). For both HM and LM pectins, the gel modulus is determined by the number of effective junction zones formed between pectin chains. HM pectin forms good gels at low pH and the presence of solutes, generally sucrose. LM pectin forms firm gels in the presence of calcium ions in a manner very similar to alginates. Pectins with low levels of methyl esterification are commonly used for edible coating developments. Pectin gels have high water vapor permeability; thus they can prevent dehydration only by acting as sacrificial agents. Pectin coatings are generally not as strong as other films. Some researchers have reported that edible coating made from fruit puree containing pectin in its formulation can be used to extend the shelf life of fresh-cut fruits and vegetables, as well as to enhance their nutritional value and increase their consumer appeal.

Starch is one of the most abundant natural polysaccharides used as food hydrocolloid (Narayan, 1994), because it is inexpensive, abundant, biodegradable, and easy to use. Starch is available from different botanical sources, including wheat, corn, rice, potato, cassava, yam, and barley, but corn represents the major commercial source (Riaz, 1999). Coatings made from starch are often transparent or translucent, odorless, tasteless, and colorless, and have low permeability to oxygen at low-to-intermediate relative humidity (Myllarinen et al., 2002). Usually, these films are formed by the drying of a gelatinized dispersion, as hydrogen bonds form between hydroxyl groups (Lourdin et al., 1995).

Alginates are another polysaccharide with a high potential to cast films. They are extracted from brown seaweeds of the *Phaephyceae* class and are the salts of alginic acid, which is a linear copolymer of D-mannuronic and L-guluronic acid monomers. Alginates possess good film-forming properties, producing uniform, transparent, and water-soluble films. Alginates' gel-forming properties are mainly due to their capacity to bind a number of divalent ions like calcium and are strongly correlated with the proportion and length of the guluronic acid blocks (G-blocks) in their polymeric chains. On the other hand, gellan is a microbial polysaccharide secreted by the bacterium *Sphingomonas elodea* (formerly referred to as *Pseudomonas elodea*). The functionality of gellan gum depends on its degree of acylation. High acyl gellan forms soft, very elastic, transparent, and flexible gels, while low acyl gellan forms hard, non-elastic, brittle gels (Sworn, 2000). The mechanism of gelation involves the formation of a three-dimensional network, which in turn is formed by double helical junction segments that are complexated with cations and hydrogen bonds (Takahashi et al., 2004). Both polysaccharides are increasingly finding use in the food industry as texturizing and gelling agents (Yang and Paulson, 2000).

A polysaccharide commonly used in the formulation of edible coatings is chitosan, which is mainly obtained from crab and shrimp shells (Hirano, 1999). This coating material has excellent film-forming properties, broad antimicrobial activity, and compatibility with other substances, such as vitamins, minerals, and antimicrobial agents (Park and Zhao, 2004; Durango et al., 2006; Chien et al., 2007a, 2007b; Ribeiro et al., 2007). The main drawback of chitosan is that this coating can affect the taste and odor of coated products. In fact, the use of chitosan-based coatings may generate slight flavor modifications because of its typical astringent/bitter taste.

Carrageenan is a complex mixture of several polysaccharides. In fact, carrageenan is a generic term for polysaccharides extracted from certain species of red seaweed of the family, *Rhodophycae* (De Ruiter and Rudolph, 1997). The three main commercial carrageenans are ι-, κ-, and λ-carrageenan, whose names specify the major substitution pattern present in the galactan backbone. Refined carrageenans produce clear solutions and, therefore, clear films. However, plasticizers are needed in order to improve their mechanical and structural properties.

Mucilages are generally hetero-polysaccharides obtained from plant stems (Trachtenberg and Mayer, 1981). McGarvie and Parolis (1979) determined that the mucilage extracted from the stems contains residues of D-galactose, D-xylose, L-arabinose, L-rhamnose, and D-galacturonic acid. The complex polysaccharide is part of dietary fiber and has the capacity to absorb large amounts of water, dissolving and dispersing itself and forming viscous or gelatinous colloids (Dominguez-López, 1995). Recently, some authors have proposed the use of mucilage gels as coatings for fruits and vegetables.

11.3.2 PROTEIN-BASED COATINGS

Proteins that can be used in the formulation of edible coatings for fresh-cut fruits and vegetables include both those derived from animal sources, such as casein and whey protein, and those obtained from plant sources like zein, wheat gluten, and soy protein (Gennadios et al., 1994). Like polysaccharides, proteins exhibit excellent oxygen, carbon dioxide, and lipid-barrier properties, particularly at low relative humidity, as well as outstanding strength and structural integrity. However, protein films and coatings are a relatively poor water barrier which can be attributed to the inherent hydrophilicity of proteins as well as to the hydrophilic plasticizers incorporated into the film matrix to impart adequate flexibility (Kester and Fennema, 1986; Gennadios et al., 1994; Sothorvit and Krochta, 2000; Baldwin and Baker, 2002). In fact, protein films are generally brittle and susceptible to cracking, and the addition of compatible plasticizers is necessary to improve their extensibility (Lim et al., 2002).

Although the use of protein-based coatings on fresh-cut fruits and vegetables may increase the nutritional value of these products, protein coatings have been explored less extensively than polysaccharides for their use on fresh-cut commodities. In addition, when proteins of animal origin are employed in the formulation, fresh-cut products may become less appealing to certain groups of consumers, such as vegetarians or vegans, also introducing the risk of allergenicity and intolerance (Baldwin and Baker, 2002).

11.3.2.1 Animal Proteins

Milk proteins, such as whey proteins and caseins, have been extensively studied due to their excellent nutritional value and several of their functional properties, which are important for the formation of edible films (McHugh and Krochta, 1994a). According to Chen (1995), edible films based on milk proteins are flavorless, tasteless, and flexible, and, depending on the formulation, their appearance varies from transparent to translucent.

Casein is the major dairy protein group, representing 80% of the total composition of milk proteins, with a mean concentration of 24 to 29 g/l in bovine or goat milk (Audic et al., 2003). Four principal fractions, α_{s1}, α_{s2}, β-, and κ-caseins have been identified. Each of the four protein fractions has unique properties that affect their ability to form films. Among them, β-casein is the most interesting to produce films of weak permeability to water vapor (Lacroix and Cooksey, 2005). Caseins can form films from aqueous solutions without further treatment based on their random-coil nature and the ability to form extensive intermolecular hydrogen, electrostatic, and hydrophobic bonds, resulting in an increase of interchain cohesion (Avena-Bustillos and Krochta, 1993; Gennadios et al., 1994).

Among protein-based coatings, whey-protein coatings have been the subject of intense investigations over the past decade. Whey proteins are soluble proteins present in milk serum after caseinate is coagulated during cheese processing, representing

20% of total milk proteins. Whey protein is composed of several individual proteins, with β-lactoglobulin, α-lactalbumina, bovine serum albumin, and immunoglobins being the main proteins (Kinsella, 1984). Whey proteins, when appropriately processed, produce transparent, flavorless, and flexible films, similar to caseinate films (Lin and Zhao, 2007). Formation of intact and insoluble whey-protein films, as a result of formation of intermolecular disulfide bonds, can be produced by heat denaturation of the proteins (Gennadios et al., 1994; McHugh and Krochta, 1994a).

11.3.2.2 Plant Proteins

Soy, zein, and wheat gluten proteins are the main plant-origin proteins studied as coating materials for fruit and vegetable applications.

Soy protein is a viable and renewable recourse for producing edible coatings, and soy protein concentrate (SPC) and soy protein isolate (SPI) are the most studied as coating materials for fresh-cut products. Both are extracted from defatted protein meal and contain 65% to 72% and 90% protein on a dry basis, respectively (Mounts et al., 1987). Soy protein coatings are typically prepared from SPI with the addition of a plasticizer for improving flexibility (Gennadios et al., 1994). However, the substantial amounts of hydrophilic plasticizers used to impart flexibility may increase the poor moisture resistance observed in these coatings.

Zein composes a group of prolamins found in corn endosperm and includes approximately 45% to 50% of the protein in corn (Shukla and Cheryan, 2001). Zein is one of a few proteins, such as collagen and gelatin, already commercially used as an edible coating (Buffo and Han, 2005). Corn zein protein is insoluble in water, a characteristic that affects the barrier properties of its films. The ability of zein and its resins to form tough, glossy, and hydrophobic grease-proof coatings and their resistance against microbial attack have been of commercial interest (Lin and Zhao, 2007). Zein-based films have similar or lower water vapor permeabilities than other protein-based coatings (Krochta et al., 1997). Park (1999) indicated that zein coatings improve the shelf life of vegetables because they limit exposure of the vegetables to ambient oxygen and increase the internal carbon dioxide concentration.

Wheat proteins account for 8% to 15% of the dry weight of wheat kernels (Kasarda et al., 1976). Commercially, it is an industrial by-product of wheat starch production via wet milling. There are four wheat-protein classes, based on solubility in different solvents, namely, albumins, globulins, gliadins, and glutenins. Wheat gluten proteins are insoluble in water and require a complex solvent system with basic or acidic conditions in the presence of alcohol and disulfide bond-reducing agents. In addition, wheat gluten films are very brittle, so the addition of plasticizers in the formulation is necessary to induce film flexibility.

11.3.3 COMPOSITE COATINGS AND EMULSIONS

As mentioned, polysaccharides and proteins are polymeric and hydrophilic in nature, and thus are good film-formers with excellent oxygen, aroma, and lipids barriers at low relative humidity, though they are poor moisture barriers. Each material has some unique, though limited, functions. The integration of proteins, polysaccharides, and lipids in a single coating can substantially improve its functionality (Lin and

Zhao, 2007). Due to the presence of microscopic pores and elevated solubility and diffusivity, lipids offer limited oxygen barrier properties. However, lipid films generally exhibit low water vapor barrier permeability, due to their low polarity (Kester and Fennema, 1986), but are usually opaque and relatively inflexible (Guilbert et al., 1996). Generally, lipids contribute to the improvement of the water vapor resistance, whereas hydrocolloids confer selective permeability to O_2 and CO_2, as well as durability, structural cohesion, and integrity (Krochta et al., 1997).

Composite coatings can be applied as a bilayer or a stable emulsion. Plasticizers, such as glycerol, polyethylene glycol, or sorbitol, may be added to modify film/coating mechanical properties and provide increased flexibility (Guilbert et al., 1996). Lipids generally form an additional layer when applied on a polysaccharide or protein layer, thus forming bilayer composite film/coatings, while the lipid in the emulsion composite films/coatings is dispersed and entrapped in the matrix of protein or polysaccharide (Lin and Zhao, 2007). Some authors have reported that emulsified coatings are less efficient than bilayer coatings due to the nonhomogeneous distribution of the lipid phase. Nonetheless, they have the advantage of needing only one application step instead of the two needed for bilayer coatings. The improved moisture-barrier properties of composite coatings have made them promising candidates for coating fresh-cut fruits and vegetables.

11.3.4 PLASTICIZERS

Plasticizers are an important component in the formulation of edible films and coatings. These are low molecular weight compounds that can be added to a hydrocolloid solution to improve the flexibility, resilience, barrier, and mechanical properties of the film matrix. They must be compatible with film-forming polymers, reduce intermolecular forces, and increase mobility of polymer chains (Donhowe and Fennema, 1994). Although plasticizers can improve the flexibility and elongation of hydrophilic films, they also affect their permeability (McHugh and Krochta, 1994b). As a general rule, the addition of a plasticizer causes an increase in the permeability of a film or coating.

Hydrophilic compounds such as polyols (glycerol, propylene glycol, polypropylene glycol, sorbitol, and sucrose) and polyethylene glycol are commonly used as plasticizers in hydrophilic formulations (Guilbert et al., 1996). Lipophilic compounds, such as vegetable oils, lecithin, and, to a lesser extent, fatty acids, may also act as emulsifiers and plasticizers (Kester and Fennema, 1986; Donhowe and Fennema, 1994). The effectiveness of a plasticizer is dependent upon three factors: size, shape, and compatibility with the film matrix (Sothornvit and Krochta, 2005).

11.4 PROPERTIES OF EDIBLE COATINGS

Edible coatings are known to improve shelf life and maintain quality of minimally processed foods (Baldwin et al., 1995). As can be seen in Figure 11.1, edible coatings have the potential to provide a moisture barrier on the surface of cut produce. They also are selective barriers to oxygen, retarding respiration rates and ethylene production as well as sealing from loss of flavor volatiles. In

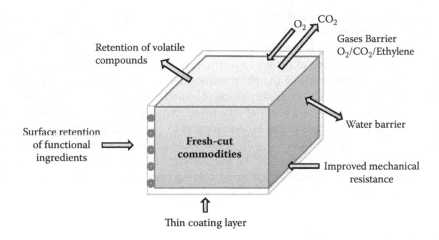

FIGURE 11.1   Potential properties of edible coatings on fresh-cut fruits and vegetables.

addition to increased barrier properties, edible films and coatings control adhesion, cohesion, and durability, improving the appearance of coated foods and maintaining quality during shipping and storage. Nonetheless, the major benefit of edible coating is that they can be consumed along with food, can provide additional nutrients, may enhance sensory characteristics, and may include quality-enhancing antimicrobials.

11.4.1   Moisture Barrier

Water loss leads to a loss of turgor and crispness, and occurs rapidly in fresh-cut products due to the absence of a cuticle and subepidermal layers and the exposure of internal tissues (Shackel et al., 1991). However, water loss can be greatly retarded by appropriate packaging (Toivonen and Brummell, 2008). In this way, edible coatings decrease the water vapor transmission rate by forming a barrier on the fruit or vegetable surface. The ability of coatings to function as barriers to water vapor relies on external conditions, including temperature and relative humidity; the characteristics of the commodity, such as type of product, variety, maturity, and water activity; and the characteristics of the coating, such as composition, concentration of solids, viscosity, chemical structure, polymer morphology, degree of cross-linking, and type of plasticizer used (Olivas and Barbosa-Cánovas, 2005).

11.4.2   Gas Barrier

Edible coatings are also used as a protective barrier to reduce respiration and transpiration rates through the tissue surface. The coating acts as a gas barrier around each fruit or vegetable piece and creates a modified atmosphere inside the piece (Rojas-Graü et al., 2008). The coating can regulate gas exchange between the fresh produce and its surrounding atmosphere, which would slow respiration and delay deterioration (Lin and Zhao, 2007). However, although gas transfer reduction between the food and the environment is desirable, extremely impermeable coatings may induce anaerobic conditions that can lead to a decrease in the production of characteristic aroma volatile compounds (Mattheis and Fellman, 2000; Perez-Gago et al., 2003). An excessive modification of the internal atmosphere (too high carbon dioxide or too low oxygen concentrations) of the plant tissues can develop ethanol and alcoholic flavors as a result of anaerobic fermentation (Lin and Zhao, 2007). The formation of fermentative metabolites as a result of anaerobic respiration is often associated with off-flavors, and its presence might be detrimental to quality (Day, 2001). The selection of an edible coating material with appropriate permeability and the control of environmental temperature and relative humidity are critical to modify the internal environment of fresh produce by adjusting coating permeability to the produce respiration (Lin and Zhao, 2007).

11.4.3 Aroma Barrier

Mass transfer of volatile compounds can occur between a food and its surrounding medium, leading to physicochemical changes and food deterioration (Karel and Lund, 2003). These phenomena can be successfully reduced with an appropriate coating. Edible coatings prevent the loss of natural volatile flavor compounds from fresh produce and the acquisition of foreign odors (Lin and Zhao, 2007). Some researchers studied the volatile barrier properties of various edible films (Miller and Krochta, 1997; Hambleton et al., 2008, 2009a, 2009b), but specific studies on fresh-cut fruits or vegetables are rather limited.

11.4.4 Carrier Properties

Quality, stability, safety, and functionality of fresh-cut commodities can be improved by incorporating antioxidants, antimicrobials, and active ingredients into edible coatings (Rojas-Graü et al., 2009). Fresh-cut fruits and vegetables processing operations can induce undesirable changes in color and appearance of these products during storage and marketing. Application of antioxidant treatments by immersing the peeled or cut product is the most common way to control browning of these commodities. However, antioxidant agents can be added into the coating matrix to protect the cut surface against enzymatic browning, which is usually caused by the enzyme polyphenol oxidase. Ascorbic acid, some thiol-containing compounds (cysteine, N-acetylcysteine, and reduced glutathione), carboxylic acids (citric acid and oxalic acid), and several resorcinol derivatives (4-hexylresorcinol) have been used as antibrowning agents in coating formulations. On the other hand, microorganisms can rapidly grow on the surface of fresh-cut fruits and vegetables during storage. Microbial stability, related to postprocessing contamination or multiplication, can be controlled by antimicrobial agents incorporated into the coatings (Cagri et al., 2004). Dipping of the product into aqueous solutions containing antimicrobials is the most practical way to extend the microbial stability of fresh-cut fruits. Nevertheless, application of antimicrobial agents directly on the food surface may have limited benefits, because the active substances are rapidly neutralized or diffused from the surface into the food product (Min and Krochta, 2005). Antimicrobial edible films and coatings could contribute to maintaining effective concentrations of the active compounds on the food surfaces (Gennadios and Kurth, 1997). Several types of antimicrobials have been incorporated into edible coatings for extending the shelf-life of fresh commodities, but their use in fresh-cut fruits and vegetables is yet limited. At the moment, organic acids and plant essential oils are the main antimicrobial agents incorporated into edible coatings used on fresh-cut produce. Despite the good results achieved so far, the incorporation of certain antibrowning and antimicrobial agents into formulations may have detrimental consequences on the flavor of the coated product. Some authors indicated that high concentrations of sulfur-containing compounds such as *N*-acetylcysteine and glutathione may produce an unpleasant odor in fruits and vegetables (Richard et al., 1992; Iyidogan and Bayindirli, 2004; Rojas-Graü et al., 2006). In the case of essential oils, the major drawback is their strong flavor that could change the original taste of foods.

In addition, minimal processing operations may result in a dramatic loss of firmness in fruit tissues due to the action of pectic enzymes (Toivonen and Brummell, 2008). Treatments with calcium salts are the most common way to control softening phenomena in fresh-cut produce (Garcia, Herrera, and Morilla, 1996). Texture enhancers such as calcium chloride can also be incorporated into the edible coating formulation. Hence, the use of calcium chloride for cross-linking some polymers could minimize softening phenomena. Edible films and coatings are also an excellent vehicle to enhance the nutritional value of fruits and vegetables by carrying basic nutrients that are lacking or present in low amounts in fruits and vegetables.

11.5 APPLICATIONS OF EDIBLE COATINGS ON FRESH-CUT COMMODITIES

Edible coatings may reduce the deleterious effects concomitant with minimal processing when they are used to coat fresh-cut fruits and vegetables. Some of the most recent reports on applications of edible coatings for improving the quality and extending the shelf life of fresh-cut fruits and vegetables are summarized in Table 11.1.

The first documented coating application on a fresh-cut product was reported by Bryan (1972), who observed that a carrageenan-based coating applied on cut grapefruit halves resulted in less shrinkage, leakage, or deterioration of taste after 2 weeks of storage at 4°C. Since then, many edible coatings have been evaluated in order to improve the shelf life of fresh-cut fruits. Rojas-Graü et al. (2008) observed that the use of alginate and gellan edible coatings effectively prolonged the shelf life of apple wedges by 2 weeks of storage. Recently, Eissa (2007) suggested that the use of a coating containing chitosan (2%) could be beneficial for extending shelf life, maintaining quality, and to some extent, controlling decay of fresh-cut mushroom. Olivas et al. (2007) reported that alginate coatings extended the shelf life of fresh-cut 'Gala' apples without causing anaerobic respiration. Baldwin et al. (1996) enhanced the storage life of cut apples with a CMC-based edible coating. By contrast, Sothornvit and Rodsamran (2008) observed an important increase in translucency of fresh-cut mango coated with a mango edible film, which was higher when cut fruits were stored at room temperature. Durango et al. (2006) and Devlieghere

TABLE 11.1

Some Applications of Edible Coatings (ECs) to Fresh-Cut Fruits and Vegetables

Commodity	EC Matrix	Additives	Relevant Results	References
Melon	Alginate/gellan/pectin	0.6–1.5% Gly, 2% CaCl$_2$	Prevented dehydration and maintained firmness	Oms-Oliu et al. (2008a)
	Soy protein	0.9–3.5% Gly, 2.6% MA, 2.6% LA	Good carrier properties without sensory changes	Eswaranandam et al. (2006)
	Alginate	1.5% Gly, 2.5% MA, 0.3% Poil, 2% CaL	Maintained firmness and inhibited the microbial growth	Raybaudi-Massilia et al. (2008b)
Apple	Apple puree/pectin	3% Gly, 0.5% AA, 0.5% CA, 5–10% Lipids	Delayed browning and reduced moisture loss	McHugh and Senesi (2000)
	Alginate	1.5% Gly, 2.5% MA, 1% NAC, 1% GSH, 0.7% LG or 0.3% Coil, 2% CaL	Inhibited native microbiota and maintained the initial texture	Raybaudi-Massilia et al. (2008a)
	Carrageenan/WPC	1% AA, 1% AC, 0.05% OA, 1% CaCl$_2$	Inhibited the loss of firmness and maintained the original color during storage without changes in sensory properties	Lee et al. (2003)
	Alginate/gellan	1.5% Gly, 1% NAC, 2% CaCl$_2$	Maintained the original color and firmness by 2 weeks of storage	Rojas-Graü et al. (2008)
	Alginate/apple puree	1.5% Gly, 1% NAC, 2% CaCl$_2$, 0.3–0.6% VII or 1–1.5% LG or 0.1% to 0.5% Ooil	Reduced native psychrophilic aerobes, molds, and yeast; lemongrass (1–1.5%) and oregano (0.5%) reduced >4 log CFU/g of inoculated *Listeria inocua*	Rojas-Graü et al. (2007b)
	WPC-BW	0.5–1% AA, 0.1–0.5% Cys, 0.005–0.02 4-HR	Reduced surface browning, but 4-HR was less effective	Perez-Gago et al. (2006)
Papaya	Alginate/gellan	0.25–2% Gly, 1% AA, 2% CaCl$_2$	Preserved the natural ascorbic acid content of the fruit	Tapia et al. (2008)
Pear	Methylcellulose	1% AA, 0.1% PS, 0.25% CaCl$_2$	Prolonged shelf life by retarding browning	Olivas et al. (2003)

Product	Coating	Additives/concentration	Effect	Reference
	Alginate/gellan/pectin	1.5% Gly, 0.75% NAC, 0.75 GSH, 2% CaCl2	Prevented browning for 2 weeks without affecting fruit firmness	Oms-Oliu et al. (2008b)
Pineapple	Alginate	1% CA, 1% AA, 2% CaCl$_2$	Helped to retain internal liquids	Montero-Calderón et al. (2008)
Mango	Chitosan	—	Retarded water loss and inhibited the growth of microorganisms	Chien et al. (2007a)
Carrot	Xanthan gum	5% GluconalCal, 0.2% Vit E	Good carrier properties without sensory changes	Mei et al. (2002)
	Alginate	0.1% CA, 2% CaCl$_2$	Improved the surface color of carrots	Amanatidou et al. (2000)
	Calcium caseinate/WPI/pectin/CMC	2.5% Gly	Retained fresh product characteristics for at least 8 d	LaFortune et al. (2005)
			Protected carrots against dehydration during storage	
Lettuce	Alginate	1% CaCl$_2$	Increased crispness	Tay and Perera (2004)
	Chitosan	2% lactic acid/Na-lactate	Limited antimicrobial effect	Devlieghere et al. (2004)
Broccoli	Corn zein	Oleic acid (1g/1g zein)	Maintained their original firmness and color	Rakotonirainy et al. (2001)
Mushroom	Chitosan	—	Extended shelf life and inhibited native microbiota	Eissa (2007)

Notes: Gly, glycerol; BW, beeswax; AA, ascorbic acid; MA, malic acid; CaCl$_2$, calcium chloride; CaL, calcium lactate; CA, citric acid; GSH, glutathione; NAC, N-acetylcysteine; Cys, cysteine; PS, potassium sorbate; WPC, whey protein concentrates; WPI, whey protein isolate; LG, lemongrass; Coil, cinnamon oil; Ooil, oregano oil; VlI, vanillin; Poil, palmarosa oil; LA, lactic acid; Vit E, vitamin E; CMC, carboxy-methylcellulose; 4-HR, 4-hexylresorcinol.

et al. (2004) used a chitosan-based coating to cover carrots and lettuce, respectively, observing a reduction in the respiration rate and ethylene production, as well as a decrease in firmness loss. Eswaranandam et al. (2006) extended the shelf life of fresh-cut cantaloupe using a soy protein coating containing malic acid and lactic acid in the formulation. Garcia et al. (1998) found a significant effect of starch-based coating on the color, weight loss, firmness, and shelf life of coated strawberries. Later, Garcia et al. (2001) observed that a starch-based coating containing potassium sorbate, plasticizer, and sunflower oil improved the water vapor barrier, reduced microbial growth, and exhibited selective gas permeability, thus extending the storage life of strawberries. Shon and Haque (2007) observed a decrease in browning of cut apples and potatoes when using an edible coating containing sour whey flour. Pennisi (1992) observed a reduction of browning and water loss of fresh-cut apple slices covered with a chitosan-lauric acid composite coating. Del-Valle et al. (2005) improved the shelf life of strawberries using a cactus-mucilage edible coating that better maintained the fruit physical and sensorial properties. Valverde et al. (2005) and Martínez-Romero et al. (2006) proposed aloe vera gel–based edible coatings for preventing moisture loss, reducing texture decay, and controlling respiratory rate of table grapes and sweet cherries, respectively, while reducing microbial proliferation. In addition, Martínez-Romero et al. (2006) maintained sweet cherries without any detrimental effect on taste, aroma, or flavor during storage using an aloe vera-based coating. Lafortune et al. (2005) protected carrots against dehydration and maintained their firmness during storage using calcium caseinate and whey protein isolate edible coatings. Chien et al. (2007a) reported the effectiveness of chitosan in maintaining quality and extending shelf life of sliced mango. Assis and Pessoa (2004) and Han et al. (2005) also proposed chitosan for extending the shelf life of sliced apples and fresh strawberries, respectively. Howard and Dewi (1995) and Li and Barth (1998) reported that the use of a cellulose-based edible coating prevented the development of white surface discoloration in fresh-cut carrots. Similar results were reported by Avena-Bustillos et al. (1994a) and Mei et al. (2002), who treated minimally processed carrots with an emulsion of sodium caseinate/stearic acid and a xanthan gum solution, respectively. Xu et al. (2001) reported that soy protein coatings can retard the senescence process of kiwifruit. Rakotonirainy et al. (2001) maintained the original firmness and color of broccoli florets using a zein-based coating. Wong et al. (1994) employed a bilayer of an acetylated monoglyceride and ascorbate buffer containing calcium ions for controlling gas diffusion through coated cut apples, and attributed the large reductions in the rate of gas evolution to this bilayer. LeTien et al. (2001) achieved reduced browning rates in apple slices coated with a combination of whey protein and CMC. Cellulose-based edible coatings on sliced mushrooms have been shown to significantly reduce enzymatic browning (Nisperos-Carriedo et al., 1991). Plotto et al. (2004) coated fresh-cut mangoes with several edible coatings, and they observed that a CMC-based formulation containing maltodextrin presented the highest scores for visual quality and flavor. Sonti et al. (2003) coated apple cubes with whey protein concentrate and whey protein isolate, obtaining a delay in browning and texture decay. Pen and Jiang (2003) reported that chitosan retarded the development of browning, maintained sensory quality, and retained levels of total soluble solids, acidity, and ascorbic acid in sliced Chinese water chestnuts.

Edible coatings have also been extensively used to protect fresh-cut fruits and vegetables from surface dehydration and texture loss. Tanada-Palmu and Grosso (2005) reported that wheat gluten with lipid-based (beeswax, stearic acid, and palmitic acid) bilayer coatings considerably retained firmness and reduced weight loss of fresh strawberries. Avena-Bustillos et al. (1997) reduced water loss of apple pieces using an emulsion containing calcium caseinate and an acetylated monoglyceride. Montero-Calderón et al. (2008) reported that the use of alginate coatings significantly improved the shelf-life of fresh-cut pineapple, as reflected in higher juice retention in contrast with the substantial juice leakage observed in other evaluated packaging conditions. McHugh and Senesi (2000) reduced moisture loss of fresh-cut apples when applying wraps made from apple puree and pectin containing various concentrations of lipids. Similarly, Wong et al. (1994) reported that a cellulose/lipid bilayer edible film reduced between 12 and 14 times water loss of apple slices. Likewise, Olivas et al. (2003) found that the incorporation of stearic acid into methylcellulose-based coatings played an important role in reducing weight loss of pear wedges, whereas coatings without lipid addition showed poor moisture barrier. Baldwin et al. (1995) indicated that a coating formed by a milk protein (casein) and a lipid (acetylated monoglyceride) was effective to provide protection from moisture loss and oxidative browning for up to 3 days in fresh-cut apples. Han et al. (2004) reported that a chitosan-based coating containing calcium resulted in at least a 24% reduction in the drip loss of frozen-thawed raspberries and increased their firmness by about 25% in comparison with uncoated fruits. Similar results were reported by Ribeiro et al. (2007), who observed a decrease in firmness loss of fresh strawberries coated with a calcium-enriched carrageenan coating when compared to noncoated fruit. Oms-Oliu et al. (2008a) reported that the use of calcium chloride, as a crosslinking agent of polysaccharide-based edible coatings (alginate, gellan, and pectin), helped to maintain firmness of fresh-cut melon during storage. Similar results were obtained by Rojas-Graü et al. (2008), who observed that apple wedges coated with alginate or gellan edible coatings and immersed into a calcium chloride solution maintained their initial firmness during refrigerated storage (Figure 11.2). Hernández-Muñoz et al. (2008) observed that the addition of calcium gluconate to the chitosan (1%) coating formulation increased the firmness of strawberries during refrigerated storage. Lee et al. (2003) indicated that incorporating 1% of calcium chloride within the whey protein concentrate coating formulation helped to maintain firmness of fresh-cut apple pieces. Olivas et al. (2007) maintained firmness of fresh-cut 'Gala' apples using a calcium chloride solution dipping and, subsequently, an alginate edible coating. Avena-Bustillos et al. (1994b, 1997) reduced water loss of apples, celery sticks, and zucchini by using an emulsion containing calcium caseinate and an acetylated monoglyceride.

The use of edible coatings to deliver functional compounds in order to improve the safety, nutritional, and quality properties of fresh-cut fruits has been proposed recently. The incorporation of antibrowning agents into edible coatings applied on fresh-cut fruits has been studied by various authors. Brancoli and Barbosa-Cánovas (2000) decreased surface discoloration of apple slices by coating slices with MC, maltodextrin, ascorbic acid, and calcium chloride. Similarly, Olivas et al. (2003) preserved fresh-cut pear wedges from surface browning by applying a MC-based coating

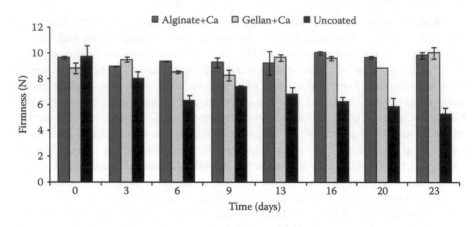

FIGURE 11.2 Changes in firmness of fresh-cut apples coated with alginate, gellan containing calcium chloride (Ca), or uncoated during storage. (Adapted from Rojas-Graü, M. A., Tapia, M. S., and Martin-Belloso, O. 2008. Using polysaccharide-based edible coatings to maintain quality of fresh-cut Fuji apples. *Lebensmittel Wissenschaft und Technologie* 41:139–147.)

containing ascorbic and citric acids. Pérez-Gago et al. (2006) reduced browning of cut apples by using a whey protein concentrate–beeswax coating containing ascorbic acid, cysteine, or 4-hexylresorcinol. Lee et al. (2003) studied the effect of whey protein and carrageenan concentrate edible coatings in combination with antibrowning agents on fresh-cut apple slices and observed that the incorporation of ascorbic, citric, and oxalic acids was advantageous in maintaining color during 2 weeks. Rojas-Graü et al. (2007a) applied alginate- and gellan-based coatings to fresh-cut 'Fuji' apples, proving that these coatings were good carriers of antioxidant agents such as cysteine and glutathione, which helped to maintain the color of cut fruits during storage. Similarly, Tapia et al. (2008) successfully applied alginate- and gellan-based coatings with added cysteine, glutathione, and ascorbic acid to fresh-cut papayas.

Furthermore, the incorporation of antimicrobial agents into edible coatings is gaining importance as a mode of reducing the deleterious effects induced by minimal processing on fresh-cut produce. In this way, Rojas-Graü et al. (2007b) achieved a 4 log reduction the inoculated population of *Listeria innocua* in fresh-cut apple when lemongrass or oregano oils were incorporated into an apple puree–alginate edible coating. Likewise, Raybaudi-Massilia et al. (2008a) demonstrated that the addition of cinnamon, clove, or lemongrass oils or their active compounds (citral, cinnamaldehyde, and eugenol) into an alginate-based coating reduced the population of *Escherichia coli* O157:H7 by more than 4 log CFU/g and extended the microbiological shelf life of Fuji apples for at least 30 days (Table 11.2). Later, Raybaudi-Massilia et al. (2008b) reported that the incorporation of 0.3% v/v palmarosa oil into the alginate coating inhibited the growth of the native microbiota and reduced the population of inoculated *Salmonella enteritidis* in fresh-cut melon. Park et al. (2005) reported a reduction of 2.5 and 2 log CFU/g in the counts of *Cladosporium* sp. and *Rhizopus* sp., respectively, on strawberries coated with a chitosan-based edible film, just after the coating application. A reduction in the counts of aerobic and

TABLE 11.2

Survival of *Escherichia coli* O157:H7 in Inoculated Fresh-Cut Apples Coated with Alginate Edible Coatings (AEC) with or without Incorporation of Plant Essential Oils or Their Active Compounds and Stored at 5°C for 30 Days

Treatments	Survival (log CFU/g) at Storage Day					
	0	3	7	14	21	30
Uncoated	6.26	5.75	5.31	4.9	3.78	3.49
AEC	5.03	4.7	3.8	3.3	2.92	2.15
AEC + Cin (0.7%)	3.14	<2	<2	<2	<2	<2
AEC + Cy (0.5%)	3.26	<2	<2	<2	<2	<2
AEC + Clo (0.7%)	2.89	2.63	<2	<2	<2	<2
AEC + Eu (0.5%)	3.36	3.11	2.4	2.2	<2	<2
AEC + Lem (0.7%)	2.24	<2	<2	<2	<2	<2
AEC + Cit (0.5%)	2.65	<2	<2	<2	<2	<2

Notes: Cin, cinnamon; Cy, cinnamaldehyde; Clo, clove; Eu, eugenol; Lem, lemongrass; Cit, citral.

coliform microorganisms during storage has been also reported. Garcia et al. (2001) extended the storage life of fresh strawberries to more than 28 days using a starch-based coating containing potassium sorbate and citric acid. Franssen and Krochta (2003) significantly reduced the populations of *Salmonella montevideo* on tomatoes when incorporating citric, sorbic, or acetic acids in HPMC coatings. The coating by itself resulted in a 2 log CFU/g reduction in the counts of the pathogenic strain, but addition of 0.4% sorbic acid led to a significantly higher inactivation. Krasaekoopt and Mabumrung (2008) observed that the incorporation of 1.5% and 2% chitosan in the methylcellulose coating applied on fresh-cut cantaloupe produced a better microbiological quality in the final product.

As commented before, edible coatings offer the potential to improve the nutritional quality and antioxidant properties of fresh-cut produce. Oms-Oliu et al. (2008b) maintained the vitamin C and total phenolic content in pear wedges coated with alginate, gellan, or pectin edible coatings. Oms-Oliu et al. (2008a) also observed that the use of an alginate coating may contribute to reducing the wounding stress induced by processing in fresh-cut 'Piel de Sapo' melon. Tapia et al. (2008) reported that the addition of ascorbic to the alginate edible coating helped to preserve the natural ascorbic acid content in fresh-cut papaya, maintaining its nutritional quality throughout storage. Serrano et al. (2006) maintained total phenolics, ascorbic acid, and high retention of total antioxidant activity in table grape coated with aloe vera gel coatings. Chien et al. (2007b) maintained the ascorbic acid content of sliced red pitayas (dragonfruit) coated with low molecular weight chitosan. Hernández-Muñoz et al. (2006) indicated that chitosan-coated strawberries retained more calcium gluconate (3,079 g/kg dry matter) than strawberries dipped into calcium solutions

(2,340 g/kg). Likewise, Han et al. (2004) observed that chitosan-based coatings had the capability to hold high concentrations of calcium gluconate or vitamin E in fresh and frozen strawberries and red raspberries, thus significantly increasing their content in both fruits. Mei et al. (2002) developed xanthan gum coatings containing high concentrations of calcium and vitamin E with the purpose of enhancing the nutritional and sensory qualities of fresh baby carrots. The results of this study showed that calcium and vitamin E contents in the coated carrots were increased from 2.6% to 6.6% and from 0 to about 67% of the Dietary Reference Intake (DRI) values per serving (85 g), respectively, without affecting the fresh aroma, fresh flavor, sweetness, crispness, and β-carotene level. Additionally, baby carrots coated with xanthan gum exhibited less white surface discoloration and greater orange color intensity ratings than uncoated samples. Tapia et al. (2007) maintained values higher than 10^6 cfu/g *Bifidobacterium lactis* Bb-12 in papaya and apple pieces coated with alginate or gellan film-forming solutions during refrigerated storage (10 days), demonstrating the feasibility of these polysaccharide coatings to carry and support viable probiotics on fresh-cut fruit.

11.6 INDUSTRIAL CONSIDERATIONS

Edible films and coatings can be applied by different methods, such as brushing, wrapping, spraying, dipping, casting, panning, or rolling. The type of food product determines the best coating method. Spray application is the conventional method for applying most coatings to whole fruits and vegetables. However, spraying requires that the bottom surface of the food product be coated in a subsequent step after application of the initial coating and drying.

The simplest way to apply a coating is directly by immersion into a solution. Dipping is advantageous when a product requires several applications of a coating to obtain uniformity on an irregular surface, such as fresh-cut fruits and vegetables. Depending on the concentration of the coating solution, the product will absorb an appropriate amount of coating material, necessary to form the desired layer. Coatings formed by immersion may be less uniform than coatings applied by other methods, and multiple dipping may be necessary to ensure full coverage. Coating integrity is a critical factor that depends on surface tension, adhesion to the food substrate, and flexibility of the coating. Better uniformity can be promoted by adding surfactants to the solution to reduce surface tension (Pavlath and Orts, 2009). In addition, some plasticizer (glycerol, mannitol, sorbitol, sucrose, and so on) needs to be added to the coating solution to keep the developing film from becoming brittle. Usually, after dipping and draining the excess of coating material, the film is allowed to set or solidify on the product (Donhowe and Fennema, 1994).

Casting is another technique in which coating-forming solutions are poured onto a level surface and allowed to dry, usually within a confined space. Casting produces freestanding films that exhibit a specified thickness, smoothness, and flatness (Donhowe and Fennema, 1994). Depending upon firmness and flexibility, cast films can then be used to wrap surfaces. This technique allows the films to be cut to any size and supplies an innovative and easy method for carrying and delivering a wide variety of ingredients such as flavorings, colorants, and vitamins, which can be used later to cover foods. In fact, some researchers have reported that fruit and vegetable wraps can be used to extend the shelf life of fresh-cut fruits and vegetables, as well as to enhance their nutritional value and increase their consumer appeal. McHugh and Senesi (1999) patented the first edible films made from fruit or vegetable purees. These authors developed a novel method (apple wraps) to extend the shelf life and improve the quality of fresh-cut apples. These wraps were made from apple puree containing various concentrations of fatty acids, fatty alcohols, beeswax, and vegetal oil and were then applied on apple pieces as preformed films (McHugh and Senesi, 2000). According to these authors, wraps were significantly more effective than coatings. Despite the good results obtained with the use of wraps, coatings are more popularly used than preformed films. Nowadays, wax coatings on whole fruits and vegetables and cellulose-based coatings on fresh-cut fruits and vegetables are the most common commercial applications of edible coatings. Nevertheless, the development of new technologies to improve the application of edible films and coatings on fresh-cut commodities is a major issue for future research. At the moment, most studies on fresh-cut fruits and vegetables applications have been conducted at a laboratory scale. Hence, further research should be focused on a commercial scale with the purpose of providing more realistic information that can be used to commercialize fresh-cut products coated with edible films or coatings. In spite of these limitations, food industries are looking for edible films and coatings to be used on a broad spectrum of foods and add value to their products, while increasing their shelf life.

11.7 REGULATORY STATUS

Because edible films and coatings are an integral part of the edible portion of food products, they ought to observe all regulations required for food ingredients. To maintain edibility, all film-forming components, as well as any functional additives in the film-forming materials, should be food-grade nontoxic materials, and all process facilities should meet high standards of hygiene (Guilbert and Gontard, 1995; Guilbert et al., 1996; Han, 2002; Nussinovitch, 2003). According to the European Directive (ED, 1995, 1998) and U.S. regulations (FDA, 2006), edible films and coatings can be classified as food products, food ingredients, food additives, food contact substances, or food packaging materials. The U.S. Food and Drug Administration (FDA) stated that any compound to be included in the formulation should be generally recognized as safe (GRAS) or regulated as a food additive and used within specified limitations (FDA, 2006). With the exception of chitosan, polysaccharides including cellulose and its approved derivatives, starches and approved derivatives, and seaweed extracts (agar, alginates, carrageenan), as well as beeswax, carnauba wax, candelilla wax, steric acid, and some glycerols are GRAS substances (Ustunol, 2009). In Europe, the ingredients that can be incorporated into edible coating formulations are mostly regarded as food additives and are listed within the list of additives for general purposes, although pectins, Acacia and karaya gums, beeswax, polysorbates, fatty acids, and lecithin are mentioned apart for coating applications (ED, 1995). In any case, the use of these coating forming substances is allowed, provided that the *quantum satis* principle is observed. This directive was complemented recently by the introduction of specific purity criteria for food additives (ED, 2008). With respect to protein-based coatings, the most commonly used proteins such as corn zein, wheat gluten, soy protein, and milk proteins also have GRAS status. However, some concerns have been raised due to the allergenicity or intolerance some consumers have to wheat proteins or lactose. The presence of a coating with a known allergen on a food must be clearly labeled, because many edible coatings are made with ingredients that could cause allergic reactions. Within these allergens, milk, soybeans, fish, peanuts, nuts, and wheat are the most important. The relationship of wheat and other cereal proteins to celiac disease, particularly, should be mentioned in the label. In fact, the use of any edible coating based on cereal proteins should be accompanied by proper labeling, especially when films or coatings are applied on fresh fruits and vegetables, because this could deprive celiac patients of their dietary needs.

11.8 FINAL REMARKS

Edible coatings are promising systems for the improvement of quality, shelf life, safety, and functionality of fresh-cut fruits and vegetables. However, more research is still needed to gather information about coating formulations, especially when new ingredients are incorporated. Moreover, scientific research has to be carried out to identify safety issues related to the potential toxicity or allergenicity of some edible coating materials, as well as their sensory implications. Finally, most studies on fresh-cut applications have been conducted at a laboratory scale. Further research should be focused on a commercial scale with the purpose of providing more realistic information that can be used to commercialize fresh-cut products coated with edible films or coatings.

REFERENCES

Assis, O. B., and Pessoa, J. D. 2004. Preparation of thin films of chitosan for use as edible coatings to inhibit fungal growth on sliced fruits. *Brazilian Journal of Food Technology* 7:7–22.

Audic, J. L., Chaufer, B., and Daufin, G. 2003. Non-food applications of milk components and dairy co-products: A review. *Lair* 83:417–438.

Avena-Bustillos, R. J., Cisneros-Zevallos, L. A., Krochta, J. M., and Saltveit, M. E. 1994a. Application of casein-lipid edible film emulsions to reduce white blush on minimally processed carrots. *Postharvest Biology and Technology* 4:319–329.

Avena-Bustillos, R. J., and Krochta, J. M. 1993. Water vapor permeability of caseinate-based edible films as affected by pH, calcium crosslinking and lipid content. *Journal of Food Science* 58:904–907.

Avena-Bustillos, R. J., Krochta, J. M., and Saltveit, M. E. 1997. Water vapour resistance of Red Delicious apples and celery sticks coated with edible caseinate-acetylated monoglyceride films. *Journal of Food Science* 62:351–354.

Avena-Bustillos, R. J., Krochta, J. M., Saltveit, M. E., Rojas-Villegas, R. J., and Sauceda-Perez, J. A. 1994b. Optimization of edible coating formulations on zucchini to reduce water loss. *Journal of Food Engineering* 21:197–214.

Baldwin, E. A., and Baker, R. A. 2002. Use of proteins in edible coatings for whole and minimally processed fruits and vegetables. In *Protein Based Films and Coatings*, ed. A. Gennadios, 501–515. Boca Raton, FL: CRC Press.

Baldwin, E. A., Nisperos-Carriedo, M. O., and Baker, R. A. 1995. Use of edible coatings to preserve quality of lightly (and slightly) processed products. *Critical Reviews in Food Science and Nutrition* 35:509–524.

Baldwin, E. A., Nisperos, M. O., Chen, X., and Hagenmaier, R. D. 1996. Improving storage life of cut apple and potato with edible coating. *Postharvest Biology and Technology* 9:151–163.

Banks, N. H. 1984. Studies of the banana fruit surface in relation to the effects of TAL prolong coating on gaseous exchange. *Scientia Horticulturae* 24:279–286.

Bauchot, A. D., John, P., Soria, Y., and Recasens, I. 1995. Sucrose ester-based coatings formulated with food-compatible antioxidants in the preservation of superficial scald in stored apples. *Journal of the American Society for Horticultural Science* 120:491–496.

Brancoli, N., and Barbosa-Cánovas, G. V. 2000. Quality changes during refrigerated storage of packaged apple slices treated with polysaccharide films. In *Innovations in Food Processing*, ed. G. V. Barbosa-Cánovas, and G. W. Gould, 243–254. Lancaster, PA: Technomic.

Bryan, D. S. 1972, December 26. Prepared citrus fruit halves and method of making the same. U.S. patent 3,707,383.

Buffo, R. A., and Han, J. H. 2005. Edible films and coatings from plant origin proteins. In *Innovations in Food Packaging*, ed. J. H. Han, 277–300. New York: Academic Press.

Cagri, A., Ustunol, Z., and Ryser, E. 2004. Antimicrobial edible films and coating. *Journal of Food Protection* 67:833–848.

Carrillo-Lopez, A., Ramirez-Bustamante, F., Valdez-Torres, J. B., Rojas-Villegas, R., and Yahia, E. M. 2000. Ripening and quality changes in mango fruit as affected by coating with an edible film. *Journal of Food Quality* 23:479–486.

Chen, H. 1995. Functional properties and applications of edible films made of milk proteins. *Journal of Dairy Science* 78:2563–2583.

Chien, P. J., Sheu, F., and Lin, H. R. 2007b. Quality assessment of low molecular weight chitosan coating on sliced red pitayas. *Journal of Food Engineering* 79:736–740.

Chien, P. J., Sheu, F., and Yang, F. H. 2007a. Effects of edible chitosan coating on quality and shelf life of sliced mango fruit. *Journal of Food Engineering* 78:225–229.

Choi, W. Y., Park, H. J., Ahn, D. J., Lee, J., and Lee, C. Y. 2002. Wettability of chitosan coating solution on 'Fuji' apple skin. *Journal of Food Science* 67:2668–2672.

Cuq, B., Gontard, N., and Guilbert, S. 1995. Edible films and coatings as active layers. In *Active Food Packaging*, ed. M. Rooney, 111–142. Glasgow: Blackie.

Day, B. 2001. Modified atmosphere packaging of fresh fruits and vegetables—an overview. *Acta Horticulturae* 553:585–590.

Debeaufort, F., and Voilley, A. 2009. Lipid-based edible films and coatings. In *Edible Films and Coatings for Food Applications*, ed. M. E. Embuscado and K. C. Huber, 135–168. New York: Springer.

Del-Valle, V., Hernández-Muñoz, P., Guarda, A., and Galotto, M. J. 2005. Development of a cactus-mucilage edible coating (*Opuntia ficus indica*) and its application to extend strawberry (*Fragaria ananassa*) shelf-life. *Food Chemistry* 91:751–756.

De Ruiter, G. A., and B. Rudolph. 1997. Carrageenan biotechnology. *Trends in Food Science and Technology* 8:389–429.

Devlieghere, F., Vermeulen, A., and Debevere, J. 2004. Chitosan: antimicrobial activity, interactions with food components and applicability as a coating on fruit and vegetables. *Food Microbiology* 21:703–714.

Dhalla, R., and Hanson, S. W. 1988. Effect of permeable coatings on the storage life of fruits. II. Pro-long treatment of mangoes (*Mangifera indica* L. cv. Julie). *International Journal of Food Science and Technology* 23:107–112.

Dominguez-López, A. 1995. Review: use of the fruit and stems of the prickly pear cactus (*Opuntia* spp.) into human food. *Food Science and Technology International* 1:65–74.

Donhowe, I. G., and Fennema, O. 1994. Edible films and coatings: characteristics, formation, definitions, and testing methods. In *Edible Coatings and Films to Improve Food Quality*, ed. J. M. Krochta, E. A. Baldwin, and M. O. Nisperos-Carriedo, 1–24. Lancaster, PA: Technomic.

Drake, S. R., Fellman, J. K., and Nelson, J. W. 1987. Postharvest use of sucrose polyesters for extending the shelf-life of stored Golden Delicious apples. *Journal of Food Science* 52:1283–1285.

Durango, A. M., Soares, N. F., and Andrade, N. J. 2006. Microbiological evaluation of an edible antimicrobial coating on minimally processed carrots. *Food Control* 17:336–341.

ED-European Parliament and Council Directive No. 95/2/EC. 1995. On food additive other than colors and sweeteners. http://ec.europa.eu/food/fs/sfp/addit_flavor/flav11_en. pdf. Accessed September 25, 2008.

ED-European Parliament and Council Directive No. 98/72/EC. 1998. On food additive other than colors and sweeteners. http://ec.europa.eu/food/fs/sfp/addit_flavor/flav11_en. pdf. Accessed September 25, 2008.

ED-European Parliament and Council Directive No. 2008/84/EC. 2008. Laying down specific purity criteria on food additives other than colours and sweeteners. http://eur-lex.europa. eu/LexUriServ/LexUriServ.do?uri=OJ:L:2008:253:0001:0175:EN:PDF. Accessed October 30, 2008.

Eissa, H. A. A. 2007. Effect of chitosan coating on shelf life and quality of fresh-cut mushroom. *Journal of Food Quality* 30:623–645.

Eswaranandam, S., Hettiarachchy, N. S., and Meullenet, J. F. 2006. Effect of malic and lactic acid incorporated soy protein coatings on the sensory attributes of whole apple and fresh-cut cantaloupe. *Journal of Food Science* 71:S307–S313.

FDA, U.S. Food and Drug Administration. 2006. Food additives permitted for direct addition to food for human consumption 21CFR172, subpart C. Coatings, Films and Related Substances.

Franssen, L. R., and Krochta, J. M. 2003. Edible coatings containing natural antimicrobials for processed foods. In *Natural Antimicrobials for Minimal Processing of Foods,* ed. S. Roller, 250–262. Boca Raton, FL: CRC Press.

García, J. M., Herrera, S., and Morilla, A. 1996. Effects of postharvest dips in calcium chloride on strawberry. *Journal of Agricultural and Food Chemistry* 44:30–33.

García, M. A., Martino, M. N., and Zaritzky, N. E. 1998. Starch-based coatings: effect on refrigerated strawberry (*Fragaria ananassa*) quality. *Journal of the Science of Food and Agriculture* 76:411–420.

Garcia, M. A., Martino, M. N., and Zaritzky, N. E. 2001. Composite starch-based coatings applied to strawberries (*Fragaria ananassa*). *Nahrung-Food* 45:267–272.

Gennadios, A., and Kurth, L. B. 1997. Application of edible coatings on meats, poultry and seafoods: a review. *Lebensmittel Wissenschaft und Technologie* 30:337–350.

Gennadios, A., McHugh, T. H., Weller, G. L., and Krochta, J. M. 1994. Edible coatings and films based on proteins. In *Edible Coatings and Films to Improve Food Quality*, ed. J. M. Krochta, E. A. Baldwin, and M. O. Nisperos-Carriedo, 201–277. Lancaster, PA: Technomic.

Guilbert, S., and Gontard, N. 1995. Edible and biodegradable food packaging. In *Foods and Packaging Materials—Chemical Interactions,* ed. P. Ackermann, M. Jägerstad, and T. Ohlsson, 159–168. London, UK: The Royal Society of Chemistry.

Guilbert, S., Gontard, N., and Gorris, L. G. M. 1996. Prolongation of the shelf life of perishable food products using biodegradable films and coatings. *Lebensmittel Wissenschaft und Technologie* 29:10–17.

Hambleton, A., Debeaufort, F., Beney, L., Karbowiak, T., and Voilley, A. 2008. Protection of active aroma compound against moisture and oxygen by encapsulation in biopolymeric emulsion-based edible films. *Biomacromolecules* 9:1058–1063.

Hambleton, A., Debeaufort, F., Bonnotte, A., and Voilley, A. 2009a. Influence of alginate emulsion-based films structure on its barrier properties and on the protection of micro-encapsulated aroma compound. *Food Hydrocolloids* 23:2116–2124.

Hambleton, A., Fabra, M. -J., Debeaufort, F., Dury-Brun, C., and Voilley, A. 2009b. Interface and aroma barrier properties of iota-carrageenan emulsion-based films used for encapsulation of active food compounds. *Journal of Food Engineering* 93:80–88.

Han, C., Lederer, C., McDaniel, M., and Zhao, Y. 2005. Sensory evaluation of fresh strawberries (*Fragaria ananassa*) coated with chitosan-based edible coatings. *Journal of Food Science* 70:S172–S178.

Han, C., Zhao, Y., Leonard, S. W., and Traber, M. G. 2004. Edible coatings to improve storability and enhance nutritional value of fresh and frozen strawberries (*Fragaria × ananassa*) and raspberries (*Rubus ideaus*). *Postharvest Biology and Technology* 33:67–78.

Han, J. 2002. Protein-based edible films and coatings carrying antimicrobial agents. In *Protein-Based Films and Coatings,* ed. A. Gennadios, 485–498. Boca Raton, FL: CRC Press.

Han, J. H. 2000. Antimicrobial food packaging. *Food Technology* 54:56–65.

Hardenburg, R. E. 1967. Wax and related coatings for horticultural products. A bibliography. *Agricultural Research Bulletin* 51–15, USDA, Washington, DC.

Hernández-Muñoz, P., Almenar, E., Ocio, M. J., and Gavara, R. 2006. Effect of calcium dips and chitosan coatings on postharvest life of strawberries (*Fragaria × ananassa*). *Postharvest Biology and Technology* 39:247–253.

Hernández-Muñoz, P., Almenar, E., Valle, V. D., Velez, D., and Gavara, R. 2008. Effect of chitosan coating combined with postharvest calcium treatment on strawberry (*Fragaria × ananassa*) quality during refrigerated storage. *Food Chemistry* 110:428–435.

Hirano, S. 1999. Chitin and chitosan as novel biotechnological materials. Art was here, but was deleted. *Polymer International* 48:732–734.

Howard, L. R., and Dewi, T. 1995. Sensory, microbiological and chemical quality of mini-peeled carrots as affected by edible coating treatment. *Journal of Food Science* 60:142–144.

Iyidogan, N. F., and Bayindirli, A. 2004. Effect of L-cysteine, kojic acid and 4-hexylresorcinol combination on inhibition of enzymatic browning in Amasya apple juice. *Journal of Food Engineering* 62:299–304.

Karel, M., and Lund, D. B. 2003. *Physical Principles of Food Preservation.* New York: Marcel Dekker.

Kasarda, D. D., Bernardin, J. E., and Nimmo, C. C. 1976. Wheat proteins. In *Advances in Cereal Science and Technology,* ed. Y. Pomeranz, 158–236. St. Paul, MN: American Association of Cereal Chemists.

Kester, J. J., and Fennema, O. 1986. Edible films and coatings: a review. *Food Technology* 40:47–59.

Kinsella, J. E. 1984. Milk proteins: physicochemical and functional properties. *Critical Reviews in Food Science and Nutrition* 21:197–262.

Krasaekoopt, W., and Mabumrung, J. 2008. Microbiological evaluation of edible coated fresh-cut cantaloupe. *Kasetsart Journal—Natural Science* 42:552–557.

Krochta, J. M., and De Mulder-Johnston, C. 1997. Edible and biodegradable polymer films: challenges and opportunities. *Food Technology* 51:61–74.

Lacroix, M., and Cooksey, K. 2005. Edible films and coatings from animal-origin proteins. In *Innovations in Food Packaging,* ed. J. H. Han, 301–317. New York: Academic Press.

Lafortune, R., Caillet, S., and Lacroix, M. 2005. Combined effects of coating, modified atmosphere packaging, and gamma irradiation on quality maintenance of ready-to-use carrots (*Daucus carota*). *Journal of Food Protection* 68:353–359.

Lee, J. Y., Park, H. J., Lee, C. Y., and Choi, W. Y. 2003. Extending shelf-life of minimally processed apples with edible coatings and antibrowning agents. *Lebensmittel Wissenschaft und Technologie* 36:323–329.

Le Tien, C., Vachon, C., Mateescu, M. A., and Lacroix, M. 2001. Milk protein coatings prevent oxidative browning of apples and potatoes. *Journal of Food Science* 66:512–516.

Li, P., and Barth, M. M. 1998. Impact of edible coatings on nutritional and physiological changes in lightly processed carrots. *Postharvest Biology and Technology* 14:51–60.

Lim, L. T., Mine, Y., Britt, I. J., and Tung, M. A. 2002. Formation and properties of egg white protein films and coatings. In *Protein-Based Films and Coatings*, ed. A. Gennadios, 233–252. Boca Raton, FL: CRC Press.

Lin, D., and Zhao, Y. 2007. Innovations in the development and application of edible coatings for fresh and minimally processed fruits and vegetables. *Comprehensive Reviews in Food Science and Food Safety* 6:60–75.

Lin, S. Y. D., and Krochta, J. M. 2005. Whey protein coating efficiency on surfactant-modified hydrophobic surfaces. *Journal of Agricultural and Food Chemistry* 53:5018–5023.

Lourdin, D., Valle, G. D., and Colonna, P. 1995. Influence of amylase content on starch films and foams. *Carbohydrate Polymers* 27:261–270.

Martin-Belloso, O., Rojas-Graü, M. A., and Soliva-Fortuny, R. 2009. Delivery of flavor and active ingredients using edible films and coatings. In *Edible Films and Coatings for Food Applications*, ed. M. E. Embuscado and K. C. Huber, 295–313. New York: Springer.

Martínez-Romero, D., Alburquerque, N., Valverde, J. M., Guillén, F., Castillo, S., Valero, D., and Serrano, M. 2006. Postharvest sweet cherry quality and safety maintenance by Aloe Vera treatment: a new edible coating. *Postharvest Biology and Technology* 39:93–100.

Mattheis, J., and Fellman, J. K. 2000. Impacts of modified atmosphere packaging and controlled atmospheres on aroma, flavor, and quality of horticultural commodities. *HortTechnology* 10:507–510.

McGarvie, D., and Parolis, H. 1979. The mucilage of *Opuntia ficusindica*. *Carbohydrate Research* 69:171–179.

McHugh, T. H., and Krochta, J. M. 1994a. Water vapor permeability properties of edible whey protein-lipid emulsion films. *Journal of the American Oil Chemists' Society* 71:307–312.

McHugh, T. H., and Krochta, J. M. 1994b. Permeability properties of edible films. In *Edible Coatings and Films to Improve Food Quality*, ed. J. M. Krochta, E. A. Baldwin, and M. O. Nisperos-Carriedo, 139–187. Lancaster, PA: Technomic.

McHugh, T. H., and Senesi, E., inventors. 1999. USDA-ARS-WRRC assignee. Filed 1999 June 11. Fruit and vegetable edible film wraps and methods to improve and extend the shelf life of foods. U.S. Patent Application Serial No. 09/330,358.

McHugh, T. H., and Senesi, E. 2000. Apple wraps: a novel method to improve the quality and extend the shelf life of fresh-cut apples. *Journal of Food Science* 65:480–485.

Mei, Y., Zhao, Y., Yang, J., and Furr, H. C. 2002. Using edible coating to enhance nutritional and sensory qualities of baby carrots. *Journal of Food Science* 67:1964–1968.

Miller, K.S., and Krochta, J.M. 1997. Oxygen and aroma barrier properties of edible films: a review. *Trends in Food Science and Technology* 8:228–237.

Min, S., and Krochta, J. M. 2005. Inhibition of Penicillium commune by edible whey protein films incorporating lactoferrin, lactoferrin hydrolysate, and lactoperoxidase systems. *Journal of Food Science* 70:M87–M94.

Montero-Calderón, M., Rojas-Graü, M. A., and Martin-Belloso, O. 2008. Effect of packaging conditions on quality and shelf-life of fresh-cut pineapple (*Ananas comosus*). *Postharvest Biology and Technology* 50:182–189.

Mounts, T. L., Wolf, W. J., and Martinez, W. H. 1987. Processing and utilization. In *Soybeans: Improvement, Production, and Uses,* 820–866. Madison, WI: American Society of Agronomy Inc., Crop Science Society of America Inc., and Soil Science Society of America, Inc.

Myllarinen, P., Buleon, A., Lahtinen, R., and Forssell, P. 2002. The crystallinity of amylase and amylopectin films. *Carbohydrate Polymers* 48:41–48.

Narayan, R. 1994. Polymeric materials from agricultural feedstocks. In: *Polymers from Agricultural Coproducts,* ed. M. L. Fishman, R. B. Friedman, and S. J. Huang, 2–28. Washington, DC: American Chemical Society.

Nieto, M. B. 2009. Structure and function of polysaccharide gum-based edible films and coatings. In *Edible Films and Coatings for Food Applications,* ed. M. E. Embuscado and K. C. Huber, 57–112. New York: Springer.

Nisperos-Carriedo, M. O., Baldwin, E. A., and Shaw, P. E. 1991. Development of an edible coating for extending postharvest life of selected fruits and vegetables. *Proceedings of the Florida State Horticultural Society* 104:122–125.

Nussinovitch, A. 2003. *Water Soluble Polymer Applications in Foods.* Oxford: Blackwell Science.

Olivas, G. I., and Barbosa-Canovas, G. V. 2005. Edible coatings for fresh-cut fruits. *Critical Review in Food Science and Nutrition* 45:657–670.

Olivas, G. I., Mattinson, D. S., and Barbosa-Canovas, G. V. 2007. Alginate coatings for preservation of minimally processed 'Gala' apples. *Postharvest Biology and Technology* 45:89–96.

Olivas, G. I., Rodriguez, J. J., and Barbosa-Cánovas, G. V. 2003. Edible coatings composed of methylcellulose, stearic acid, and additives to preserve quality of pear wedges. *Journal of Food Processing and Preservation* 27:299–320.

Oms-Oliu, G., Soliva-Fortuny, R., and Martín-Belloso, O. 2008a. Using polysaccharide-based edible coatings to enhance quality and antioxidant properties of fresh-cut melon. *LWT—Food Science and Technology* 41:1862–1870.

Oms-Oliu, G., Soliva-Fortuny, R., and Martín-Belloso, O. 2008b. Edible coatings with antibrowning agents to maintain sensory quality and antioxidant properties of fresh-cut pears. *Postharvest Biology and Technology* 50:87–94.

Park, H. J. 1999. Development of advanced edible coatings for fruits. *Trends in Food Science and Technology* 10:254–260.

Park, S. I., Stan, S. D., Daeschel, M. A., and Zhao, Y. Y. 2005. Antifungal coatings on fresh strawberries (*Fragaria* × *ananassa*) to control mold growth during cold storage. *Journal of Food Science* 70:M202–M207.

Park, S., and Zhao, Y. 2004. Incorporation of a high concentration of mineral or vitamin into chitosan-based films. *Journal of Agricultural and Food Chemistry* 52:1933–1939.

Pavlath, A. E., and Orts, W. 2009. Edible films and coatings: why, what, and how? In *Edible Films and Coatings for Food Applications,* ed. M. E. Embuscado and K. C. Huber, 1–23. New York: Springer.

Pen, L. T., and Jiang, Y. M. 2003. Effects of chitosan coating on shelf life and quality of fresh-cut Chinese water chestnut. *LWT—Food Science and Technology* 36:359–364.

Pennisi, E. 1992. Sealed in (plastic) edible film. *Science News* 141:12.

Perez-Gago, C., Rojas, C., and Del Rio, M. A. 2003. Effect of hydroxypropyl methylcellulose-lipid edible composite coating on plum (cv. Autumn giant) quality during storage. *Journal of Food Science* 68:879–883.

Perez-Gago, M. B., Serra, M., and del Rio, M. A. 2006. Color change of fresh-cut apples coated with whey protein concentrate-based edible coatings. *Postharvest Biology and Technology* 39:84–92.

Plotto, A., Goodner, K. L., and Baldwin, E. A. 2004. Effect of polysaccharide coating on quality of fresh cut mangoes (*Mangifera indica*). *Proceedings of the Florida State Horticultural Society* 117:382–388.

Rakotonirainy, A. M., Wang, Q., and Padua, G. W. 2001 Evaluation of zein films as modified atmosphere packaging for fresh broccoli. *Journal of Food Science* 66:1108–1111.

Raybaudi-Massilia, R. M., Mosqueda-Melgar, J., and Martín-Belloso, O. 2008b. Edible alginate-based coating as carrier of antimicrobials to improve shelf-life and safety of fresh-cut melon. *International Journal of Food Microbiology* 121:313–327.

Raybaudi-Massilia, R. M., Rojas-Graü, M. A., Mosqueda-Melgar, J., and Martín-Belloso, O. 2008a. Comparative study on essential oils incorporated into an alginate-based edible coating to assure the safety and quality of fresh-cut Fuji apples. *Journal of Food Protection* 71:1150–1161.

Riaz, M. N. 1999. Processing biodegradable packaging material from starches using extrusion technology. *Cereal Food World* 705–709.

Ribeiro, C., Vicente, A. A., Teixeira, J. A., and Miranda, C. 2007. Optimization of edible coating composition to retard strawberry fruit senescence. *Postharvest Biology and Technology* 44:63–70.

Richard, F. C., Goupy, P. M., and Nicolas, J. J. 1992. Cysteine as an inhibitor of enzymatic browning. 2. Kinetic studies. *Journal of Agricultural and Food Chemistry* 40:2108–2114.

Ridley, B. L., O'Neill, M. A., and Mohnen, D. 2001. Pectins: structure, biosynthesis, and oligogalacturonide-related signaling. *Phytochemistry* 57:929–967.

Rojas-Graü, M. A., Raybaudi-Massilia, R. M., Soliva-Fortuny, R. C., Avena-Bustillos, R. J., McHugh, T. H., and Martín-Belloso, O. 2007b. Apple puree-alginate edible coating as carrier of antimicrobial agents to prolong shelf-life of fresh-cut apples. *Postharvest Biology and Technology* 45:254–264.

Rojas-Graü, M. A., Sobrino-López, A., Tapia, M. S., and Martín-Belloso, O. 2006. Browning inhibition in fresh-cut 'Fuji' apple slices by natural antibrowning agents. *Journal of Food Science* 71:S59–S65.

Rojas-Graü, M. A., Soliva-Fortuny, R., and Martín-Belloso, O. 2009. Edible coatings to incorporate active ingredients to fresh-cut fruits: a review. *Trends in Food Science and Technology* 20:438–447.

Rojas-Graü, M. A., Tapia, M. S., and Martin-Belloso, O. 2008. Using polysaccharide-based edible coatings to maintain quality of fresh-cut Fuji apples. *Lebensmittel Wissenschaft und Technologie* 41:139–147.

Rojas-Graü, M. A., Tapia, M. S., Rodríguez, F. J., Carmona, A. J., and Martín-Belloso, O. 2007a. Alginate and gellan based edible coatings as support of antibrowning agents applied on fresh-cut Fuji apple. *Food Hydrocolloids* 21:118–127.

Serrano, M., Valverde, J. M., Guillen, F., Castillo, S., Martínez-Romero, D., and Valero, D. 2006. Use of Aloe vera gel coating preserves the functional properties of table grapes. *Journal of Agricultural of Food Chemistry* 54:3882–3886.

Shackel, K. A., Greve, C., Labavitch, J. M., and Ahmadi, H. 1991. Cell turgor changes associated with ripening in tomato pericarp tissue. *Plant Physiology* 97:814–816.

Shon, J., and Haque, Z. U. 2007. Efficacy of sour whey as a shelf-life enhancer: use in antioxidative edible coatings of cut vegetables and fruit. *Journal of Food Quality* 30:581–593.

Shukla, R., and Cheryan, M. 2001. Zein: the industrial protein from corn. *Industrial Crops and Products* 13:171–192.

Sonti, S., Prinyawiwatkul, W., No, H. K., and Janes, M. E. 2003. Maintaining quality of fresh-cut apples with edible coating during 13-days refrigerated storage. In *IFT Annual Meeting Book of Abstracts,* 45. New Orleans, LA: Institute of Food Technologists.

Sothornvit, R., and Krochta, J. M. 2000. Plasticizer effect on oxygen permeability of β-lactoglobulin films. *Journal of Agricultural and Food Chemistry* 48:6298–6302.

Sothornvit, R., and Krochta, J. M. 2005. Plasticizers in edible films and coatings. In *Innovations in Food Packaging,* ed. J. H. Han, 403–433. New York: Academic Press.

Sothornvit, R., and Rodsamran, P. 2008. Effect of a mango film on quality of whole and minimally processed mangoes. *Postharvest Biology and Technology* 47:407–415.

Sumnu, G., and Bayindirli, L. 1995. Effects of coatings on fruit quality of Amasya apples. *Lebensmittel-Wissenschaft und-Technologie* 28:501–505.

Sworn, G. 2000. Gellan gums. In *Handbook of Hydrocolloids,* ed. G. O. Phillips and P. A. Williams, 117–135. Boca Raton, FL: CRC Press/Woodhead.

Takahashi, R., Tokunou, H., Kubota, K., Ogawa, E., Oida, T., Kawase, T., and Nishinari, K. 2004. Solution properties of gellan gum: change in chain stiffness between single- and double-stranded chains. *Biomacromolecules* 5:516–523.

Tanada-Palmu, P. S., and Grosso, C. R. F. 2005. Effect of edible wheat gluten-based films and coatings on refrigerated strawberry (*Fragaria ananassa*) quality. *Postharvest Biology and Technology* 36:199–208.

Tapia, M. S., Rojas-Graü, M. A., Carmona, A., Rodriguez, F. J., Soliva-Fortuny, R., and Martin-Belloso, O. 2008. Use of alginate and gellan-based coatings for improving barrier, texture and nutritional properties of fresh-cut papaya. *Food Hydrocolloids* 22: 1493–1503.

Tapia, M. S., Rojas-Graü, M. A., Rodríguez, F. J., Ramírez, J., Carmona, A., and Martin-Belloso, O. 2007. Alginate- and gellan-based edible films for probiotic coatings on fresh-cut fruits. *Journal of Food Science* 72:E190–E196.

Tasdelen, O., and Bayindirli, L. 1998. Controlled atmosphere storage and edible coating effects on storage life and quality of tomatoes. *Journal of Food Processing and Preservation* 22:303–320.

Tay, S. L., and Perera, C. O. 2004. Effect of 1-methylcyclopropene treatment and edible coatings on the quality of minimally processed lettuce. *Journal of Food Science* 69:131–135.

Toivonen, P. M. A., and Brummell, D. A. 2008. Biochemical bases of appearance and texture changes in fresh-cut fruit and vegetables. *Postharvest Biology and Technology* 408:1–14.

Trachtenberg, S., and Mayer, A. M. 1981. Composition and properties of Opuntia Ficus-indica mucilage. *Phytochemistry* 20:2665–2668.

Ustunol, Z. 2009. Edible films and coatings for meat and poultry. In *Edible Films and Coatings for Food Applications*, ed. M. E. Embuscado and K. C. Huber, 245–268. New York: Springer.

Valverde, J. M., Valero, D., Martinez-Romero, D., Guillen, F., Castillo, S., and Serrano, M. 2005. Novel edible coating based on aloe vera gel to maintain table grape quality and safety. *Journal of Agricultural and Food Chemistry* 53:7807–7813.

Vargas, M., Pastor, C., Chiralt, A., McClements, D. J., and González-Martínez, C. 2008. Recent advances in edible coatings for fresh and minimally processed fruits. *Critical Reviews in Food Science and Nutrition* 48:496–511.

Wong, W. S., Tillin, S. J., Hudson, J. S., and Pavlath, A. E. 1994. Gas exchange in cut apples with bilayer coatings. *Journal of Agricultural and Food Chemistry* 42:2278–2285.

Xu, S., Chen, X., and Sun, D. W. 2001. Preservation of kiwifruit coated with an edible film at ambient temperature. *Journal of Food Engineering* 50:211–216.

Yang, L., and Paulson, A. T. 2000. Effects of lipids on mechanical and moisture barrier properties of edible gellan film. *Food Research International* 33:571–578.

12 Hazard Analysis and Critical Control Point and Hygiene Considerations for the Fresh-Cut Produce Industry

Peter McClure

CONTENTS

12.1 INTRODUCTION

Food safety management systems have evolved from the surveillance-based approaches that considered reported foodborne disease statistics, surveillance of foods, food handlers, food-processing operations, and training and education. This evolution progressed to assessment of producer's risk (chance that a product will cause foodborne disease in the marketplace) and assessment of hazards in the mid-1970s. At the time, it was concluded that the only practical approach, with potential for immediate application, was hazard assessment that considered severity of the hazard and specifying adequate safety margins to minimize the hazard.

In food-processing businesses, this early work led to the development of risk assessment systems, with the objective of providing clearer insight into the requirements for the most effective controls necessary to assure consumer safety. The philosophy behind the approach taken and the techniques applied at the time (including what is now termed dose-response and predictive mathematical models used in exposure assessment) were early examples of the *safety by design* approach now adopted by many food producers. The developments evolved into a systems approach involving multidisciplinary teams (including microbiologists, process engineers, and process scientists) that were able to carry out hazard identification, semiquantification and ranking of risks, and identification of effective controls and monitoring procedures. At this stage, HAZOP (HAZard and OPerability) or fault tree analysis was applied, and the HACCP (Hazard Analysis and Critical Control Points system) was in the early stages of being recognized as the preferred approach for food safety management in operational practice. The HACCP system is recognized as having significant advantages over "inspectional" approaches because of the inherent weaknesses in microbiological sampling and testing, the time required to obtain results, and the associated costs.

The continuing trend for fresher, less heavily processed or preserved and more nutritious foods and the all-year-round availability of fresh foods have led to this group of products becoming an increasing component of the human diet. The fresh-cut sector of the fresh produce market is the fastest-growing segment in the United States (FDA, 2007). The sourcing of this produce from regions all over the globe also means that hazards associated with these materials are introduced into areas where they may not normally occur, and they may be spread over wide geographic regions. Unfortunately, the increase in consumption of contaminated fresh produce has led to a significant rise in foodborne disease caused by these vehicles. In the 1970s, these products were responsible for 1% of foodborne disease cases in the United States, and in the 1990s, this figure rose to 12% (Sivapalasingam et al., 2004). The period between 1990 and 2003 saw fresh fruit and vegetables overtake poultry in the number of foodborne disease outbreaks in the United States, where they numbered 554. Of 72 outbreaks associated with fresh produce in the United States between 1996 and 2006, 25% of these were linked to fresh-cut produce (FDA, 2007). In Europe, a total of 77 outbreaks associated with fruit or vegetables were reported between 1997 and 2007, involving six different microorganisms (Heaton and Jones, 2008). There is evidence that changes in production, processing, and distribution practices have led to some of these outbreaks (Beuchat, 2002), but it is also the case that pathogens such as *Escherichia coli* O157 are now more prevalent in the environment than they used to be.

The reasons or causes for these outbreaks are not always identified, and the very nature of fresh produce means that control of hazards may be limited or restricted by the intervention techniques or approaches available because of the impact on product characteristics. The application of HACCP to fresh produce at the farm level is a relatively recent development, partly stimulated by the recent increases in foodborne disease associated with these products and the recognition by national and international food safety authorities that control of hazards at the farm level is critical for assuring consumer safety. In some parts of the fresh produce sector, the adoption of HACCP has been slow, and it is considered by some groups to be inappropriate due to the lack of science and understanding that enables the identification of reliable critical control points and their corrective actions. Other groups, such as the International Fresh-Cut Produce Association (IFPA), have recognized the importance of HACCP and have provided guidelines to the industry for designing specific HACCP plans for fresh-cut produce. In a survey of members in 1997, the IFPA reported that 61% of respondents had implemented verified HACCP plans (NACMCF, 1999).

This chapter describes the principles of HACCP and supporting food safety management systems, such as Good Agricultural Practice (GAP) and Good Hygienic Practice (GHP). It is recognized that small and medium-sized producers of fresh produce may not have the resources or expertise to carry out HACCP studies in the same depth that the larger food processors currently do, so this chapter also provides sources of information that should aid small businesses in managing the risks associated with produce they market.

12.2   HISTORY AND PRINCIPLES OF HAZARD ANALYSIS AND CRITICAL CONTROL POINT

The HACCP system was originally devised by Pillsbury, the National Aeronautics and Space Administration (NASA), and the U.S. Army Laboratories at Natick, Massachusetts, to ensure that foods consumed by astronauts were safe (Bauman, 1990). Since then, HACCP has become the principal internationally recognized and accepted method for food safety assurance. It was originally targeted at microbiological safety, but it now encompasses chemical and physical hazards in foods.

The HACCP system is based on seven principles:

1. Hazard analysis (identify hazards and assess their severity and risk).
2. Determine critical control points.
3. Specify criteria to ensure control (establish critical limits).
4. Monitor critical control points (establish a system to monitor control points).
5. Take corrective action whenever monitoring indicates criteria are not met (establish the corrective action to take when a critical control point [CCP] is not under control).

6. Verify that the system is functioning as planned (establish procedures for verification to confirm that HACCP is working effectively).
7. Establish documentation concerning all procedures and records appropriate to these principles and their applications.

The main aims of HACCP are to identify relevant hazards and control points that are critical for assuring safety. If a hazard must be controlled but no CCPs are identified, redesign of the operation should be considered. The HACCP plan should be reviewed when the process or equipment is changed, when consumer use is different, and when new hazards appear that may be relevant to consumer safety.

12.3   APPLICATION OF HACCP

There are a number of steps involved in the application of HACCP and development of a HACCP plan. These are summarized in Figure 12.1. The following sections describe these steps in more detail.

12.3.1   ASSEMBLE HACCP TEAM

The fresh produce operation needs to ensure that the relevant specific knowledge and expertise are used for the development of an effective HACCP plan. This is often achieved by assembling a multidisciplinary team, with up to six team members, but in many cases there will be smaller numbers of "skill sets" available, and some experience and knowledge may not be available in-house. Where such expertise is not available, advice should be sought from other sources. Expert information for fresh produce is available from trade associations such as the Chilled Food Association in the United Kingdom (http://www.chilledfood.org/index.htm), the FDA (http://www.cfsan.fda.gov/~dms/prodgui3.html), the International Sprout Growers Association (http://www.isga-sprouts.org/), and the IFPA (http://www.unitedfresh.org/about/history). In this step, the scope of the HACCP study is also identified, which includes the types/classes of hazards to be considered and the extent of control a particular business is able to exert over the operations covered. For fresh produce, all three classes of hazards (see Table 12.1) may be relevant and important, and the operations covered may well include consideration of the planting and growing areas (proximity to sources of biological, chemical, and physical hazard contamination, such as farm animal waste and runoff that may be contaminated); preharvest, harvesting, and postharvest processing; transport; handling; and storage. Some materials used (e.g., seed, fertilizer, pesticides) may not be under direct control of the business, but any hazards associated with these should be considered in the HACCP study.

12.3.2   DESCRIBE THE PRODUCT

In the case of fresh-cut produce, the product may be a single fruit, vegetable, or herb, or it may be a mixture of different produce. The product description also includes the packaging material and gas used for the product, other ingredients used (e.g., those used to control hazards and some materials that may constitute hazards in their own right), intended

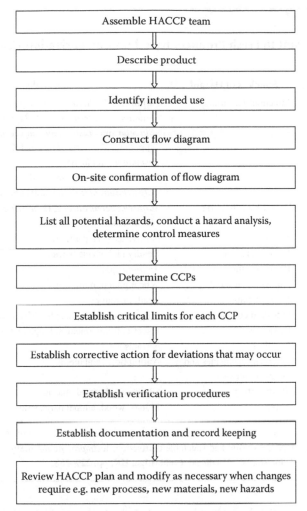

FIGURE 12.1 Process map for application of HACCP. World Health Organization (WHO) 1997 HACCP—Introducing the Hazard Analysis and Critical Control Point System, Food Safety Unit, Programme of Food Safety and Food Aid, WHO, Geneva. Page 23, 1997. With permission.

storage conditions, and shelf life of the product. The description should also consider any aspects of preharvest or postharvest processing that are relevant to the product, such as residues that may be present, washing, and precooking of the ingredients or some of the ingredients. This information will aid in one of the subsequent steps that analyzes the hazards to be considered and subsequent CCPs that have to be established.

12.3.3 IDENTIFY THE INTENDED USE OF THE PRODUCT

The intended use of the product may have a significant impact on the hazards present on fresh produce at the time of release from the business operation. It may also give

TABLE 12.1
Hazards Relevant to Fresh Produce, Based on Categorization Used by Early in 2005

Hazard Class	Subclass of Hazard	Examples
Biological	Infectious bacteria	Pathogenic *Escherichia coli* (e.g., *E. coli* O157:H7), *Salmonella enterica* (e.g., *S. Typhimurium*), *Shigella* spp., *Listeria monocytogenes*, *Yersinia* spp. (e.g., *Y. enterocolitica* and *Y. pseudotuberculosis*)
	Toxico-infectious bacteria	**Bacillus cereus** and other pathogenic **Bacillus** spp., **Clostridium botulinum** (infant botulism)
	Toxinogenic bacteria	**Staphylococcus aureus, C. perfringens**, *C. botulinum*
	Viruses	Noroviruses, Hepatitis A
	Poisonous plants	Deadly nightshade berries
	Poisonous fungi	*Amanita* spp.
	Allergenic materials	Wheat gluten, nuts/seeds
Chemical	Heavy metals	Lead, mercury
	Industrial contaminants	Polychlorinated biphenyls (PCBs), dioxins
	Pesticides	Insecticides, herbicides, fungicides, rodenticides
	Fertilizers	Nitrates
	Veterinary medicines and growth promoters	Antibiotics, clenbuterol
Physical	Sharp/cutting objects	Glass, metal, wood, plastic
	Choking	Stones, wood, animal parts, string, nuts, seed-stones

Note: Bold text represents additional examples to those provided by Early, R. 2005. Good agricultural practice and HACCP in fruit and vegetable cultivation. In *Improving the safety of fresh fruit and vegetables,* ed. Jongen, W., Woodhead, Cambridge, UK, pp. 229–268.

an opportunity for other hazards to be introduced into the product and should include all the steps, such as transport, storage, and handling by the food producer, retailer, and consumer or customer (in the case of food-service operators), and preparation prior to consumption. It is important that the producer takes account of consumer uses that may be different than those intended. For example, herbs and vegetables may be intended for use in foods intended to be cooked, but consumers may use these in ready-to-eat foods such as salads. The intended use also requires consideration of the likely consumers of the product and susceptible individuals who may be more vulnerable when exposed to particular hazards. The intended use should consider how the product will be stored and include temperature of storage and shelf life. Realistic storage temperatures and handling practices should be considered here. Often, fresh produce is intended to be stored at refrigeration temperatures, and the temperature of retail and consumer fridges can vary. Data on temperature distribution in fridges are available from various sources (Laguerre et al., 2002; Kosa et al., 2007; James and Evans, 2008).

There are examples of outbreaks of foodborne illness from fresh produce which were not considered by the producers. A number of outbreaks of botulism were linked to baked potatoes some years ago (Angulo et al., 1998; Bhutani et al., 2005), and more recently, an outbreak of botulism was linked to fresh carrot juice manufactured in the United States (FDA, 2007). It is thought that the product was temperature abused either during distribution or in the homes of consumers affected, allowing growth of proteolytic *Clostridium botulinum*, which would not occur at chill storage temperatures.

12.3.4 Construct the Flow Diagram

The flow diagram should contain information on each step being considered by the producer. For fresh products, this will often start with receipt of materials such as seeds, fertilizers, pesticides, and field or cultivating medium preparation. These stages are then typically followed by sowing, cultivation, harvesting, rinsing cut surfaces, packing, and transport. Product processing will often involve chilling, storage, further cutting, washing, rinsing, centrifugation/drying, packaging and labeling, secondary packaging, and chilled storage and transport. Refrigerated storage during further distribution and storage is an essential component for fresh produce in providing required shelf life. It may be appropriate to separate the flow diagram by hazard classes to simplify consideration of the overall HACCP plan. The information contained in the flowchart needs to be sufficiently detailed to permit consideration of how hazards may be introduced and how their concentration may change through further processing. The information used to build the flowchart may come from various sources, including suppliers, manufacturers of equipment, regulatory authorities, trade associations, in the form of guidelines or best practice documents, as well as from members of the HACCP team who are familiar with the cultivation, harvesting, and postharvest processing of the product. Examples of flow diagrams for fresh produce are available for field crop production (Early, 2005), endive production (Willocx et al., 1994), sprouted seeds (International Sprout Growers Association), shredded lettuce (IFPA), and tomatoes (Rushing et al., 1996). It is important to use these only as guides, as each HACCP study will be specific to the business operation under consideration. At this stage, the flowchart is a theoretical description of the process.

12.3.5 Confirm the Flow Diagram

Because the previous stage is a theoretical depiction of the process involved for the product in question, this next stage should provide confirmation of that information. Where practical, it is important to provide this confirmation by following the process on-site, with a working line, although this may be difficult if the product or parts of the line are new and not yet commissioned. Details of the operations conditions may be added or amended at a later stage. Ideally, the flow diagram is confirmed throughout all operational periods, including day, night, and weekend shifts, and should also consider line stoppages that may be anticipated.

12.3.6 Hazard Analysis

Hazard analysis considers all the procedures involved with production and the subsequent supply chain (e.g., field preparation, seed germination, drilling and planting, growing, fertilizer/pesticide application, irrigation, harvesting, processing, storage, handling, and distribution). This involves the identification of all relevant hazards, their sources and points of contamination, and the probability and extent of their survival or growth during each stage of "processing." It also includes consideration of the severity of the hazards identified and the risk they pose for consumer safety. The hazards considered should include biological agents, such as infectious, toxico-infectious, and toxigenic bacteria, viruses, protozoal parasites, fungal pathogens, and mycotoxins; poisonous plants and plant materials; poisonous fungi; allergenic materials; toxic chemicals, such as those used as pesticides; antibiotics that may be used in agriculture; and physical hazards, such as stones, glass, insects, and rodents. Table 12.1 provides examples of the various hazards that could be present on raw produce and is based on the categorization used by Early (2005).

Biological hazards commonly associated with fresh produce are described in more detail in Chapter 3 and also by Beuchat (2002), Dawson (2005), Koopmans and Duizer (2004), ICMSF (2005), NACMCF (1999), WHO (1998), Heaton and Jones (2008), and the FDA Guide to Minimize Microbial Food Safety Hazards for Fresh Fruits and Vegetables (FDA, 1998). Pesticide residues in relation to fresh produce are discussed by Winter (2005) and Chiang et al. (2005), and other chemical and physical hazards relevant to fresh produce are described by Rhodehamel (1992). For chemical and physical hazards, their concentration is unlikely to change unless there are dilution or concentration effects. For biological hazards, growth of these may occur during processing, despite fresh produce, such as fruit, having relatively low pH. Examples of survival and growth characteristics for pathogens associated with fresh produce are provided by NACMCF (1999). The principal determinants for growth are pH, temperature of storage, and water availability. Predictive models and independent growth and survival studies can provide valuable information to estimate changes in levels of microorganisms during storage. A recent summary of predictive models relevant to chilled foods is provided by McClure and Amézquita (2008). In many cases, if multiple pathogens may be introduced at different stages (e.g., through raw materials), they may be grouped together if they have common characteristics (e.g., susceptibility to decontamination/washing regimes), and the most resistant type is chosen as the target because other less resistant forms will therefore be automatically taken into account.

Other sources of information for hazard analysis include foodborne disease surveillance databases and records, including the Rapid Alert System for Food and Feed (RASFF) in Europe (http://ec.europa.eu/food/food/rapidalert/index_en.htm) and the *Morbidity and Mortality Weekly Report* (MMWR) in the United States (http://www.cdc.gov/mmwr/). It may be the case that emerging hazards are relevant to a product and may not have been identified in the original HACCP study. Although that agent may not have been associated with any foodborne disease cases caused by fresh produce, there may sufficient information to indicate that it should be identified as a relevant hazard. For example, *E. coli* O157:H7 was not associated with fresh produce a number of years ago, but there was sufficient information on the farm animal reservoirs of this serious pathogen and information on its low infectious dose, to indicate that it should be considered a risk in these products. Often, it is the case, however, that a new pathogen is not taken seriously until it has been linked to the food type in question.

More recently, inspection agencies responsible for ensuring proper implementation of HACCP have been looking for evidence to support nonidentification of hazards in HACCP plans. This is done to ensure that hazards are not overlooked by producers.

12.3.7 IDENTIFY CRITICAL CONTROL POINTS

Hazards may be controlled at a number of points in a process, and the intention of a CCP is to provide control sufficient to eliminate a hazard or reduce it to an acceptable level. A process that can be used to identify critical control points is shown in Figure 12.2. It is important that points that contribute to control of hazards are not confused with CCPs, and that there is a manageable number of CCPs identified. For example, there may be points at which good or best practice may contribute to minimizing presence or levels of particular hazards, but these are not CCPs.

The identification of CCPs in fresh produce production may be problematic due to the limited choice of intervention steps that can effectively reduce or eliminate hazards. For example, the application of heat treatment, such as pasteurization, to eliminate infectious pathogens, is often not possible because of the deleterious effects on the organoleptic properties of fresh produce. Consequently, controls points that prevent contamination of raw produce and growth of pathogenic microorganisms are of particular importance in fresh produce operations. The hazards that tend to pose the greatest risk, because of their characteristics and their presence in the farm environment or on food handlers, tend to be infectious agents that can cause disease even if they are present at low levels. In evaluating the field production system for fresh-market tomatoes, a critical point where microbial contamination would be likely to occur was not identified. Soil contact was minimized during production by holding vines off the ground and using plastic mulch. Fertilizer was fed through irrigation tubes, and no organic manures were used. The control points identified were in harvesting and handling.

It is necessary to distinguish between controls points that are required for food safety (CCPs) and those that are required to maintain quality. These are often referred to as Quality Control Points (QCPs), and noncritical points are referred to as Quality Points (QPs). The same process/rationale used to identify CCPs is used to identify QCPs and QPs. For fresh produce, agents that impact on quality can also impact on safety. For example, Wells and Butterfield (1996) showed that the rate of *Salmonella* isolation from bacterial soft-rot produce was twice as high compared to intact produce, and that *Salmonella* co-cultured with soft-rot-inducing organisms such as *Erwinia carotorova* or *Pseudomonas viridiflava* on potato, carrot, and green pepper grew more quickly. It is also the case that physically damaged produce is more susceptible to microbiological contamination and colonization, due to the availability of nutrient sources from within the fresh produce.

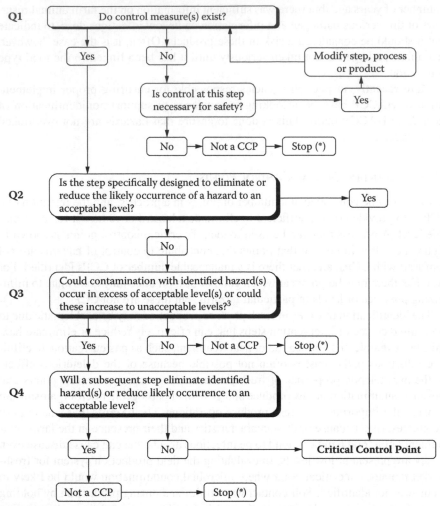

FIGURE 12.2 Example of questions used to identify CCPs. World Health Organization (WHO) 1997 HACCP—Introducing the Hazard Analysis and Critical Control Point System, Food Safety Unit, Programme of Food Safety and Food Aid, WHO, Geneva. Page 23, 1998. With permission.

12.3.8 Set Critical Limits

The limits set at CCPs should be directly or indirectly measurable and properly validated. A limit distinguishes between acceptable and unacceptable levels of a control point. For some hazards, such as metal fragments, it may be relatively straightforward to set limits (e.g., absence of detectable metal fragments in all materials processed passing through a particular stage in processing). Others may be defined by national or international regulatory agencies or bodies. For biological hazards, a limit set at a particular point in a process may need to anticipate subsequent increases in levels of the hazard, if this can occur in the process identified.

Examples of limits that may be set for fresh produce include absence of specific pathogens in a set quantity of product; maximum levels of pesticides or fertilizers in produce; maximum temperature/humidity during storage; and no glass, metal, animal, stone, or wood fragments in product at particular points in processing.

12.3.9 ESTABLISH MONITORING SYSTEMS

Monitoring systems are used to verify that controls are working as expected and to trigger alerts when control is lost. These systems are often automated and take measurements on a regular basis. The measurement system should provide information in sufficient time to make adjustments to prevent breaching of critical limits. It is often possible to trigger process adjustments when results indicate a trend toward loss of control at a CCP, before they reach that point. It is important that the data derived from monitoring are assessed by personnel with the awareness and authority to carry out corrective actions. Monitoring procedures for CCPs in fresh produce processing often rely on rapid measurements because of the limited shelf life of these products. Physical and chemical testing are often preferred to microbiological analysis because of the speed with which these can be carried out, and they will be indirect indicators of control of microbiological hazards. It is important to keep accurate records of results to provide evidence for verification, and in case outside parties seek evidence that a process is operating effectively. These records can also be very useful when deviations or failures occur during processing. Using statistical tools such as Cusum analysis (see Montgomery, 2005), it is often possible to identify when a process has lost control and also when control has been regained. This allows accurate identification of product batches that have been affected by the deviation, and these can be put on hold or destroyed, minimizing any losses for the producer.

12.3.10 ESTABLISH CORRECTIVE ACTION PROCEDURES

The purposes of corrective actions are to ensure that the CCP has been brought back under control and to deal effectively with any product that may be defective or substandard, sometimes identified in the previous step. To identify appropriate corrective actions, it is good practice to anticipate failures at CCPs and consider realistic "what if" scenarios.

12.3.11 VALIDATE HACCP PLAN, IMPLEMENT AND VERIFY HACCP SYSTEM

When an HACCP plan is completed, the next step is to validate this plan before implementation and subsequent verification. The validation step involves the gathering of evidence to show that the HACCP plan is effective (i.e., that it works when put into practice). A number of validation activities have been identified by ILSI (1999), and these include evidence to confirm the following:

- Skills in the HACCP are team adequate.
- The flow diagram adequately describes the process.

- Significant hazards have been considered and appropriate preventative measures identified to minimize levels and presence of hazards.
- Appropriate CCPs have been identified and are at the correct stages in the process.
- Critical limits are established for each hazard, and these are appropriate for the CCPs identified.
- Monitoring systems can demonstrate the efficacy of control and preventative measures.
- Corrective actions have been identified for each CCP, nonconforming product can be prevented from reaching the consumer, and persons responsible for these actions are identified and understand their responsibilities.
- Procedures are in place to verify plans and documented evidence describing the whole HACCP system is available and records are retained.

Verification is the application of methods, procedures, tests, and other evaluations in addition to monitoring, to determine compliance with the HACCP plan. This step provides confirmation that the relevant actions and outcomes identified in the HACCP plan have been carried out and achieved. The verification step needs to cover the whole production and processing cycle. The factors that can impact on fresh produce production are many and varied, and the verification of HACCP for these products can be more complex and require more iterations than for other product types.

12.3.12 ESTABLISH AND MAINTAIN DOCUMENTATION

Good record maintenance and documentation are key requirements for HACCP and provide the point of reference for the HACCP team, new members of the team, risk managers, and auditors who will examine evidence to demonstrate compliance with internal and external (e.g., regulatory) requirements and standards. The documentation includes specifications for raw materials and final product, material safety data sheets, procedures for preparation and storage of materials applied before planting/sowing, during cultivation, and subsequent processing, process flow diagram; control chart (including a list of identified hazards, control and preventative measures, and CCPs), monitoring procedures; corrective actions; records of measurements taken, validation and verification data.

12.3.13 REVIEW HACCP PLAN

Review of HACCP plans is often overlooked and should be undertaken on a regular basis, particularly when changes are made to the process or materials used in production. Any changes that could lead to the introduction of new hazards in production should trigger the HACCP team to review the HACCP plan and consider any knock-on effects for CCPs that were previously identified. In many cases, if new hazards are introduced, they may well be controlled by the existing control measures.

12.4   GOOD AGRICULTURAL PRACTICE

Good agricultural practice (GAP) and other procedures, such as hygienic practice, are sometimes referred to as prerequisite programs and should not be confused with HACCP. These prerequisite programs provide the underpinning procedures that must be in place in addition to any procedures identified in the HACCP plan.

Good agricultural practice is the application of quality assurance and management at the farm level. GAP guidance documents have been produced by the U.S. Food and Drug Administration (FDA)/United States Department of Agriculture (USDA) and are entitled "Guide to Minimize Microbial Food Safety Hazards for Fresh Fruits and Vegetables" (FDA/USDA, 1998) and "Guide to Minimize Microbial Food Safety Hazards of Fresh-Cut Fruits and Vegetables" (FDA/USDA, 2008). The first of these provides recommendations for growers, packers, and shippers to use good agricultural and good manufacturing practices in areas where they have control to prevent or minimize microbial food safety hazards in fresh produce. This guidance is based on eight basic principles and practices, as shown in Table 12.2. The elements considered in these guidelines are shown in Table 12.3. The second guidance document primarily addresses microbiological hazards and appropriate control measures for these, although some chapters also discuss physical and chemical hazards. The key areas covered in the second guidance document are personnel health and hygiene, training, building and equipment, and sanitation operations.

Other organizations have also provided guidance on GAP, to establish "best practice." The Food and Agriculture Organization (FAO, 2007) advocates a nonprescriptive method that considers the environmental, economic, and social sustainability of farm production and postproduction processes for production of safe and quality foods. This guidance is based on ten elements of agricultural practice, many of which are shared with the eight principles identified by the FDA/USDA. These ten elements are soil, water, crop/fodder production, crop protection, animal production, animal health and welfare, harvest and on-farm processing and storage, energy and waste management, human welfare, health, and safety, and wildlife and landscape. Although it may appear that aspects of animal production and welfare may have no relevance to fresh produce production, this is not necessarily the case. Many of the microbiological hazards that have caused disease and have been linked with fresh produce have animal reservoirs, and farm animals are major sources of these agents. Clearly, fresh produce producers must take account of these sources and put in place procedures that will minimize the possibility of these hazards coming into contact with produce they are growing.

The private sector has also developed standards that originated from the Euro-Retailer Produce Working Group (EUREP) and has now evolved into the GLOBALGAP standard. This standard covers crops, livestock, and aquaculture, with the crops including fruit and vegetables. More recently, GLOBALGAP and the Safe Quality Food (SQF) program in the United States have agreed to a harmonization of GAP and HACCP-based approaches for food safety management at the farm

TABLE 12.2
Basic Principles of Microbiological Food Safety in Relation to Growing, Harvesting, Packing, and Transporting Fresh Produce

Principle	Description
1	Prevention of microbiological contamination of fresh produce is favored over reliance on corrective actions once contamination has occurred.
2	To minimize microbiological food safety hazards in fresh produce, growers, packers, or shippers should use good agricultural and management practices in those areas over which they have control.
3	Fresh produce can become microbiologically contaminated at any point along the farm-to-table food chain. The major source of microbiological contamination with fresh produce is associated with human or animal feces.
4	Whenever water comes into contact with produce, its source and quality dictate the potential for contamination. Minimize the potential of microbiological contamination from water used with fresh fruits and vegetables.
5	Practices using animal manure or municipal biosolid wastes should be managed closely to minimize the potential for microbiological contamination of fresh produce.
6	Worker hygiene and sanitation practices during production, harvesting, sorting, packing, and transport play a critical role in minimizing the potential for microbiological contamination of fresh produce.
7	Follow all applicable local, state, federal laws and regulations, or corresponding or similar laws, regulations, or standards for operators outside the United States, for agricultural practices.
8	Accountability at all levels of the agricultural environment (farm, packing facility, distribution center, and transport operation) is important to a successful food safety program. There must be qualified personnel and effective monitoring to ensure that all elements of the program function correctly and to help track produce through the distribution channels to the producer.

Source: FDA/USDA, 1998.

level (Freshplaza, 2008). GLOBALGAP identified a number of control points and compliance criteria for fruit and vegetables that address many of the aspects that would be covered by prerequisite programs. These points and criteria cover soil and substrate management, irrigation, harvesting, and produce handling.

Legislation in the United Kingdom and Europe has moved responsibility for fresh produce away from government into the supply chain, where trade associations and retailers are driving the food safety agenda through development of their own standards (Monaghan, 2006). This recent study describes a code of practice developed by Marks and Spencer, called "Field to Fork." This code of practice (COP) covers all aspects of growing and packing of fresh produce, and includes sections on pesticide management, food safety, organic produce, traceability, environment, packing, and genetically modified organisms. This COP was developed following the principles of HACCP but acknowledges, in many cases, that it is not possible to eliminate risk at a CCP, only to minimize it. Recently updated examples of the requirements for site

TABLE 12.3
Good Agricultural Practice—
Areas for Consideration in
Produce Production and Harvest
Considered by FDA/USDA

Area

Water quality
Land history and surrounding properties
Soil amendments
Field sanitation
Pest control
Agricultural chemicals
Worker sanitation facilities
Worker health and hygiene
Containers and packing material
Tools and equipment
Transport
Post-harvest cooling
Storage
Product traceability

Source: Sperber, W.H. 2005. HACCP does not
work from farm to table. *Food Control,*
16, 6, 511–514.

selection and environmental impact management of organic material inputs in the current COP (Marks and Spencer, "Field to Fork: Fresh Produce Code of Practice") are shown in Table 12.4 and Table 12.5. The COP also provides specific instructions covering water, workers, personal protective clothing, and wildlife.

12.5 RECENT DEVELOPMENTS IN FOOD SAFETY MANAGEMENT

Quantitative microbiological risk assessment (QMRA) is a relatively recent development (Hatheway, 1997) aimed fundamentally at the protection of consumers, used in decision making on food safety issues and in helping responsible authorities to meet public health goals. To date, these have mostly focused on risks associated with single hazards and broad groups of products enabling government agencies to identify products of greatest concern to public health and to identify key aspects of their processing and handling that impact most on risk. More recently, QMRA has been applied within the food industry and is fast becoming an indispensable part of product/process design, and in the evaluation of control measures along the steps of manufacturing and the supply chain. The main purpose of QMRA is to quantify risk or to determine the effects of process or product interventions on risk. With a number of potential applications, the first step is to clearly determine the purpose of

TABLE 12.4

Site Selection and Environmental Input Management, Marks and Spencer's "Field to Fork" Code of Practice

	Question	Answer	Crop Category 1 and 2	Crop Category 3 and 4
1	Has grazing of land occurred in previous 18 months?	Yes	Do not crop until 18 months from date last grazed.	Risk assess.
		No	Go to Question 2	
2	Are active landfill/municipal waste disposal operations within 5 km?	Yes	Risk assess with specific reference to prevailing wind and bird flight paths. Do not crop within 1 km.	Risk assess with specific reference to prevailing wind and bird flight paths.
		No	Go to Question 3.	
3	Is the land at risk of flooding?	Yes	Do not crop if there is evidence of annual flooding. Do not crop for a period of 24 months post flooding by potentially contaminated water.	Risk assess impact of potential flood risk upon crop safety. Do not crop if there is evidence of flooding within the past 12 months by water contaminated with animal feces.
		No	Go to Question 4.	
4	Is there public access to the growing areas, such as public footpaths or roads?	Yes	Take all reasonable measures to control potential crop contamination with specific reference to dogs.	Take all reasonable measures to control potential crop contamination with specific reference to dogs.
		No	Go to Question 5.	
5	Are there any risks from neighboring agricultural business?	Yes	Do not crop if there is a risk of contamination by run off from livestock operations, spray application of slurries occurs within 1 km, or composting of animal wastes occurs within 1 km.	Risk assess potential for contamination.
		No		
6	Is there any animal (domestic, farm, or wildlife) activity that may pose a risk?	Yes	Risk assess and employ all reasonable measures to discourage animal activity in and around the crop/irrigation sources.	Risk assess and employ all reasonable measures to discourage animal activity in and around the crop/irrigation sources.

Notes: Category 1 represents crops eaten raw with no protective skin removed before eating or that have some risk or history of pathogen contamination; Category 2 represents crops that can be eaten raw and either have no protective skin removed before eating or have some risk or history of pathogen contamination; Category 3 represents crops that can be eaten raw, either have a protective skin or grow clear of the ground or have no significant risk or history of pathogen contamination; and Category 4 represents crops that are always cooked before eating.

Source: Mark's and Spencer's "Field to Fork" Code of Practice. Personal communiation. With permission.

TABLE 12.5

Management of Organic Material Inputs, Marks and Spencer's "Field to Fork" Code of Practice

	Question	Answer	Category 1 and 2 Crops	Category 3 and 4 Crops
1	Has raw manure been applied to land in past 24 months?	Yes	Do not crop for 24 months from date of last application.	Do not crop Category 3 crops for 18 months or Category 4 crops for 12 months from last application.
		No	Go to Question 2.	
2	Has composted, pelleted, or heat-treated animal manure been applied to land within 24 months?	Yes	Go to Question 3.	
		No	Go to Question 6.	
3	Has all animal manure applied/proposed for application to soil been fully composted to the standard set out in Marks and Spencer's composting protocol or satisfactorily heat treated to the specified acceptable standard?	Yes	No restriction.	No restriction.
		No	Go to Question 4.	
4	Has all animal manure applied/proposed for application been fully composted or processed and material certified as clear of pathogens by appropriate sampling and microanalysis, but control points are not fully verified?	Yes	Do not crop for 24 months unless granted a written derogation from a trained F2F Auditor who may reduce the period to 18 months.	If manure was from a nonintensive source assessed as free of *Escherichia coli* O157, no restriction.
		No	Treat as raw manure. Go to Question 1.	
5	Has Green waste material been applied to land in 3 months prior to earliest anticipated cropping?	Yes	Go to Question 6.	
		No	Go to Question 7.	
6	Has applied Green waste been composted in compliance with Marks and Spencer's Composting Protocol?		No restrictions.	No restrictions.
		No	Go to Question 7.	
7	Has applied Green waste intended for application been assessed as fully composted, with confidence that the constituents were free from animal manure/derivatives, and verified free of pathogens but control points were not verified?	Yes	Do not crop for 3 months.	Risk assess.
		No	Treat as raw manure (Q2)	

the QMRA—that is, to define what questions will be answered and identify which data and calculations are required. The first draft of a risk profile includes a detailed description of the product formulation and the manufacturing process, as well as the control measures implemented in the process. It should also include a review of available data in order to identify critical gaps that need to be filled before the QMRA can be conducted. It is noteworthy that much of the information gathered for HACCP plans is the same or similar to that used in QMRA. However, the purposes and objectives of these two activities should not be confused.

Generally, QMRAs are very data-demanding, and a combination of different sources is usually necessary. For example, data can be generated from experiments (e.g., challenge tests), retrieved from databases (e.g., ComBase) or scientific publications, obtained from models (e.g., temperature distributions from a heat transfer model), or come from expert elicitation. The more relevant data are available, the lower the uncertainty of risk estimates will be, thereby offering valuable guidance to risk managers for decision making. When data are missing, it is necessary to make assumptions during the QMRA process. These assumptions, along with all the data available, need to be clearly documented in order to make the conclusions of the risk assessment as transparent as possible.

The well-known QMRA structure established by the Codex Alimentarius (CAC, 1999) includes four major elements:

- *Hazard identification*: Hazards (pathogens or their toxins) of concern with the food product are identified as well as their association with adverse health effects. This part is predominantly qualitative, and sources of information are typically scientific literature, public databases, epidemiological data, relevant government bodies, and expert elicitation.
- *Exposure assessment*: Estimates the probability of intake of the pathogenic agent and the cell numbers consumed. Exposure assessments must indicate the portion size or the unit used to determine the amount of food to be consumed. The main outcome of exposure assessments is an estimation of the level of the pathogenic agent (with some level of uncertainty) and the likelihood of occurrence in foods at the point of consumption. In processing of fresh produce, this includes the initial prevalence and concentration of the pathogenic agent, the effect of any intervention steps, the response of the microorganism (growth, survival, or inactivation) along the manufacturing process, the frequency of recontamination, the growth during storage and distribution, and the handling practices by the final consumer (e.g., temperatures in the household refrigerator).
- *Hazard characterization*: Describes the health impact (severity and duration of adverse effects) of consuming a specified number of cells of the pathogenic agent on individuals, and it is frequently determined by assessing pathogen–host relationships, and ideally, dose–response relationships. It is difficult to obtain dose–response data of highly virulent pathogens, because human voluntary feeding studies are not possible. Furthermore, correlating animal responses to the responses of susceptible human populations has considerable uncertainties. Relationships can be derived from

epidemiological studies, again, with high levels of uncertainty in many cases.

- *Risk characterization*: Provides a risk estimate as a result of combining the information and analyses performed in the previous three elements. The output is an indication of the level of disease resulting from a given exposure. It must include a description of the uncertainties associated with the estimate. The level of confidence of this estimate depends on the assumptions made, and on the variability and uncertainty identified in all the previous steps. It is important to differentiate between variability and uncertainty for subsequent risk management decisions, and this is an area of growing interest in QMRA. A risk assessor must understand that predictive microbiology models cannot provide exact predictions of the response of microorganisms in food products due to natural variability of both microorganisms and foods, as well as lack of knowledge and data (i.e., uncertainty) when developing a model.

When QMRA is used as part of product/process design of chilled foods, such as fresh produce, it is not uncommon to focus primarily on exposure assessment. The target outcome of the assessment is reduction of the exposure so that the levels of the pathogenic agent are below those required by regulatory agencies at the point of consumption. In this approach, mitigation and control strategies are designed for the exposure only. The quantitative nature of exposure assessment makes it the most complex part of the QMRA process. Thus, exposure assessments are usually focused on specific food categories and processes in order to reduce complexity and facilitate risk management decisions. In the exposure assessment part of the QMRA process, predictive microbiology and process modeling play a fundamental role. For that reason, a properly conducted QMRA requires a multidisciplinary team of people, involving expertise in food safety and microbiology, food processing and engineering, and mathematical/statistical and computing skills.

So far, there are relatively few examples of risk assessments applied to fresh produce. Bassett and McClure (2008) recently reported a qualitative risk assessment approach for fresh fruit. This approach was based on the framework described above and considered multiple hazards and a variety of fresh fruits. A review of potential risk management options was carried out, and the authors concluded that washing to a recommended protocol was an appropriate risk management action for many of the fresh fruits considered, particularly when GAP and GHP are followed, with the addition of refrigerated storage for low-acid fruit. In study considering the microbiological standards of irrigation water, Stine (2005) described a quantitative approach to calculate the concentration of infectious pathogens necessary to achieve a risk of 1 in 10,000 annual risk of infection. This is the accepted level of risk used for drinking water by the U.S. Environmental Protection Agency.

12.5.1 Food Safety Metrics

Public health targets set by responsible authorities have to be interpreted by other parties involved in the provision of food. The World Trade Organization introduced the concept of appropriate level of protection (ALOP) as a public health target, and other concepts have been proposed and introduced to translate these targets into meaningful and tangible objectives for the food industry. Among these are Food Safety Objectives (FSOs), Performance Objectives (POs), and Performance Criteria (PC) proposed by the International Commission on Microbiological Specifications for Foods (ICMSF) and adopted by the Codex Alimentarius Food Hygiene Committee. The discussion of how FSOs translate to ALOPs is not yet agreed, and so far, only example FSOs have been discussed and considered. It will be interesting to follow developments over the next few years to see if accepted levels of risk are agreed upon for foodborne diseases, as they currently are for waterborne diseases in the United States.

12.6   CONCLUSIONS/REMARKS/FUTURE TRENDS

There is evidence of fresh produce causing increasing disease through contamination from organic wastes applied to agricultural land to fertilize crops, fecally contaminated waters used for irrigation, direct fecal contamination from livestock, wild animals, and birds, contaminated sprays, and postharvest contamination by poor worker hygiene. The control of food safety for fresh produce has lagged behind other (e.g., processed) foods in the application of food safety management systems, such as HACCP, due partly to poor understanding of these environments and also to the absence of prerequisite programs, such as GAP, that are necessary to underpin HACCP. It is also the case that some texts suggest there may not be a need to apply HACCP if hazards cannot be identified. Sperber (2005) expressed concern about the application of HACCP to primary production processes because of the lack of definitive critical control points, emphasizing the importance of prerequisite programs (see Table 12.6). The key message from Sperber is that food safety is HACCP *plus* prerequisite programs.

The recent emphasis on development of GAP was promoted partly through government agencies and also through the private sector. Continuing outbreaks of foodborne disease associated with fresh produce indicate that there are still improvements that need to be made. The recently updated guidance on fresh-cut produce issued by the FDA/USDA (2008) is a very useful reference point for producers in this sector. The guidance refers to HACCP and states that although HACCP is not currently required for processing fresh-cut produce, FDA encourages fresh-cut produce processors to take a proactive role in minimizing microbial food safety hazards, recommending consideration of a preventative control program (e.g., HACCP) to build safety into processing operations.

There are a few examples of HACCP being applied to fresh produce and also of trade associations such as the UK CFA promoting the application of HACCP to fresh produce. In a study looking at the application of HACCP in apple juice production, Senkel et al. (1999) concluded that pathogen-free product could only be achieved where GMP and sanitation procedures were used together with HACCP. In the few studies describing implementation of HACCP for fresh produce, there have been notable successes. For example, Peters (1998) reported on HACCP plans that were developed for a number of businesses, including very small businesses (single-person operations, family businesses, and small enterprises) as well as large

TABLE 12.6
Prequisite Programs Commonly Used in Food Processing

Prerequisite Programs Commonly Used in the Food Processing Industry
Cleaning and sanitation
Purchasing requirements
Pest control
Labeling
Rework
Facility and equipment design
Supplier approval
Employee training
Foreign material control
Good agricultural practices
Personal hygiene
Water/ice/air control
Maintenance
Transportation
Product retrieval
Allergen control
Chemical control
Product specifications
Products storage control

Source: Sperber, W.H. 2005. HACCP does not work from farm to table. *Food Control*, 16, 6, 511–514.

integrated operations. This study also concluded that prerequisite programs need to be in place to support detailed HACCP plans. HACCP plans have also been developed for sprouted seeds, shredded lettuce, and tomatoes by trade associations and other groups in the United States. Rushing (2000) also described the application of HACCP for production of fresh tomatoes and concluded that this was an effective food safety management system, involving public agencies and the industry, suggesting it should be used as a model for other produce industries to follow. The implementation of this program followed outbreaks of salmonellosis linked to fresh market tomatoes in 1990 and 1993.

Critical information required to carry out HACCP studies includes knowledge of sources of pathogens and control of these sources, their persistence, and transmission. Although there have been major advances in many of these areas in recent years, there is still much to learn that can be incorporated or considered in future HACCP studies. Perhaps the greatest challenge at present is the integration of HACCP with prerequisite programs and the implementation of these in all regions where fresh produce is grown, including developing countries. The absence of effective intervention methods, apart from irradiation (see Molins et al., 2001), that can be used without damaging the structure and decreasing the vitamin content of fresh produce means that control points must rely on factors that may have limited effects in reducing numbers of pathogens or restricting their growth. For this reason, it is essential that the numbers present on raw materials are minimized (hence the importance of GAP) and that pathogens are not introduced postharvest, for example, by worker hygiene (importance of GHP). Although irradiation is not accepted by consumers as a preferred method of preservation, it should be considered, particularly where vulnerable groups are considered.

The application of GAP, GHP, and HACCP should eventually result in better control in the fresh produce sector, with involvement of regulatory agencies, the industry, consumers, and food scientists in academia, and it is important that all the relevant stakeholders contribute to these activities. There is one final point on HACCP training that needs to be addressed in the future. A recent study from the United Kingdom (Jones et al., 2007) concluded that catering businesses causing foodborne disease outbreaks were not less likely to have staff with formal training or less likely to use HACCP. This indicates that HACCP training methods should be reviewed and optimized to ensure that those who are trained actually use the training and change behavior.

12.7   SOURCES OF INFORMATION

The Chilled Food Association in the United Kingdom published best practice guidelines for the production of chilled foods which include information on hazards, control measures, and HACCP systems relevant to chilled foods. The CFA also issued guidelines for good hygienic practice in the manufacture of chilled foods, which include a detailed account of Good Manufacturing Practice relevant to fresh produce. Information from small businesses that have implemented HACCP systems and useful information related to HACCP training and the future of HACCP, are provided by Mayes and Mortimore (2000). For a more detailed account of recently developed food safety metrics, see ICMSF (2002). The microbiological hazards associated with fresh produce and foodborne disease outbreaks associated with these products are covered in more detail by ICMSF (2005). A review of recent research and reevaluation of guidelines for food-service and restaurant operators, regulatory agencies, and consumers for handling fresh-cut leafy green salads was published (Palumbo et al., 2007). A recent review of quality assurance systems applied to fresh produce is provided by da Cruz et al. (2006), and commodity-specific food safety guidelines for lettuce and leafy greens can be found at http://www.cfsan.fda.gov/~acrobat/lettsup.pdf.

NOMENCLATURE

Control measure: Any action and activity that can be used to prevent or eliminate a food safety hazard or reduce it to an acceptable level.

Corrective action: Any action to be taken when the results of monitoring at the CCP indicate a loss of control.

Critical Control Point (CCP): A step at which control can be applied and is essential to prevent or eliminate a food safety hazard or reduce it to an acceptable level.

Critical limit: A criterion that separates acceptability from unacceptability.

Deviation: Failure to meet a critical limit.

Dose-response assessment: The determination of the relationship between the magnitude of exposure (dose) to a chemical, biological, or physical agent and the severity or frequency of associated adverse health effects (response).

Exposure assessment: The qualitative or quantitative evaluation of the likely intakes of biological, chemical, and physical agents via food, as well as exposure from other sources, if relevant.

Flow diagram: A systematic representation of the sequence of steps or operations used in the production or manufacture of a particular food item.

HACCP: Hazard Analysis and Critical Control Point—A system that identifies, evaluates, and controls hazards that are significant for food safety.

HACCP plan: A document prepared in accordance with the principles of HACCP to ensure control of hazards that are significant for food safety in the segment of the food chain under consideration.

Hazard: A biological, chemical, or physical agent in, or condition of, food with the potential to cause an adverse health effect.

Hazard analysis: The process of collecting and evaluating information on hazards and conditions leading to their presence to decide which are significant for food safety and should be addressed in the HACCP plan.

Hazard characterization: The qualitative or quantitative evaluation of the nature of the adverse health effects associated with the hazard. For the purpose of microbiological risk assessment, the concerns relate to microorganisms and their toxins.

Hazard identification: The identification of biological, chemical, and physical agents capable of causing adverse health effects and which may be present in a food or particular group of foods.

Monitor: The act of conducting a planned sequence of observations or measurements of control parameters to assess whether a CCP is under control.

Risk: A function of the probability of an adverse health effect, and the severity of that effect, consequential to a hazard in a food.

Risk assessment: A scientifically based process consisting of the following steps: hazard identification, hazard characterization, exposure assessment, and risk characterization.

Step: A point, procedure, operation, or stage in the food chain, including raw materials, from primary production to final consumption.

Validation: Obtaining evidence that the elements of the HACCP plan are effective.

Verification: The application of methods, procedures, tests, and other evaluations, in addition to monitoring, to determine compliance with the HACCP plan.

REFERENCES

Angulo, F.J., Getz, J., Taylor, J.P., Hendricks, K.A., Hatheway, C.L., Barth, S.S., Solomon, H.M., Larson, A.E., Johnson, E.A., Nickey, L.N., and Ries, A.A. 1998 A large outbreak of botulism: the hazardous baked potato. *Journal of Infectious Diseases*, 178, 172–177.

Bassett, J., and McClure, P.J. 2008 A risk assessment approach for fresh fruits. *Journal of Applied Microbiology*, 104, 925–943.

Bauman, H. 1990 HACCP: concepts, development and application. *Food Technology*, 44, 156–158.

Beuchat, L.R. 2002 Ecological factors influencing survival and growth of human pathogens on raw fruits and vegetables. *Microbes and Infection*, 4, 413–423.

Bhutani, M., Ralph, E., and Sharpe, M.D. 2005 Acute paralysis following a "bad potato": a case of botulism. *Canadian Journal of Anesthetics* 52, 433–436.

Codex Alimentarius Commission (CAC). 1999 *Principles and Guidelines for the Conduct of Microbiological Risk Assessment*, CAC/GL-30-1999. Available at: http://www.who.int/foodsafety/publications/micro/cac1999/en/, accessed June 2008.

Chiang, J.-M., Chen, T.-H., and Fang, T.J. 2005 Pesticide residue monitoring in marketed fresh vegetables and fruits in cental Taiwan (1999–2004) and an introduction to the HACCP system. *Journal of Food and Drug Analysis*, 13, 368–376.

da Cruz, A.G., Cenci, S.A., and Maia, M.C.A. 2006 Quality assurance requirements in produce processing. *Trends in Food Science and Technology*, 17, 406–411.

Dawson, D. 2005 Foodborne protozoan parasites. *International Journal of Food Microbiology*, 103, 207–227.

Early, R. 2005 Good agricultural practice and HACCP in fruit and vegetable cultivation. In *Improving the safety of fresh fruit and vegetables,* ed. Jongen, W., Woodhead, Cambridge, UK, pp. 229–268.

Food and Agriculture Organisation (FAO). 2007. Good Agricultural Practices—An FAO Working Concept, http://fao.org/prods/gap/resources/keydocument_en.htm, accessed June 2010.

Food and Drug Administration, United States Department of Agriculture (FDA/USDA). 1998. Guide to minimize microbial food safety hazards for fresh fruits and vegetables. Available at http://www.fda.gov/downloads/food/GuidanceComplianceInformation/GuidanceDocuments/ProduceandPlanProducts/UCM169112.pdf, accessed June 2010.

Food and Drug Administration (FDA). 2007 Guidance for Industry on Refrigerated Carrot Juice and Other Refrigerated Low-Acid Juices. http://www.cfsan.fda.gov/~dms/juicgu15.html, accessed June 2008.

Food and Drug Administration (FDA). 2008 Guide to Minimize Microbial Food Safety Hazards of Fresh-cut Fruits and Vegetables, http://www.cfsan.fda.gov/~dms/prodgui4.html, accessed June 2008.

Freshplaza. 2008 GLOBALGAP and FMI/SQF to work in partnership on audit checklist and standards for growers. www.freshplaza.com/news_detail.asp?id=11585, accessed June 2008.

Hathaway S.C. 1997 Development of food safety risk assessment guidelines for foods of animal origin in international trade. *Journal of Food Protection*, 60, 1432–1438.

Heaton, J.C., and Jones, K. 2008 Microbial contamination of fruit and vegetables and the behaviour of enteropathogens in the phyllosphere: a review. *Journal of Applied Microbiology*, 104, 613–626.

International Commission on Microbiological Specifications for Foods (ICMSF). 2002 *Microorganisms in Foods 7: Microbiological testing in food safety management,* Kluwer Academic/Plenum, New York.

International Commission on Microbiological Specifications for Foods (ICMSF). 2005 *Microorganisms in Foods 6: Microbial ecology of food commodities,* Kluwer Academic/Plenum, New York.

International Life Sciences Institute (ILSI). 1999 Validation and verification of HACCP. Available at: http://europe.ilsi.org/publications/Report+Series/haccpvalidation.htm, accessed June 2010.

James, S.I., and Evans, J. 2008 A review of the performance of domestic refrigerators. *Journal of Food Engineering* 87, 2–10.

Jones, S.L., Parry, S.M., O'Brien, S.J., and Palmeri, S.R. 2007 Are staff management practices and inspection risk ratings associated with foodborne disease outbreaks in the catering industry in England and Wales? *Journal of Food Protection*, 71, 550–557.

Jongen, W. 2005 *Improving the safety of fresh fruit and vegetables,* Woodhead, Cambridge, UK.

Koopmans, M., and Duizer, E. 2004 Foodborne viruses: an emerging problem. *International Journal of Food Microbiology*, 90, 23–41.

Kosa, K.M., Cates, S.C., Karns, S., Godwin, S.L., and Chambers, D. 2007 Consumer home refrigeration practices: results of a Web-based survey. *Journal of Food Protection*, 70, 1640–1649.

Laguerre, O., Derens, E., and Palagos, B. 2002 Study of domestic refrigerator temperature and analysis of factors affecting temperature: a French survey. *International Journal of Refrigeration-Revue Internationale Du Froid* 25: 653–659.

Mayes, T., and Mortimore, S. 2000 *Making the most of HACCP: learning from others' experience,* Woodhead, Cambridge, UK.

McClure, P.J., and Amézquita, A. 2008 Predicting the behaviour of microorganisms in chilled foods. In *Chilled foods: a comprehensive guide,* ed. Brown, M.H., Woodhead, Cambridge, UK. pp. 477–528.

Molins, R.A., Motarjemi, Y., and Kaferstein, F.K. 2001 Irradiation: a critical control point in ensuring the microbiological safety of raw foods. *Food Control,* 12, 347–356.

Monaghan, J.M. 2006 United Kingdom and European approach to fresh produce food safety and security. *Horttechnology,* 16, 559–562.

Montgomery, D.C. 2005 *Introduction to statistical quality control,* 5th edition, John Wiley and Sons, New York.

National Advisory Committee on Microbiological Criteria for Foods (NACMCF). 1999 Microbiological safety evaluations and recommendations on fresh produce. *Food Control,* 10, 117–143.

Palumbo, M.S., Gorny, J.R., Gombas, D.E., Beuchat, L.R., Bruhn, C.M., Cassens, B., Delaquis, P., Fraber, J.M., Harris, L.J., Ito, K., Osterholm, T., Smith, M., and Swanson, K.M.J. 2007 Recommendations for handling fresh-cut leafy green salads by consumers and retail foodservice operators. *Food Protection Trends,* 27: 892–898.

Peters, R.E. 1998 The broader application of HACCP concepts to food quality in Australia. *Food Control,* 9, 83–89.

Rhodehamel, E.J. 1992 Overview of biological, chemical and physical hazards. In *HACCP: principles and applications,* eds. Pierson, M.D., and Corlett, D.A. Jr., Van Nostrand Reinhold, New York, pp. 8–28.

Rushing, J.W. 2000 A case study of salmonellosis associated with consumption of fresh-market tomatoes and the development of a Hazard Analysis Critical Control Points (HACCP) program. *Hortscience,* 36, 29–32.

Rushing, J.W., Angulo, F.J., and Beuchat, L.R. 1996 Implementation of a HACCP program in a commercial fresh-market tomato packinghouse: a model for the industry. *Dairy Food and Environmental Sanitation,* 16, 549–553.

Senkel, I.A., Henderson, R.A., Jolbitado, B., and Meng, J. 1999 Use of hazard analysis critical control point and alternative treatments in the production of apple cider. *Journal of Food Protection,* 62, 778–785.

Sivapalasingam, S., Friedman, C.R., Cohen, L., and Tauxe, R.V. 2004 Fresh produce: a growing cause of outbreaks of foodborne illness in the United States, 1973 through 1997. *Journal of Food Protection*, 67, 2342–2353.

Sperber, W.H. 2005 HACCP does not work from farm to table. *Food Control*, 16, 6, 511–514.

Stine, S.W. 2005 Application of a microbial risk assessment to the development of standards for enteric pathogens in water used to irrigate fresh produce. *Journal of Food Protection*, 68, 913–918.

Wells, J.M., and Butterfield, J.E. 1996 *Salmonella* contamination associated with bacterial soft-rot of fresh fruits and vegetables in the marketplace. *Plant Disease*, 81, 867–872.

Willocx, F., Tobback, P., and Hendrickx, M. 1994 Microbial safety assurance of minimally processed vegetables by implementation of the hazard analysis critical control point (HACCP) system. *Acta Alimentaria*, 23, 221–238.

Winter, C.K. 2005 Pesticide residues in fruit and vegetables. In *Improving the safety of fresh fruit and vegetables,* ed. Jongen, W., Woodhead, Cambridge, UK, pp. 135–155.

World Health Organization (WHO). 1997 HACCP—Introducing the Hazard Analysis and Critical Control Point system, Food Safety Unit, Programme of Food Safety and Food Aid, WHO, Geneva.

World Health Organization (WHO). 1998 Surface decontamination of fruits and vegetables eaten raw: a review. WHO/FSF/FOS/98.2. http://www.who.int/foodsafety/publications/fs_management/surfac_decon/en/, accessed 29 June 2008.

13 Process Design, Facility, and Equipment Requirements

Alessandro Turatti

CONTENTS

13.1 FRESH-CUT PRODUCTS AND SAFETY

Food safety keeps the fresh-cut industry strong and ensures that consumers have confidence that the fruit and vegetables processed by the fresh-cut companies are safe and healthy.

Food safety is a shared responsibility, and everyone involved in the food chain from farm to fork must take responsibility for safeguarding the food supply. This especially applies to fresh-cut products. The United Fresh Produce Association (UFPA) defines fresh-cut produce as "any fresh-cut fruit or vegetable or any combination thereof that has been physically altered but remains in the fresh state. These products are items such as bagged salads, baby cut carrots, and broccoli florets."

Therefore, for fresh-cut vegetables that are eaten raw, there is no treatment that can be relied upon to significantly decrease the numbers of contaminating microorganisms. Because fresh-cut fruits and vegetables are living and breathing during and after processing, they are subject to fast deterioration and can support the growth of large populations of microorganisms. Unlike other processed foods, there is no kill step throughout processing, and there is no treatment, other than good temperature management, that will significantly retard deterioration jointly with good sanitation and Good Manufacturing Practices (GMPs).

Eliminating all the risks is very difficult, as washing with antimicrobial compounds, although significant, frequently results only in a relatively small reduction. Management of them is based on identifying and controlling those factors that are central in preventing contamination or limiting growth of pathogenic microorganisms between farm and plate. Consequently, maintaining the safety and quality of fresh-cut produce is a key challenge that is met only by doing several small things in the appropriate ways. If temperature is not managed well, adherence to GMPs and superior sanitation will not ensure either safety or quality. A poor sanitation program will prevent a Hazard Analysis and Critical Control Point (HACCP) from functioning entirely.

Due to the diversity of raw materials, processing conditions, and packaging systems used in the production of fresh-cut products, it is impossible to institute a one-size-fits-all approach to achieve microbial safety. Rather, the processor has to cautiously consider a broad variety of factors and hurdles—raw material quality, hygienic processing, temperature, water activity, acidity, modified atmosphere—in determining ways to control microbial growth. Through the selection and combination of these elements, the processor is able to determine the optimum shelf life for a product and establish conditions for its use that will ensure safe products for consumers.

Hence, safety and sanitation are a top priority for fresh-cut processors from plant design to cleaning procedures. New design elements of processing equipment are a vital part of this continuous evolution of food safety and sanitation. These elements help plant managers more effectively implement cleaning practices, and in many cases, the new designs create a more efficient overall process. This chapter will deal only with general hygiene concepts. It is not a substitute for complete familiarity with legislation and national standards.

Prior to approaching the theme of sanitary design, it is essential to briefly outline the risks involved in processing fresh-cut products.

13.1.1 FOOD HAZARDS

The U.S. Food and Drug Administration (FDA) defines food hazard as "a biological, chemical, or physical agent that is reasonably likely to cause human illness or injury in the absence of its control."

Food hazards mainly related to fresh-cut products are generally categorized as microbial, chemical, or physical. Most of the principles of safety revolve around the avoidance of entry, harborage, or buildup in the facility of anything that does not belong, be it microbial hazards, insects, rodents, plain old dirt, and even unauthorized personnel. This principle is especially important at the points in the process where the product that will be ultimately consumed is most vulnerable. Even produce that is clean and free from pathogenic microorganisms and chemical contaminants when harvested, can easily be recontaminated if it is not handled hygienically (during the process that ultimately ends with the consumer). Produce can be contaminated by unhygienic equipment and the unhygienic practices of the personnel involved in the process. High standards of hygiene will minimize the risk of produce recontamination.

Microbial, chemical, or physical hazards may be very different from one product to another:

1. Microbial Hazards. Microbial hazard means occurrence of a microorganism that has the potential to cause illness or injury.

   Processing plants might receive produce from several different farms, thus increasing the risk that different soilborne pathogens may be present, including *Escherichia coli, Salmonella* sp., and Hepatitis A. Poor handling procedures, scarce sanitation and cleaning, and contaminated water augment these risks at the processing plant level. Additionally, processors, similarly to other suppliers and retailers in the food chain, continue to remain at risk for cases of deliberate tampering and bioterrorism.

2. Chemical Hazards. Fresh-cut foods, similar to other food products, are also subject to contamination by environmental contaminants and residues from pesticides or veterinary drugs. Compliance of raw material with the relevant legislation is essential. Supplier selection, evaluation, and follow-up are the best control measures. Chemicals such as cleaning agents, lubricants, and pest control materials may also present on-site chemical hazards. The correct use of food-grade chemicals, where appropriate, and application of GMP are the best control measures.

   Those involved with sanitation and sanitary design should be aware of how to protect foods against allergens that are naturally occurring proteins with many characteristics of interest to food safety. Allergen infestation frequently occurs because of product cross-contamination through an allergen-containing product during manufacture. Allergens are organic, but not living organisms that are water insoluble, only slightly acid soluble (according to some sources), and alkali soluble. Allergens are hard to eradicate when baked onto a surface, such as wheat protein in a finish dryer, because they are heat resistant as well as resistant to proteolysis and to extremes in pH.

3. Physical Hazards (foreign bodies removal). Physical hazards might include foreign bodies such as metal, glass, wood, and plastic. Their control is ensured by raw material quality (specifications, supplier evaluation) and provisions applied during processing (e.g., metal detectors after packaging, filters in line).

Foreign matter is the main single source of customer complaints for numerous food processors, retailers, and food and safety inspectors. Preventing foreign matter contamination of foods depends on understanding the source of each potential material and taking appropriate steps to prevent contamination. The instrumental methods of detection are air separators, potentiometric (conventional metal detection, using the same principle used in airline passenger screening), X-ray, and optical sorting.

13.2 SANITATION DESIGN OF PROCESSING PLANTS

13.2.1 GENERAL OUTLINES

It is recommended that the processing facility and its structures (such as walls, ceilings, floors, windows, doors, vents, and drains) be designed in order to be simple to clean and maintain and to protect the product from microbial, physical, and chemical contamination. As an example, designing food contact surfaces to be smooth, nonabsorbent, smoothly bonded, without niches, and sealed would make these surfaces easier to clean and, thus, would prevent the harborage of microbial pathogens.

Adequate food safety controls, operating practices, and facility design can reduce the potential for contamination by using location and flow of humans, product, equipment, and air.

Hygienic processing can be summarized in a simple "3 C principle":

- Keep Clean
- Keep Cold
- Avoid Cross-contamination

If these principles cannot be followed, there will be a risk of contamination or pathogen growth, which must then be controlled by other means.

These three basic principles are reflected in the following guidance and recommendations.

13.2.2 FORWARD-ONLY MOVEMENT—"LOW CARE" AND "HIGH CARE"

This requires that there should be no "crossing over" in the processing line between the raw material and clean products. This principle does not impose a linear processing: It is possible to separate in different areas the machines according to the function. GMP dictates that packaging surrounding produce that has arrived from the field or storage area must travel in a reverse direction from the produce.

The rationale behind this principle is that process flow must ensure that produce moves through the facility from input, where there can be high levels of contamination, to output, where there should be lower levels of contamination. Contamination from earlier steps in the process must not be allowed to enter later steps in the process due to poor plant layout.

It is important that processors ensure they are supplied with good quality, safe, fresh produce. If high levels of contamination exist on the raw material, then it is unlikely that further processing will be able to reduce it to safe levels.

Directly related to the forward-only movement is the fact that different areas are separated by walls in order to progressively increase cleanliness from the trimming room to the packaging section.

Good processing design requires the segregation of "low care" and "high care" production areas (lately the terms used have been "low risk" and "high risk," underlining that care has to be considered a must). This means that processed product is washed thoroughly from a low care preparation area to a high care area. In the low care area, product is initially prepared, washed, or peeled so that it can be effectively washed through a partition from low care to high care. In the high care area, only ready-to-eat foods are handled, and this area is designed to a high standard of hygiene, where practices relating to personnel, ingredients, equipment, and environment are managed to minimize contamination by microorganisms. Therefore, in this area, the contamination risk by foreign bodies such as packaging, metal, and poor hygiene conditions, is reduced to the absolute minimum through good hygiene practice. Segregation of processing personnel within both areas is critical.

The separation is enhanced by the installation of sanitizer systems (i.e., hand and boot sanitizing stations) inside entrance doors to critical areas. Entrances and outlets to the processing areas must be kept at a minimum, and the hand washing and sanitizing stations must be located in the transit area in adequate quantity. The hand and boot sanitizing stations must be of a sufficient number to ensure ease of use and availability and must be conveniently located. If they are not convenient, they will not be used. But, they should not be so close as to present a risk of contamination of the product.

The hand-washing stations need to be constructed to prevent recontamination. A knee-activated valve, automatic electronic valve, or foot-activated valve is ideal. It is crucial that these systems force the operators, visitors, and inspectors to pass through them in order to enter and exit the processing areas, being clear that these systems must be maintained in good state. Inspections should include testing a number of hand-washing stations to ensure that they are working properly.

Proper hand washing and sanitizing can prevent contamination. By hand washing, organic matter and transient bacteria are removed, so that sanitizing can efficiently reduce and eliminate bacteria. There are several important guidelines that must be followed by employees, in order not to neutralize or cancel the effectiveness of hand washing and sanitizing: They must not wear jewelry or covering over their fingers, like duct tape or adhesive bandages. Organic matter can lodge between the skin and the jewelry or the tape, subsequently resulting in rapid microbial growth—a source of contamination. Personal items might also contribute to contamination and have to be stored away from production areas and processing rooms.

Eating, drinking, or smoking in production areas and processing rooms must be strictly forbidden. Close proximity of the hands with the nose during these practices might cause contamination as the nose harbors *Staphylococcus aureus* in about 50% of the healthy population.

Skin contaminants are also a concern; therefore, elbows, arms, and other uncovered skin surfaces should not come into contact with food or food preparation surfaces.

Generally, the tours, audit inspections, and visits operated by quality assurance (QA) personnel should be done starting with finished product rooms and ending in raw material receiving areas (therefore, in the opposite direction of the product flow).

The separation between raw and finished product must also be operated from the operational standpoint. Therefore, it is suggested that standard operating procedures be implemented regarding the product flow: Improper personnel flow or human faults might undermine even the best physical separation.

Foodborne illness is often traced not only to improper cleaning and sanitizing of food equipment and the environment but also to the lack of standard operating procedures related to the flow of the product. Consequently, employees (including maintenance and janitorial staff) working with raw materials must be barred from entering finished product rooms.

For older constructions where physical separation of low care (risk) and high care (risk) is not possible, strict processes and procedures have to be considered with the utmost importance to help avoid cross-contamination.

In the written employee hygiene programs in place, generally a color-coding system is used to separate equipment and employees. This highly assists in the dislocation of equipment in and out of the facility by maintenance team. The most common rule is to associate different areas of the facility with different colors. The coding system based on colors can concern clothing (i.e., smocks, uniforms), cleaning supplies (e.g., brushes, shovels), containers, and any other equipment.

When taking into consideration the internal space, sufficient room for people and equipment will have to be considered (thus including overhead clearance). The movement of people and equipment (fork lift, pallet jacks) will have to avoid bottlenecks and "traffic jams." The location of the required utilities (i.e., electricity, water, gas) must also be taken into consideration.

13.2.3 TEMPERATURE CONTROL

The separation of the areas is also important because the facilities are designed and equipped in such a way that the temperatures inside the rooms are in accordance with the requirements of the phase of the processing.

According to French regulations, packaged products must be immediately stored at 4°C (39.2°F) and maintained at this temperature or lower (0–4°C; 32–39.2°F).

New technologies developed from *Turatti Srl* and *Air Liquide* are available in the market in order to cool the temperature of the processed vegetable or fruit by means of a nitrogen system. Nitrogen cooling exhibits unique properties when used for cooling, providing many benefits over other systems. The process takes place in a controlled-atmosphere condition that impedes the growth of oxidation-based spoilage and microbial action. This action results in improvements of the color of the leaves, with final product freshness visibly enhanced. In addition, because of the obtainable low temperatures, the cooling time is shortened, resulting in an additional reduction in microbial activity.

13.3 DESIGN OF THE FACILITIES

Engineering design and construction of a processing facility for fresh-cut products (likewise, all the food-processing plants) requires a significant amount of planning. Information must be collected from many qualified sources and combined into documents that are simple to read and share within the organization in charge of designing and building the facilities.

All regulatory requirements will mandatorily apply to the design and construction of the facility. Additionally, recommendations from dedicated associations as *United Fresh (UFPA)* and the qualified company engaged to design and construct the building will have to be followed.

There are several basic elements to consider when designing food-processing plants. Given that each situation is unique, some elements will not apply, and some important key factors will lead to elements that might be added. Elements listed emphasize engineering design with a stringent focus on food safety, leaving aside other topics (i.e., economics, marketing, logistics, and other necessary inputs).

In order to guarantee adequate sanitation programs to obtain safe food, the facility and surroundings in which fresh-cut processing operations are operated have to be designed and constructed with stringent sanitary design principles in mind. The environment of a fresh-cut processing facility is potentially an ideal source of microbiological contamination. At this stage, it should also be pointed out that several existing facilities do not enclose optimal sanitary design and construction; therefore, a number of renovations and adjustments might be required (where feasible).

Additionally, there may be specific state or local code requirements for food handling or processing establishments in the area which specify where the operation should be located and how it should be constructed and maintained.

The main focus of this section will be on the internal part of the facility, with some general notes about the premises and location.

13.3.1 PREMISES, SURROUNDINGS, BUILDING SITE, AND LOCATION

In the fresh-cut business, the choice of a location is very often influenced by the possibility to receive the raw material in the shortest time in order to process it rapidly. At the same time, the site of the distribution facilities of the customers (i.e., retail, supermarkets, food service operators) must be sufficiently close. It is evident that both conditions are not occurring easily, thus involving sometimes compromising solutions.

It is of the utmost importance that several factors be taken in due consideration regarding site selection and its exterior surroundings. Chemical pollutants might be a potential contamination source, and harborages and infestations (i.e., insects, rodents, birds, and other pests), must be carefully considered.

It is essential that an adequate pest control management program be in place for the fresh-cut processing facility (relevant documentation and proper records must be cautiously kept). The role of insects, rodents, birds, frogs, and other pests in spreading foodborne pathogens has been documented in several studies. Additionally, sanitary landscaping will be very influential in depriving vermin of harborage.

As a general rule that applies to all food-processing facilities, the location must be far away from every contamination source (i.e., facility must not be near oil refineries, chemical plants, sewage treatment facility, or cow pasture, as these types of high-risk facilities may create smoke, dust, sewage, or odors that could impact the safety of the products).

Special safety measures are compulsory to keep contaminants and odors from entering the yard or facility, in case it is found that sources of contamination are adjacent to the building.

13.3.2 Loading Areas and Receiving Rooms—Openings

As trucks need to unload raw material or ship final products to and from the processing area, receiving rooms should be enclosed as much as possible. When designing and building these areas, proper steps must be taken in order to prevent the entrance of pests (i.e., insects, rodents, and birds). Dock leveler plates must be equipped with brush seals to keep rodents out, while dock canopies should prevent the likelihood of bird nesting. Air curtains might help to prevent insects from entering the building.

Additionally, any openings into the facility building (thus including entry and exit doors, windows, and ventilation ducts) must be properly sealed and protected.

13.4 INTERIOR BUILDING DESIGN AND CONSTRUCTION

Several different sanitary objectives come together for interior building design and construction. The main aim is to maximize the protection of fresh-cut products from contamination, reducing likely harborages of pests and microorganisms. Consequently, the focus remains on cleanability and sanitation.

It is evident (as previously discussed) that although existing facilities might be very challenging as laws, regulations, and recommendations are very often adjourned, a new facility can be easily built with sanitary design criteria in mind.

Nevertheless, it has to be firmly stated that even if a facility is designed and built to the latest and most recent sanitary specifications, this does not guarantee a safe food product if the facility is not effectively cleaned and maintained regularly on an appropriate schedule.

As a general rule, materials such as wood and aluminum that might contaminate the product must be avoided in any food-processing facility.

13.4.1 Interior Walls and Ceilings

Interior sanitary walls should have a hard, flat, smooth, and washable surface applied to a suitable base. Several properties are also related to the wall: It must be free of pits and cracks and resistant to cleaning and sanitizing chemicals and corrosion. Coves with radii sufficient to promote sanitation will be installed at the junctures of floors and walls in all rooms.

The companies operating in the manufacturing of wall materials are always developing new types that are available for food-processing and handling areas.

The inaccessibility of most ceilings means that dust, condensation, and other materials might accumulate. Therefore, ceilings should meet the same objectives mentioned for walls, being of the easy-to-clean type.

13.4.2 FLOORS

Floors must be easily cleanable and meet important requirements: Floors must be smooth, durable, resistant to corrosion, nonabsorbent, not subject to slippage, and very hard wearing. Because of the intense daily exposure to a variety of products (i.e., chemicals as sanitizers, acids, water, dust, and pieces of fresh-cut products), the floor in a fresh-cut processing facility is the most difficult surface to maintain at an acceptable level.

Additionally, floors must withstand abuses caused by the movement of equipment, forklifts, and other types of mechanical stresses (i.e., displacements of dollys and pallets, holes drilled to place machinery, etc.).

The most widely utilized base material is concrete, which is then covered with a coating or a brick material. There are many available highly acceptable materials for the surfacing of floors in a fresh-cut processing facility.

As water in a fresh-cut processing area is an important element, adequate slope for drainage and prevention of water pooling must be provided.

Even if, as stated, several materials would provide an acceptable floor surface if appropriately installed, floors must be cleaned and sanitized at the end of the production and maintained in good repair.

13.4.3 FLOOR DRAINS AND SEWAGE

As previously stated, flooring is one of the biggest concerns in a food-processing plant. In addition to having a floor that is durable, nonskid, sanitary, and easy to maintain, floor drains must be carefully considered: Typically, the floor drains are the weakest points in a flooring installation, as they might be a major source of microbial contamination inside a facility.

The floors must be sloped to the drains at a certain angle (i.e., 1–2%). Floor drains should be of adequate size and number and be properly located. The covering of the drains must be enough to prevent slipping and must be designed and installed so that they are cleanable. Additionally, floor drains must have grates or covers that are removable for cleaning.

13.4.4 INTERIOR LIGHTING

Good quality lighting of sufficient intensity is a must, not only for production efficiency but also for safety, sanitation, and inspection: This especially applies to the operations related to cleaning and sanitizing.

Suggested lighting requirements might differ between different regulations and other sources. Light fixtures (designed to be resistant to moisture and easy to sanitize) should be shielded to prevent contamination from broken bulbs and should have an unbreakable cover.

The lighting regulations are generally part of an overall glass management policy aiming to forbid the entrance of unprotected glass inside the facility.

13.4.5   HEATING, VENTILATION, AND AIR-CONDITIONING (HVAC) SYSTEMS

Heating, ventilation, and air-conditioning (HVAC) systems operate in order to maintain the temperature and humidity of the facility and prevent both condensation and circulation of dust. These systems are integral components of any industrial facility; nevertheless, in the food-processing sector (and with special relevance in a plant where fresh-cut products are processed), these systems may take on added importance due to the need to maintain strict temperature controls.

Positive air pressure differentials in critical or sensitive food handling rooms must be produced. The highest pressure will be in the packing area, flowing to the trimming areas and raw material receiving. Consequently, the facility must incorporate properly sized units and an adequate distribution system.

Environmental hygiene monitoring must include air sampling as well. Many experts observe that airborne contamination is strongly suspected as the cause of some pathogenic contamination and unfortunately this happened with pathogenic micro organisms (especially *Listeria monocytogenes*). Therefore, a comprehensive evaluation of the processing and ventilating air utilized in the plant must be undertaken.

Depending on the fresh-cut product being processed and the stage of processing, diverse levels of air handling and filtration are utilized; nevertheless, in order to guarantee the appropriate performance of the systems, daily sanitary operations are always required.

13.4.6   EMPLOYEE FACILITIES AND PERSONNEL NEEDS

These areas include training areas or classroom, lockers, break room, smoking area, restroom facilities, sinks and sanitation stations (foot bath, foam spray, hand wash, and sanitize before entering room), office area, and visitor reception area. These facilities should be arranged to minimize contamination resulting from traffic patterns.

Suitable facilities for employees must be arranged: in these areas, they can safely store their clothes and other personal items. A sufficient number of toilets must be located in compliance with regulations.

The facilities of the employees should not open straight into processing (or other critical areas); therefore, in the majority of cases, a two-door separation should be placed between the personnel facilities and fresh-cut processing areas.

13.4.7   REFRIGERATION UNITS AND COOLERS

Permanent refrigeration rooms and coolers must have adequate sanitary construction as in other rooms and areas in the fresh-cut facility.

The evaporators must be periodically checked to make sure they maintain the correct temperatures in the rooms where they are mounted.

As applies to other areas of the facilities, refrigeration units can be a source of contamination; Therefore, these units must be designed and installed with cleaning and sanitizing always in mind. In particular, scrupulous attention must be paid to the draining system of the condenser. Drip pans must be accessible to cleaning. Additionally, they must be piped to a drain system in order to remove condensate into drains, strictly avoiding critical processing areas.

13.4.8 Compressed Air Lines

Air compressors must be of the oil-free type. Additionally, compressed air lines can contain condensate and become an issue on a microbiological standpoint. Filters must be cleaned very often.

13.4.9 Electrical Boxes

Electrical boxes, control panels, and associated equipment have to be water resistant.

Control panels and electrical boxes must be continuously welded or mounted with approximate 1 inch (25 mm) stand-offs (to allow for easy cleaning and pest inspection) or properly sealed with caulking material to the frame or wall. The top side of the control panels will have a sloped shielding at a minimum of a 20-degree angle to prevent ledges that can collect dirt and dust. Light switches are also sites where moisture, pests, and microorganisms can accumulate, and they should be cleaned and sanitized regularly.

13.4.10 Final Considerations about the Facility

The aforementioned recommendations and guidelines have a general outline, but each fresh-cut facility is peculiar and of its own. For this reason, the facilities must be inspected regularly to protect raw and fresh-cut products from contamination that might result from atypical situations.

Fresh-cut processing (likewise, the whole food-processing sector) is one of the most energy-intensive industries. As energy costs rise and the supply of energy remains questionable, plant managers must search for applicable ways to reduce consumption in this energy-intensive activity. Energy efficiency results from different important decisions that range from insulation to the choice of processing and refrigeration equipment and controls.

Another essential is related to water consumption. The increasing scarcity of a good quality water supply to support food-processing processes and stricter regulatory constraints on wastewater discharge from these facilities give a sense of the drivers toward water recovery and reuse in these types of facilities. A plant survey must establish facility water savings potential by identifying areas where water is wasted or where water could be reused. The source of the water (being of the quality of drinking water) must also be tested periodically, avoiding cross-connections with nonpotable water.

Different approaches are actually being successfully applied to this problem, as the use of reverse osmosis and microfiltration to treat various recovered process water streams produces an excellent quality water for reuse within the process plant.

The quality assurance laboratory is extremely important. As mentioned, whole fruits and vegetables are living organisms whose journey to either consumption or spoilage begins at harvest. Cutting the outer surface, or skin, of produce during processing exposes many more surfaces and accelerates the rate of respiration and therefore breakdown. As a result, microbiological testing is extremely important, and the laboratory must be segregated from the processing areas. The walls and ceilings have to be constructed of acid-resistant materials, and adequate lighting should be available. The employees of the laboratory must assist in audits of the ranches, harvesters, cooler, and processing plant, and operate inspections for employee hygiene, cleanliness, record keeping, and policies.

Also, maintenance areas must be physically segregated from processing areas, allowing enough room to guarantee access for the machines or parts when they are brought in.

13.5 SANITATION DESIGN OF FRESH-CUT EQUIPMENT

13.5.1 General Food Equipment Design Recommendations

All plant equipment and utensils should be so designed and of such material and workmanship as to be adequately cleanable, and they will be properly maintained. The design, construction, and use of equipment and utensils will preclude the adulteration of food with lubricants, fuel, metal fragments, contaminated water, or any other contaminants.

All equipment should be so installed and maintained as to facilitate the cleaning of the equipment and of all adjacent spaces. Food-contact surfaces should be corrosion resistant when in contact with food. They should be made of nontoxic materials and designed to withstand the environment of their intended use and the action of food, and, if applicable, cleaning compounds and sanitizing agents. Food-contact surfaces will be maintained to protect food from being contaminated by any source, including unlawful indirect food additives.

1. *Hygienic equipment design.* The purpose of hygienic equipment design is to (Shapton and Shapton, 1991):
   a. Give maximum protection to the product.
   b. Provide product contact surfaces necessary for processing which will not contaminate the product and are readily cleanable.
   c. Provide junctures that minimize "dead" areas where chemical or microbial contamination may occur.
   d. Give access for cleaning, maintenance, and inspection.
2. *Surfaces.* The food-contact surfaces must meet the following criteria:
   a. The surface must be nontoxic to the food.
   b. The surface must be nonreactive with the food.
   c. The surface must be noncontaminating to the food.
   d. The surface must be noncorrosive.

e. The surface must be cleanable.

In addition, the surfaces must be able to withstand the environment of the process area. The most accepted food contact surface and processing equipment fabrication is stainless steel. Stainless steel of the 300 series (304 or 316) is preferred for food contact surfaces. It should be polished to a 180 grit finish.

Elements that must not be used in food contact surfaces or in food zones are antimony, cadmium, copper, copper alloys (brass and bronze), lead, monel, wood, and glass (FDA regulations 21 CFR, sections 170–190).

3. *Welding.* Welds on food contact surfaces should be ground smooth and polished to the same texture as the adjoining surfaces. Food contact surfaces should only be butt welded with no overlap welding. Equipment used to grind and polish stainless steel should be dedicated to stainless steel to avoid the development of rust by using equipment contaminated with mild steel.

13.5.2 Fresh-Cut Equipment Design

The HACCP approach of preventing microbial, chemical, and physical contamination of the product is essential in the design, fabrication, and installation of equipment for fresh-cut processing. Processing equipment should be built without points where product builds up and can fall back into the product stream.

In addition, all equipment should be able to be disassembled, cleaned, sanitized, and reassembled in the time available for the clean-up shift. There should be no special tools required to disassemble the equipment. Nothing more complicated than a screwdriver and a crescent wrench should be required.

Using the "less is more" design principle is very important in reaching a high level of sanitation on the machine. A key tenant to system design to optimize sanitation is in the simplicity of design. This means using one-piece construction wherever possible and reducing the number of hollow parts on a system. Even the legs of a system can be thoroughly scrutinized for how they fit into a cleaning procedure (a simpler leg design is a simpler leg to clean).

Simplicity is important: The more efficiently a cleaning crew can thoroughly clean a system, the better, both in terms of safety and the bottom line. A machine cannot make a plant safe, but it can make safety easier to achieve.

13.5.2.1 Process Phases

13.5.2.1.1 Field Operations

Harvesting equipment should be built using the principles of sanitary design. The materials that are used should be appropriate for food use. In addition, the equipment should be designed for ease of sanitation and the prevention of contamination by foreign objects.

Examples of sanitary design features include the following:

- Conveyor belt materials should be appropriate for food contact.

- All overhead hydraulic hoses and fittings should be protected with a catch pan.
- All night-lights should be shielded.
- All direct contact surfaces should be constructed of materials that are appropriate for food contact.

Each harvest crew member should wear appropriate clean protective garments and should be trained in the proper use of toilet facilities and hand-wash stations. Such efforts should be part of an ongoing training effort.

13.5.2.1.2 Raw Material

13.5.2.1.2.1 Transport, Storage, and Precooling   It is important that processors ensure they are supplied with good quality, safe, fresh produce. If high levels of contamination exist on the raw material, then it is unlikely that further processing will be able to reduce it to safe levels. Processors must ensure that the produce they accept into their premises has been produced in accordance with good agricultural practices.

Similarly, they must ensure that produce is not further contaminated during transport to the factory and storage prior to processing. Adherence to good hygienic practices in these areas detailed in Storage Facility Hygiene and Transport Hygiene is necessary, and these should be supplemented with additional measures as required by local facilities and working practices.

Most fresh fruits and vegetables require thorough cooling immediately after harvest in order to deliver the highest quality product to the consumer and to extend the shelf life of fresh-cut products. Proper cooling delays the inevitable quality decline of produce and lengthens its shelf life. To lower the core temperature of the produce arriving from the fields, there is a different set of ways to proceed.

The first is hydrocooling, when warm produce is cooled directly by chilled water. Produce is either immersed in cold water or cold water is sprayed onto the product (thus requiring plastic or wax-covered boxes). The size of the product determines how long it needs to stay in the water (i.e., carrots and asparagus require less than 10 minutes and produce larger than 3 inches, 7.6 cm in diameter requiring from 30 to 60 minutes).

Vacuum coolers are the most common application for produce that can lose water quickly (i.e., lettuce, leafy greens, and cauliflower). The vacuum coolers operate by lowering the temperature that water boils at, creating an evaporative cooling process. The cooling process requires the pallets of produce to be loaded into a vacuum chamber. The chamber is closed, and a vacuum is created using large vacuum pumps.

One disadvantage of vacuum coolers is that they can cause water loss (for most produce, that amounts to a 2% to 4% moisture loss). Additionally, vacuum coolers use a large amount of energy and are expensive.

The most costly method for cooling produce is the third one: ice (primarily used for broccoli). In addition to being expensive, ice packing is very inefficient; therefore, growers are looking for alternatives to cooling broccoli.

The most widely used system, the fourth, is forced air cooling: Chilled air is delivered at a high velocity over the product to remove damaging field heat.

It is extremely important to consider at first the type of produce to be cooled when selecting the cooling method. If a wide range of produce is going through the cool-down process, then a forced air cooler may be the way to go, but if it is primarily for leafy greens, then a vacuum cooler may be more efficient.

13.5.2.1.3 Trimming and Waste Removal

The required proportions of the ingredients in salad mixes are achieved during trimming. The trimming table is supplied with the final percentage of each salad, taking into account the respective processing output. All unwanted parts of the plant, including most of the outer green leaves and core area, are removed manually. This operation causes injury that could be minimized by using very sharp knives (Bolin and Huxsoll, 1991). Trimming may be partly mechanized.

Wounding of plant tissue results in leakage of enzymes and their substrates that are normally in different cell compartments. The destruction of cell microstructures leads to biochemical spoilage such as texture breakdown, off-flavor, and browning.

Dumping and unloading must be operated by means of use of systems to minimize tissue damage and contamination. This phase is ideally performed in an unloading area that is separate from the clean, processing area (high care or high risk).

The initial trim/core operation is suitable to remove loose leaves, core of lettuce, and large stems. Worker hygiene and training are significant, and operators should wear appropriate clean protective garments. In order to make sure that quality standards are consistently met, several fresh-cut processors recently installed mechanical and optical inspection systems to remove foreign bodies from the product.

Waste removal is a key challenge in all food-processing environments, but it is more of an issue in the fresh-cut fruit and vegetable industry. This labor-intensive process relies upon dozens of workers in each facility hand peeling and trimming fruits and vegetables prior to packaging. Food waste including peels, rinds, tops, tails, cores, and whole rejected pieces are generated throughout the plant.

Waste removal systems (to get rid of unwanted waste from the plant) range from very easy solutions to rather complex ones as belts, flumes, augers, and vacuum. Every fresh-cut fruit and vegetable plant needs to get cabbage leaves, onion tops and tails, carrot tops, pineapple cores, melon rinds, and similar undesirables from the trim line to the waste truck.

Each of the systems presents some advantages and disadvantages. Belts are flexible to be moved and guarantee easy changeover, but they might create dangerous dropping points and sanitation issues if not properly washed. In addition, belted conveyor waste can be collected in strategically placed bins which then need to be carried by lift truck and dumped in dedicated areas. Nevertheless, this is a labor-intensive proposition in addition to the floor space requirements and potential safety concerns.

Flume-carried and water-evacuated waste needs to be dewatered prior to collection in a trailer for disposal. Solids are separated from water and compressed; water on the other side needs to be treated and discharged in dedicated areas. The law and regulations on this topic are becoming more restrictive, raising the bar on the need to have very hygienic design and manufacturing.

The vacuum system without the use of water as a carrying medium is widely utilized. The pneumatic conveying system consists of a blower package that effectively pressurizes a stainless steel or polyvinyl chloride (PVC) waste pipe located beneath the trimming area. The pipeline is fed through a number of rotary airlocks that meter the waste into the system. Within seconds, the waste is transported to a receiver or flexible hose situated above a waste truck or bin for removal.

Reduced water use is an important issue as water cleanliness, chemical requirements, and access rights have become industry hot buttons in recent years.

13.5.2.1.4 Cutting, Shredding, and Size Reduction

Depending on the relevant produce and its application, various shredding or size reduction machines can be used. These are, in almost all instances, constructed of good quality stainless steel (grade 316 or better), but a very well-organized, competent cleaning program must be put in place to ensure that thorough cleaning is undertaken and actually achieved on each machine every day.

With all such machines, there is a potential risk of metal contamination through breakage of blades or contamination with metal. Produce should be passed through a metal detector as a final part of the processing operation, and this should detect any product that has been contaminated with metal slivers or parts.

13.5.2.1.5 Washing

Washing is a critical part of any produce preparation process, especially if a raw, processed fresh produce is sold as ready-to-eat. Washing serves three purposes, and the correct washing process must be accurately designed, controlled, and applied to the correct type of produce. Washing should

- Remove pieces of actual dirt and debris.
- Reduce the microbial and chemical load on the produce.
- Reduce the temperature of the finished product to help enhance shelf life.

An optimum washing system for any prepared vegetable process generally consists of three separate washing stages and three tanks. The first of these tanks aims to eliminate general field dirt and debris. A flotation washing system, where high volumes of air are blown into the tank through spare pipes located 10 to 12 inches beneath the surface of the water, is the preferred solution for products that float. This creates a violent "Jacuzzi" effect that causes produce to tumble around. Any accompanying dirt and debris should be loosened and washed off—if such dirt and debris are likely to float, a proper design will incorporate a system to remove floating debris. Dirt and debris that sink to the bottom of the tank should be released through a periodic drainage system with continuous water renewal.

If antibacterial chemical water treatments are used, it is recommended that, where possible, an automated chemical monitoring and dosing machine be employed for optimum control of the process. These machines are available for chlorine but may not be available for other chemicals.

There are many chemical treatments applied to water, for example, water-softening treatments, flocculation treatments, and decontamination treatments. They are used to kill microorganisms in water and on the produce in contact with water. This approach should never be substituted for practices that ensure that high-quality process water is used throughout postharvest handling activities. However, the use of antibacterial chemicals can enhance the safety of an operation adhering to good hygienic practices.

A new washing system entered the market. The innovative design utilizes the proven technology of the closed pipe flume concept without using a centrifugal pump for product movement. In this closed flume, the product contact time with the sanitizing water solution is accurately controlled, guaranteeing full submersion and treatment time. The laminar flow in the flume piping gently separates the washed products for full exposure to the treatment without creating mechanical damage. As a result, delicate products such as baby leaf and fruit can be effectively washed while maintaining high product quality. The closed flume design prevents the release of the sanitizer in the production area, improving plant safety.

13.5.2.1.6 Dewatering and Drying

On removal from the wash tank, excess water must be separated from the washed produce (depending on the produce item). This can be achieved by simple draining for the necessary time period or alternatively where leafy bulky products are involved, by spin drying, dewatering, or drying in fluidized drying tunnels.

Spin dryers are available in different shapes and capacities. The centrifugation cycle generally begins with a soft loading of the fragile leaves followed by a smooth acceleration and a discharge of the product.

The dewatering system can stand alone as a single unit or be utilized as the first stage of a patented continuous air dryer. As a stand-alone unit, the system is utilized to effectively dewater sliced, shredded, and diced food products. Additionally, the system is suitable to adjust the surface moisture on the unwashed leafy items and to recover expensive treatment liquids utilized by the fresh-cut fruit processing industry (sliced apples, diced tomatoes, etc.). The dewatering system's unique design makes it ideal for dewatering sensitive products such as diced tomatoes, sliced mushrooms, fruit, and other fresh-cut products. Because of its high dewatering efficiency, the drying process can often be eliminated, greatly reducing mechanical damage.

The air dryer is able to dewater all kinds of leafy vegetables, including baby leaf, making use of a technology developed over many years. The fluidized drying system uses the most traditional and safe method for removing water from the leafy products, making use of absolutely natural technologies and preserving the integrity of the product.

Thanks to an accurate study, the possibility to access each component of the tunnel was privileged in order to make the cleaning and maintenance operations easier. It is possible to check, by means of several samplings, the effectiveness of this system, that allows products to dry in an environment free from external contamination, thus considerably improving the performances (yet very good) of the previous drying systems. The results show a longer shelf life of the product.

Care must be taken to ensure the washed material is treated appropriately, that proper stock rotation is achieved, and that the air temperature within the drying area or room is maintained as low as possible (33.8–41°F equal to 1–5°C). If produce is left wet and microorganisms are present, then these may increase in number, making the produce unsafe (and encouraging spoilage).

13.5.2.1.7 Packing

Packaging materials and methods used throughout the process must be suitable for their intended purpose. General hygiene used in the production of packaging is obviously important, but the processor must also ensure that the packaging material is marked as food grade and that it is used as specified. This will ensure that chemicals in the packaging material will not migrate into the product.

Many different packing formats are possible, from a standard polyethylene plastic bag through to rigid plastic containers, buckets, boxes, tubs, vacuum packing, and so forth. In all instances, the type of packaging must be adequate for its use and application. Processors must remember that the product contained therein is still alive and respiring, so for some produce items with high respiration rates (e.g., mushrooms, broccoli), a packaging medium with a high permeability must be used. Permeability is the rate at which various packaging materials allow carbon dioxide and oxygen to pass through their walls, and the selection of packaging material with suitable permeability is critical to good processing and, in some aspects, critical to food safety standards.

If Modified Atmosphere Packing (MAP) is used, the processor must ensure that the packaging and gas mixtures are appropriate for the purpose. It is essential that processors wishing to extend the shelf life of produce in this way obtain suitable technical advice and carry out careful shelf-life testing to ensure that pathogenic microorganisms cannot exploit the storage conditions. Certain pathogenic microorganisms can survive and grow without oxygen and at low temperatures over extended storage periods. In addition, MAP designed to prevent produce spoilage may inadvertently remove a natural barrier to the growth of pathogenic microorganisms, namely, competitive spoilage microorganisms. For these reasons, the application of MAP should not be considered without technical support.

13.5.2.1.8 Fresh-Cut Fruit

Fresh-cut fruit products for retail and food service applications have recently appeared in the marketplace. In the coming years, it is commonly perceived that the fresh-cut fruit industry will have unprecedented growth.

State-of-the-art production facilities are designed specifically for ensuring fresh-cut fruit food safety and maximum quality. Within the premises, the same 3 C principle stated above is commonly applied. Regular microbial testing of the plant environment and all food contact surfaces is performed.

On receipt of every batch of fruit, quality control checks must be carried out by operators on random samples (as a general recommendation, approved fruit vendors follow Good Agricultural Practices). The quality control checks generally include a visual inspection of the fruit, for example, to ensure that there is no bruising or rot damage. In addition, color charts (that may be placed on the wall) help the operator assess the condition and ripeness of the fruit. All quality control checks must meet or exceed the high standards set out in the specification. Fruit is generally delivered on a daily basis, and the delivery is recorded on a computerized system in order to monitor and track it throughout production.

Very large fruit is used in processing to help increase yield (the amount of usable fruit obtainable from a fruit once skin/pips/stones are removed). It is also far quicker to prepare larger fruit than to handle tiny fruit which would be more time consuming. The fruit is stored at low temperatures, around 4–6°C (39.2–42.8°F). Storage is often short, as it is usually used soon after it has been delivered.

Commonly, to minimize contamination risk, suppliers are required to prewash and sort all raw materials. Washing and sanitizing are very important. Washing and treatment systems were conceived in order to wash and sanitize or treat the fruit once cut. In general, fresh-cut fruit pieces should be rinsed just after cutting with cold (0–1°C, equal to 32–33.8°F), chlorinated water (50 to 200 ppm total chlorine) with a pH of 7 or less. This helps extend product shelf life by reducing microbial loads, removing cellular juices at cut surfaces which may promote browning, and actually inhibiting the enzymatic reactions involved in fruit browning.

Due to the diverse variety of fruit, raw materials, and processing conditions, it is not possible to determine a general approach to achieve microbial safety. For instance, melons are washed and sanitized thoroughly to remove foreign matter and microbes from the outer skin and are then transferred, from the low-risk area to the high-risk area, by water (which keeps the melons chilled and prevents bruising). Two additional systems that are widely utilized are steam and hot water. In both cases, the outer skin of melons, pineapple, oranges, and so forth, is taken to a high temperature (i.e., 100°C or 212°F) in order to get rid of bacteria and microorganisms. A hot water system is preferable to steam as it might reach all areas of fruits with corrugated skin, such as melons or pineapple.

Melons, pineapples, or apples are placed by hand in a machine that removes the outer skin of the fruit. The working process offered by this machine provides blades to remove fruit ends that are of the self-centering type to suit each fruit dimension. Fruits can be at first longitudinally cut in halves. Different types of machines are configured to produce wedge slices, ring slices, or whole peeled fruits.

Whole pineapples are washed carefully to remove contaminants. Operators may remove the skin of the pineapple by hand. (This is a skilled procedure, requiring much concentration to remove the skin only, and not the flesh.) Automated machines are also available. The peeled pineapple is passed from low risk through to high risk. The pineapple is dipped to remove any foreign matter and microbes. In the high-risk area, the peeled pineapples are quartered, have their hard inner cores removed, and are diced. The diced pineapple is stored in a blast chiller until needed.

Steam peeling systems have appeared in the market to peel the most delicate types of fruits (i.e., tropical fruits).

Final cooling with nitrogen cooling exhibits unique properties, providing many benefits over other systems. As mentioned above, the process takes place in controlled-atmosphere conditions which impede the growth of oxidation-based spoilage and microbial action.

Little research has been performed on foodborne human pathogens on fresh-cut fruits. This is an area that will have to be investigated in the future. It was also noticed that irradiation of fresh-cut fruit products may be beneficial in reducing the number of bacteria present on the product.

The difficulties encountered with fresh-cut fruit processing, while not insurmountable, will require processors to perform at new and higher levels of technical and operational sophistication.

It is recommended that in the future, researchers, manufacturers, and the industry should work together to overcome barriers that hamper delivery of high quality fresh-cut fruits throughout the year. As the food service industry and home meal replacement expand, there will be a greater demand for fresh-cut fruits with acceptable shelf life.

13.6 CONCLUSIONS

The increased weight placed on sanitary design of fresh-cut processing facility and equipments is a rather new trend. The gap between the meat and dairy industry and this relatively new one is continuously narrowing.

Since microorganisms are ubiquitous and mutate incessantly (adapting to different types of cleaners and sanitizers), it is extremely important to develop dynamic plans of sanitary design to appropriately control microbial contamination (the most serious of the three HACCP hazards).

To all stakeholders in the fresh-cuts industry (from regulators to processors to equipment manufacturers), equipment designed with hygienic goals in mind is fast becoming an area of primary concern. For this reason, sanitation design will continue to develop as manufacturers and industry organizations apply collective knowledge and advanced engineering to create safer and more efficient processes. There is an unprecedented quantity of information related to machine design and implementation in a clean processing floor environment. Organizations such as UFPA (United Fresh Produce Association), PMA (Produce Marketing Association), 3-A, and the USDA offer plant owners and managers information on what to look for in processing equipment that will best suit their needs within a plant. These organizations also offer invaluable guidance to machine manufacturers to continually grow and evolve their easy-to-clean designs. Maximizing use of the knowledge base is the best way to continue the advance in safety and sanitation in the fresh-cut industry.

Additionally, the industry and its organizations are working with equipment manufacturers to establish sanitary equipment design guidance in a proactive effort to offer basic tenets, flowcharts, and checklists to help in the evaluation of effective sanitary design attributes. The FDA, moreover, is focusing on the cleanability of processing equipment during strengthened inspections of food-processing facilities.

REFERENCES

Anonymous. 1988. Guide des bonnes pratiques hygiéniques concernant les produits végétaux prêts a l'emploi, dits "de 4ème gamme," approuvé le 1er Août 1988 par le directeur général de la concurrence, de la consommation et de la répression des fraudes. *Bulletin officiel de la concurrence, de la consommation et de la répression des fraudes,* 17: 221–233.

Anonymous. 1993. Arrêté du 22 Mars 1993 relatif aux règles d'hygiène applicables aux végétaux et préparations de végétaux crus prêts a l'emploi a la consommation humane. *Journal Officiel de la République Française,* 75: 5586–5588.

Anonymous. 1996. Guide de bonnes pratiques hygiénique végétaux crus prêts a l'emploi. *Les éditions du Journal Officiel,* Paris, 71.

ANSI/NSF/3-A Standard 14159-1-2000, which covers the materials, design, and construction requirements of equipment used in meat and poultry processing.

Beaulieu J.C., and Gorny J.R. 2001. *Fresh-Cut Fruits USDA Handbook* 66, 3, 16–30.

Bolin H.R., and Huxsoll C.C. 1991. Effect of preparation procedures and storage parameters on quality retention of salad-cut lettuce. *J. Food Sci.,* 56: 60–62, 67.

Bolin H.R., Stafford A.C., King A.D. Jr., and Huxsoll C.C. 1977. Factors affecting the storage stability of shredded lettuce. *J. Food Sci.,* 42: 1319–1321.

Cantwell M. 2003. *Fresh-Cut Produce: Maintaining Quality and Safety,* Davis.

Current Good Manufacturing Practice (CGMP) regulations. The regulations are current through February 1994. The CGMP regulations are issued under Title 21, Code of Federal Regulations, Part 110 (21 CFR 110). Amendments to the regulations appear in the *Federal Register.*

FDA. *Commodity Specific Food Safety Guidelines for the Melon Supply Chain,* 1st ed., 2005.

FDA. *Commodity Specific Food Safety Guidelines for the Lettuce and Leafy Greens Supply Chain,* 1st ed., 2006.

FDA. Guide to Minimize Microbial Food Safety Hazards of Fresh-cut Fruits and Vegetables, Draft Final Guidance, III definitions.

Gorny J.R., and Kitinoja L. 1999. *Postharvest Technology for Small Scale Produce Marketers: Economic Opportunities, Quality and Food Safety,* University of California, Davis.

Herner R.H., and Krahn T.R. 1973. Chopped lettuce should be kept dry and cold. Yearbook Prod. Mark. Assoc., p. 130 in Bolin et al. (1977).

International Commission on Microbiological Specifications for Foods (ICMSF). 1986. *Microorganisms in Foods 2: Sampling for Microbiological Analysis: Principles and Specific Applications.* 2nd ed. Blackwell Scientific, Oxford.

International Commission on Microbiological Specifications for Foods (ICMSF). 1988. *Microorganisms in Foods 4: Applications of the Hazard Analysis Critical Control Point (HACCP) System to Ensure Microbiological Safety and Quality.* 2nd ed. Blackwell Scientific, Oxford.

International Fresh-Cut Produce Association (IFPA). 2001. *Food Safety Guidelines for the Fresh-cut Produce Industry.* 4th ed. IFPA, Alexandria, VA.

International Fresh-Cut Produce Association (IFPA). 2003. *IFPA Sanitary Equipment Design Buying Guide and Checklist.* 1st ed. IFPA, Alexandria, VA.

International Fresh-Cut Produce Association (IFPA) and Produce Marketing Association (PMA). 1999. *Fresh-cut produce handling guidelines.* 3rd ed. IFPA/PMA, Alexandria, VA, 17–26; 27–42.

Laminkara O. 2002. *Fresh-Cut Fruits and Vegetables: Science, Technology, and Market,* 21–43, CRC Press, Boca Raton, FL.

Lelieveld H.L.M., Mostert M.A., and Holah J. 2005. *Handbook of Hygiene Control in the Food Industry.* 1st ed CRC Press, Boca Raton, FL.

Nguyen-The C., and Carlin F. 2000. Fresh and processed vegetables. In *The Microbiological Safety and Quality of Food,* B. M. Lund, T. C. Baird-Parker, and G. W. Gould (Eds.), Gaithersburg, MD, Aspen, pp. 620–684.

Ohlsson T., and Bengtsson N. 2002. *Minimal Processing Technologies in the Food Industry.* Cambridge, Woodhead.

Shapton D., and Shapton N. 1991. *Principles and Practices for the Safe Processing of Foods.* Cambridge, Woodhead.

Turatti A. 1997. L'evoluzione dell'impiantistica nella lavorazione dei prodotti della IV gamma. Gli ortofrutticoli di IV gamma. Problemi e prospettive. Milano 22 aprile 1997. Seminario I.V.T.P.A. 67–82.

U.S. Department of Health and Human Services, Food and Drug Administration Center for Food Safety and Applied Nutrition (CFSAN). 1998. Guidance for Industry: Guide to Minimize Microbial Food Safety Hazards for Fresh Fruits and Vegetables, October.

Varoquaux P., Mazollier J., and Albagnac G. 1996. The influence of raw material characteristics on the storage life of fresh-cut butterhead lettuce. *Postharvest Biol. and Technol.*, 9, 127–139.

Varoquaux P., and Varoquaux F. 1990. Les fruits de quatrième gamme. *Arb. Fruit.*, 3, 35–38.

Varoquaux P., and Wiley R.C. 1996. Biological and biochemical changes in minimally processed refrigerated fruits and vegetables. In *Minimally Processed Refrigerated Fruits and Vegetables,* R.C. Wiley (Ed.), New York: Chapman & Hall, pp. 226–268.

14 Quality Assurance of Fresh-Cut Commodities

José M. Garrido

CONTENTS

14.1 INTRODUCTION: HYGIENE AND FOOD SAFETY INTERNATIONAL STANDARDS

Since the food crises that took place in Europe in the 1990s ("mad cow disease," dioxins, etc.), consumers are questioning the safety of food products. In this context, the European Union (EU) published the *White Book of Food Safety* in January 2000 with the proposal of a food safety approach that turned out to be radically new in relation to the existing approaches up to that moment. The principle that governs the White Book is the transparency at every level of the FOOD SAFETY policy, what, undoubtedly, will help to increase the confidence of the consumers.

Also in the year 2000, the Codex Alimentarius Commission (CAC) informed that denunciations were received by the accreditation bodies of the European countries questioning the inclusion of all food safety aspects on the quality systems certified on the basis of ISO 9000. For this reason, the CAC reacted by proposing a guide for the utilization and promotion of Quality Management Systems in such a way that they fulfill the food safety criteria. Both the White Book and the CAC proposed, for the whole food chain, quality systems that integrate the following three concepts:

- *Risk Assessment,* as a function of the scientific knowledge, that includes:
  - Identification and characterization of hazards
  - Assessment of the exposure
  - Characterization of the risk
- *Risk Management,* competence of the industry, of the different elements and steps of the food chain:
  - Hazard Analysis and Critical Control Points (HACCP) and prerequisites

- Good practices
- Suppliers and subcontractors
- Traceability ahead and backward
- Crisis management and withdrawal of the affected product
- *Risk Communication,* responsibility of all the stakeholders (scientists, food chain, authorities, consumers) concerning information exchange, suggestions, and knowledge

In 1993, the Public European Administration regulated the compulsory nature of the HACCP systems implementation, excluding the primary sector from its scope (Directive 93/43/CEE). The first schemes of food safety susceptible of certification arose in 1997. The fulfilling of requirements by a company provides a guarantee to the stakeholders that the HACCP system management is effective and the products are safe. These procedures of Hygiene and Food Safety have different origins:

- *Suppliers' associations and big retailers,* with the aim of covering their responsibility related to the food safety of their trademarks:
  - British Retail Consortium (BRC)
  - International Food Standard (IFS)
- *Manufacturers' associations,* to protect the image of the whole sector:
  - FAMI-QS (European Feed Additives and Premixtures—Quality System) additives and premixes for animal feeding
  - IGMP (International Good Manufacturing Practices) (packaging)
- *Food Plus GmbH* and their GLOBALGAP (Global Partnership for Good Agricultural Practices, formerly EUREPGAP, Euro-Retailer Producer Working Group—Good Agricultural Practices), covering aspects from production to postharvest manipulation of agricultural products
- *International Standard Organization (ISO),* with the aim of providing guidelines for management procedures (ISO 9001, ISO 14001, ISO 22000, etc.)

To assure the correct application and efficiency of the guidelines, the certification schemes should be credited, which implies that

- The competent national body validates the standard officially
- The certification must be issued by a certification body credited for that scope

The certification schemes for Hygiene and Food Safety offer several advantages:

- They provide criteria for the design, implementation, and operation of the management system. There is normally a lack of these aspects in the regulations of the different countries, because they have a generic character.
- The criteria are uniform among countries, which allows "speaking the same language," eliminating the barriers of marketing caused by technical problems related to the lack of safety and hygiene of the products.

- They provide elements to the organization to allow them to manage the food safety and the hazards associated with the productive processes in an effective way, so it turns into an added value to the company that can be communicated to the whole chain, thus achieving consumer confidence.

14.2   COMPARISON OF THE INTERNATIONAL STANDARD GUIDELINES

The international guidelines for food safety management have structures and elements in common, namely:

- *HACCP* system's requirements, based on the Codex criteria
- System requirements of quality management based on the *ISO 9001* guidelines
- *Product* requirements: control of the product and packaging characteristics
- Requirements of food *processing*: suppliers, hygiene, maintenance, machinery, and facilities
- *Personnel* requirements: good manufacturing practices, training and control of diseases
- All of them include the compliance of the *legal requirements*—that is, horizontal and national legislation, technical and sanitary sectorial regulations, as well as procedures applicable to the sector.

As a consequence, seven HACCP principles based on the Codex Alimentarius must be performed:

- Identify hazards and evaluate risks.
- Establish preventive measures (prerequisites).
- Identify Critical Control Points (CCPs).
- Establish critical limits and tolerances of the CCPs.
- Establish the CCP's monitoring system and the corrective actions to face deviations.
- Establish the documentation system and record keeping procedures.
- Establish the verification procedure for ensuring the system is working as intended.

The HACCP's prerequisites used to be the following:

- Cleaning and sanitation plan
- Facilities and equipment design; layout and validation
- Maintenance and calibration plan
- Pest control
- Good Manufacturing Practices and personnel training
- Water control
- Allergens control

TABLE 14.1
Main Differences between ISO 9001 and the Specific Procedures for Food Safety

Approach	ISO 9001	Food Safety
Exclusive for food safety	Implicit, but neither exclusive nor developed	Reason for these standards
Customer (expectations)	Reason for this standard	Commented in BRC/IFS, not included in ISO 22000
Traceability	Requirement	Detailed in IFS/BRC, requirement in ISO 22000
Organoleptic/sensory	Requirement (as an specification/contract)	Commented in BRC/IFS, not included in ISO 22000
Quality characteristics (nonhygienic)	Requirement (as a specification/contract)	Commented in BRC/IFS, not included in ISO 22000
Service characteristics	Requirement (as a specification/contract)	Not specifically included

- Specifications for raw materials, ingredients, and packaging materials
- Traceability
- Supplier control
- Production planning
- Product recall and withdrawal; food crisis and alerts management manual

The main differences between ISO 9001 and the specific Procedures for Food Safety are detailed in Table 14.1. In addition, the main differences between the Procedures of Hygiene and Food Safety are shown in Table 14.2.

TABLE 14.2
Main Differences between Standards of Hygiene and Food Safety

Approach	Process Certification (ISO 22000)	Process and Product Certification (IFS/BRC)
Necessary accreditation to certificate	45012 Any accreditation body	45011 Any recognized body by BRC/IFS
Control of the certification body's activity	Only according to accreditation requirements (includes auditor's qualification)	Particular requirements BRC/IFS Accreditation requirements IFS: Auditor's requirements
HACCP	Compulsory according Codex	Compulsory according to Codex, specific requirements
Quality system	Compulsory	Compulsory, specific requirements
Prerequisites	Compulsory, indicates relationship	Compulsory, indicates relationship
Detail level of prerequisites	None	Very high

IFS is a standard officially supported by German, French, and Italian distribution chains, so that it is in these countries, together with Austria and the Netherlands, where most certificates are emitted. In the case of BRC, British, Nordic, and Australian distribution chains are officially supporting it. These countries together with South Africa are those with the largest number of certifications. Trademarks as Compass or United Biscuits also support BRC. In the case of Spain, up to December 2007, there were 312 food companies certified by BRC. Most of those companies are also being certified in compliance with IFS, because it is supported by retail chains such as Carrefour, Alcampo, El Corte Inglés, Consum, and Eroski.

14.3   FROM ISO 9001 TO ISO 22000

In 1993, the European Union issued the Directive 93/43/CEE, related to the hygiene of food products, according to which food companies were required to implement self-control systems based on the HACCP model, as a prevention system for food safety. Nevertheless, because of the food crises that occurred in Europe in the 1990s, consumers lost confidence in food products. The White Book of Food Safety, published in January 2000, established the basis of a new regulatory system concerning this matter, which became applicable through the following regulations:

- Directive 95/2001, related to the general safety of products
- Regulation 178/2002: Food legislation principles and general requirements as well as the creation of the European Authority of Food Safety and the establishment of the procedures related to food safety
- "Hygiene Pack":
  - Regulation 852/2004, related to the hygiene of food products
  - Regulation 853/2004, with the specific requirements for the hygiene of food products of animal origin
  - Regulation 854/2004, which established specific requirements for the official management and control of food products of animal origin intended for human consumption
- Regulation 2073/2005, with microbiological criteria applicable to food products

The first schemes for food safety certification appeared in 1997 and can be classified into two groups:

- Voluntary standards of HACCP certification
  - Danish standard DS 3027 (1st ed. 1997)
  - Dutch standard RvA (1st ed. 1999)
- Mandatory certification standards to provide products to retailers:
  - British Retail Consortium procedure (BRC, United Kingdom)
- International Food Standard procedure (IFS, France, Germany)

Since the mid-1980s, ISO 9000 series guidelines have had a large influence on the general industry as a model for the implementation and certification of quality assurance/management systems. Nevertheless, and though its approach is directed to the prevention and detection of defective products, it provided a poor definition of specific requirements directed to guarantee the exit to the market of a safe food product. In this context, in 2001, the Technical Committee 34/WG8 within the International Standardization Organization (ISO), which was very successful in the elaboration and diffusion of ISO 9000 (Quality Management) and ISO 14000 (Environmental Management) standards, started elaborating ISO 22000 guidelines for Food Safety Management, trying to put together the strategies of ISO 9000 and the HACCP system, in order to guarantee that the products are safe and fulfill all the quality requirements.

ISO 22000 is the first Food Safety Management Standard with an international consensus, elaborated on by the ISO, which includes a net of standardization bodies that represent 148 countries. Its title is "Food Safety Management Systems: Requirements for any organization in the food chain." As every ISO guideline, it is an optional standard guideline to satisfy a market demand. It was approved and published in 2005, having been elaborated by experts and with the consensus of organizations of international reference for the food sector: Codex Alimentarius, Global Food Safety Initiative, CIAA (Confederation of the Food and Drink Industries in the EU), World Food Safety Organization, or Bureau Veritas Certification.

In contrast to the IFS or the BRC Food Standard, the ISO 22000 procedure can be applied by any type of organization involved in the food chain: producers of animal feed, primary producers, food-processing organizations, transporters, storage operators and subcontractors, retailers and distribution shops of food services, and even organizations interrelated with the food chain (equipment producers, packaging materials, cleaning agents, additives, etc.). It provides an adapted frame for the criteria specified by the White Book (EU) and the Codex for the design (Level 1—Risk evaluation), function (Level 2—Risk management), and interactive communication (Level 3—Risk communication).

ISO 22000 guidelines contain requirements for planning, execution, verification, and updating of the whole management system of food safety, as well as requirements directed to satisfy the food safety demands of the clients/consumers and safety authorities. It combines the following four key elements of food safety:

- Management system, based on the ISO 9000 procedure
- Interactive communication along the food chain
- Favorable environment (facilities and maintenance programs)
- Hazard control validated and verified by combining several measures managed for:
  - A HACCP plan
- A prerequisites program and good manufacturing practices

The ISO 22000:2005 guideline has the following structure:

- 0. Preamble and introduction
- 1. Scope

- 2. Normative references
- 3. Terms and definitions
- 4. Food Safety Management System (FSMS)
- 5. Management responsibility
- 6. Resource management
- 7. Planning and realization of safe products
- 8. Validation, verification, and improvement of the FSMS

It also features a series of informative Annexes:

- Annex A: Cross-references between ISO 22000:2005 and ISO 9001:2000
- Annex B: Cross-references between HACCP (CODEX, 2001) and ISO 22000:2005
- Annex C: Codex references for specific sectors providing examples of control measures, including prerequisites programs and a guide for their selection and use

Among the main benefits of the ISO 22000 implementation in the organizations, the following can be pointed out:

- Communication is organized and directed to the same aim among the commercial parts
- More efficient and dynamic control of hazards
- Resources optimization along the whole chain
- Documentation, planning, and verification process improvement
- Resources saving: audit overlapping is reduced
- Confidence earned from the organization to its customers

The ISO 22000 standard gives the following added value to organizations:

- An internationally recognized standard that harmonizes different requirements from customers or countries
- An auditable standard that can be used for third-party certification bodies with clear requirements
- A filling of the gap between ISO 9001:2000 and HACCP
- A contribution to a better understanding and to a greater development of the HACCP
- A good reference model for safety authorities

14.4 BRITISH RETAIL CONSORTIUM (BRC) GLOBAL STANDARD FOOD GUIDELINES

BRC is the product certification standard promoted by the British Retail Consortium, which represents 90% of British distributors, with more than

290,000 shops and 3 million workers, and its main objective is to defend the interests of retailers.

BRC has established several standards for trademarks owned by retailers (private brands): BRC GS *Storage and Distribution* (version 1), BRC IOP GSF *Packaging* (version 3), and BRC GS *Consumer Product* (version 2), for nonfood products manufacturers; nevertheless, the *BRC Global Food Standard* (version 5), aimed at *food products* manufacturers, is the most known and extended. The BRC members demand that the manufacturers of their private brands be certified by one of the previous standards. The aim is to assure, at low cost, the responsibility they are assuming with the ownership of their trademarks regarding safety, quality, and legality, because the certification is assumed by the manufacturer.

In 1990, the food legislation of the United Kingdom coined the concept of "due diligence," or "co-responsibility." As a consequence of the food crises of the 1990s, in addition to the weakness of ISO 9000 regarding food safety and the absence of an international standard in the matter, the first version of the BRC Food Standards arose on October 1998 and has had successive revisions in June 2000, April 2002, January 2005, and, the last one (version 5), December 2007, applicable from July 2008. Therefore, BRC Food Standard is a regulation of compulsory fulfillment by suppliers of food products to big and medium retailers of the United Kingdom. It includes requirements and recommendations applicable to the manipulation of any type of food product in any phase after the primary production: processing, packaging, storage, transport, distribution, manipulation, and supply to the consumer. In addition to UKAS (United Kingdom Accreditation Service) and BRC, certification bodies, research stakeholders, trade entities, food industry representatives, and distributors take part in its elaboration and revision.

The standard requires

- The implementation of the HACCP
- A documented and effective quality management system
- Prerequisites (food hygiene elements): control of design requirements and maintenance of facilities, processes, products, and staff

Every section begins with a declaration of intentions as the outstanding paragraph, followed by the requirements that develop the contents of the section. In the last version (the fifth), seven requirements called *Fundamental* were established. A nonconformity (NC) facing a fundamental requirement supposes the inability to obtain the certificate until the NC is corrected and the corrective action is verified by the certification body in a complete audit. Three levels of NCs exist:

- *Critical:* Critical deviation that could result in hazardous or unsafe conditions for individuals or is likely to prevent performance of a legal requirement. If the deviation is related to a nonfundamental requirement, the certification is not obtained until the NC is corrected and the corrective action is verified by the certification body in a follow-up visit.

- *Major:* Basic deviation in the requirements of a declaration of intention or any clause of the standard. If the deviation corresponds to a nonfundamental requirement, the certificate is not obtained until the evidence of its correction is demonstrated within 28 days, although in this case, upon decision of the certification body, a personal checking or at least documentary evidence with verification in the following evaluation is needed.
- *Minor:* Basic deviation from the requirements but being demonstrated that there are no doubts on the conformity of the product. The certification is not obtained until documentary evidence of its correction within 28 days is demonstrated. The certification body verifies this fact in the following evaluation.

The BRC standard is structured in seven chapters:

1. Senior Management commitment
2. Food Safety Plan—HACCP
3. Quality Management System
4. Installation standard
5. Product Control
6. Processes Control
7. Staff

It contains ten fundamental requirements, which must be established, implemented, continuously supported, and controlled by the organization. The lack of implementation of a fundamental requirement can lead to damages in the product's legality, its quality, or its safety, so the certificate cannot be granted. These fundamental requirements are as follows:

- Senior Management commitment and Continual Improvement § 1
- Food Safety Plan—HACCP Clause 2
- Internal Audits § 3.5
- Corrective and Preventive Action § 3.8
- Traceability § 3.9
- Layout, Product Flow, and Segregation § 4.3.1
- Housekeeping and Hygiene § 4.9
- Handling Requirements for Specific Materials—Materials Containing Allergens and Identity Preserved Materials § 5.2
- Control of Operations § 6.1
- Training § 7.1

The last version 5 introduces important changes from the previous versions that go into the definition and scope of the requirements in depth. Among all of them, the most remarkable may be the following:

- Higher emphasis in the *Senior Management Commitment*: establishment, control and revision of the Food Safety Objectives ($\approx$ ISO 22000); reinforcement of the review's accomplishment by the direction; support commitment to the food safety team ($\approx$ ISO 22000); continuous improvement requirement of the management system
- Requirements of the *Food Safety Plan—HACCP*: relevant information should be compiled before performing the HACCP analysis ($\approx$ ISO 22000); detailed information on the expected use of the product, taking into account the sensible population groups (children, allergics, elders); detailed flow diagrams; proofs of acceptable levels for each detected hazard in the final product ($\approx$ ISO 22000); validation of CCPs by the Food Safety Team ($\approx$ ISO 22000); verification activities to prove that the HACCP is effective ($\approx$ ISO 22000); revision of the HACCP plan following the established procedures ($\approx$ ISO 22000)
- Requirements in the *Management System*: internal annual audits; running effective corrective actions; manufacturing procedures to fulfill the costumer's requirements ($\approx$ IFS); accomplishment of annual exercises of traceability that include quantities checkups; strict requirements for the management of incidents, product recall, and withdrawal
- Major detail in *the facilities requirements*: safety and prevention of access to production areas; aptitude certification for using materials in contact with food; higher detail for staff's clothes; presence sheets with specification and safety of the chemical products ($\approx$ IFS); detailed requirements for the glass and hard plastic manipulation; procedures of housekeeping and hygiene that must be confirmed in case of changes ($\approx$ ISO 22000); detailed requirements of pest control
- Demanding requirements in the *product control*: detailed measures in allergens control; organoleptic tests and self-life studies ($\approx$ IFS); detailed requirements to assure the reliability of the analytical methods
- Some changes in *process control*: Validation needed based on the HACCP of any change in the process (changes in the formulation, methods of production, equipment, packaging) ($\approx$ ISO 22000)
- Some changes in the *staff requirements*: Detailed and rigorous requirements in the training and capability to carry out the activity; graphs of personnel circulation on a plan ($\approx$ IFS); more flexibility in the use of face masks based on the risk analysis

The certification process is detailed in the standard. The certification bodies must be credited for the certification of product and be in the directory of BRCs Web site (http://www.brc.org.uk/standards/). The duration of the audit depends on the size of the organization and the processes it performs. The frequency of audits is about 6 to 12 months, depending on the nonconformities, and the organization decides if they can be published or not. All the requirements have to be revised in the follow-up audits, and the reports must be in English.

14.5 INTERNATIONAL FOOD STANDARD (IFS) V5 GUIDELINES

The presence of trademarks owned by the food distribution companies (private brands) is an increasingly extended phenomenon in Europe. The concept was created in 1869 by Sainsbury, and it now approaches two segments of the market:

- Products of equivalent quality to the first class marks of the manufacturer
- Low price or hard discount products.

The consequences for the distributors owning a private mark can be summarized as

- Increased communication with the final consumer
- More legal responsibilities that can have repercussions on their image
- The consumer's perception that the manufacturer is the distributor of the product or, at least, that it is responsible for the quality, legality, and safety of the products with its trademark

It was already discussed how these circumstances were the origin of the edition of the BRC schemes within the British distribution. In the same direction, the association of food distributors of Germany (HDE, since 2002), those of France (FDC, from the version v4 of 2004 and the actual version v5), and those of Italy (Federdistribuzione), stand the edition and revision of the International Food Standard (IFS) (http://www.food-care.info/).

The current version v5 of IFS was the result of the revision done by all the stakeholders: distributors' associations, manufacturers, Accreditation and Certification Bodies, and IFS's expert staff. It has the following characteristics:

- A unique control list, without distinguishing among basic and higher requirements, and the removal of the named *Suggestions*
- More requirements that include the *Risk Analysis* approach
- More emphasis in *Processes* and *Procedures*
- A new punctuation system that allows a more simple comparison of results
- Change in the audit frequency to 12-month cycles (validity period of the certificate)
- Determination of more requirements "Knock Out" (KO) focused on fundamental aspects of food safety
- More detailed requirements for accreditation and certification bodies and auditors

The standard structure is divided into four parts: Protocol of audit (Part 1); Standard requirements (Part 2); Requirements for accreditation and certification bodies and auditors, accreditation process and certification IFS (Part 3); and a Part 4, related with the models of audit report, results, corrective actions plan, and certification.

Part 2 requirements are divided into five areas:

1. Senior Management Responsibility
2. Quality Management System
3. Resource Management
4. Production Process
5. Measurements, Analysis, and Improvements

The 251 requirements are evaluated individually using a punctuation scale that goes from 20 points that represent the "full compliance," to 0 points, meaning "criteria not satisfied," passing through the 15 points of "small deviation from the requirements" and the five of "partial implantation of the requirements." Achieving 95% or more of the maximum possible points grants the top level of the certificate, being the basic level that obtained by the fulfillment of 75% and 95% of the above-mentioned maximum points.

In addition, two high levels of nonconformity are established:

- *Major NC,* which represents 15% of the maximum punctuation and implies not receiving the certificate until a corrective action plan is sent within 14 days and audit in less than 6 months. They are established by the auditor when a substantial deficiency exists among the requirements, including the regulation of production and destination country, or when it could cause a serious hazard for the health of the consumer.
- *K.O.* for 10 NC of the standard. The fact of not implementing some of them has as a consequence not obtaining the certificate.

The KO requirements are defined specifically in the standard and refer to the following aspects:

- Senior Management Responsibilities § 1.2.4
- HACCP Analysis § 2.1.3.8
- Staff Hygiene § 3.2.1.2
- Raw material specifications, legal requirements § 4.2.2
- Final products specifications § 4.2.3
- Foreign body detection § 4.9.1
- Traceability System, including OGMs and allergens control § 4.16.1
- Internal audits § 5.1.1
- Product Recall and Withdrawal § 5.9.2
- Corrective actions § 5.11.2

14.6 GLOBALGAP STANDARD

The Euro-Retailer Produce—Good Agricultural Practices (EUREP-GAP) Standard was published in 1997 as an initiative of the retail sector and named EUREP Working Group (http://www.globalgap.org). The engine behind the initiative included European retailers, in response to the increasing concern of the consumers toward food safety, environmental and labor standards, and with the idea of harmonizing their own standards, often very different. It was also intended to answer the interest of the producers in the development of common certification procedures. In September 2007, in coherence with the world extension of the bodies implied in the system and with vocation of being a global reference, it changed its name EUREP into *GLOBALGAP*.

GLOBALGAP is a reference with scope for the whole primary production process (crop fields, aquaculture, and animal husbandry), with Food Safety requirements, labor safety, and environmental criteria, which are gaining global acceptance as a valuable scheme to assure the quality of the products. It establishes the frame for the development of Good Agricultural Practices (GAP) in farms, defining fundamental elements for the best practice in the primary global production, acceptable for the principal retail worldwide groups. Nevertheless, the procedures adopted for some individual retailers and some producers can exceed those of GLOBALGAP. The main benefits that it offers to the producers are the following:

1. Reduction of risks related to food safety in the world production
2. Reduction of the conformity costs, avoiding multiple audits and proliferation of requirements for the buyers
3. Increasing the integrity of the insurance programs of farms worldwide: common criterion, checking independent reports and actions to be taken

The producers receive GLOBALGAP's approval from an independent verification organization approved by GLOBALGAP and credited according to ISO 45011.

Three options can be registered in the program:

- *Option 1*: A farm with one or multiple areas of production or "establishments," which are a property or are managed by an individual or an organization, without constituting separated legal entities.
- *Option 1 with Quality Management System*: This one must be solid enough to assure (and it must remain demonstrated in the audits) that the registered producers/production areas fulfill uniformly the requirements of the standard.
- *Option 2, Group of Producers*: Group of several farmers with legally separated entities who want to accede to the certification in a unitary way, in order to support the integrity of the whole set.

The documents of the program are:

1. *General regulations*: Establish the rules for the management of the standard.
2. *Control Points and Compliance Criteria (CPCC)*: Establish the requirements that the producer must fulfill, leaving specific details for each requirement.
3. *Checklist*: Is the base for the external audit to the producer and must be used to fulfill the annual internal audit requirement.

The CCP are structured in:

- Major obligations
- Minor obligations
- Recommendations

Though when the country has more restrictive regulations than those of the GLOBALGAP standard, the legislation of the country where the producer operates should be observed; only when laws do not exist or are more permissive does GLOBALGAP specify a minimal acceptable level of fulfillment.

The CPCC, which it is the section object of audit, is divided into areas and subareas:

- An "All Farm" (AF) Base Module
- Three areas subdivided into different subareas:
  - Crops Base (CB) Module
    - Fruit and Vegetables (FV)
    - Combinable Crops (CC)
    - Coffee Green (CO)
    - Tea (TE)
    - Flower and Ornamentals (FO)
  - Livestock (LS) Base Module: Cattle and Sheep (CS); Dairy (DV), Pig (PG), Poultry (PY)
- Aquaculture Base (AB) Module: Salmonids (SN)

In this way, the chapters applicable to a crop of fruits and vegetables are the following:

- Section AF: Base Module for All Farm:
  - AF.1 Record keeping and internal self-assessment/internal inspection
  - AF.2 Site history and site management
  - AF.3 Workers health, safety, and welfare
  - AF.4 Waste and pollution management, recycling, and reuse
  - AF.5 Environment and conservation
  - AF.6 Complaints
  - AF.7 Traceability
- Section CB: Crops Base Module:
  - CB.1 Traceability
  - CB.2 Propagation material
  - CB.3 Site history and site management
  - CB.4 Soil Management
  - CB.5 Use of fertilizers
  - CB.6 Irrigation
  - CB.7 Integrated Pest Management
  - CB.8 Plant protection products
- Section FV: Fruit and Vegetables Module:
  - FV.1 Propagation Material
  - FV.2 Soil and Substrate Management

- FV.3 Irrigation
- FV.4 Harvesting
- FV.5 Produce Handling (not applicable if Produce Handling in a packing facility on farm is excluded from certification)

15 Future Trends in Fresh-Cut Fruit and Vegetable Processing

Gemma Oms-Oliu and Robert Soliva-Fortuny

CONTENTS

15.1 CONSUMER DEMANDS AND MARKET DEVELOPMENT

In the last few decades, a tremendous revolution has occurred within the family structure. The inclusion of women in the labor force has caused a radical change in lifestyles, characterized by a dramatic reduction in the times for meal preparation. An increasing number of people have at least one meal away from home, making use of public or private food services. In this context, industrial kitchens need to prepare and cook large numbers of meals in short periods of time, often with limited staff and equipment. At the same time, consumers have become more health conscious about food choices and have developed interest in both fresh and convenience products (Rocha and Morais, 2007).

European nutrition experts agree that consumption of sufficient amounts of fruits and vegetables is key to a healthy diet and can play an integral role in reducing cardiovascular diseases, certain types of cancer, obesity, and diabetes. They are low-caloric food items, but at the same time, they contain remarkable amounts of some minor functional constituents in foods, such as fiber, vitamins, and minerals. They also contain phytonutrients that offer protection against degenerative diseases, leading to lower mortality and increased life expectancy and quality. Encouraging consumers to increase their intake of fruits and vegetables is a worldwide issue. Nutrition experts have proposed lowering the price of healthy foods to increase consumption. However, increased health information may be a more efficient policy tool than price decrease to increase the consumption of fruit and vegetables. In this regard, the international movement of "five a day" promotes the consumption of fruits and vegetables worldwide and is present in over 40 countries on five continents. The claim for five servings of fruits and vegetables is based on the daily minimum consumption recommended by the scientific and medical community in a healthy diet. Great importance is being placed on consuming fresh products as part of a broader emphasis on health. Despite increasing knowledge about the health benefits of diets high in fruits and vegetables, many consumers' diets are still deficient in the recommended intake. Americans consume only half as much as recommended by the Food Guide Pyramid. In Southern European countries, trends show that diets are moving away from the traditional "Mediterranean diet" based on fruit and vegetables, bread and other cereals, olive oil, and fish consumption (Rodrigues and de Almeida, 2001).

This scenario created the challenge, and also the opportunity, for the introduction to the markets of new products like fresh-cut fruits and vegetables as a way to increase the consumption of fruits and vegetables to the recommended levels for a healthy diet. Fresh-cut produce meets the expressed consumer desire for convenience, quality of appearance, and healthy nutrition. A considerable number of fresh-cut commodities are already available in the markets of many developed countries, thus becoming a useful tool for attracting consumers and boosting the intake of fruit and vegetables servings. Minimally processed products are one of the major growing segments in food retail establishments (Soliva-Fortuny and Martín-Belloso, 2003). At the top of the freshness tree, products labeled as "organic" stand out from the rest, corresponding to produce that has been grown without the aid of chemicals and delivered free from preservatives, with emphasis on the purity of the product and the effectiveness of the packaging that protects it. Once sold in specialty stores, organic products now rate special sections in a growing number of retail stores. Sales of organic packaged salad mixes, one of the fastest-growing categories, have grown at a rate of 200% over the past three years and show no sign of slowing down.

The market for fresh-cut fruits and vegetables is consolidated in the United States, as well as in some European countries such as the United Kingdom, France, or The Netherlands, whereas in countries such as Spain and Italy, its development is still moderate. The evolution of the North American market is due to the wide array of products and presentations offered, the increase in exhibition space, and the increase in shelf-life of up to 10–16 days for fresh salads. Such a reality differs from that of Spain or Italy, where the shelf-life of fresh-cut salads is around a week. This time difference may be related to technology issues, but especially to the logistic development that allows maintenance of the cold chain. The marketing of fresh-cut fruits and vegetables requires the appropriate combination of technologies for extending the shelf-life of the products, maintaining the sensory and organoleptic characteristics of the original fresh product.

15.2 NEW APPROACHES TOWARD THE CONTROL OF QUALITY AND SAFETY OF FRESH-CUT FRUITS AND VEGETABLES

Much research is still to be done in order to develop technologies that render fresh-cut fruits and vegetables products with high sensory quality, microbiological safety, and nutritional value. In general, it is currently possible to reach a shelf-life of at least 1 week for most refrigerated (5°C) products. However, some commodities would require a shelf-life of more than 2 weeks, so that success in their commercialization can be attained. Variations in quality and shelf-life, safety through the control of temperature and hygiene conditions in a "from farm to fork" approach, and quality of raw materials are some of the main points that may raise concerns by consumers (Artés et al., 2007). Maintaining the correct product temperatures through the entire chill chain is often the most important factor in ensuring quality and safety of fresh-cut fruits and vegetables. These products are often packaged under modified atmospheres, having a relatively short shelf-life and exhibiting high vulnerability to temperature abuses. Thus, maintenance of an adequate temperature as close to 0°C as possible is required to keep the product safe for consumption.

A characteristic feature of fresh-cut produce is the need for an integrated approach, where different aspects, such as raw materials, handling, processing, packaging, and distribution, must be properly managed to make shelf-life extension possible. The intelligent selection of different preservation techniques, without obviating the intensity of each treatment and the sequence of application to achieve a specified outcome, is expected to have significant prospects for the future of minimally processed fruit and vegetables. Unit operations such as peeling and shredding need further development to make them gentle. There is no sense in disturbing the quality of produce by rough treatment during processing, and then trying to limit the damage by subsequent use of preservatives.

15.2.1 DISINFECTION TECHNOLOGIES

In order to minimize microbiological spoilage, and at the same time provide safe and high-quality fresh-cut fruit and vegetables, the industry needs to implement improved strategies for commodities disinfection. Although chlorine is still the most commonly used sanitizer due to its efficacy, cost-effectiveness ratio, and simple use, future regulatory restrictions are likely and will require the development of functional alternatives. In some European countries including Germany, the Netherlands, Switzerland, and Belgium, the use of chlorine in fresh-cut products is forbidden (Carlin and Nguyen-the, 1999). As a consequence, several innovative approaches have been explored for the decontamination of minimally processed fruits and vegetables. These alternatives must satisfy consumers and, at the same time, be compatible with the sensory characteristics of the products to be treated. However, different studies have demonstrated that decontamination treatments such as hydrogen peroxide or acidic electrolyzed water can even enhance the microbial growth rate depending on the product and applied conditions (Gómez-López et al., 2008). Consequently, after some days of refrigerated storage, the benefit of the decontamination can be lost. Hence, microbial populations in decontaminated produce could reach higher levels than those found in their non-decontaminated counterparts. The use of chlorine dioxide and the cyclic exposure to ozone gas outstand among the alternative treatments for decontamination with a higher potential (Cooksey, 2005; Aguayo et al., 2006). However, the practical implementation of gaseous treatments during storage of minimally processed vegetables packaged in retail bags can pose a problem, because gases have to diffuse to any surface and could be degraded by the food before accomplishing their desired effect. Physical treatments for fresh-cut fruit and vegetables are being considered as alternatives to chemical preservation techniques. Low-dose gamma irradiation is very effective in reducing bacterial, parasitic, and protozoan pathogens in raw foods. Its effectiveness has been proven in fresh-cut carrots, lettuce, and cantaloupe. Ultraviolet light (UV) is a relatively inexpensive and easy-to use technique for food preservation. The application of high UV dose levels requires continuous UV sources that can operate during long time periods, which could compromise quality due to the consequent damage of the treated tissues. Pulsed light (PL) could be an alternative technology for the decontamination of food surfaces and food packages. This technology consists of the release of short intense pulses of broad-spectrum light. It has been suggested that short pulse widths and high peak intensities may have a competitive advantage over continuous UV treatment systems, especially in those situations where rapid disinfection is required, because the energy density can be multiplied manyfold (Dunn et al., 1995; FDA, 2000). PL can be used in the final steps of minimal processing; however, treatments that effectively penetrate packaging materials are still a challenge to this technology. Gómez-López et al. (2007), Elmnasser et al. (2007), and Oms-Oliu et al. (2010) summarized the main limitations of the PL systems for food applications. According to them, one of the most important limitations of PL treatments is the control of heating, which could substantially impair the quality of fresh-like commodities.

15.2.2 PACKAGING TECHNOLOGIES

Packaging also has a determinant role in the preservation and quality retention of fresh-cut commodities. In recent years, packaging has developed due to increased demands on product safety, shelf-life extension, cost efficiency, environmental issues, and consumer convenience. Different from other fresh foods such as meat and fish, fruits and vegetables continue to actively metabolize during postharvest periods. By matching appropriate film permeation rates for O_2 and CO_2 with the respiration rate of the packaged fresh-cut commodities equilibrium, modified atmosphere packaging (MAP) can be established inside the package. Products are often packaged after flushing with different mixtures of gases (O_2, CO_2, and N_2). The use of low O_2 concentrations (1–5%) and high CO_2 concentrations (5–10%) in combination with storage at refrigeration temperatures (optimally 4°C), is proposed as optimal storage conditions for fresh-cut vegetables to maintain sensory and microbial quality. Polyvinyl chloride (PVC), used primarily for overwrapping, and polypropylene and polyethylene, used for bags, are the films most widely used for packaging minimally processed foods. Multilayered films, often with ethylene vinyl acetate, are manufactured with differing gas transmission rates. However, the most difficult task in manufacturing fresh-cut fruits and vegetable products of good quality with a shelf life of several days is that only a few packaging materials on the market are permeable enough to match the respiration of fruit and vegetables. Packaging films currently available for fresh-cut produce do not have sufficient O_2 and CO_2 transmission rates, especially when the produce has high respiration, and as a consequence, too low O_2 levels and excessive amounts of CO_2 in package headspace are often detrimental to fresh-cut fruit. In addition, most MAP systems are designed for a specific temperature, and films with adequate O_2 permeability, adequate response to temperature variations, or both are rare (Cameron et al., 1995). Thus, changes in the environmental temperature create a specific problem in MAP systems because the respiration rate is more influenced by temperature changes than is the film permeability used to obtain the modified atmosphere. Packaged fruits and vegetables are usually exposed to varying surrounding temperatures during handling or retail display, resulting in decreased O_2 and increased CO_2 levels inside the package due to a rise in the respiration rate of the product. Due to this fact, it is difficult to maintain an optimum atmosphere inside a package when the surrounding temperature does not remain constant. In designing MAP systems for fresh-cut commodities, it would be prudent to realistically evaluate the time and temperature conditions that the products will likely encounter along the postharvest chain, as well as the likelihood of mixed load conditions. It then will become possible to design systems such as a combination MAP that can maintain optimum atmospheres throughout the postharvest handling chain (Brecht et al., 2003).

To address some of the limitations of using polymeric films for MAP, active packages are being developed. An active package will respond to environmental changes such as temperature or atmosphere composition, or to physiological changes in the product, which may be indicated by the evolution of volatile compounds such as ethanol or ethylene. Some films allow increasing gas transmission when temperature increases due to a reversible melting of the side-chains in the polymers. These films can be tailor made to match changes in permeation properties to the temperature response of respiration rates of a specific commodity. Other permeation patches may consist of other highly permeable films, microperforation, or a combination of the two. Incorporation of patches may facilitate more precise control of permeation properties of packages. Sachets also can be incorporated into packages (Forney, 2007). Other features are being developed that respond to the environment to modulate gas transmission properties of the package (Cameron et al., 1995). Examples include pores that open to increase gas transmission in response to a rise in temperature. Sachets may contain a variety of substances that can absorb or release gases and provide another mechanism for regulating atmosphere composition and product quality (Ozdemir and Floros, 2004).

The application of superatmospheric O_2 concentrations has also been suggested to overcome limitations of traditional MAP atmospheres. Some researchers have claimed that superatmospheric O_2 concentrations (≥ 70 kPa) can be an alternative to low O_2 modified atmospheres in order to prevent undesired anoxic respiration, inhibit the growth of naturally occurring spoilage microorganisms, and maintain fresh-like sensory quality of fresh-cut produce (Amanatidou et al., 1999; Jacxsens et al., 2001; Van der Steen et al., 2002). Results by Allende et al. (2004) showed that high O_2 atmospheres (80–100 kPa) alleviated tissue injury, reduced microbial growth, and were beneficial in maintaining quality of fresh-cut baby spinach.

Edible coatings can be applied as either a complement or an alternative to MAP in order to improve the shelf-life of fresh-cut fruits. Edible films and coatings may help to reduce the deleterious effects concomitant with minimal processing, not solely retarding food deterioration and enhancing its quality, but also improving its safety due to their natural biocide activity or by incorporating antimicrobial compounds (Petersen et al., 1999). The application of edible coatings to deliver active substances is one of the major advances reached so far in order to increase the shelf life of fresh-cut produce. The functionality of edible coatings can be improved by incorporating antimicrobial agents (chemical preservatives or antimicrobial compounds obtained from a natural source), antioxidants, and functional ingredients such as minerals and vitamins. A technique that can potentially be used to incorporate functional ingredients and antimicrobials into edible coatings for fruits is micro- and nanoencapsulation. Micro- and nanoencapsulation is defined as a technology for packaging solids, liquids, or gaseous substances in miniature (micro- and nanoscale) sealed capsules that can release their contents at controlled rates under specific conditions. Release can be solvent activated or signaled by changes in pH, temperature, irradiation, or osmotic shock (Vargas et al., 2008).

The main problem when applying the coatings to fresh-cut fruits is the low adherence presented by the highly hydrophilic cut surface fruit. Recent studies in the field of edible coatings have focused on the development of new technologies that allow for a more efficient control of coating properties and functionality. To this end, new methodologies have been developed, most based on composite or multilayered systems. Nevertheless, applications to food products are still scarce. One of these new methodologies consists of the development of multilayered coatings by means of the layer-by-layer (LbL) electrodeposition (Bernabé et al., 2005; Marudova et al., 2005; Krzemiski et al., 2006; Weiss et al., 2006). The LbL technique could be used to coat highly hydrophilic food systems such as fresh-cut fruits and vegetables. In the near future, multilayered edible coatings will receive more attention (than single-layer coatings) as they could be specially engineered to incorporate and allow the controlled release of vitamins and other functional or antimicrobial agents (Vargas et al., 2008). A possible multilayered structure could include three layers: a matrix layer (e.g., biopolymer based) that contains the functional substance; an inner control layer to govern the rate of diffusion of the functional substance by allowing its controlled release; and a barrier layer that prevents the migration of the active agent from the coated food as well as controls the permeability to gases. Another approach to improve coating properties is to make nanocomposites by incorporating nanosized clay materials such as layered silicates into biopolymer-based matrices. Rhim et al. (2006) incorporated different types of nanoparticles (montmorillonites, nano-silver, and silver-zeolite) into chitosan matrix, obtaining composites with better mechanical, water vapor barrier, and antimicrobial properties than the traditional chitosan coating. Cellulose nanofibers have also shown good possibilities as reinforcements in composite coatings for food packaging. However, even if these studies seem to be promising, the major concern of the scientific community when incorporating these nanomaterials into edible coatings or food is still unsolved: the lack of studies into their possible toxicity.

15.2.3 BIOTECHNOLOGICAL APPROACHES

Biotechnological approaches have been developing in order to extend the shelf life of fresh-cut fruits and vegetables. Molecular methods for detecting human pathogens or plant genetic transformation may create fruits or vegetables best suitable for minimally processed foods (Rodov, 2007). Fresh-cut fruit and vegetables are typically eaten raw, without thermal sterilization procedures like cooking or pasteurization, and therefore represent a significant food safety challenge (Bhagwat, 2006). Modern biotechnology provides fast and sensitive methods for detecting foodborne pathogens on fresh-cut produce. Real-time polymerase chain reaction (RT-PCR) is probably one of the most popular tools for detecting foodborne pathogens. Various commercially available RT-PCR systems have been tested for recognition and quantification of *Listeria monocytogenes* (Liming et al., 2004) and *Salmonella* spp. (Cheung et al., 2004) on fresh-cut products such as fresh-cut cantaloupe, mixed salads, and cilantro leaves. In addition to PCR techniques, other biotechnological methods were employed for detecting various foodborne pathogens in fresh-cut fruit or vegetable products. Among those methods, enzyme-linked fluorescence immunoassay and immunostrip test (Huang et al., 2005), pulsed-field gel electrophoresis (Francis and O'Beirne, 2006), random amplified polymorphic DNA (RAPD), and restriction endonuclease analyses (REA) (Aguayo et al., 2004). In spite of the impressive progress made in the sphere of molecular diagnostics of human pathogens, the problem of efficient detection and elimination of microbial hazards in fresh-cut foods is still far from its practical solution. Neither sensitivity nor reaction times of the available methods allow online monitoring of raw materials or final products for the presence of pathogens. Its application for fresh-cut produce needs additional revolutionary technological changes. These changes are expected to come from the progress of nanotechnology. Development of inexpensive disposable nanobiosensors will improve food safety control in food chain management, in particular, the rapid detection of foodborne pathogens (Rodov, 2007).

On the other hand, genetic transformation may create fruits or vegetables best suitable for fresh-cut processing. Desirable traits for such genotypes would include inhibited enzymatic browning, firm texture, slow tissue degradation, inhibited senescence, and protection against microbial proliferation. However, prospects of practical implementation of these genotypes depend on their acceptance by consumers. So far, just a few gene-engineering projects have been directly and intentionally oriented to the needs of the fresh-cut industry. Antisense inhibition of polyphenol oxidase (PPO) gene expression suppressed browning potential in apple tissues (Murata et al., 2000, 2001). In lettuce, isolation and characterization of a wound inducible phenylalanine ammonia-lyase (PAL) gene from Romaine lettuce (Campos et al., 2004) may bring more efficient biotechnological control of enzymatic browning in this crop. Inhibition of cell wall degrading enzymes, which are involved in tissue softening, may also positively affect the texture of fresh-cut produce. Antisense suppression of the tomato β-galactosidase codifying genes (TBG4) resulted in fruits that at red-ripe stage were 40% firmer than the wild-type control (Smith et al., 2002). Such tomatoes might be expected to preserve good texture as fresh-cut products and be good candidates for minimal processing. Some benefits can be reached by using transgenic fruits or vegetables with inhibited ethylene production or sensitivity.

Slices of Vedrantais melon with antisense gene for the key ethylene biosynthesis enzyme ACC-oxidase demonstrated better storage performance than those of the wild-type fruit. The advantages of the transgenic fresh-cut melon were expressed as higher firmness, soluble solids content, and acidity; preferable flavor; sweetness; texture; and visual quality (Fonseca et al., 2001). Biotechnology can also render the fresh-cut produce an additional protection against microbial colonization. Modern genomic methods may provide a tool for determining the specific pathogen mechanisms involved in the interaction between plant tissue and pathogenic microorganisms. A study carried out by Palumbo et al. (2005) in fresh-cut cabbage has been one of the first steps on the way to biotechnological modulation of the fresh-cut produce–pathogen interaction.

15.3 SUMMARY

The fresh-cut produce market has experienced a dramatic transformation during the past decade, and the expectations for forthcoming years are encouraging. Vegetables have made up the majority of fresh-cut produce sales, and it is expected that this will continue to be the case. However, the fresh-cut fruit sector started to gain ground during the last decade because technical limitations that precluded the industrialization of fruits are being overcome. However, growth will not happen at the same rate if increased innovation through new products development is not conducted appropriately. The adaptation of the distribution chains to these changing requirements is of vital importance to allow for growth of the fresh-cut industry. Development of tailor-made crops through biotechnology as well as research on novel preservation treatments and packaging strategies are required to continue to boost the progress of the fresh-cut sector in the next years.

REFERENCES

Aguayo, E., Escalona, V.H., and F. Artés. 2006. Effect of cyclic exposure to ozone gas on physicochemical, sensorial and microbial quality of whole and sliced tomatoes. *Postharvest Biology and Technology* 39: 169–177.

Aguayo, V., Vitas, A.I., and I. Garcia-Jalon. 2004. Characterization of *Listeria monocytogenes* and *Listeria innocua* from a vegetable processing plant by RAPD and REA. *International Journal of Food Microbiology* 90: 341–347.

Allende, A., Luo, Y., McEvoy, J.L., Artés, F., and C.Y. Wang. 2004. Microbial and quality changes in minimally processed baby spinach leaves stored under superatmospheric oxygen and modified atmosphere conditions. *Postharvest Biology and Technology* 33: 51–59.

Amanatidou, A., Smid, E.J., and L.G.M. Gorris. 1999. Effect of elevated oxygen and carbon dioxide on the surface growth of vegetable-associated microorganisms. *Journal of Applied Microbiology* 86: 429–438.

Artés, F., Gómez, P.A., and F. Artés-Hernández. 2007. Physical, physiological and microbial deterioration of minimally fresh processed fruits and vegetables. *Food Science and Technology International* 13(3): 177–188.

Bernabé, P., Peniche, C., and W. Argüelles-Monal. 2005. Swelling behaviour of chitosan/pectin polyelectrolyte complex membranes. Effect of thermal crosslinking. *Polym. Bull.*, 55: 367–375.

Bhagwat, A.A. 2006. Microbiological safety of fresh-cut produce: where are we now? In: K.R. Matthews (ed.), *Microbiology of Fresh Produce,* ASM Press, Washington, DC, pp. 121–147.

Brecht, J.K., Chau, K.V., Fonseca, S.C., Oliveira, F.A.R., Silva, F.M., Nunes, M.C.N., and R.J. Bender. 2003. Maintaining optimal atmosphere conditions for fruits and vegetables throughout the postharvest handling chain. *Postharvest Biology and Technology* 27: 87–101.

Cameron, A.C., Talasila, P.C., and D.W. Joles. 1995. Predicting film permeability needs for modified-atmosphere packaging of lightly processed fruits and vegetables. *HortScience* 30(1): 25–34.

Campos, R., Nonogaki, H., Suslow, T., and M.E. Saltveit. 2004. Isolation and characterization of a wound inducible phenylalanine ammonia-lyase gene (LsPAL1) from Romaine lettuce leaves. *Plant Physiology* 121: 429–438.

Carlin, F., and C. Nguyen-the. 1999. Minimally processed produce-microbiological issues. In *Proceeding of the international conference on fresh-cut produce,* 9–10 September, 1999. Chipping Campden, UK: Campden and Chorleywood Food Research Association (CCFRA).

Cheung, P.Y., Chan, C.W., Wong, W., Cheung, T.L., and K.M. Kam. 2004. Evaluation of two real-time polymerase chain reaction pathogen detection kits for *Salmonella* spp. in food. *Letters in Applied Microbiology* 39: 509–515.

Cooksey, K. 2005. Effectiveness of antimicrobial food packaging materials. *Food Additives and Contaminants* 22: 980–987.

Dunn, J., Ott, T., and W. Clark. 1995. Pulsed light treatment of food and packaging. *Food Technology* 49(9): 95–98.

Elmnasser, N., Guillou, S., Leroi, F., Orange, N., Bakhrouf, A., and M. Federighi. 2007. Pulsed-light system as a novel food decontamination technology: a review. *Canadian Journal of Microbiology* 53, 813–821.

FDA, U.S. Food and Drug Administration. 2000. Kinetics of microbial inactivation for alternative food processing technologies: pulsed light technology. Available at: http://vm.cfsan.fda.gov/~comm/ift-puls.html. Accessed 2 May 2008.

Fonseca, R.M., Goularte, M.A., Silva, J.A., Lucchetta, L., Marini, L., Zanuzo, M.R., Antunes, P.L., and C.V. Rombaldi. 2001. Conservabilidade de meloes transgenicos, cv. Vedrantais, minimamente processados e refrigerados. *Revista Brasileira de Agrociencia* 7: 149–151.

Forney, C.F. 2007. New innovations in the packaging of fresh-cut produce. *Acta Horticulturae* 746: 53–60.

Francis, G.A., and D. O'Beirne. 2006. Isolation and pulsed-field gel electrophoresis typing of *Listeria monocytogenes* from modified atmosphere packaged fresh-cut vegetables collected in Ireland. *Journal of Food Protection* 69: 2524–2528.

Gómez-López, V.M., Ragaert, P., Debevere, J., and F. Devlieghere. 2007. Pulsed light for food decontamination: a review. *Trends in Food Science and Technology* 18: 464–473.

Gómez-López, V.M., Ragaert, P., Debevere, J., and F. Devlieghere. 2008. Decontamination methods to prolong the shelf-life of minimally processed vegetables, state-of-the-art. *Critical Reviews in Food Science and Nutrition* 48(6): 487–495.

Huang, C.C., Yang, Y.R., Liau, S.M., Chang, P.P., and C.Y. Cheng. 2005. Development of a modified enrichment method for the rapid immunoassay of *Escherichia coli* O157 strains in fresh-cut vegetables. *Journal of Food and Drug Analysis* 13: 64–70, 99.

Jacxsens, L., Devlieghere, F., Van der Steen, C., and J. Debevere. 2001. Effect of high oxygen atmosphere packaging on microbial growth and sensorial qualities of fresh-cut produce. *International Journal of Food Microbiology* 71: 197–210.

Krzemiski, A., Marudova., M., Moffat, J., Noel, T. R., Parker, R., Welliner, N., and S.G. Ring. 2006. Deposition of pectin/poly-L-lysine multilayers with pectins of varying degrees of esterification. *Biomacromolecules*, 7: 498–506.

Liming, S.H., Zhang, Y., Meng, J., and A.A. Bhagwat. 2004. Detection of *Listeria monocytogenes* in fresh produce using molecular beacon–real-time PCR technology. *Journal of Food Science* 69: M240–M245.

Marudova, M., Lang, S., Brownsey, G.J., and S.G. Ring. 2005. Pectin-chitosan multilayer formation. *Carbohydr. Res.*, 340: 2144–2149.

Murata, M., Haruta, M., Murai, N., Tanikawa, N., Nishimura, M., Homma, S., and Y. Itoh. 2000. Transgenic apple (*Malus × domestica*) shoot showing low browning potential. *Journal of Agricultural and Food Chemistry* 48: 5243–5248.

Murata, M., Nishimutra, M., Murai, N., Haruta, M., Homma, S., and Y. Itoh. 2001. A transgenic apple callus showing reduced polyphenol oxidase activity and lower browning potential. *Bioscience, Biotechnology and Biochemistry* 65: 383–388.

Oms-Oliu, G., Martín-Belloso, O., and R. Soliva-Fortuny. 2010. Pulsed light treatments for food preservation. A review. *Food and Bioprocess Technology* 3: 13–23.

Ozdemir, M., and J.D. Floros. 2004. Active food packaging technologies. *Critical Reviews in Food Science and Nutrition* 44: 185–193.

Palumbo, J.D., Kaneko, A., Nguyen, K.D., and L. Gorski. 2005. Identification of genes induced in *Listeria monocytogenes* during growth and attachment to cut cabbage, using different display. *Applied Environmental Microbiology* 71: 5236–5243.

Petersen, K., Nielsen, P.V., Bertelsen, G., Lawther, M., Olsen, M.B., Nilsson, N.H., and G. Mortensen. 1999. Potential of bio-based materials for food packaging. *Trends in Food Science and Technology* 10: 52–68.

Rhim, J.W., Hong, S.I., Park, H.M., and K.W. Ng Perry. 2006. Preparation and characterization of chitosan-base nanocomposite films with antimicrobial activity. *Journal of Agricultural and Food Chemistry* 54: 5814–5822.

Rocha, A.. and A.M.M.B. Morais. 2007. Role of minimally processed fruit and vegetables on the diet of the consumers in the XXI century. *Acta Horticulturae* 746: 265–272.

Rodov, V. 2007. Biotechnological approaches to improve quality and safety of fresh-cut fruit and vegetable products. *Acta Horticulturae* 746: 181–193.

Rodrigues, S.S., and M.D.V. de Almeida. 2001. Portuguese household food availability in 1990 and 1995. *Public Health Nutrition* 4: 1167–1171.

Smith, D.L., Abbott, J.A., and K.C. Gross. 2002. Down-regulation of tomato beta-galactosidase 4 results in decreased fruit softening. *Plant Physiology* 129: 1755–1762.

Soliva-Fortuny, R.C., and O. Martín-Belloso. 2003. New advances in extending the shelf-life of fresh-cut fruits: a review. *Trends in Food Science and Technology* 14: 341–353.

Van der Steen, C., Jacxsens, L., Devlieghere, F., and J. Debevere. 2002. Combining high oxygen atmospheres with low oxygen modified atmosphere packaging to improve the keeping quality of strawberries and raspberries. *Postharvest Biology and Technology* 26: 49–58.

Vargas, M., Pastor, C., Chiralt, A., McClements, D.J., and C. González-Martínez. 2008. Recent advances in edible coatings for fresh and minimally processed fruits. *Critical Reviews in Food Science and Nutrition* 48: 496–511.

Weiss, J., Takhistov, P., and D.J. McClements. 2006. Functional materials in food nanotechnology. *J. Food Sci.*, 71: 107–116.

Index